ECOPHYSIOLOGY OF HIGH SALINITY TOLERANT PLANTS

Tasks for Vegetation Science 40

SERIES EDITORS

The titles published in this series are listed at the end of this volume.

Ecophysiology of High Salinity Tolerant Plants

Edited by

M. AJMAL KHAN

University of Karachi,
Pakistan

and

DARRELL J. WEBER

Brigham Young University,
Provo, UT, U.S.A.

Springer

A C.I.P. Catalogue record for this book is available from the Library of Congress.

ISBN-10 1-4020-4017-2 (HB)
ISBN-13 978-1-4020-4017-7 (HB)
ISBN-10 1-4020-4018-0 (e-book)
ISBN-13 978-1-4020-4018-4 (e-book)

Published by Springer,
P.O. Box 17, 3300 AA Dordrecht, The Netherlands.

www.springer.com

Printed on acid-free paper

Printed in the Netherlands

Dedication of this volume

We wish to dedicate this volume to the many scientists who are involved in the research of high saline tolerant plants. It is through their effort that our knowledge of the mechanisms of saline tolerance is being understood. They are the ones who find high saline tolerant plants and evaluate their possible value in reclamation and economic use.

Contents

Contributing Authors

Abdelly, Chedly, INRST, Laboratoire d'Adaptation des Plantes aux Contraintes Abiotiques, BP 95, Hammam-Lif 2050, Tunisia
Akhtar, K., Nuclear Institute for Agriculture and Biology, Faisalabad, Pakistan
Arshad, M., Cholistan Institute of Desert Studies, Islamia University, Bahawalpur, Pakistan
Ashraf, M. Y., Nuclear Institute for Agriculture and Biology (NIAB), P.O. Box 128, Jhang Road, Faisalabad, Pakistan
Ashraf, M., Department of Botany, University of Agriculture, Faisalabad 38040, Pakistan
Attran, Elham, Biology Department, Teacher Training University, Tehran, Iran
Barth, Hans-Jörg, Dept. of Physical Geography, University of Regensburg, Germany
Baslar, S., Dokuzeylul University, Buca-Izmir, Turkey
Beltran-Morales, A. Universidad Autónoma de Baja California Sur, La Paz, B.C.S. México
Biel, Karl Y., Institute of Basic Biological Problems of Russian Academy of Sciences, Pushchino Moscow Region 142290, Russia
Böer, Benno, UNESCO Regional Office in the Arab States of the Gulf Doha, P. O. Box 3945, State of Qatar
Bujar, M., Biology Department, Teacher Training University, Tehran, Iran
Carter, Christy, George E. Brown, Jr., U.S. Salinity Laboratory, USDA/ARS,
Grieve, Catherine M.T., George E. Brown, Jr., U. S. Salinity Laboratory, USDA/ARS, 450 W. Big Springs Road, Riverside, California 92507, USA
Choudhary, R., Botany Department, J.N. Vyas University, Jodhpur, India 342001
Debez, Almed, INRST, Laboratoire d'Adaptation des Plantes aux Contraintes Abiotiques, BP 95, Hammam-Lif 2050, Tunisia
DePew, M. W., Agronomist/Soil Scientist with Environmental Technical Services, LC
Duan, Deyu, Research Center for Agricultural Resources, Institute of Genetics and Developmental Biology, Chinese Academy of Sciences, 286 Huaizhong Road, Shijiazhuang, Hebei 050021, P. R. China
Dinesh, R., Botany Department, J.N. Vyas University, Jodhpur, India 342001
Dogan, Y., Dokuzeylul University, Buca-Izmir, Turkey
Egan, Todd P., Elmira College, 1 Park Place, Elmira, NY, 14901, USA
Escobar-Hernandez, A., Cibnor. La Paz, B.C.S. México
Fenech-Larios, L., Universidad Autónoma de Baja California Sur, La Paz, B.C.S. México
Gallagher, J.L., Halophyte Biotechnology Center, College of Marine Studies, 700 Pilottown Road, University of Delaware, Lewes, Delaware, 19958, USA

Garcia-Hernandez, J.L., Centro de Investigaciones Biológicas del Noroeste (CIBNOR), S.C. Mar Bermejo # 195, Col. Playa Palo Santa Rita, La Paz, B. C. S., México CP: 23090

Gaylord, Barrett, Elmira College, 1 Park Place, Elmira, NY 14901, USA

Gehlot, H. S., Botany Department, J.N. Vyas University, Jodhpur, India 342001

Geissler, Nicole, Institute for Plant Ecology, Justus-Liebig-University of Giessen, D-35392 Giessen, Germany

Ghars, M. A., INRST, Laboratoire d'Adaptation des Plantes aux Contraintes Abiotiques, BP 95, Hammam-Lif 2050, Tunisia

Grignon, Claude, B & PMP, Agro-M INRA, 34060 Montpellier CEDEX 1, France

Gul, B., Department of Botany, University of Karachi, Karachi-75270, Pakistan

Gulzar, S., Department of Botany, Government Superior Science College, Shah Faisal Colony, Karachi-75230, Pakistan

Hanks, Joseph, Dept. of Integrated Biology, Brigham Young University, Provo, Utah 84602, USA

Hanson, Lee D., Department of Chemistry and Biochemistry, Brigham Young University, Provo, Utah 84602, USA

Harris, Lyneen C., Department of Chemistry and Biochemistry, Brigham Young University, Provo, Utah 84602, USA

Hernandez-Contreras, H., Universidad Autónoma de Baja California Sur, La Paz, B.C.S. México

Huchzermeyer, Bernhard, Botany Institute, Plant Developmental Physiology and Bioenergetics, Hannover University, D-30419 Hannover, Germany

Hussin, Sayed, Institute for Plant Ecology, Justus-Liebig-University of Giessen, D-35392 Giessen, Germany

Joshi, S., Botany Department, J.N. Vyas University, Jodhpur, India 342001

Keller, Emily, A., Department of Plant and Animal Science, Brigham Young University, Provo, UT 84602, USA

Khan, M. A., Department of Botany, University of Karachi, Karachi-75270, Pakistan

Khan, M. Athar, Nuclear Institutes of Agriculture, Tandojam, Pakistan

Khavari-Nejad, R. A. Biology Department, Teacher Training University, Tehran, Iran

Koyro, Hans-Werner, Institute for Plant Ecology, Justus-Liebig-University of Giessen, D-35392 Giessen, Germany

Li, Weiqiang, Research Center for Agricultural Resources, Institute of Genetics and Developmental Biology, Chinese Academy of Sciences, 286 Huaizhong Road, Shijiazhuang, Hebei 050021, P. R. China

Li, Xianggan, Syngenta, P.O. Box 12257, 3054 Cornwallis Road, Durham, NC 27709-2257, USA

Liu, Xiaojing, Research Center for Agricultural Resources, Institute of Genetics and Developmental Biology, Chinese Academy of Sciences, 286 Huaizhong Road, Shijiazhuang, Hebei 050021, P. R. China
Lopez-Aguilar, R., Centro de Investigaciones Biológicas del Noroeste (CIBNOR), S.C. Mar Bermejo #195, Col. Playa Palo Santa Rita, La Paz, B. C. S., México CP: 23090
Marcum, Kenneth B., Department of Applied Biological Sciences, Arizona State University, USA
Murillo-Amador, B., Centro de Investigaciones Biológicas del Noroeste (CIBNOR), S.C. Mar Bermejo # 195, Col. Playa Palo Santa Rita, La Paz, B. C. S., México CP: 23090
Naidoo, G., School of Biological and Conservation Sciences, University of KwaZulu-Natal, P/Bag X54001, Durban, 4000, South Africa
Naidoo, Y., School of Biological and Conservation Sciences, University of KwaZulu-Natal, P/Bag X54001, Durban, 4000, South Africa
Özturk, M., Ege University, Bornova-Izmir, Turkey
Sakcali, M. S., Marmara University, Institute of Sciences, Istanbul, Turkey
Sankhla, Narendra, Texas A&M University, Agriculture Research Center, Dallas (TX), 75252 USA
Sarwar, G., Nuclear Institute for Agriculture and Biology (NIAB), P.O. Box 128, Jhang Road, Faisalabad, Pakistan
Seliskar, Denise M., Halophyte Biotechnology Center, College of Marine Studies, 700 Pilottown Road, University of Delaware, Lewes, Delaware, 19958, USA
Shirazi, M. U., Nuclear Institute of Agriculture, Tandojam, Pakistan
Smaoui, Abderrazzak, INRST, Laboratoire de Caractérisation et de la Qualité de l'Huile d'Olive, BP 95, Hammam-Lif 2050, Tunisia
Smith, Bruce N., Department of Plant and Animal Science, Brigham Young University, Provo, UT 84602, USA
State University, Tempe, Arizona, USA
Tadano, T., Faculty of Applied Biology, Tokyo University of Agriculture, 1-1-1 Sakuraoka, Setagaya-ku, Tokyo, 156-8502, Japan
Tillman, P. H., Engineer/Agronomist with Barbuda Farms International, Ltd.
Troyo-Dieguez, E. Centro de Investigaciones Biológicas del Noroeste (CIBNOR), S.C. Mar Bermejo # 195, Col. Playa Palo Santa Rita, La Paz, B. C. S., México CP: 23090
Valdez-Cepeda, R., Universidad Autónoma de Baja California Sur, La Paz, B.C.S. México
Weber, D. J., Dept. of Integrated Biology, Brigham Young University, Provo, Utah 84602, USA
Yensen, Nicholas P., NyPa International, Tucson AZ 85705, USA
Zarrouk, Moktar, INRST, Laboratoire de Caractérisation et de la Qualité de l'Huile d'Olive, BP 95, Hammam-Lif 2050, Tunisia

Preface

The halophytes are highly specialized plants, which have greater tolerance to salt. They can germinate, grow and reproduce successfully in saline areas which would cause the death of regular plants. Most halophytic species are found in salt marsh systems along seashores or around landlocked inland lakes and flat plains with high evaporation. The halophytes play very significant role in the saline areas specially in the coast by overcoming the salinity in different ways, viz. with regulating mechanisms in which excess salts are excreted and with out regulating mechanism, which may include succulents or cumulative types. Besides that they protect coast from erosion and cyclones, provide feeding ground and nursery for fish, shrimps and birds. Halophytes get increasing attention today because of the steady increase of the salinity in irrigation systems in the arid and semi-arid regions where the increasing population reaches the limits of freshwater availability. In many countries, halophytes have been successfully grown on saline wasteland to provide animal fodder and have the potential for rehabilitation and even reclamation of these sites. The value of certain salt-tolerant grass species has been recognized by their incorporation in pasture improvement programs in many salt affected regions throughout the world. There have been recent advances in selecting species with high biomass and protein levels in combination with their ability to survive a wide range of environmental conditions, including salinity.

Our limited understanding of how halophytes work, as this may well be our future as our limit of fresh water is reached. It is important that we preserve these unusual plants and their habitats, not just for their aesthetic beauty, but also as a resource for the development of new salt tolerant and halophyte crop of economic importance. Over the last ten years much new information has become available about the genetics, molecular biology, ecology, physiology, and physiological ecology of high salinity tolerant plants. A binational US-Pakistan workshop with the support of National Science Foundation, USA was organized in Brigham Young University, Provo, Utah to discuss the currents trends on High Salinity Tolerant Plants. This volume was put together primarily based on the presentations made in that meeting. However, papers were also invited from those who were unable to attend the workshop due to unavoidable circumstances and other imminent halophyte biologists.

This volume is a good collection of papers on the halophytes. Effect of soil salinity in combination of other soil factors were discussed by Barrett Gaylord, Todd Egan, Ajmal Khan, Bilquees Gul and group from Tunis led by Prof. Chedly Abdelly. These papers discussed various germination strategies employed by halophytes to be successful and the role of calcium in promoting seed germination under saline conditions. Salt tolerance of halophytes during the mature vegetative stage were discussed by a group from University of Agriculture Faisalabad, Pakistan lead by Prof. M. Ashraf, Prof. Xiaojing from Shijiazhuang, China, Jack Gallagher from USA, Yasin Ashraf, from Nuclear Institute for Agriculture and Biology, Pakistan. They indicated there are plants which can survive on salt concentration approaching seawater and produced considerable biomass. Dr. Lee Hansen has been instrumental in developing the calorimetric responses to determine the level of stress in plants and have another excellent paper to indicate the efficacy of the techniques used. There are several contributions dealing with the physiology of salt tolerant shrubs and grasses and the mechanism of the salt tolerance.

Halophytic vegetation of the most regions is poorly understood. There are several contributions describing the halophytic vegetation of Great Basin Utah, Saudi Arabia etc. Several papers were included on economic utilization of halophytes. Nick Yensen and Michael DePew are the leading name for successful utilization of halophytes as grain crop, turf and forage in the highly saline areas and have made a significant contribution in introducing this technology from laboratory stage to a profit making enterprise. Similarly US Salinity Laboratory has added a new dimension in saline agriculture by growing floriculture crops. Nick Yensen and Benno Böer look ahead in the 21st century and the necessity for the halophytic research, its potential and possible contribution in the world with limited supply of fresh water.

We wish to acknowledge the support provided by the National Science Foundation (Grant no. NSF INT-0220495) that made it possible to hold the High Saline Tolerant Plants Symposium in Provo, Utah. We appreciated the support and co-operation of Brigham Young University in providing services in connection with the symposium. We especially want to express gratitude to Dr. W.M. Hess who was involved with the planning of the symposium. We would also like to thank Mr. Zaheer Ahmed for his invaluable contribution in editing and making it camera ready for the publication.

Foreword

Volume 40 of the T:VS series, edited by Professors Khan and Weber comes as a present for the 25th birthday of the series. With many other volumes of the series, it deals with problems of saline ecosystems and halophytes. Both editors are long time experts in that field. They were able to attract many other experts in salinity research to present interesting papers from various aspects of saline systems.

Several volumes in this series have made it possible to include papers from colleagues working in developing countries. This is also the case in this volume. While this brought some criticism about problems with the English language, I am still convinced that we should continue the inclusion of such papers from otherwise neglected areas. We realize that this will be criticized, but under the present financial constraints it is not possible to obtain the editorial help for a better presentation. All who were involved in discussions, reviewing and editing, helped to make the contents of the papers as clear as possible. First priority, however, remained the scientific value of the papers, which is sometimes difficult to detect, because the research priority and level varies to some extent in the countries from where papers are included. This book, like several others before in the series, is a good example for our efforts.

This volume contains papers ranging from seed physiology to presentations of new concepts to understand saline systems. In between are papers dealing with the present level of ecophysiological work and the description of halophytic species present in regions of interest. The latter is very welcome in the context of this book series, because we have several papers on that topic and may soon approach a complete picture of halophyte species diversity from the entire world.

With regards to the contributors of this volume, it is rewarding for the series editor to see that many papers come from colleagues with which I have had contacts for a long time. It is a pleasure for me to help them publish their research work here and in the future. I hope that the two volume editors and the book as a whole receive the attention in the scientific community that they deserve.

I thank the Publisher for the continued interest in the series and hope that it gains similar standing in the new Springer environment as it had in the previous Kluwer publishing house.

Helmut Lieth
Osnabrück, July 2005

CHAPTER 1

HOW SALTS OF SODIUM, POTASSIUM, AND SULFATE AFFECT THE GERMINATION AND EARLY GROWTH OF ATRIPLEX ACANTHOCARPA (CHENOPODIACEAE)

BARRETT GAYLORD AND TODD P. EGAN

Elmira College, 1 Park Place, Elmira, NY 14901, USA

Abstract. *Atriplex acanthocarpa* is a desert shrub that grows in soils containing many different types of salts. The effects of two sodium and two potassium salts on the germination and early growth of *Atriplex acanthocarpa* (Torr.) Wats. (armed saltbush) were investigated. Seeds were germinated in petri dishes with solutions of 0, 85, 170, and 340 mM NaCl, KCl, Na_2SO_4, and K_2SO_4 in an incubator with 16 hour days and 8 hour nights with a constant temperature of 24°C. Seeds were allowed to germinate for 12 days. After this time seeds were rinsed of their solutions and allowed to germinate in distilled water for 12 more days to test for specific ion toxicity. Germination and shoot growth were inhibited primarily by salt concentration as opposed to salt type. Recovery germination of rinsed, ungerminated seeds was high indicating that inhibition of seed germination was due to an osmotic affect as opposed to a specific ion affect. In contrast, the congener *Atriplex prostrata*, which native to oceanic shores, may demonstrate specific ion toxicity to salts other than sodium chloride.

1. INTRODUCTION

Atriplex acanthocarpa (Torr.) Wats. (armed saltbush) is a shrub that grows to 1 meter tall. Male flowers are terminal, whereas female flowers are axillary. Flower color is pinkish. Seeds are found in a hard bracteole that is 6-12 mm long and crested. Leaves are 2-6 cm long and are ovate to lanceolate and silvery white (Jones, 1977). Leaf margins can be dentate or undulating with a base that is hastately lobed (Welsh & Crompton, 1995).

Atriplex acanthocarpa occurs in the U.S. from New Mexico east to Texas and south into Mexico (Welsh & Crompton, 1995). In Texas it can be found growing on brackish or saline clay soils along the shore from Baffin to Oso bays (Jones, 1977). Two varieties occur in the U.S. The northern variety, *Atriplex acanthocarpa* var. *acanthocarpa*, occurs from southeast Arizona, and southern New Mexico into western Texas. The southern variety which is represented in this study is the variety *coahuilensis* and occurs in southern Texas and into Mexico (Welsh & Crompton, 1995).

M.A. Khan and D.J. Weber (eds.), Ecophysiology of High Salinity Tolerant Plants, 1-9.

A major taxonomic revision of *A. acanthocarpa* was performed by Henrickson (1988). In this revision Henrickson divides the species into four varieties located in the Southwestern U.S. and Northern Mexico based on morphological variation (Henrickson, 1988). In addition to the varieties discussed earlier (*acanthocarpa* and *coahuilensis*), Henrickson includes two additional varieties. The variety *stewartii* occurs in Mexico, but approximately at the same latitude as *coahuilensis* which occurs in Mexico and Texas. The most southerly occurring variety, *pringlei*, occurs only in northern Mexico (Henrickson, 1988).

Few ecological or physiological experiments have studied armed saltbush, but a foraging quality study by Garza and Fulbright (1988) compared the quality of forage from armed saltbush to fourwing saltbush (*Atriplex canescens* (Pursh) Nutt.). This first study of plant chemistry on armed saltbush found that armed saltbush generally had higher amounts of crude protein compared to fourwing saltbush. Magnesium, potassium, and calcium levels were about the same between armed and fourwing saltbush. Sodium levels in armed saltbush were much higher than in fourwing saltbush. Differences in phosphorous content varied between these species depending on the month. In vitro organic matter digestibility was higher in armed saltbush for most months except November, and this was probably due to fourwing saltbush being woodier than armed saltbush. Therefore, on saline soils deer and cattle can use both of these species as nutritious vegetation (Garza & Fulbright, 1988).

Because armed saltbush can grow in saline soils there is interest in using it to re-vegetate saline rangelands (Garza & Fulbright, 1988) and it is therefore an economically important species. Globally, soils may contain a variety of several different ions. Aside from sodium and chloride, soils in Texas can contain sulfate, magnesium, and potassium ions. These ions naturally occur in the soil (Provin & Pitt, 2001). Soils containing sulfate ions are very common in northern prairie soils (Curtin et al., 1993). Sodium sulfate can also be found in high proportions in the soils of western Canada (Warne et al., 1989). In India it was found that chloride and sulfate ions are common in saline areas where barley and wheat are important crop species (Manchanda et al., 1982).

Because different salts occur in soils worldwide, and the genus *Atriplex* is important as grazing fodder, we decided to study how well different species of *Atriplex* germinate and grow in varying concentrations of different ions. In a previous study, Egan et al. (1997) examined the effects of different salts of sodium and potassium on the germination of *Atriplex prostrata*, an annual herb of temperate climates commonly found growing in soils containing sodium chloride (Gleason & Cronquist, 1991). They found that inhibition of germination was due mostly to salt concentration, and that there was a high percentage of germination after seed treatments were flushed with distilled water. This supports the hypothesis that there was no specific ion effect, and that seeds prevented from germinating were acted upon by osmotic stress (Egan et al., 1997). If seeds did not germinate after being flushed with distilled water, it would be very likely that the salt concentration of the solution was too high and the embryos in the seeds died of a specific ion effect (Ungar, 1991).

There are many ways different ions can affect seeds and mature plants exposed to varying salt concentrations (Ungar, 1991). Therefore, the purpose of this study was to determine 1) if germination of *A. acanthocarpa* was inhibited by an osmotic effect or a specific ion effect, 2) the effect of osmotic stress on the rate of seed germination, and 3) how well seeds can recover germination after they have been treated with different ions. Our hypothesis is that the desert shrub *A. acanthocarpa* will germinate and grow better in potassium and sulfate salts compared to the sea-side annual *A. prostrata*.

2. MATERIALS AND METHODS

Seeds of *Atriplex acanthocarpa* were collected in Webb County, TX five miles east of Loredo, TX on State Route 359 along an open roadside in ranch lands (27°28.968'N, 99°20.808'W). In April of 2001 seeds were removed from their bracteoles with clippers. Four replicates of ten seeds were tested at 0, 85, 170, and 340 mM NaCl, KCl, Na_2SO_4, and K_2SO_4. Seeds were placed in petri dishes with 5 mL of each salt type and concentration level and covered with tight fitting lids to prevent evaporation.

The dishes were placed in an incubator with 16 hour days and 8 hour nights with a constant temperature of 24°C. Four species of halophytes (*Atriplex acanthocarpa*, *Atriplex canescens*, *Blutaparon vermiculare*, and *Rumex chrysocarpus*) were tested at this temperature regime, and *Atriplex acanthocarpa* had high germination. The number of germinated seeds in each dish was recorded daily. After twelve days, seeds in treatment solutions were no longer germinating, so all germinated seedlings were removed and their shoot length was measured. All ungerminated seedlings were rinsed with distilled water and placed back in their dishes with 5 ml of distilled water for 12 more days. If seeds germinated after being rinsed with distilled water then seed germination was assumed to have been inhibited by an osmotic effect as opposed to a specific ion effect (Ungar, 1991). An osmotic effect is caused by solutes in the environment that lower the osmotic potential to a point where germination or growth is inhibited. Enforced dormancy and growth inhibition due to osmotic stress can be alleviated after seeds are removed from a saline environment. A specific ion effect is due to the chemical influence/toxicity of a given ion, and not an osmotic stress caused by that ion. Calcium ions can, in some instances, ameliorate the specific ion effects of other ions (Ungar, 1991). Germination and growth of the previously ungerminated seedlings was monitored daily and recorded as before for twelve more days.

A modified Timson's index was used to determine the rate of seed germination in the saline and recovery groups. The sum of the number of germinated seeds each day was divided by the number of days that seeds were allowed to germinate in each treated dish (Timson, 1965). Therefore if all of the seeds germinated in one day, the highest Timson's index would be ten.

Statistical analyses were performed using NCSS (1995). A two-way analysis of variance of salt type and salt concentration was used to determine differences among treatment group means. Bonferroni post-hoc tests were used to determine significant differences between treatment groups.

3. RESULTS

3.1. Timson's Index

Timson's index values ranged from a mean high value for the control groups (0 mM) of 7.51 (S.E. ± 0.44) to a low for the 340 mM salt groups (all salt types) of 2.62 (S.E. ± 0.71) out of a possible high value of 10. There was a significant effect of salt concentration on the rate of seed germination ($F = 25.51$, $P < 0.001$). There was also a significant effect of salt type ($F = 5.65$, $P = 0.002$) on germination, but there was no interaction of concentration and salt type ($F = 0.99$, $P = 0.46$). Bonferroni post-hoc tests indicated that as salt concentrations increased the rate at which seeds germinated decreased. A trend found for each concentration was that the sodium salts often had lower Timson's index values compared to the potassium salts, but this was only shown to be statistically significant for potassium chloride at the 340 mM level where the KCl-treated seeds had a mean Timson's index of 4.63 (± 1.71), but NaCl and Na_2SO_4-treated seeds had values of 2.56 (± 0.50) and 1.81 (± 0.56), respectively (Figure 1).

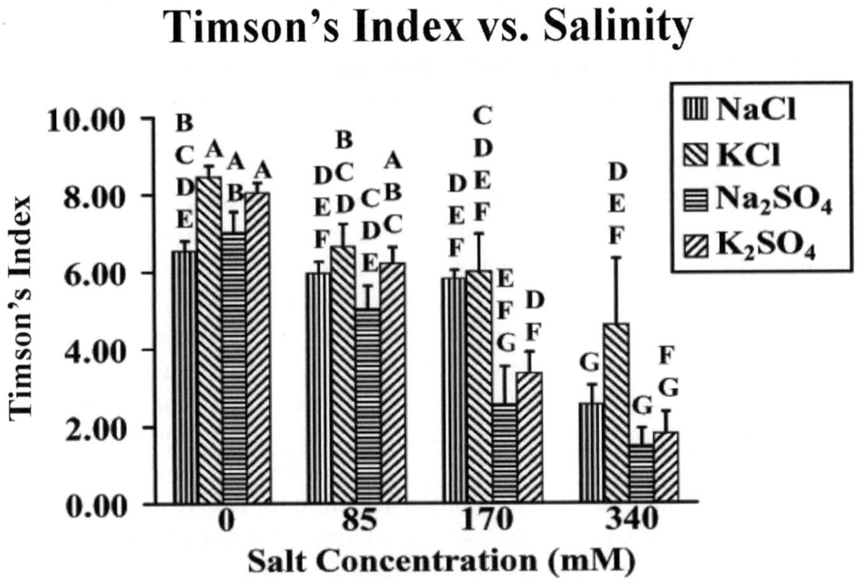

Figure 1. *Modified Timson's index (± S.E) of germination velocity for* A. acanthocarpa *for 0, 85, 170, and 340 mM NaCl, KCl, Na_2SO_4, and K_2SO_4 treatment groups. Different letters denote significant differences among all groups indicated by Bonferroni tests.*

3.2. Percent Germination

Mean percent germination ranged from a mean high value in the control groups (0 mM) of 86% (S.E. ± 3) to a low for the 340 mM salt groups (all salt types) at 20 (S.E. ± 10). Percent seed germination was significantly affected by both concentration ($F=8.33$, $P<0.001$) and salt type ($F=4.41$, $P=0.01$), but not by an interaction between the two ($F=1.21$, $P>0.05$). Although there was an overall decrease in percent germination as salt concentration increased, the Bonferroni test indicated few differences between salt types or salt concentrations. This decrease is seen at the 340 mM salt concentration where the KCl-treated seeds had a higher germination rate of 65% (± 18) compared to the next highest germination rate of 43% (± 3) for NaCl-treated seeds (Figure 2).

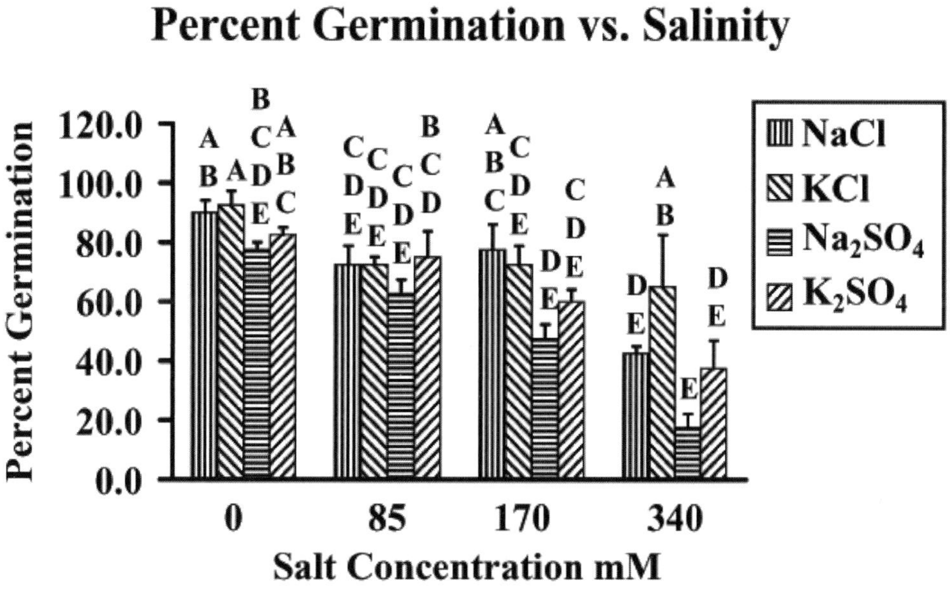

Figure 2. *Percent germination (± S.E) for* A. acanthocarpa *for 0, 85, 170, and 340 mM NaCl, KCl, Na_2SO_4, and K_2SO_4 treatment groups. Different letters denote significant differences among all groups indicated by Bonferroni tests.*

3.3. Seedling Length

Seedling length ranged from a mean high value in the control groups (0 mM) of 9.1 mm (S.E. ± 1.4) to a low in the 340 mM salt groups (all salt types) of 1.0 mm (S.E. ± 0.0). There was a significant effect of salt concentration on shoot length ($F=3.69$, $P>0.02$). However, there was no effect of salt type on shoot length ($F=0.57$, $P>0.05$). There was no interaction between concentration and type ($F=0.46$, $P>0.05$). The main differences indicated by the Bonferroni tests were between the control groups and the 170 mM and 340 mM treatment groups. In general, control groups differed little in seedling length from the 85 mM treatment group, but more from the higher salinity treatments (Figure 3).

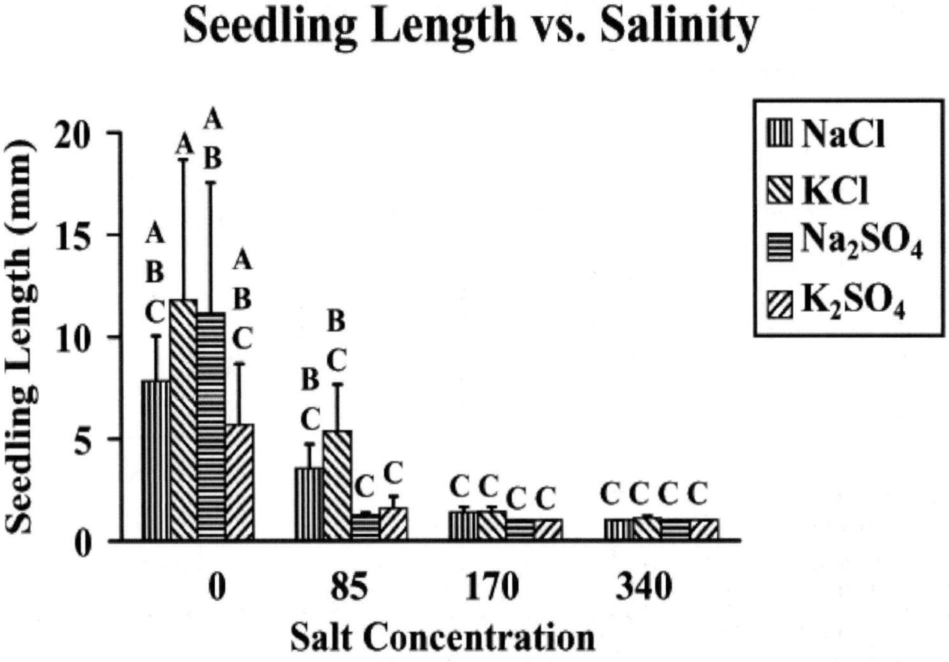

Figure 3. *Seedling length (± S.E) for* A. acanthocarpa *for 0, 85, 170, and 340 mM NaCl, KCl, Na_2SO_4, and K_2SO_4 treatment groups. Different letters denote significant differences among all groups indicated by Bonferroni tests.*

3.4. Recovery

Different numbers of seeds for each group were left after the salt treatment in each group, so maximum Timson's index values would be different for each group and comparisons between groups would not be meaningful. Therefore, Timson's index values are not included as part of the recovery data. A similar issue of few data points arises with percent germination, so no statistical analyses were performed.

Mean percent recovery germination does demonstrate good recovery after the treatment solutions were rinsed from the seeds and replaced with distilled water, indicating that enforced seed dormancy was due to an osmotic effect as opposed to a specific ion effect. This data should be interpreted keeping in mind that the mean number of ungerminated seeds remaining in each dish ranged from 1-9 seeds with a mean number of seeds equal to 3.8. However, a general trend in the data shows higher percent germination for seeds at 170 mM and 340 mM salt concentrations because there were more seeds to germinate in the recovery phase of this experiment as opposed to the 0 mM and 85 mM groups where most of the seeds germinated in the salt treatment phase of this experiment (Table 1).

4. DISCUSSION

Several studies have compared the effects of chloride and sulfate salts on different plant species to determine if our knowledge of the effects of NaCl on plants can be

Table 1. *Percent germination recovery of* Atriplex acanthocarpa *seeds 12 days after salt solutions were replaced with distilled water.*

	Salt Concentration (millimolar)			
Salt Type	*0*	*85*	*170*	*340*
NaCl	50.00	25.00	41.67	16.67
KCl	0.00	37.50	70.83	58.73
Na_2SO_4	33.33	5.00	49.58	42.76
K_2SO_4	12.50	27.78	44.17	39.44

extrapolated onto the effects of other salt types (Onnis et al., 1981; Bal & Chattopadhyay, 1985; Romo & Haferkamp, 1987; Egan et al., 1997), but the results have been too variable to formulate a general answer this question (Curtin et al., 1993).

Egan et al. (1997) tested the effects of sodium chloride, potassium chloride, sodium sulfate, and potassium sulfate on *Atriplex prostrata* seed germination and early growth and found results similar to those of the present study of *A. acanthocarpa*. We found that germination of *A. acanthocarpa* was inhibited by an osmotic effect as opposed to specific ion toxicity (Figure 1). This is reflected in the high recovery germination rates (Table 1). It should be repeated that at 340 mM, the sodium salt-treated seeds of *A. acanthocarpa* had a significantly lower Timson's index than KCl-treated seeds at 340 mM. This is interesting because in the study of *A. prostrata* at the highest salt concentration sodium also had a statistically significant effect on lowering seed germination (Egan et al., 1997). These results are interesting in light of Glenn et al. (1996) and their work with the annual *Atriplex canescens*. They found that growth was negatively correlated with initial amounts of potassium in the leaf, but not sodium. They also found that *A. canescens* accumulated more sodium than potassium (Glenn et al., 1996).

Atriplex prostrata, an annual species, treated with these same salts in a similar manner also showed good recovery, usually near 100% even for seeds at high salt concentrations (Egan et al., 1997). Keiffer and Ungar (1995) observed that *A. prostrata* germination can occur even after seeds have been treated with a 10% NaCl solution for 2 years. Comparing germination of *A. acanthocarpa* to another woody species, *Atriplex griffithii*, was interesting because Ungar and Khan (2001) demonstrated that once the bracteole was removed from *A. griffithii* seeds there was greater germination. It is possible that the bracteole of *A. griffithii* prevented germination because it was causing an osmotic stress due to the high salt concentration in the bracteole and that the seed was able to recover from once the bracteole was removed (Ungar & Khan, 2001).

Atriplex prostrata percent germination generally decreased as salt concentration increased. With KCl and K_2SO_4 the two lower salt concentration treatment groups (-0.75 and -1.00 MPa) did not differ from each other, but did differ from the highest salt concentration (-1.50 MPa). For NaCl, percent germination for the highest and lowest salt concentration treatment groups differed from each other, but neither differed from the intermediate concentration. For Na_2SO_4 all three concentration treatment groups differed from each other (Egan et al., 1997). In the present study with *A. acanthocarpa*, percent germination differences between salt types were not as pronounced (Figure 2) which may be due to *A. acanthocarpa* being native to western deserts (Henrickson, 1988) where many different ions occur in the soil (Provin & Pitt, 2001), and *A. prostrata* being commonly found along sea beaches (Gleason & Cronquist, 1991) where NaCl is the predominant salt.

Seedling growth for *A. acanthocarpa* was similar to *A. prostrata* (Egan et al., 1997). For *A. prostrata* there was a significant effect of salt concentration and salt type even though a Bonferroni post-hoc test did not demonstrate differences among salts at the same concentration (Egan et al., 1997). When comparing the effect of salt types on seedling growth for *A. acanthocarpa* at the same salt concentrations there was no significant effect of salt type. Once again we see that *A. prostrata*, which is native to habitats where NaCl is the predominant soil salt, was more affected by ions other than Na^+ and Cl^- compared to *A. acanthocarpa* which is indigenous to soils with a variety of ions. Al-Jibury and Clor (1986) demonstrated similar results with bitter lentil (*Securigera securidaca*). They found that for seedling growth osmotic pressure had a greater effect on seed germination than the different types of salts used (Al Jibury & Clor, 1986).

It was shown that for mature plants of *A. prostrata*, potassium salts had a stronger inhibitory effect than sodium salts (Egan & Ungar, 1998) on both plant growth and survival. It would therefore be interesting to test the effects of these salts on mature *A. acanthocarpa*.

Percent germination recovery for *A. prostrata* was 100% for chloride salts, but less for SO_4^{2-} salts (Egan et al., 1997). Percent recovery germination for *A. acanthocarpa* did not reflect differences in salt type which lends further support to the hypothesis that different ions have less of an affect on desert shrubs.

5. CONCLUSIONS

The effects of different salts on *A. prostrata* and *A. acanthocarpa* germination and early growth were similar; however, there were some differences. The results of this experiment support our hypothesis that the desert shrub, *A. acanthocarpa*, was more tolerant of a variety of ions that may naturally occur in a desert environment compared to its seaboard congener *A. prostrata*. A further study comparing the growth of mature *A. acanthocarpa* to *A. prostrata* plants might strengthen this hypothesis and tell us more about the relationships between ions and the genus *Atriplex*.

6. ACKNOWLEDGEMENTS

The authors thank Ross McCauley for sending seeds of *A. acanthocarpa*, and J. Forrest Meekins for her helpful comments on manuscript preparation. We also thank Rebecca Mattano for the many hours of helping to remove bracteoles from *A. acanthocarpa* seeds.

7. LITERATURE CITED

Al-Jibury, L.K. & Clor, M.A. 1986. Interaction between sodium, calcium and magnesium chlorides affecting germination and seedling growth of *Securigera securidaca* Linn. Annals of Arid Zone 25: 105-110.

Bal, A.R. & Chattopadhyay, N.C. 1985. Effect of NaCl and PEG 6000 on germination and seedling growth of rice (*Oryza sativa* L.). Biologia Plantarum 27: 65-69.

Curtin, D.H., Steppuhn, H. & Selles, F. 1993. Plant responses to sulfate and chloride salinity: growth and ionic relations. Soil Science Society of America Journal 47: 1304-1310.

Egan, T.P. & Ungar, I.A. 1998. Effect of different salts of sodium and potassium on the growth of *Atriplex prostrata* (Chenopodiaceae). Journal of Plant Nutrition 21: 2193-2205.

Egan, T.P., Ungar, I.A. & Meekins, J.F. 1997. The effect of different salts of sodium and potassium on the germination of *Atriplex prostrata* (Chenopodiaceae). Journal of Plant Nutrition 20: 1723-1730.

Garza Jr., A. & Fulbright, T.E. 1988. Comparative chemical composition of armed saltbush and fourwing saltbush. Journal of Range Management 41: 401-403.

Gleason, H.A. & Cronquist, A. 1991. Manual of the Vascular Plants of Northern United States and Adjacent Canada (2nd edn). New York, Bronx: New York Botanical Garden, 910 pp.

Glenn, E., Pfister, R., Brown, J.J., Thompson, L.L. & O'Leary, J. 1996. Na and K accumulation and salt tolerance of *Atriplex canescens* (Chenopodiaceae) genotypes. American Journal of Botany 83: 997-1005.

Henrickson, J. 1988. A revision of the *Atriplex acanthocarpa* complex (Chenopodiaceae). The Southwestern Naturalist 33: 451-463.

Jones, F.B. 1977. Flora of the Texas Coastal Bend. Corpus Christi, Texas: Mission Press. 262 pp.

Keiffer, C.H. & Ungar, I.A. 1995. Germination responses of halophyte seed exposed to prolonged hypersaline conditions. In: Khan, M.A. & Ungar, I.A. (Ed.), Biology of Salt Tolerant Plants, pp. 45-50. Chelsea, Michigan: Book Crafters.

Manchanda, H.R., Sharma, S.K. & Bhandari, D.K. 1982. Response of barley and wheat to phosphorous in the presence of chloride and sulphate salinity. Plant and Soil 66: 233-241.

NCSS. 1995. NCSS 6.0 Statistical System for Windows User's Guide. Number Cruncher Statistical Systems, Kaysville, Utah.

Onnis, A., Pelosini, F. & Stefani, A. 1981. *Puccinellia festucaeformis* (Host) Parl.: Germinazione e crescita iniziale in funzione della salinita del substrato. Giornale Botanico Italiano 115: 103-116.

Provin, T. & Pitt, J.L. 2001. E-60 Managing soil salinity. Texas A and M University, Texas: Texas Agricultural Extension Service. 5 pp.

Romo, J.T. & Haferkamp, M.R. 1987. Effects of osmotic potential, potassium chloride, and sodium chloride on the germination of greasewood (*Sarcobatus vermiculatus*). Great Basin Naturalist 47: 110-116.

Timson, J. 1965. New method for recording germination data. Nature 207: 216-217.

Ungar, I.A., 1991. Ecophysiology of Vascular Halophytes. Boca Raton, Florida: CRC Press. 209 pp.

Ungar, I.A. & Khan, M.A. 2001. Effect of bracteoles on seed germination and dispersal of two species of *Atriplex*. Annals of Botany 87: 233-239.

Warne, P., Guy, R.D., Rollins, L. & Reid, D.M. 1990. The effects of sodium sulfate and sodium chloride on growth, morphology, photosynthesis, and water use efficiency of *Chenopodium rubrum*. Canadian Journal of Botany 68: 999-1006.

Welsh, S.L. & Crompton, C. 1995. Names and types in perennial *Atriplex Linnaeus* (Chenopodiaceae) in North America selectively exclusive of Mexico. Great Basin Naturalist 55: 322-334

CHAPTER 2

HALOPHYTE SEED GERMINATION

M. AJMAL KHAN AND BILQUEES GUL

Department of Botany, University of Karachi, Karachi-75270, Pakistan

Abstract. Halophyte seed germination, although displays a high degree of inter- and intra-specific variability, shows some patterns in response to various environmental factors. Seeds of stem succulent species germinated better in highly saline conditions. The salt tolerance decreased progressively from leaf succulent, secreting to non-secreting grass halophytes. Seed germination of cold desert halophytes progressively increased with an increase in temperature while seeds of warm desert halophytes showed better germination at cooler temperatures. Halophytes from moist temperate regions germinated better at cooler temperatures. The percentage of un-germinated seeds that recovered when they were transferred to distilled water varied significantly with variation in salinity and temperature regimes in different species. Seeds of some species failed to germinate when exposed to high salinity and temperature stress. While seeds of other halophytic species showed various levels of recovery ranging from 20% to complete recovery of germination. There are some species where recovery of germination is higher then untreated control. Higher temperature inhibited germination recovery for most of the species reported. Germination regulating chemicals like GA_3, kinetin, ethylene, fusicoccin, proline, betaine, thiourea and nitrate released the innate dormancy in seeds of some sub-tropical species while GA_3 was most effective. Germination regulating chemicals had better effect in releasing innate dormancy of Great Basin halophytes and fusicoccin appeared to be more effective. Alleviation of salinity induced dormancy using different chemicals was more successful in Great Basin species in comparison to sub-tropical species. Fusicoccin and ethephon succeeded with subtropical species while all chemicals alleviated either partially or completely the salinity-induced dormancy in Great Basin halophytes except for proline, betaine and nitrate.

1. INTRODUCTION

Halophytes, plants capable of growing and reproducing in saline conditions, as a group have several physiological adaptations that facilitate their survival in saline environments. Of these, most common among halophytes is the ability of osmotic adjustment i.e., allowing for the uptake of water into the plant despite the salt content. In other words, the plants essentially become “saltier” than the soil water. For the plant to adjust osmotically and yet still function physiologically, the plant adjusts both by absorbing and sequestering salts and also by the manufacture of organic plant – derived osmotica. The ability of many halophytes to sequester and compartmentalize ionic compounds (salt, including metals) even when these ionic compounds are in high concentration in the soil, make them ideal candidates for saline agriculture. Halophytes are distributed in a variety of climatic conditions

M.A. Khan and D.J. Weber (eds.), Ecophysiology of High Salinity Tolerant Plants, 11-30.

from coast to mountain valleys (Khan, 2003b). Generally, most tropical halophytes are perennials while majority of the moist temperate halophytes are annuals. However, in cold deserts like Great Basin, USA a large number of perennial halophytes are also reported. Halophytes utilize a broad range of physiological adaptations to salinity and may be stem succulents, leaf succulents, secreting forbs and grasses, annual and perennial grasses, pseudohalophytes and non-halophytes (Breckle, 1983).

The success of halophyte populations is greatly dependent on the germination response of their seeds particularly in temperate conditions while seed germination in subtropical habitats confers an ultimate advantage (Khan, 2003a). The soils where halophytes normally grow become more saline due to rapid evaporation of water particularly during summer, therefore, the soil surface tends to have higher soil salinity and higher water potentials (Khan & Gul, 1998; Khan & Ungar, 1998b). Seed germination in arid and semi-arid regions usually occurs after the rains by reducing surface soil salinity (Khan, 1999). The germination of halophytes could be inhibited under saline conditions due to: i) a complete inhibition of germination process at salinities beyond the tolerance limit of species, ii) delaying the germination of seeds at salinities that cause some stress to seeds but do not prevent germination, iii) causing the loss of viability of seeds due to high salinity and temperature and iv) upsetting growth regulator balance in the embryo to prevent successful initiation of germination process. There is a great deal of variability in the response of halophytes to increasing salinity, moisture, light, and temperature stresses and their interactions (Khan, 2003a).

The information available on the germination of halophytic seeds is far from complete (Khan, 1999). From a total of about 2400 species reported (Lieth et al., 1999); patchy data is available for only about a few hundred species (Baskin & Baskin, 1995; Ungar, 1995). Several factors determine the germination responses of halophytic seeds including salinity, temperature, light, habit, life form, habitat, water etc (Khan, 2003a). It would be interesting to determine patterns of germination if any due to these factors. Present study is an attempt to look for patterns if there are any based on the characteristics mentioned above.

2. SALINITY EFFECTS

Seeds of many halophytic species are reported to germinate best under fresh water conditions or at salinities below 100 mM NaCl (Ungar, 1991) indicating that seeds do not require low water potential for germination (Ungar, 1995). Halophytes vary considerably in their ability to tolerate salt (Table 1). The seawater concentration varies from about 0.6 M NaCl (temperate) to 0.7 M NaCl (arid sub-tropical). The highest salinity concentration at which a seed is reported to germinate is 1.7 M NaCl. Chapman (1960) reported that few seeds of a stem succulent halophyte *Salicornia herbacea* germinated at 1.7 M NaCl and seeds of *Haloxylon ammodendron* and *H. persicum* germinated at 1.3 M NaCl (Tobe et al., 2000). This is followed by a leaf succulent species, *Kochia americana* (1.2 M NaCl;

Clarke & West, 1969) and a grass *Spartina alterniflora* (1.02 M NaCl; Mooring et al., 1971). The first three highly salt tolerant species are stem succulent followed closely by a grass and a leaf succulent (Table 1).

Table 1. *Sodium chloride concentration at which seed germination of halophytes was reduced from 75 – 100% to about 10%.*

Species	*NaCl (M)*	*References*
Salicornia herbacea	1.70	Chapman, 1960
Haloxylon ammodendron	1.30	Huang et al., 2003
Haloxylon persicum	1.30	Tobe et al., 2000
Kochia Americana	1.20	Clark & West, 1969
Spartina alterniflora	1.03	Mooring et al., 1971
Sarcocornia perennis	1.03	Redano et al., 2004
Kochia scoparia	1.00	Khan et al., 2001b
Salsola iberica	1.00	Khan et al., 2002c
Halogeton glomeratus	1.00	Khan et al., 2001c
Arthrocnemum macrostachyum	1.00	Khan & Gul, 1998
Sarcobatus vermiculatus	1.00	Khan et al., 2002a
Salicornia bigelovii	1.00	Rivers & Weber, 1971
Suaeda moquinii	1.00	Khan et al., 2001a
Salicornia rubra	1.00	Khan et al., 2000
Suaeda japonica	0.90	Yokoishi &Tanimoto, 1994
Cressa cretica	0.86	Khan, 1991
Salicornia pacifica	0.86	Khan & Weber, 1986
Halocnemum strobilaceum	0.86	Pujol et al., 2001
Suaeda depressa	0.85	Ungar, 1962
Salicornia europaea	0.85	Ungar, 1962, 1967
Tamarix pentandra	0.85	Ungar, 1967
Allenrolfea occidentalis	0.80	Gul & Weber, 1999
Halosarchia pergranulata	0.80	Short & Colmer, 1999
Salsola imbricata	0.80	Khan, unpublished data

2.1. Salinity tolerance during seed germination of stem succulent halophytes

This indicates that high salinity tolerance during germination is quite independent of plant life form (Table 1). Salt tolerance at germination stage for stem succulent species is reported in the (Table 2). About 60% of the species reported could germinate above seawater salinities (Chapman, 1960; Ungar, 1962, 1967; Rivers & Weber, 1971; Joshi & Iyengar, 1982; Khan & Weber, 1986; Patridge & Wilson, 1987; Khan & Gul, 1998; Gul & Weber, 1999; Khan et al., 2000; Tobe et al., 2000). However, seeds of species like *Halopeplis perfoliata*, *Salicornia brachystachya*, *S. dolistachya* and *Arthrocnemum australacium* failed to germinate at concentrations above 0.25 M NaCl (Clarke & Hannon, 1971; Mahmoud et al., 1983; Huiskes et al., 1985). The data presented in (Table 2) clearly indicates that a large percentage of the stem succulent halophytes are highly salt tolerant, however, some species could not germinate at salinities above 0.3M NaCl (Table 2).

Table 2. *Sodium chloride concentration at which seed germination of stem succulent halophytes was reduced from 75 – 100% to about 10%.*

Species	NaCl (M)	References
Salicornia herbacea	1.70	Chapman, 1960
Haloxylon ammodendron	1.30	Huang et al., 2003
Haloxylon persicum	1.30	Tobe et al., 2000
Sarcocornia perennis	1.03	Redano et al., 2004
Sarcocornia fruticosa	1.03	Redando et al., 2004
Salicornia bigelovii	1.00	Rivers & Weber, 1971
Salicornia rubra	1.00	Khan et al., 2000
Salicornia europaea	0.85	Ungar, 1962, 1967
Allenrolfea occidentalis	0.80	Gul & Weber, 1999
Halosarchia pergranulata	0.80	Short & Colmer, 1999
Sarcocornia quinquifolia	0.69	Patridge & Wilson, 1987
Salicornia pacifica	0.68	Khan & Weber, 1986
Salicornia brachiata	0.60	Joshi & Iyengar, 1982
Salicornia virginica	0.60	Zedler & Beare, 1986
Haloxylon stocksii	0.50	Khan & Ungar, 1996
Arthrocnemum halocnemoides	0.40	Malcolm, 1964
Halopeplis amplexicaulis	0.40	Tremblin & Binet, 1982
Salicornia patula	0.34	Berger, 1985
Halopeplis perfoliata	0.25	Mahmoud et al., 1983
Salicorni brachystachya	0.24	Huiskes et al., 1985
Salicornia dolistachya	0.24	Huiskes et al., 1985
Arthrocnemum australacicum	0.23	Clarke & Hannon, 1970

2.2. Salinity tolerance during seed germination of leaf succulent halophytes

Twenty seven percent of leaf succulent halophytes are reported to germinate at or above seawater salinity (Table 3). However, and about 26% failed to germinate at concentrations above 0.2 M NaCl (Kingsbury et al., 1976; Ungar, 1962, 1967, 1991; Joshi & Iyengar, 1982; Khan et al., 1987; Rozema, 1975; Bakker et al., 1985). Leaf succulents share equal distribution among halophytes at all salinity tolerance levels (Table 3).

2.3. Salinity tolerance during seed germination of secreting halophytes

Few secreting halophytes (only 19%) which could germinate above seawater salinity (Table 4, Woodell, 1985; Khan, 1991; Ungar, 1967; Binet, 1965; Ignaciuk & Lee, 1980; Raccuia et al., 2004). Most secreting halophytes show germination at NaCl concentrations ranging from 0.34 to 0.52 M NaCl (Prado et al., 2000; Carter & Ungar, 2003). While few of them have low salt tolerance during germination (Mahmoud et al., 1983; Ladiges et al., 1981; Fernandes et al., 1985).

Table 3. *Sodium chloride concentration at which seed germination of leaf succulent halophytes was reduced from 75 – 100% to about 10%.*

Species	*NaCl (M)*	*References*
Kochia americana	1.20	Clark & West, 1969
Kochia scoparia	1.00	Khan et al., 2001b
Salsola iberica	1.00	Khan et al., 2002c
Sarcobatus vermiculatus	1.00	Khan et al., 2001a
Suaeda moquinii	1.00	Khan et al., 2001a
Suaeda japonica	0.90	Yokoishi & Tanimoto, 1994
Suaeda depressa	0.85	Ungar, 1962
Salsola imbricata	0.80	Mehrunnisa (unpublished data)
Cakile maritima	0.60	Barbour, 1970
Plantago lanceolata	0.60	Bakker et al., 1985
Salsola kali	0.60	Woodell, 1985
Suaeda maritima	0.60	Boucaud & Ungar, 1976
Suaeda fruticosa	0.50	Khan & Ungar, 1998b
Cakile maritima	0.50	Debez et al., 2004
Cochelaria danica	0.43	Bakker et al., 1985
Rumex crispus	0.43	Bakker et al., 1985
Ceratoides lanata	0.34	Workman & West, 1967
Cotula cornopifolia	0.34	Patridge & Wilson, 1987
Plantago maritima	0.34	Macke & Ungar, 1971
Sperglaria media	0.34	Ungar & Binet, 1975
Silene maritima	0.30	Binet, 1968
Spergularia rupicola	0.30	Okusanya, 1979
Samolus valerandi	0.25	Schat & Scholten, 1985
Lasthenia glabrata	0.20	Kingsbury et al., 1976
Sperglaria marina	0.17	Ungar, 1991
Suaeda limearis	0.17	Ungar, 1962
Suaeda nudiflora	0.17	Joshi & Iyengar, 1982
Iva annua	0.13	Ungar, 1967
Chrysothamnus nauseosus	0.09	Khan et al., 1987
Glaux maritima	0.09	Rozema, 1975
Sperglaria salina	0.09	Bakker et al., 1985

2.4. Salinity tolerance during seed germination of monocotyledonous halophytes

Grass species which could tolerate above 0.6 M salinity but their germination was significantly reduced to 20% (Table 5)

2.5. Comparative salinity effects on seed germination.

It seems that when we compare the salinity tolerance of halophytes from different groups they differ significantly in their salt tolerance above seawater levels (Table 6).

Table 4. *Sodium chloride concentration at which seed germination of secreting dicotyledonous halophytes was reduced from 75 – 100% to about 10%.*

Species	*NaCl (M)*	*References*
Limonium vulgare	1.40	Woodell, 1985
Atriplex rosea	1.00	Khan et al., 2004b
Cressa cretica	0.85	Khan, 1991
Tamarix pentandra	0.85	Ungar, 1967
Atriplex tornabeni	0.77	Binet, 1965
Atriplex laciniata	0.60	Ignaciuk & Lee, 1980
Atriplex nummularia	0.52	Uchiyama, 1987
Atriplex triangularis	0.51	Khan & Ungar, 1984
Atriplex prostrata	0.50	Katembe et al., 1998
Atriplex canescense	0.40	Mikheil et al., 1992
Atriplex lentiformis	0.40	Mikheil et al., 1992
Atriplex polycarpa	0.40	Mikheil et al., 1992
Limonium stocksii	0.40	Zia & Khan, 2004
Atriplex stocksii	0.35	Khan & Rizvi, 1994
Atriplex halimus	0.34	Zid & Boukhris, 1977
Atriplex patula	0.34	Ungar, 1996
Mesembryanthemum australe	0.34	MacKay & Chapman, 1954
Atriplex glabriuscula	0.24	Ignaciuk & Lee, 1980
Limonium axillare	0.17	Mahmoud et al., 1983
Melulaca ericifolia	0.17	Ladiges et al., 1981
Atriplex rependa	0.09	Fernandez et al., 1985

Table 5. *Sodium chloride concentration at which seed germination of secreting monocotyledonous halophytes was reduced from 75 – 100% to about 10%.*

Species	*NaCl (M)*	*References*
Spartina alterniflora	1.03	Mooring et al., 1971
Puccinellia fastucaeformis	0.80	Onnis & Miceli, 1975
Ruppia maritima	0.68	Koch & Seelinger, 1988
Puccinellia lemmoni	0.60	Harivandi et al., 1982
Puccinellia nuttalliana	0.51	Macke & Ungar, 1971
Aeluropus lagopoides	0.50	Gulzar & Khan, 2001
Sporobolus ioclados	0.50	Khan & Gulzar, 2003ab
Urochondra setulosa	0.50	Gulzar et al., 2001
Desmostachya bipinnata	0.50	Gulzar (unpublished data)
Hordeum marinum	0.45	Onnis & Bellattato, 1972
Distichlis spicata	0.43	Cluff & Roundy, 1988
Sporobolus airoides	0.38	Hyder & Yasmin, 1972
Sporobolus virginicus	0.38	Breen et al., 1977
Hordeum jubatum	0.32	Ungar, 1974
Puccinellia distans	0.30	Harivandi et al., 1982
Halopyrum mucronatum	0.20	Khan & Ungar, 2001b

Fifty percent stem succulent species could germinate above seawater followed by 27% in leaf succulent and about 19% both in secreting and grass species (Table 6). It is also interesting to note that the salt tolerance of stem succulent halophytes at germination is higher than 0.2 M NaCl (Table 6). Most halophytes belonging to other groups have germination tolerance ranges between 0.2 to 0.6 M NaCl except for leaf succulents where one fourth could not germinate at or above 0.2 M NaCl (Table 6).

Table 6. *Sodium chloride concentration at which seed germination of halophytes was reduced from 75 – 100% to about 10%.*

Adaptations	*Number of species*	*<0.2*	*0.21-0.40*	*0.41-0.60*	*>0.61*
Stem Succulents	20	0	35	15	50
Leaf Succulents	30	26	13	24	27
Secreting	21	6	26	46	19
Grasses	16	0	31	50	19

3. RECOVERY

The enforced dormancy response for halophyte seeds to saline conditions is of selective advantage to plants growing in highly saline habitats because seeds could withstand high salinity stress and provide a viable seed bank for recruitment of new individuals, but seed germination would be limited to periods when soil salinity levels were within the species tolerance limits (Ungar, 1982). However, halophyte seeds differ in their ability to recover from salinity stress and germinate after being exposed to hyper-saline conditions (Table 7 & 8).

3.1. Recovery of germination of temperate halophytes from salinity stress

Halophytes from the Great Basin desert (a cool temperate area) are highly salt tolerant to salinity (Table 7). Halophytes like *Allenrolfea occidentalis* (Gul & Weber, 1999), *Kochia scoparia* (Khan et al., 2001a), *Salicornia rubra* (Khan et al., 2000) and *Salsola iberica* (Khan et al., 2002c) had 80% or higher recovery of germination when exposed to 1000 mM NaCl (Table 7).

A substantial recovery from germination occurred at the NaCl concentrations up to 600 mM NaCl in *Halogeton glomeratus* (Khan et al., 2001b), *Sarcobatus vermiculatus* (Khan et al., 2002a), *Suaeda moquinii* (Khan et al., 2001a) and *Triglochin maritima* (Khan & Ungar, 1999). This data showed that seeds of Great Basin halophytes have the ability to tolerate high salinity when present in the seed bank. All the species reported here recovered substantially up to 600 mM NaCl but some could almost completely recover from the NaCl concentration of 1000 mM NaCl (Table 7).

3.2. *Recovery of germination of sub-tropical halophytes from salinity stress*

Recovery of germination of sub-tropical halophytes also showed some variability (Table 8) and they appeared to be less salt tolerant while in the seed bank when compared with temperate desert species (Table 7).

Arthrocnemum macrostachyum showed a substantial recovery at 1000 mM NaCl (Khan & Gul, 1998) while all others recovered in up to 600 mM NaCl (Table 8). *Aeluropus lagopoides* (Gulzar & Khan, 2001), *Atriplex stocksii* (Khan, 1999), *Limonium stocksii* (Zia & Khan, 2004) and *Urochondra setulosa* (Gulzar et al., 2001) showed about 75% recovery at 600 mM NaCl (Table 8). While *Cressa cretica* (Khan, 1999), *Haloxylon stocksii* (Khan & Ungar, 1996), *Salsola imbricata* (Khan, unpublished data), *Suaeda fruticosa* (Khan & Ungar, 1998b) and *Sporobolus ioclados* (Khan & Gulzar, 2003a) showed poor recovery responses.

Table 7. *Percentage recovery of germination of temperate halophytes at various NaCl concentrations (mM).*

Name of species	*NaCl (mM)*					
	0	*200*	*400*	*600*	*800*	*1000*
Kochia scoparia	0	85	88	100	100	100
Salsola iberica	1	2	22	37	60	82
Halogeton glomeratus	100	85	72	52	22	8
Allenrolfea occidentalis	0	82	83	98	98	98
Salicornia rubra	0	1	23	38	60	78
Sarcobatus vermiculatus	0	0	61	47	22	0
Suaeda moquinii	0	0	62	50	25	8
Triglochin maritima	15	36	80	65	-	-

Table 8. *Percentage recovery of germination of sub-tropical halophytes at various NaCl concentrations (mM).*

Name of species	*NaCl (mM)*			
	0	*200*	*400*	*600*
Aeluropus lagopoides	0	60	82	89
Arthrocnemum macrostachiyum	0	19	83	96
Atriplex stocksii	23	38	71	75
Cressa cretica	4	76	72	28
Haloxylon stocksii	20	6	58	50
Limonium stocksii	0	82	98	98
Salsola imbricata	0	1	17	19
Sporobolus ioclados	40	19	21	39
Suaeda fruticosa	70	40	0	0
Urochondra setulosa	38	88	75	60

4. TEMPERATURE AND SALINITY EFFECT ON SEED GERMINATION

Several factors (water, temperature, light and salinity) interact in the soil interface, which regulate seed germination. They may even co-act with the seasonal variation in temperature to determine the temporal pattern of germination. Variation in temperature under saline conditions has differential effects on the germination of halophytes (Ungar, 1995; El-Keblawy and Al-Rawai 2005) and this variation could be due to ecological regions of the world where they belong.

4.1. Germination of sub-tropical halophytes under various temperature regimes under saline conditions

Sub-tropical halophytes predominantly showed optimal germination at 20-30°C (Table 9) and any further increase or decrease in temperatures affected seed germination (Khan & Rizvi, 1994; Khan & Ungar, 1996, 1997, 1998b, 1999, 2000, 2001a; Gulzar & Khan, 2001; Gulzar et al., 2001).

Table 9. *Percent germination of subtropical halophytes at different temperatures.*

Species	*10/20*	*15/25*	*20/30*	*25/35*
Aeluropus lagopoides	+	+	+++	++
Arthrocnemum macrostachyum	++	+++	+++	++
Atriplex stocksii	+++	++	++	++
Cressa cretica	+++	++	++	+
Halopyrum mucronatum	+	+	++	+++
Haloxylon stocksii	+++	+++	+++	++
Limonium vulgare	++	++	+++	++
Salsola imbricata	++	++	+++	++
Sporobolus ioclados	++	++	+++	++
Suaeda fruticosa	+	+++	+++	++
Urochondra setulosa	+	++	+++	++
Zygophyllum simplex	+	+	+++	++

4.2. Germination of Great Basin desert halophytes under various temperature regimes under saline conditions

All halophytic species studied from the cold Great Basin desert progressively modified their seed germination with changes in temperature (Khan & Weber, 1986; Khan et al., 1987; Gul & Weber 1999; Khan et al., 2001abc). Germination increased with an increase in temperature (Table 10) and optimal germination was obtained at temperature regime of 25–35°C (Khan & Weber, 1986; Gul & Weber, 1999; Khan, 1999; Khan et al., 2000, 2001abc).

4.3. Germination of moist temperate region halophytes under various temperature regimes under saline conditions

Germination of halophytes from moist temperate regions usually shows better germination at lower temperature (5-15°C) regime (Table 11, Khan & Ungar, 1984; Badger & Ungar, 1989; Khan & Ungar, 1998a; Ungar, 1977; Okusanya & Ungar, 1983; Ungar & Capilupo, 1969). Seed germination of halophytes under natural conditions is regulated by variation in soil salinity and ambient thermoperiod (Khan & Ungar, 1984; Badger & Ungar, 1989; Ungar, 1995). The salt tolerance of seeds appears to be affected by thermoperiod (Morgan & Myers, 1989; Khan & Ungar, 1996). Seeds of halophytes are known to tolerate high salinity during their presence in the soil and are known to germinate when soil salinities are reduced (Khan & Ungar, 1996; Ungar, 1995).

Table 10. *Percent germination of Great Basin halophytes at different temperatures.*

Species	*5/15*	*10/20*	*15/25*	*20/30*	*25/35*
Allenrolfea occidentalis	-	+	++	+++	+++
Atriplex rosea	+	++	+++	+++	+++
Chrysothamnus nauseosus	+	+	++	+++	+++
Halogeton glomeratus	++	++	+++	+++	+++
Kochia scoparia	++	++	+++	+++	+++
Salicornia rubra	+	++	++	+++	+++
Salicornia utahensis	++	?	++	++	++
Salsola iberica	+	++	+++	+++	+++
Sarcobatus vermiculatus	++	++	+++	+++	++
Suaeda moquinii	++	++	+++	+++	+++
Triglochin maritima	- - -	++	++	+++	+++

Table 11. *Percent germination of moist temperate halophytes at different temperatures.*

Species	*5/15*	*10/20*	*5/25*	*20/30*
Atriplex prostrata	+	+	+++	++
Cochlearia anglica	+++	++	+	+
Crithimum maritimum	+++	++	++	-
Hordeum jubatum	+++	+++	+++	+
Polygonum aviculare	+++	+++	++	+
Salicornia europaea	++	++	+++	+
Salicornia stricta	+	++	+++	+
Spergularia marina	+++	++	+	-
Suaeda depressa	+	++	+++	-

5. TEMPERATURE EFFECTS ON THE RECOVERY OF SEED GERMINATION UNDER SALINE CONDITIONS

Role of temperature in the recovery of germination was poorly reported and most studies only focused on the recovery of seed germination based on the variation in salinity (Ungar, 1962, 1967; Clarke & Hannon, 1971; Ungar & Capilupo, 1969; Boorman, 1967; 1968; Woodell, 1985; Keiffer & Ungar, 1995). Khan and Ungar (1996) reported that variation in the recovery responses of *Haloxylon stocksii* seeds with the change in thermoperiod under various NaCl salinity treatments. They reported better recovery of germination at warmer thermoperiod. A number of studies on the effect of temperature regimes on the recovery of germination have since been conducted on the various kinds of halophytes from many parts of the world (Table 12).

5.1. Germination of Great Basin halophytes under various temperature regimes under saline conditions

Best seed germination of temperate desert halophytes occurred at 25–35°C (Khan & Gul, 2002), however optimal recovery of germination of temperate desert halophytes occurred at various temperature regimes (Table 12). Optimal seed germination of *Allenrolfea occidentalis* (Gul & Weber, 1999), *Halogeton glomeratus* (Khan et al., 2001b), *Sarcobatus vermiculatus* (Khan et al., 2002a), *Salsola iberica* (Khan et al., 2002c) were reported in 25–35°C while at 10–20°C and 15–25°C temperature regimes seeds of *Suaeda moquinii* (Khan et al., 2001a), *Salicornia rubra* (Khan et al, 2000), *Kochia scoparia* (Khan et al., 2001b) recovered better while *Triglochin maritima* showed a better recovery at 5–25°C (Khan & Ungar, 1999). *Polygonum aviculare*, a native of moist temperate region showed best recovery at colder temperature regimes (5–15°C) (Khan & Ungar, 1998a).

Table 12. *Percent recovery of germination of temperate halophytes in 400 mM NaCl various thermoperiods (°C) .*

Name of the species	*5-15*	*10-20*	*15-25*	*20-30*	*25-35*
Allenrolfea occidentalis	39	5	51	100	98
Suaeda moquinii	66	100	93	26	15
Salicornia rubra	80	98	99	58	58
Kochia scoparia	79	93	94	57	57
Sarcobatus vermiculatus	69	52	46	40	61
Salsola iberica	46	36	38	30	81
Triglochin maritima	-	10	30	-	-

5.2. Germination of Sub-tropical halophytes under various temperature regimes under saline conditions

Recovery of seed germination of subtropical halophytes does not show any pattern (Table 13). Few halophytes (*Aeluropus lagopoides* and *Limonium stocksii*) showed almost complete recovery at all temperature regimes studied (Gulzar & Khan 2001, Zia & Khan, 2004). *Atriplex stocksii* and *Suaeda fruticosa* showed about 70% recovery at 20–30°C and 15–25°C respectively (Khan & Ungar, 1998b; Khan, 1999), While most other halophytes showed a recovery response about 50% or less like *Arthrocnemum macrostachyum* (Gul & Weber, 1998), *Haloxylon stocksii* (Khan & Ungar, 1996), *Salsola imbricata* (Khan, unpublished data), while still other made little recovery at any temperature regime, *Cressa cretica* (Khan, 1999), *Sporobolus ioclados* (Gulzar & Khan, 2001), *Urochondra setulosa* (Gulzar et al., 2001). It appears from the published data that the recovery of germination of most subtropical halophytes are poor in comparison to temperate halophytes and they do not show any consistent pattern of recovery of germination responses with the change in temperature.

Table 13. *Percentage recovery of germination of sub-tropical halophytes in 400 mM NaCl at various thermoperiods (°C).*

Name of the species	*10-20*	*10-30*	*15-25*	*20-30*	*25-35*
Aeluropus lagopoides	42	-	65	89	88
Arthrocnemum macrostachyum	34	39	42	-	45
Atriplex stocksii	15	38	-	75	0
Cressa cretica	4	-	17	17	12
Haloxylon stocksii	55	30	40	-	6
Limonium stocksii	95	-	98	92	98
Salsola imbricata	03	-	4	15	3.2
Sporobolus ioclados	11	-	4	25	8
Suaeda fruticosa	30	38	71	-	51
Urochondra setulosa	11	-	27	57	29
Zygophyllum simplex	3	4	15	-	3.2

6. GROWTH REGULATORS

Growth regulator theory of dormancy has attracted a great deal of attention (Bewley & Black, 1994). This attributes the control of dormancy to various growth regulators – inhibitors, such as ABA, and promoters, such as gibberellins, cytokinins and ethylene. According to the theory, dormancy is maintained (or induced) by inhibitors such as ABA, and it can be released only when the inhibitors are removed or when promoters overcome it. A second concept is that important metabolic changes occur by the action of dormancy-breaking factors such as the synthesis of RNA, and protein or the operation of the pentose phosphate pathway (Bewley & Black, 1994). If all of the dormant conditions, including after-ripening, light or stratification requirements, and dormancies by endogenous inhibitors, are related to the balance of growth regulators (Khan, 1971; Ungar, 1991); this line of experimentation with halophytes should yield some basic information on the dormancy mechanisms of halophytes. Effect of various germination regulating chemicals like proline (Pr.), betaine (Bet.), gibberellic acid (GA), kinetin (Kin.), fusicoccin (Fc.), ethephon (Et.), thiourea (Tu.) and nitrate (Nit.) were studied on the innate and salinity induced seed dormancy of a number of sub-tropical and Great Basin halophytes.

6.1. Effect of germination regulating chemicals in alleviating innate dormancy of sub-tropical halophytes

The results presented in Table 14 show that except for proline all the germination regulating chemicals were effective in alleviating either partially or completely the innate dormancy in the annual *Zygophyllum simplex* and a perennial *Atriplex stocksii* (Khan & Rizvi, 1994; Khan & Ungar, 1997, 2000, 2002). Fusicoccin and Thiourea treatment caused a complete loss of seed dormancy (Khan & Ungar, 2002). Among seeds of perennial species like *Aeluropus lagopoides, Halopyrum mucronatum, Limonium stocksii*, *Salsola imbricata, Sporobolous ioclados*, and *Urochondra setulosa* germination regulating chemicals either had no effect or a negative effect on germination (Gulzar & Khan, 2002; Khan & Ungar, 2001b, Zia & Khan, 2004). *Arthrocnemum macrostachyum* (Bet., FA. and FC), *Haloxylon stocksii* (Bet. GA), *Suaeda fruticosa* (Pr. and Bet.) and *Sporobolous arabicus* (Fc, Et., Tu. and Nit.) have some effect (Khan et al., 1998; Khan and Ungar, 1996, 2000, 2001b).

Table 14. *Response of different germination regulating chemicals on the innate dormancy of sub-tropical halophytes.*

Species	*Pr*	*Bet.*	*GA*	*Kin*	*FC*	*Et*	*Tu.*	*Nit.*
Aeluropus lagopoides	0	0	0	0	0	0	0	0
Arthrocnemum macrostachyum	0	+++	+++ +	0	++	0	0	0
Atriplex stocksii	+	+	++	+++	+	+	++	+
Halopyrum mucronatum	-	-	0	-	-	-	0	0
Haloxylon stocksii	-	+	++	-	?	?	?	?
Limonium stocksii	-	-	-	0	?	0	0	-
Salsola imbricata	0	0	0	0	0	0	0	0
Sporobolus ioclados	0	-	0	0	0	0	0	0
Suaeda fruticosa	+	++++	0	-	0	0	0	0
Urochondra setulosa	0	0	0	0	0	0	0	-
Zygophyllum simplex	0	++	+++	+++	++++	+++	++++	+++
Sporobolus arabicus	0	0	0	0	+++	+	+++	++

6.2. Effect of germination regulating chemicals in alleviating innate dormancy of Great Basin halophytes

Germination regulating chemicals have no effect on the innate dormancy of temperate halophytes like *Allerolfea occidentalis* (Gul & Weber, 1998) *Atriplex rosea* (Khan et al., 2004b). Whereas most of them have some effect on releasing innate dormancy in the case of *Ceratoides lanata, Kochia scoparia, Salicornia rubra, Salicornia utahensis, Suaeda moquinii,* and *Triglochin maritima* (Khan & Ungar, 2001c, Khan et al., 2002b; Gul & Khan, 2003; Khan et al., 2004a). Among chemicals, fusicoccin alleviated the innate dormancy of 80% of the temperate species studied while proline, betaine, kinetin, ethephon and thiourea alleviated 50% of the temperate halophytes (Table 15).

6.3. Effect of germination regulating chemicals in alleviating salinity effects on seed germination of sub-tropical halophytes

Effect of salinity on the seed germination of subtropical halophytes could not be alleviated by many chemicals (Table 16). *Halopyrum mucronatum, Haloxylon stocksii, Salsola imbricata, Sporobolous ioclados, Suaeda fruticosa* and *Urochondra setulosa* had little or no effect of the germination regulating chemicals in alleviating salinity induced dormancy (Khan et al., 1998, Khan & Ungar, 2000, 2001ab; Gulzar & Khan, 2002). Most chemicals had some effect on alleviating salinity induced effects on the germination of *Atriplex stocksii* and *Zygophyllum simplex*. Thiourea, Ethephon and Fusicocccin were most effective (60 %) in alleviating the salinity effects on germination, followed by GA, kinetin and nitrate (Table 16). Osmotica like proline and betaine alleviated seed germination of few species (Khan et al., 1998, Khan & Ungar, 2000, 2001abc, Gulzar & Khan, 2002).

Table 15. *Response of different germination regulating chemicals on the innate dormancy of Great Basin halophytes.*

Species	*Pr*	*Bet.*	*GA*	*Kin*	*FC*	*Et*	*TU*	*Nit.*
Allenrolfea Occidentalis	0	0	0	0	0	0	0	0
Atriplex rosea	0	0	-	0	+	0	+	-
Halogeton glomeratus	+	+	0	+	++++	-	0	+++
Kochia scoparia	-	+	+	++	++	++	0	-
Salicornia rubra	0	-	0	0	++	+	+	0
Salicornia utahensis	++	++	-	++++	++++	0	++	-
Salsola iberica	+++	+	-	-	+++	+	0	-
Sarcobatus vermiculatus	0	0	0	0	0	0	0	-
Suaeda moquinii	+	++	++++	+++	+	++	++	+
Triglochin maritima	++	0	0	+	++	++	++	0

6.4. Effect of germination regulating chemicals in alleviating salinity effects on seed germination of Great Basin halophytes

Contrary to the sub-tropical halophytes, seed germination of Great Basin halophytes was substantially alleviated by the application of various chemicals (Table 17). All chemical used almost completely alleviated the salinity effects on the seed germination of *Allenrolfea occidentalis* (Gul & Weber, 1998, Gul et al.,

Table 16. *Response of different germination regulating chemicals on the salinity induced dormancy of sub-tropical halophytes.*

Species	*Pr*	*Bet.*	*GA*	*Kin*	*FC*	*Et*	*TU*	*Nit.*
Aeluropus lagopoides	+	-	0	++	+++	++++	+++	0
Arthrocnemum macrostachyum	+	+	+	+	+	+	+	0
Atriplex stocksii	++	+	+++	+++	++	0	+	+
Halopyrum mucronatum	-	-	+	0	?	?	+	+
Haloxylon stocksii	0	0	-	-	0	0	0	0
Limonium vulgare	-	-	-	-	?	++++	-	-
Salsola imbricata	-	-	-	-	-	-	-	-
Sporobolus ioclados	-	-	+	0	0	+	-	-
Suaeda fruticosa	0	+	-	0	+	++	0	-
Urochondra setulosa	0	0	0	0	0	0	0	-
Zygophyllum simplex	0	0	+++	+++	++++	++	++	+
Sporobolus arabicus	0	0	0	0	+++	0	+	+++

Table 17. *Response of different germination regulating chemicals on the salinity Induced dormancy of Great Basin halophytes.*

Species	*Pr*	*Bet.*	*GA*	*Kin*	*FC*	*Et*	*TU*	*Nit.*
Allenrolfea occidentalis	+++	++++	+++	+++	++++	++++	+++	++++
Atriplex rosea	0	0	-	+	0	+	0	-
Halogeton glomeratus	+	0	+	++	+++	-	++	+++
Kochia scoparia	-	+	++	+++	+++	+++	+++	-
Salicornia rubra	0	-	+	+	+	++	+	+
Salicornia utahensis	+	0	0	+++	+++	+++	+++	0
Salsola iberica	+	0	+	0	+	0	+	-
Sarcobatus vermiculatus	0	-	+	++	++	++	+	-
Suaeda moquinii	0	0	++	+++	+	++	+	0
Triglochin maritima	+	0	0	+++	++	+	++	+++

2000). GA, kinetin, fusicoccin, ethephon, and thiourea had substantial effect in alleviating the salinity effects on seed germination. While Nitrate, proline and betaine were effective in some species (Gul et al., 2000, Khan & Ungar, 2001c,

Gul & Khan, 2003, Khan et al., 2000, 2002c, 2003, 2004ab). Physiology of halophyte seed germination is not properly understood. There is considerable variation in the seed germination responses with different factors involved. The physiological response of seed germination is perhaps evolved to make the particular halophyte adapt to specific environmental conditions.The clue for better understanding of the causes of seed dormancy would come from identifying environmental cues that are translated to specific physiological signals. The data based on seed germination at large and halophytic seed germination in particular is too small to make tangible ecological arguments for physiology and biochemistry of halophyte seed dormancy.

7. REFERENCES

Badger, K.S. & Ungar, I.A. 1989. The effects of salinity and temperature on the germination of the inland halophyte *Hordeum jubatum*. Canadian Journal of Botany 67: 1420-1425.

Bakker, J.P., Dijkstra, M. & Russchen, P.T. 1985. Dispersal, germination and early establishment of halophytes and glycophytes on a grazed and abandoned salt-marsh gradient. New Phytologist 101: 291-308.

Barbour, M.G. 1970. Germination and early growth of the strand plant *Cakile maritima.* Bulletin of Torrey Botanical Club 97: 13-22.

Baskin, C.C. & Baskin, J.M. 1995. Dormancy types of dormancy-breaking and germination requirements in seeds of halophytes. In: M.A. Khan & I.A. Ungar (Eds.), Biology of Salt Tolerant Plants. Department of Botany, University of Karachi. pp. 23-30. Karachi, Pakistan: Department of Botany, University of Karachi

Berger, A. 1985. Seed dimorphism and germination behavior in *Salicornia patula.* Vegetatio 61: 137-143.

Bewley, J.D. & Black, M. 1994. Seeds: Physiology of Development and Germination. New York, New York: Plenum Press.

Binet, P. 1965. Action de divers rythmes thermiques journaliers sur la germination de semences de *Triglochin maritimum* L. Bulletin Societe. Normandie (Caen) 6: 99-102.

Binet, P. 1968. Dormances et aptitude a germer en milieu sale chez les halophytes. Bulletin de la Societe France Physiologie Vegetale 14: 125-132.

Breckle, S.W. 1983. Temperate deserts and semi-deserts of Afghanistan and Iran. In: D.W. Goodall & N. West. (Eds.), Ecosystems of the world. pp 271-319. Amsterdam, Netherlands: Elsevier.

Breen, C.M., Everson, C. & Rogers, K. 1977. Ecological studies on *Sporobolus virginicus* (L) Kunth with particular reference to salinity and inundation. Hydrobiologia 54: 135-140.

Boorman, L.A. 1967. Experimental studies in the genus *Limonium.* Ph. D. dissertation, University of Oxford, Oxford.

Boorman, L.A. 1968. Some aspects of the reproductive biology of *Limonium vulgare* Mill. and *L. humile* mill. Annals of Botany 32: 803-824.

Boucaud J, & Ungar IA. 1976. Hormonal control of germination under saline conditions of three halophytic taxa in the genus *Suaeda. Physiologia Plantarum* 3**7:** 143-147.

Carter, C.T. & Ungar, I.A. 2003. Germination response of dimorphic seeds of two halophyte species to environmentally controlled and natural conditions. Canadian Journal of Botany 81: 918-926.

Chapman, V.J. 1960 .Salt marshes and salt deserts of the world. New York, New York:Interscience Publishers.

Clarke, L.D. & Hannon, N.J. 1971. The mangrove swamp and salt marsh communities of the Sydney District. IV. The significance of species interaction. Journal of Ecology 59: 535-553.

Clarke, L.D. & West, N.E. 1969. Germination of *Kochia americana* in relation to salinity. Agronomy Journal 20: 286-287.

Cluff, G.J., Roundy, B.A. 1988. Germination responses of desert salt grass to temperature and osmotic potential. Journal of Range Management 41: 150-154.

Debez, A., Hamed, K.B., Grignon, C. & Abdelly, C. 2004. Salinity effect on germination, growth, and seed production of the halophyte *Cakile maritime*. Plant and Soil 262: 179-189.

El-Keblawy, A., Al-Rawai, A. 2005. Effects of salinity, temperature and light on germination of invasive *Prosopis juliflora* (SW.) D.C. Journal of Arid Environment 61: 555-565.

Fernandez, G., Johnston, M. & Olivares, P.A. 1985. Rol del pericarpio de *Atriplex repanda* en la germinacion. III. Estudio histological y quimico del pericarpio. Phyton 45: 165- 169.

Gul, B & Khan M.A. 2003. Effect of growth regulators and osmotica in alleviating salinity effects on the germination of *Salicornia utahensis.* Pakistan Journal of Botany 36: 877-886.

Gul, B., Khan, M.A. & Weber, D.J. 2000. Alleviation salinity and darkness-enforced dormancy in *Allenrolfea occidentalis* seeds under various thermoperiod. Australian Journal of Botany 48: 745-752.

Gul, B., & Weber, D.J. 1998. Role of dormancy relieving compounds and salinity on the seed germination of *Allenrolfea occidentalis*. Annals of Botany 82: 555-562.

Gul, B. & Weber, D.J. 1999. Effect of salinity, light, and thermoperiod on the seed germination of *Allenrolfea occidentalis*. Canadian Journal of Botany 77: 1-7.

Gulzar, S. & Khan, M.A. 2001. Seed germination of a halophytic grass *Aeluropus lagopoides*. Annals of Botany 87: 319-324.

Gulzar, S. & Khan, M.A. 2002. Alleviation of salinity-induced dormancy in perennial grasses. Biologia Plantarum 45: 617-619.

Gulzar, S., Khan, M.A. & Ungar, I.A. 2001. Effect of temperature and salinity on the germination of *Urochondra setulosa*. Seed Science & Technology 29: 21-29.

Harivandi, M.A., Butler, J.D. & Soltanpour, P.N. 1982. Effects of sea water concentrations on germination and ion accumulation in Alkaligrass (*Puccinellia* spp.). Communication in Soil Science and Plant Analysis 13: 507-517.

Huang, Z., Zhang, X., Zheng, G. & Gutterman, Y. 2003. Influence of light, temperature, salinity and stirage on seed germination of *Haloxylon ammodendron*. Journal of Arid environment 55: 453-464.

Huiskes, A.H.L., Stienstra, A.W., Koustaal, B.P., Markusse, M.M. & van Soelen, J. 1985. Germination ecology of *Salicornia dolichostachya* and *Salicornia brachystachya*. Acta Botanica Neerlandica 34: 369-380.

Hyder, S.Z. & Yasmin, S. 1972. Salt tolerance and cation interaction in alkali sacaton at germination. Journal of Range Management 25: 390-392.

Ignaciuk, R. & Lee, J.A. 1980. The germination of four annual strand-line species. New Phytologists 84: 581-587.

Joshi, A.J. Iyengar, E.R.R. 1982. Effect of salinity on the germination of *Salicornia brachiata* Roxb. Indian Journal of Plant Physiology 25: 65-70.

Katembe, W.J., Ungar, I.A. & Mitchell, J.P. 1998. Effect of salinity on germination and seedling growth of two *Atriplex* species (Chenopodiaceae). Annals of Botany 82: 167-175.

Keiffer, C.W., Ungar, I.A. 1995. Germination responses of halophyte seeds exposed to prolonged hypersaline conditions. In: M.A. Khan & I.A. Ungar. (Eds). Biology of Salt Tolerant Plants. Department of Botany, University of Karachi, Karachi, Pakistan: Department of Botany, University of Karachi. 43-50 pp.

Khan, A.A. 1971. Cytokinins: Permissive role in seed germination. Science 171: 853-859.

Khan, M.A. 1991. Studies on germination of *Cressa cretica* L. seeds. Pakistan Journal of Weed Science Research 4: 89-98.

Khan, M.A. 1999. Comparative influence of salinity and temperature on the germination of subtropical halophytes. In: H. Lieth, M. Moschenko, M. Lohman, H.W. Koyro & A. Hamdy. (Eds.), Halophyte Uses in different climates I: Ecological and Ecophysiological Studies. Progress in Biometeriology. Leiden, Netherlands: Backhuys Publishers. 77-88 pp.

Khan, M.A. 2003a. Halophyte seed germination: Success and Pitfalls. In: A.M. Hegazi, H.M. El-Shaer, S. El-Demerdashe, R.A. Guirgis, A. Abdel Salam Metwally, F.A. Hasan, & H.E. Khashaba (Eds.), International symposium on optimum resource utilization in salt affected ecosystems in arid and semi arid regions Cairo, Egypt: Desert Research Centre. 346-358 pp.

Khan, M.A. 2003b. An ecological overview of halophytes from Pakistan. In: H. Lieth and M. Moschenko. (Eds.), Cash crop halophytes: Recent Studies: 10 years after the Al-Ain meeting (Tasks for Vegetation Science, 38). Dordrecht, Netherlands: Kluwer Academic Press. 167-188 pp.

Khan, M.A. & Gul, B. 1998. High salt tolerance in germinating dimorphic seeds of *Arthrocnemum indicum*. International Journal of Plant Science 159: 826.832.

Khan, M.A. and Gul. B. 2002. *Arthrocnemum macrostachyum*: a potential case for agriculture using above seawater salinity. In R. Ahmed and K.A. Malik. Prospects of Saline Agriculture. Kluwer Academic Press, Netherlands. 353 - 364 pp.

Khan, M.A. & Gulzar, S. 2003a. Germination responses of *Sporobolus ioclados*: a potential forage grass. Journal of Arid Environment 53: 387-394.

Khan, M.A. & Gulzar. S. 2003b. Light, salinity and temperature effects on the seed germination of perennial grasses. American Journal of Botany 90: 131-134.

Khan, M. A. & Rizvi Y. 1994. The effect of salinity, temperature and growth regulators on the germination and early seedling growth of *Atriplex griffithii* Moq. var. *stocksii* Boiss. Canadian Journal of Botany 72: 475-479.

Khan, M.A. & Ungar, I.A. 1984. Effects of salinity and temperature on the germination and growth of *Atriplex triangularis* Willd. American Journal of Botany 71: 481- 489.

Khan, M.A. & Ungar, I.A. 1996. Influence of salinity and temperature on the germination of *Haloxylon recurvum*. Annals of Botany 78: 547-551.

Khan, M.A. & Ungar, I.A. 1997. Alleviation of seed dormancy in the desert forb *Zygophyllum simplex* L. from Pakistan. Annals of Botany 80: 395-400.

Khan, M.A. & Ungar, I.A. 1998a. Seed germination and dormancy of *Polygonum aviculare* L. as influenced by salinity, temperature, and gibberellic acid. Seed Science & Technology 26: 107-117.

Khan, M.A. & Ungar, I.A. 1998b. Germination of salt tolerant shrub *Suaeda fruticosa* from Pakistan: Salinity and temperature responses. Seed Science & Technology 26: 657-667.

Khan, M.A. & Ungar, I.A. 1999. Seed germination and recovery of *Triglochin maritima* from salt stress under different thermoperiods. Great Basin Naturalist 59: 144-150.

Khan, M.A. & Ungar, I.A. 2000. Alleviation of salinity-enforced dormancy in *Atriplex griffithii* Moq. var. *stocksii* Boiss. Seed Science & Technology 25: 83-91.

Khan, M.A. & Ungar, I.A. 2001a. Role of dormancy regulating chemicals in release of innate and salinity-induced dormancy in *Sporobolus arabicus* Boiss. Seed Science & Technology 29: 209-306.

Khan, M.A. & Ungar, I.A. 2001b. Alleviation of salinity stress and the response to temperature in two seed morphs of *Halopyrum mucronatum* (Poaceae). Australian Journal of Botany 49: 777-7783.

Khan, M.A. & Ungar, I.A. 2001c. Effect of dormancy regulating chemicals on the germination of *Triglochin maritima*. Biologia Plantarum 44: 301-303.

Khan, M.A. & Ungar, I.A. 2002. Role of dormancy-relieving compounds and salinity on the germination of *Zygophyllum simplex* L. Seed Science and Technology 30: 507-514.

Khan, M.A. & Weber, D.J. 1986. Factors influencing seed germination in *Salicornia pacifica* var. *utahensis*. American Journal of Botany 73:1163-1167.

Khan, M.A., Gul, B. & Weber, D.J. 2000. Germination responses of *Salicornia rubra* to temperature and salinity. Journal of Arid Environments 45: 207-214.

Khan, M.A., Gul, B. & Weber, D.J. 2001a. Germination of dimorphic seeds of *Suaeda moquinii* under high salinity stress. Australian Journal of Botany 49: 185-192.

Khan, M.A., Gul, B. & Weber, D.J. 2001b. Effect of salinity and temperature on the germination of *Kochia scoparia*. Wetland Ecology & Management 9: 483-489.

Khan, M.A., Gul, B. & Weber, D.J. 2001c. Seed germination characteristics of *Halogeton glomeratus*. Canadian Journal of Botany 79: 1189-1194.

Khan, M.A., Gul, B. & Weber, D.J. 2002a. Effect of temperature, and salinity on the germination of *Sarcobatus vermiculatus*. Biologia Plantarum 45: 133-135.

Khan, M.A., Gul, B. & Weber, D.J. 2002b. Improving seed germination of *Salicornia rubra* (Chenopodiaceae) under saline conditions using germination regulating chemicals. Western North American Naturalist 62: 101-105.

Khan, M.A., Gul, B. & Weber, D.J. 2002c. Seed germination in the Great Basin halophyte *Salsola iberica*. Canadian Journal of Botany 80: 650-655.

Khan, M.A., Gul, B. & Weber, D.J. 2004a. Action of plant growth regulators and salinity on the seed germination of *Ceratoides lanata*. Canadian Journal of Botany 82: 37-42.

Khan, M.A., Gul, B. & Weber, D.J. 2004b. Temperature and high salinity effect in germinating dimorphic seeds of *Atriplex rosea*. Western North American Naturalist. 64: 193-201.

Khan, M.A, Ungar, I.A. & Gul, B. 1998. Action of compatible osmotica and growth regulators in alleviating the effect of salinity on the germination of dimorphic seeds of *Arthrocnemum indicum* L. International Journal of Plant Sciences 159: 313-317

Khan, M.A., Ungar, I.A. & Gul, B. 2003. Alleviation of salinity-enforced seed dormancy in *Atriplex prostrata*. Pakistan Journal of Botany 36: 907-912.

Khan, M.A., Sankhla, N., Weber, D.J. & McArthur, E.D. 1987. Seed germination characteristics of *Chrsothamnus nauseosus* ssp *viridulus* (Astereae, Asteraceae). Great Basin Naturalist 47: 220-226.

Kingsbury, R.W., Radlow, A., Mudie, P.J., Rutherford , J. & Radlow, R. 1976. Salt stress responses in *Lasthenia glabrata*, a winter annual composite endemic to saline soils. Canadian Journal of Botany 54: 1377-1385.

Koch, E.W., Seelinger, U. 1988. Germination ecology of two *Ruppia maritima* L. populations in southern Brazil. Aquatic Botany 31: 321-327.

Ladiges, P.Y., Foord, P.C. & Willis, R.J. 1981. Salinity and water logging tolerance of some populations of *Melaleuca ericifolia* Smith. Australian Journal of Ecology 6: 203-215.

Lieth, H., Moschenko, M., Lohmann, M., Koyro, H.W. & Hamdy, A. 1999. Halophyte uses in different climates I: Ecological and Ecophysiological Studies. In: H. Lieth. (Ed.), Progress in Biometeriology Leiden. Netherlands: Backhause Publishers.

Macke, A. & Ungar, I.A. 1971. The effect of salinity on germination and early growth of *Puccinellia nuttalliana*. Canadian Journal of Botany 49: 515-520.

Mackay, J. B. & Chapman, V.J. 1954. Some notes on *Suaeda australis* Moq. var. *nova zelandica* and *Mesembryanthemum australe* Sol. Ex Forst.f. Trans Royal Society of New Zealand 82: 41-47.

Mahmoud, A., El Sheikh, A.M. & Abdul Baset, S. 1983. Germination of two halophytes: *Halopaplis perfoliata* and *Limonium axilare* from Saudi Arabia. Journal of Arid Environment 6: 87-98.

Malcolm, C.V. 1964. Effect of salt, temperature and seed scarification on germination of two varieties of *Arthrocnemum halocnemoides*. Journal of Royal Society of Western Australia 47: 72-75.
Mikheil, G.S., Meyer, S.E. & Pendleton, R.L. 1992. Variation in germination response to temperature and salinity in shrubby *Atriplex* species. Journal of Arid Environment 22: 39-49.
Mooring, M.T., Cooper, A.W. & Seneca, E.D. 1971. Seed germination response and evidence for height of ecophenes in *Spartina alterniflora* from North Carolina. American Journal of Botany 58: 48-56.
Morgan, W.C. & Myers, B.A. 1989. Germination characteristics of the salt tolerant grass *Diplachne fusca*. I. Dormancy and temperature responses. Australian Journal of Botany 37: 225-37.
Okusanya, O.T. 1979. An experimental investigation into the ecology of some maritime cliff species. II. Germination studies. Journal of Ecology 67: 293-304
Okusanya, O.T. & Ungar, I.A. 1983. The effects of time of seed production on the germination response of *Spergularia marina*. Physiologia Plantarum 59: 335-342.
Onnis, A. & Bellettato, R. 1972. Dormienza e alotolleraza in due specie spontanee di Hordeum. Giornal Botanik Italiano 106: 101-109.
Onnis, A. & Miceli, P. 1975. *Puccinellia festucaeformis* (Host) Parl.: Dormienza e influenza della salinia sulla germinazione. Giornale Botanico Italiano 109: 27-37.
Patridge, T.R. & Wilson, J.B. 1987. Germination in relation to salinity in some plants of salt marshes in Otago, New Zealand. New Zealand Journal of Botany 25: 255-261.
Prado, F.E., Boero, C., Gallardo, M. & Gonzales, J.A. 2000. Effect of NaCl on germination, growth, and soluble sugar content in *Chenopodium quinoa* Willd. seeds. Botanical Bulletin of Academy Sinica 41: 27-34.
Pujol, J.A., Calvo, J.F. & Ramirez-Diaz, L. 2001. Seed germination, growth and osmotic adjustment in response to salinity in a rare succulent halophyte from southeastern Spain. Wetland*s* 21: 256-264.
Raccuia, S.A., Cavallaro, V. & Melilli, M.G. 2004. Intraspecific variability in *Cynara cardanculus* L. var. *sylvestris* Lam. Sicilian populations: seed germination under salt and moisture stresses. Journal of Arid Environment 56: 107-116.
Redano, S., Rubio-Casal, A.E., Castillo, J.M., Luque, C.J., Alvarez, A.A., Luque, T. & Figueroa, M.E. 2004. Influence of salinity and light on germination of three taxa with contrasted habitats. Aquatic Botany 78: 255-264.
Rivers, W.G. & Weber, D.J. 1971. The influence of salinity and temperature on seed germination in *Salicornia bigelovii*. Physiologia Plantarum 24**:** 73-75.
Rozema, J. 1975. The influence of salinity, inundation and temperature on the germination of some halophytes and non-halophytes. Oecologia Plantarum 10: 341-353.
Schat, H. & Scholten, M. 1985. Comparative population ecology of dune slack species: The relation between population stability and germination behaviour in brackish environment. Vegetatio 61: 189-195.
Short, D.C. & Colmer, T.D. 1999. Salt tolerance in the halophyte *Halosarchia pergranulata* subsp. *pergranuylata*. Annals of Botany 83: 207-213.
Tobe, K., Li, X. & Omasa, K. 2000. Effect of sodium chloride on seed germination of two Chinese desert shrub *Haloxylon ammodendron* and *H. persicum* (Chenopodiaceae). Australian Journal of Botany 48: 455-460.
Tremblin, G. & Binet, P. 1982. Installation d'Halopeplis amplexicaulis (Vahl.) Ung. dans une sebkha algerienne. Acta Oecologia 3: 373-379.
Uchiyama, Y. 1987. Salt tolerance of *Atriplex nummularia*. Technical Bulletin of Tropical Agriculture Research Center 22 1-69.
Ungar, I.A. 1962. Influence of salinity on seed germination in succulent halophytes. Ecology 3: 329-335.
Ungar, I.A. 1967. Influence of salinity and temperature on seed germination. Ohio Journal of Science 67: 120-123.
Ungar, I.A. 1974. Inland halophytes of the United States. In: R. Reimold and W. Queen (Eds.), Ecology of halophytes, pp. 203-305. New York, New York: Academic Press.
Ungar, I.A. 1977. Salinity, temperature and growth regulator effects on the seed germination of *Salicornia europaea* L. Aquatic Botany 3: 329-335.
Ungar, I.A. 1982. Germination ecology of halophytes. In: D.N. Sen & K.S. Rajpurohit. (Eds.), Contribution to the ecology of halophytes. pp. 143-154. The Hague, Netherlands: Junk.
Ungar, I.A. 1995. Seed germination and seed-bank ecology of halophytes. In: J. Kigel and G. Galili. (Eds.), Seed Development and Germination. New York, New York: Marcel and Dekker Inc.
Ungar, I.A. 1991. Ecophysiology of Vascular Halophytes. Boca Raton. Louisiana: CRC Press.
Ungar, I.A. 1996. Effect of salinity on seed germination, growth and ion accumulation of *Atriplex patula* (Chenopodiaceae). American Journal of Botany 83: 604-607.
Ungar, I.A. & Binet, P. 1975. Factors influencing seed dormancy in *Spergularia media* (L.) C. Presl. Aquatic Botany 1: 45-55.
Ungar, I.A. & Capilupo, F. 1969. An ecological life history study of *Suaeda depressa* (Pursh.) Wats. Advancing Frontiers of Plant Sciences 23: 137-158.
Woodell, S.R.J. 1985. Salinity and seed germination patterns in coastal plants. Vegetatio 61: 223-229.

Workman, J.P. & West, N.E. 1967. Germination of *Eurotia lanata* in relation to temperature and salinity. Ecology 48: 659-661.

Yokoishi, T., & Tanimoto, S. 1994. Seed germination of the halophyte *Suaeda japonica* under salt stress. Journal of Plant Research 107: 385-388.

Zedler, J.B. & Beare, P.A. 1986. Temporal variability of salt marsh vegetation: the role of low-salinity gaps and environmental stress. In: D.A. Wolfe. (Ed.), Estuarine variability. pp 295-306. New York, New York: Academic Press.

Zia, S. & Khan, M.A. 2004. Effect of light, salinity and temperature on the germination of *Limonium stocksii*. Canadian Journal of Botany 82: 151-157

Zid, E. & Boukhris, M. 1977. Quelque aspects de la tolerance de l'*Atriplex halimus* L. au chlorure de sodium. Multiplication, croissance, composition minerale. Oecologia Plantarum 12: 351-362.

CHAPTER 3

SALT TOLERANCE OF SOME POTENTIAL FORAGE GRASSES FROM CHOLISTAN DESERT OF PAKISTAN

MOHAMMAD ASHRAF[1], MANSOOR HAMEED[1], MOHAMMAD ARSHAD[2], YASIN ASHRAF[3] AND K. AKHTAR[3]

[1]*Department of Botany, University of Agriculture, Faisalabad 38040, Pakistan*
[2]*Cholistan Institute of Desert Studies, Islamia University, Bahawalpur, Pakistan*
[3]*Nuclear Institute for Agriculture and Biology, Faisalabad, Pakistan*

Abstract. Cholistan desert located in the southeast of the Punjab province, Pakistan, covering an area of 25,800 km^2 is a part of Greater Thar desert. The soil types characteristically include: sand dunes, sandy soils with patches of non-saline non-sodic loamy soils and sodic clayey soils. Vegetation structure and density are greatly influenced by the rainfall. During low rainfall years even drinking water gets scarce and both the plant and animal communities are adversely affected. Biodiversity assessment survey was carried out during 1997-98. The dominant species of the lesser Cholistan among grasses were *Aristida adscensionis, Ochthochloa compressa, Lasiurus scindicus, Cymbopogon jwarancusa, Cenchrus biflorus, Sporobolus ioclados and Aeluropus lagopoides*, whereas that of Greater Cholistan *Aeluropus lagopoides, Aristida adscensionis, Cenchrus biflorus, Cenchrus pennesetiformis, Cymbopogon jwarancusa, Lasiurus scindicus, Panicum antidotale* and *Panicum turgidum*. Like other deserts, the major problem in the area is the scarcity of good quality water. The subsoil water in most places of Cholistan is brackish and unfit for normal plant growth. It is highly probable that by adopting biological approach the vast area of Cholistan can be economically exploited. Salinity tolerance of four potential forage grass species, *Cenchrus pennesetiformis, Panicum turgidum, Pennisetum divisum* and *Leptochloa fusca*, including one highly tolerant exotic grass species *Puccinellia distans* was assessed after 6 weeks growth at four salinity treatments, 2.4 (control), 8, 16 and 24 dSm^{-1}. Shoot biomass production in *Leptochloa fusca* and *Puccinellia distans* was not affected by any of the salinity levels and these grasses had greater shoot fresh and dry matter than the other species at all salinity treatments. *Pennisetum divisum* was the worst affected, whereas *Cenchrus pennesetiformis* and *Panicum turgidum* were intermediate in biomass production. Every species showed a specific accumulation pattern for different ions (Na^+, Cl-, K^+ and Ca^{2+}) in the shoots or roots. Levels of leaf monosaccharides (glucose + fructose) were similar in *Cenchrus pennesetiformis* and *Panicum turgidum* under both control and salt treatments. *Leptochloa fusca* accumulated lesser amounts of monosaccharides, although its monosaccharides doubled in the salt treatment. Disaccharides (sucrose), in the leaves of *Cenchrus pennesetiformis*, decreased markedly due to the addition of salts in the rooting medium. Trisaccharides (raffinose) were very low in relation to monosaccharides and disaccharides. In another experiment six accessions of *Panicum antidotale*, eight of *Cenchrus ciliaris* and one of *Lasiurus scindicus* from Cholistan were examined for salinity tolerance under five NaCl levels viz. 1.25 (control), 10, 15, 20 and 25 dSm^{-1}. Although significant reduction in growth attributes of all accessions was recorded, accessions KS 1/2 and Local-2 of *Cenchrus ciliaris*, LS 3/6 of *Lasiurus scindicus*, and KS 1/1 of *Panicum antidotale* were superior to the other accessions in all growth parameters measured. In conclusion, these accessions could be directly used on the salt affected soils of Cholistan using subsoil brackish water irrigation. Grasses like *Panicum turgidum, P. antidotale* and *Lasiurus scindicus* though moderately salt tolerant, could be grown on sandy soils with irrigation with moderately saline subsoil water.

M.A. Khan and D.J. Weber (eds.), Ecophysiology of High Salinity Tolerant Plants, 31-54.

1. INTRODUCTION

Pakistan lies between latitude of 23°35' and 37°05' north and longitude of 60°50' and 77°50' east, covering an area of 296,096 km^2 (Muzaffar, 1997). Heterogeneity of climatic conditions is a unique feature of Pakistan's landscape that ranges from the extremely cold mountains in the north to pleasantly warm weather near the coast of Arabian Sea in the south. The climate is semi-arid to arid subtropical in general, however, with under 250 mm of annual rainfall, while some of the driest regions receive rainfall less than 123 mm annually. The southwest monsoon arrives during June to September associated with rainfall. Temperatures are influenced by altitude and in the period just before the monsoon temperatures in the central plains reach 35-40°C. Deserts may reach up to 45°C. Winter temperatures in the northern mountains remain below freezing for several months of the year.

On the basis of climate, altitude and types of plants, vegetation of Pakistan can be divided into eleven major categories, i.e., permanent snow fields & glaciers, dry alpine & cold desert zone, alpine scrub & moist alpine, Himalayan dry coniferous, Himalayan moist temperate forest, sub-tropical pine forest, sub-tropical dry mixed deciduous scrub forest, Balochistan scrub forest, dry sub-tropical and temperate semi-evergreen scrub forest, tropical thorn forest and sand dune desert, and mangrove forest (Hussain, 1992). Sand dune deserts include areas of Cholistan, Thal, Thar deserts and coastal sandy tracts of Makran and Sind. Soil of these sandy deserts is alluvial in nature covered with sand dunes of several kilometers together, supporting a typical xeric vegetation.

Cholistan desert stretching along the southern border of the Punjab, Pakistan (area 25,800 km^2), is a part of the world's seventh largest desert (Rao et al., 1989). Cholistan merges with the Indus Valley that has been the home of one of the world's oldest civilizations, Mohenjo Daro and Harappa, about 4000 to 5000 years ago, on its western side; possibly at that time it was sharing monsoon downpours on the Indus Valley and was not so arid as today. A gradual drift in the monsoon winds declined the area into a desert (Leopold, 1963).

The desert is classified into Greater Cholistan (18,030 km^2) characterized with comparatively frequent and high sand dunes and Lesser Cholistan (7,770 km^2) with comparatively less high sparse sand dunes. Rainfall is erratic, varying from 100 mm in the north to 200 mm in the south annually and may reach 250 mm near the Pakistan-India border. Average winter temperature ranges between 14 and 16°C, with December and January being the coldest months when the temperature frequently drops below zero. Mean summer temperatures range between 34 to 37°C with maximum temperature in May/June when it may rise up to 50°C or even more. The relative humidity varies from 50 to 65%. During the summers, strong winds frequently result in sandstorms and shift sand dunes from one place to the other. The huge land mass of Cholistan comprises hard textured clayey flat grounds 'dahars' interspersed among semi-stable and unstable sand dunes that may attain a height of 100 m. Dahar soils vary in texture, structure, salinity and sodicity.

Human population, as projected for the year 1991, based on 1981 census was 97,000 with a population density of 3.73 individuals per km^2. Deep interior of the desert is much thinly populated than the peripheral zone. Semi-permanent and nomad inhabitants roam about measuring length and breadth of the desert continuously looking for forage for their livestock and more strongly for drinkable water. Total livestock in Cholistan (recorded in 1994) was 262,430 including 63,095 cattle, 114,421 sheep, 72,726 goats, and 12,188 camels. Prolonged droughts in combination with high temperatures, evaporating the last drops of water from the 'tobas' force the inhabitants to leave the region and move towards irrigated areas until the next showers. Erratic rainfall influences grass cover, which in spite of its great sprouting potentials, does not match masses of livestock dependent on it and therefore, mostly is present in the form of over-grazed stubbles unable to stabilize the sand dunes properly.

Cholistan area faces the scarcity of water for irrigation and even for drinking. The only source of water for human and animals in Cholistan is 'tobas' i.e. pits for the collection of rainwater, and wells dug at different places where subsoil water is sometimes sweet. Sub-soil water in most places is brackish where electrical conductivity ranges from 1-30 dS m^{-1}, pH from 7.2-9.0 and sodium adsorption ratio (SAR) from 1.0-210 (Ashraf & Bokhari, 1987). In view of these data the subsoil water can be considered as highly saline and thus is not fit for irrigating normal crops.

Despite recording the plant biodiversity, the research regarding the biological approach of economic utilization of the Cholistan area was based on three major objectives. Firstly, native grasses with high forage value were assessed for their degree of tolerance to brackish water with the view that if any species shows high tolerance to brackish water it would be of great value in terms of forage production on saline or normal sandy soils of Cholistan with irrigation with subsoil brackish water. Secondly, the plant species, which do not normally occur in Cholistan, but are highly tolerant to high levels of salts whether exotic or native, could be introduced into the area. Thirdly, assessment of the germplasm of different species adapted to saline and drought-hit regions of Cholistan could be an important means of identifying highly tolerant accessions to both salinity and drought. The accessions so identified could be of direct use on such vast barren areas and they could be the useful colonizers of the desert provided they are also tolerant to other environmental factors of the area.

2. MATERIALS AND METHODS

Several studies were conducted to assess the plant biodiversity of Cholistan and to explore appropriate germplasm for enhancing the productivity of the vast rangeland of Cholistan, especially for large saline/sodic patches of interdunal flats.

2.1. Experiment 1: Assessment of biodiversity in Cholistan

Cholistan was thoroughly explored for the biodiversity assessment during the summer and winter seasons. Vegetation of the area was sampled by quadrats laid along a transect line in three distinct habitat types viz. sand dunes (height of dunes at some areas is up to 100 m but in lesser Cholistan it is only up to 25 m), interdunal flats or 'Dahars' (flat clayey pieces of land measuring several km^2) and saline patches (thick saline surfaces occurring in small patches of about 100 m^2 or more). Each transect was separated from the previous one by 20 km, traveled by four-wheel jeep (see map). Along each transect line 10 quadrats, 5 x 5 m each were taken perpendicular to the transect line, with a 5 m distance in between two consecutive quadrats. Frequency, density and percent cover of all the species were recorded and their importance values were calculated following Hussain (1983).

2.2. Experiment 2: Response of some arid zone grasses to brackish water

Salinity tolerance of three native grasses, *Cenchrus pennesetiformis* Hochst. & Steud. ex Steud., *Panicum turgidum* Forssk. and *Pennisetum divisum* (Gmel.)Henr. from Cholistan and Thal deserts were assessed for salinity tolerance after six weeks growth at four salinity treatments, 2.4 (control), 8. 16 and 24 dSm^{-1} which was prepared by mixing four salts, $NaHCO_3$, $MgSO_4$ $7H_2O$, $CaCl_2$ $2H_2O$ and NaCl in 1:5:10:30 ratio in half strength Hoagland's nutrient solution. In this experiment a highly salt tolerant grass *Leptochloa fusca* Kunth from plains of the Punjab and one exotic salt tolerant species *Puccinellia distans* (L.) Parl. was also included.

2.3. Experiment 3: Assessment of salt tolerance in germplasm of some potential native grasses

Seeds of 26 accessions of four different native grass species were obtained from the Cholistan Institute of Desert Studies (CIDS) i.e., nine accessions of *Panicum antidotale* Retz., twelve of *Cenchrus ciliaris* L., four of *Lasiurus scindicus* Henr. and one of *Sporobolus ioclados* (Nees ex Trin.) Nees. They were screened for salinity resistance at Faisalabad. Fifteen of them were further selected on the basis of their salinity tolerance, including six accessions of *Panicum antidotale*, eight of *Cenchrus ciliaris* and one of *Sporobolus ioclados*. The material was then tested under five NaCl salinity levels viz., 1.25 (control), 10, 15, 20 and 25 dS m^{-1} for their performance and tolerance against salinity.

3. RESULTS

3.1. Experiment 1: Assessment of biodiversity in Cholistan

Vegetation of the Cholistan desert varies from place to place depending upon the texture and structure of the soil. Vegetation structure and density are greatly influenced by rainfall. Vegetation was studied at 26 different sites in Lesser Cholistan (Figure 1) along jeepable tracks. Dominant species among grasses *Aristida adscensionis, Ochthochloa compressa, Lasiurus scindicus, Cymbopogon jwarancusa, Cenchrus biflorus* and *Aeluropus lagopoides* (Table 1). Among herbs/undershrubs *Suaeda fruticosa, Salsola baryosma, Dipterygium glaucum, Crotalaria burhia, Haloxylon recurvum* and *Haloxylon salicornicum* and among shrubs *Calligonum polygonoides* and *Capparis decidua* were the dominant species.

Sand dune vegetation (Table 2) comprised among dicots *Aerva javanica, Capparis decidua, Dipterygium glaucum, Citrullus colocynthis, Mukia maderaspatana, Prosopis cineraria, Mollugo cerviana, Crotalaria burhia, Calligonum polygonoides, Fagonia indica* and *Tribulus longipetalus* and among grasses/sedges *Cyperus conglomeratus, Aristida adscensionis, Cenchrus biflorus, Cenchrus pennesetiformis, Lasiurus scindicus, Panicum antidotale* and *Panicum turgidum*.

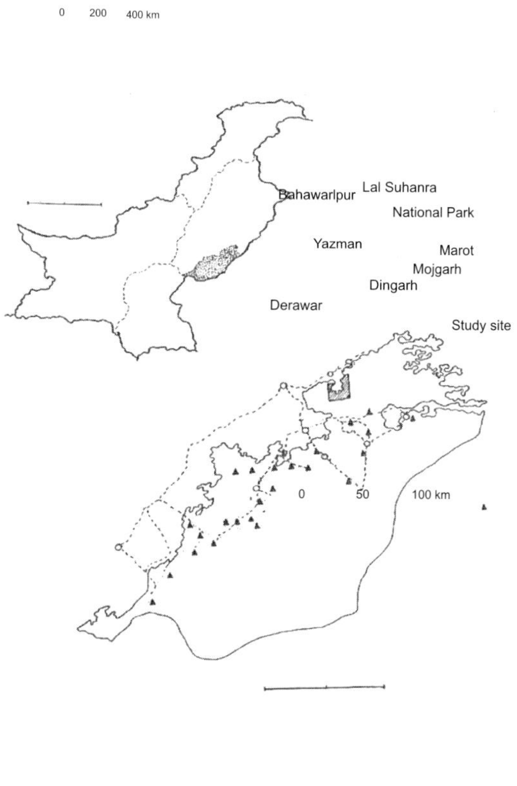

Figure1. *Map of Cholistan showing vegetation study sites.*

Table 1. *Plant species recorded from Lesser Cholistan.*

Family	*Plant species*
Aizoaceae	*Aizoon canariense, Gisekia pharnaceoides, Limeum indicum, Sesuvium sesuvioides, Trianthema triquetra, Zaleya pentandra*
Amaranthaceae	*Achyranthes aspera* var. *aspera, Aerva javanica* var. *bovei*
Asclepiadaceae	*Calotropis procera* ssp. *hamiltonii, Leptadenia pyrotechnica*
Boraginaceae	*Heliotropium crispum, Heliotropium strigosum*
Brassicaceae	*Farsetia hamiltonii*
Capparidaceae	*Capparis decidua, Cleome brachycarpa, Cleome scaposa, Dipterygium glaucum*
Chenopodiaceae	*Haloxylon recurvum, Haloxylon salicornicum, Salsola baryosma, Suaeda fruticosa*
Convolvulaceae	*Cressa cretica*
Cucurbitaceae	*Citrullus colocynthis, Mukia maderaspatana*
Cyperaceae	*Cyperus conglomerates*
Euphorbiaceae	*Euphorbia prostrata*
Malvaceae	*Abutilon muticum*
Mimosaceae	*Prosopis cineraria, Prosopis glandulosa*
Molluginaceae	*Mollugo cerviana*
Nyctaginaceae	*Boerhavia procumbens*
Papilionaceae	*Crotalaria burhia, Indigofera argentea*
Poaceae	*Aeluropus lagopoides, Aristida adscensionis, Aristida hystricula, Aristida mutabilis, Cenchrus biflorus, Cenchrus ciliaris, Cenchrus pennesetiformis, Cenchrus prieurii, Cymbopogon jwarancusa* ssp. *olivieri, Echinochloa colona, Eragrostis barrelieri, Eragrostis ciliaris, Lasiurus scindicus, Ochthochloa compressa, Panicum antidotale, Panicum turgidum, Schoenefeldia gracilis, Sporobolus ioclados, Stipagrostis plumosus*
Polygalaceae	*Polygala erioptera*
Portulacaceae	*Trianthema triquetra*
Polygonaceae	*Calligonum polygonoides*
Rhamnaceae	*Ziziphus mauritiana* var. *spontanea*
Tamaricaceae	*Tamarix androssowii, Tamarix aphylla*
Tiliaceae	*Corchorus depressus*
Zygophyllaceae	*Fagonia indica* var. *indica, Fagonia indica* var. *schweinfurthii, Tribulus longipetalus*. ssp. *longipetalus, Zygophyllum simplex*

Interdunal non-saline flats vegetation comprised among dicots *Trianthema triquetra, Calotropis procera, Haloxylon salicornicum, Haloxylon recurvum, Salsola baryosma, Suaeda fruticosa, Euphorbia prostrata* and among grasses *Cymbopogon jwarancusa* and *Sporobolus ioclados*. Only few species were recorded in highly saline patches, which among dicots were *Trianthema triquetra, Haloxylon recurvum, Salsola baryosma,* and *Suaeda fruticosa* and among grasses *Aeluropus lagopoides* and *Ochthochloa compressa* among grasses.

Table 2. *Dominant plant species recorded from different habitat types in Cholistan.*

		Habitat Types		
Family	*Species*	*Sand dune*	*Interdunal Flats*	*Saline Patches*
Aizoaceae	*Trianthema triquetra*			
Amaranthaceae	*Aerva javanica*			
Asclepiadaceae	*Calotropis procera*			
Capparidaceae	*Capparis decidua*			
	Dipterygium glaucum			
Chenopodiaceae	*Haloxylon recurvum*			
	Haloxylon salicornicum			
	Salsola baryosma			
	Suaeda fruticosa			
Cucurbitaceae	*Citrullus colocynthis*			
	Mukia maderaspatana			
Cyperaceae	*Cyperus conglomeratus*			
Euphorbiaceae	*Euphorbia prostrata*			
Mimosaceae	*Prosopis cineraria*			
Molluginaceae	*Mollugo cerviana*			
Papilionaceae	*Crotalaria burhia*			
Poaceae	*Aeluropus lagopoides*			
	Aristida adscensionis			
	Cenchrus biflorus			
	Cenchrus pennesetiformis			
	Cymbopogon jwarancusa			
	Lasiurus scindicus			
	Ochthochloa compressa			
	Panicum antidotale			
	Panicum turgidum			
	Sporobolus ioclados			
Polygonaceae	*Calligonum polygonoides*			
Zygophyllaceae	*Fagonia indica*			
	Tribulus longipetalus			

Vegetation in Greater Cholistan (Table 3) is more or less similar to that of Lesser Cholistan, but vegetation density is more sparse and dispersed. Dominant species were *Aerva javanica, Calotropis procera, Leptadenia pyrotechnica, Dipterygium glaucum, Capparis decidua, Haloxylon recurvum, Haloxylon salicornicum*, *Suaeda fruticosa*, *Mollugo cerviana*, *Neurada procumbens*, *Calligonum polygonoides*, *Fagonia indica* and *Tribulus longipetalus* among dicots and *Cyperus conglomeratus*, *Aeluropus lagopoides*, *Aristida adscensionis*, *Cenchrus biflorus*, *Cenchrus pennesetiformis*, *Cymbopogon jwarancusa*, *Lasiurus scindicus*, *Panicum antidotale* and *Panicum turgidum* among grasses and sedges. A list of major potential forage grasses is presented in Table 4.

Table 3. *Plant species recorded from Greater Cholistan.*

Family	*Plant species*
Aizoaceae	*Sesuvium sesuvioides, Zaleya pentandra, Trianthema triquetra*
Amaranthaceae	*Aerva javanica, Aerva pseudotomentosa, Amaranthus viridis*
Asclepiadaceae	*Calotropis procera, Leptadenia pyrotechnica*
Brassicaceae	*Dipterygium glaucum*
Capparidaceae	*Capparis decidua, Cleome brachycarpa, Cleome scaposa*
Chenopodiaceae	*Haloxylon recurvum, Haloxylon salicornicum, Salsola baryosma, Suaeda fruticosa*
Convolvulaceae	*Convolvulus deserti*
Cucurbitaceae	*Citrullus colocynthis*
Cyperaceae	*Cyperus conglomeratus, Cyperus rotundus*
Euphorbiaceae	*Euphorbia prostrata*
Mimosaceae	*Acacia jacquemontii, Prosopis cineraria*
Molluginaceae	*Mollugo cerviana*
Neuradaceae	*Neurada procumbens*
Orobanchiaceae	*Cistanche tubulosa*
Papilionaceae	*Crotalaria burhia, Indigofera argentea*
Poaceae	*Aeluropus lagopoides, Aristida adscensionis, Cenchrus biflorus, Cenchrus ciliaris, Cenchrus pennesetiformis, Cenchrus prieurii, Cymbopogon jwarancusa, Cynodon dactylon, Eragrostis barrelieri, Ochthochloa compressa, Lasiurus scindicus, Leptathrium senegalense, Panicum antidotale, Panicum turgidum, Stipagrostis plumosus*
Polygalaceae	*Polygala erioptera*
Polygonoides	*Calligonum polygonoides*
Tamaricaceae	*Tamarix aphylla*
Tiliaceae	*Corchorus depressus*
Zygophyllaceae	*Fagonia indica, Seetzenia lanata, Tribulus longipetalus*

3.2. Experiment 2: Response of some arid zone grasses to brackish water.

The five grass species (*Leptochloa fusca, Cenchrus pennesetiformis, Panicum turgidum, Pennisetum divisum* and *Puccinellia distans*) responded differently to increasing salinity level of the rooting medium. Increasing salinity treatment markedly reduced shoot and root dry weights in *Cenchrus pennesetiformis*, *Panicum turgidum* and *Pennisetum divisum*, whereas those of *Leptochloa fusca* and *Puccinellia distans* showed stability at all salinity treatments (Figure 2, Table

5). *Leptochloa fusca* had significantly greater shoot fresh and dry matter than the other grass species at 16 and 24 dS m^{-1}.

Root dry weight of *Pennisetum divisum* and *Cenchrus pennesetiformis* decreased with increase in external salt level, whereas that of *Panicum turgidum, Leptochloa fusca* and *Puccinellia distans* remained unaffected (Table 5). *Panicum turgidum* produced significantly greater root dry matter than the other species at all salinity levels.

Table 4. *Some major potential forage grasses of Cholistan desert.*

Grass species	*Forage value*
Aristida adscensionis	Probably grazed by cattle
Cenchrus biflorus	Palatable when young
Cenchrus ciliaris	A valuable forage grass
Cenchrus pennesetiformis	An extremely valuable fodder grass
Cenchrus prieurii	Not common but excellent fodder grass
Cynodon dactylon	A first-class fodder grass
Lasiurus scindicus	A valuable fodder grass, particularly relished by camel, sheep and cattle
Ochthochloa compressa	A good fodder grass for cattle and horses
Panicum antidotale	Palatable when quite young, acquire a saltish and bitter taste on maturity
Panicum turgidum	Very palatable, heavily grazed by cattle and camel
Stipagrostis plumosus	Has a great value as a fodder

Table 5. *Fresh and dry weights of shoot and dry weights of roots of five grass species after six weeks of growth in different concentrations of salinity in sand culture.*

	Salinity as electrical conductivity – dS m^{-1}								
	8	*16*	*24*	*8*	*16*	*24*	*8*	*16*	*24*
Species	*Percent shoot fresh weight*			*Percent shoot dry weight*			*Percent root dry weight*		
Cenchrus pennesetiformis	111	92	58	104	77	62	89	98	57
Leptochloa fusca	157	129	117	117	92	97	86	61	69
Panicum turgidum	77	73	53	80	80	74	113	146	99
Pennisetum divisum	68	35	19	67	33	18	93	37	38
Puccinellia distans	136	129	116	108	105	105	142	68	95
	LSD 5% = 28.6			LSD 5% = 23.2			LSD 5% = 29.4		

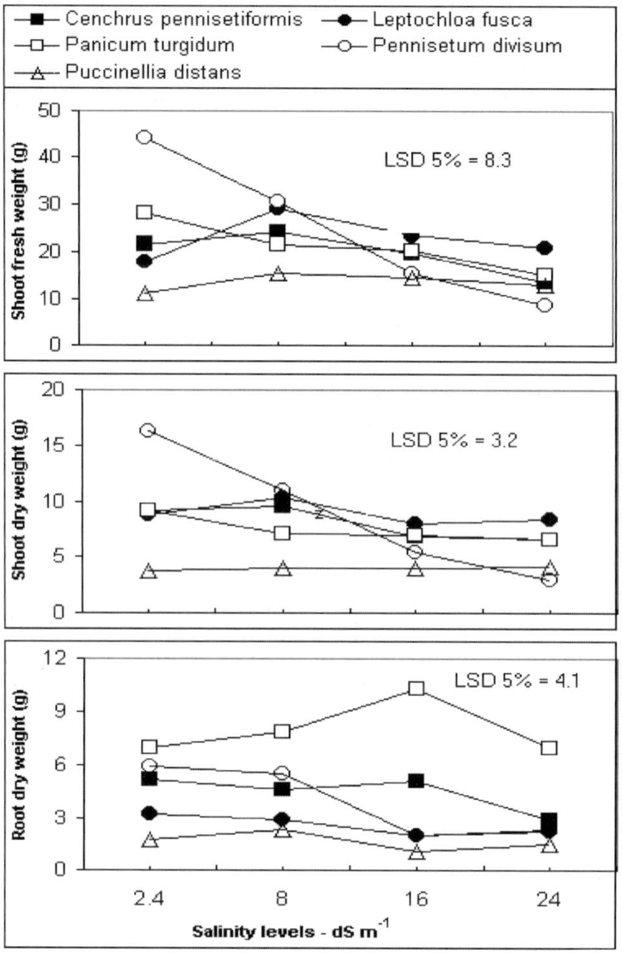

Figure 2. *Fresh and dry weights of shoots and roots of five grass species after six weeks growth at varying salinity levels in sand culture.*

Percent fresh and dry matters of *Leptochloa fusca* and *Puccinellia distans* remained unaffected at all salt treatments (Table 5). In contrast, percent biomass production in *Pennisetum divisum* was severely inhibited by salinity. Percent root biomass was inhibited in all the species except *Panicum turgidum*. Increasing salt treatments had no significant effect on shoot/root ratio of all five-grass species.

Number of tillers/plant of *Puccinellia distans* and *Cenchrus pennesetiformis* increased only at 8 dSm^{-1}, while at the other treatments this variable remained unaffected (Table 6). Significant reduction in *Panicum turgidum* regarding number of tillers/plant was recorded due to salinity, but in *Leptochloa fusca* it remained stable at all the treatments. *Puccinellia distans* showed the highest, while

Panicum turgidum and *Pennisetum divisum* the lowest number of tillers/plant of all species at all salt treatments.

Table 6. *Shoot/root ratio and number of tillers/plant of five grass species after six weeks of growth in different concentrations of salinity in sand culture.*

	Salinity as electrical conductivity – dS m⁻¹							
	2.4	*8*	*16*	*24*	*2.4*	*8*	*16*	*24*
Species	*Shoot/root ratio*				*Number of tillers/plant*			
Cenchrus pennesetiformis	1.74	2.04	1.37	1.90	15.2	26.2	14.7	13.7
Leptochloa fusca	2.44	3.32	3.64	3.40	19.2	26.2	27.5	21.0
Panicum turgidum	1.29	0.91	0.71	0.97	37.0	10.0	11.0	9.0
Pennisetum divisum	2.75	1.96	2.50	1.30	10.3	11.3	5.0	5.3
Puccinellia distans	2.11	1.59	3.23	2.33	57.0	75.2	62.2	55.2
				LSD 5%=NS				

Increasing external salt concentration affected shoot Na+ concentration non-significantly (Figure 3). *Cenchrus pennesetiformis* and *Pennisetum divisum* had significantly higher and *Leptochloa fusca* and *Puccinellia distans* lower shoot Na+ at 16 dSm^{-1}, but at higher salinity (24 dSm^{-1}) the difference in Na^+ concentrations in all the species was non-significant. Root Na^+ of all five species increased with increase in salt level. *Panicum turgidum* had significantly higher concentration than the other species at all salt treatments. At 8 dS m^{-1} *Pennisetum divisum* and *Leptochloa fusca* had intermediate and *Cenchrus pennesetiformis* and *Puccinellia distans* the lowest root Na^+.

Chloride concentrations in the shoots and roots in all five species increased with increase in salt treatment except in shoot of *Cenchrus pennesetiformis*, which had almost equal to its control at 8 and 24 dSm^{-1}, respectively. *Puccinellia distans* and *Pennisetum divisum* had significantly higher shoot Cl^- than the other species at 8 and 16, and 24 dSm^{-1}, respectively. *Panicum turgidum* had the highest root Cl^- concentration of all five species at all salt treatments.

Shoot K^+ concentrations of all species decreased significantly with increase in salinity level, whereas root K^+ of all species remained unaffected. *Pennisetum divisum* and *Leptochloa fusca* contained relatively higher shoot K^+ at 16 and 24 dS m^{-1}, respectively. *Puccinellia distans* showed relatively lower shoot K^+ at 8 and 24 dSm^{-1}. *Panicum turgidum* had the highest root K^+ of all five species at 16 and 24 dSm^{-1}.

Shoot Ca^{2+} concentrations of *Panicum turgidum* and *Puccinellia distans* increased with increase in salinity level, whereas the remaining three species did not show any consistent pattern of increase or decrease in shoot Ca^{2+}. *Cenchrus pennesetiformis* and *Leptochloa fusca* accumulated relatively greater amount of

Ca^{2+} in the shoots than the other species at 8 and 16, and 24 dSm^{-1}, respectively. Increasing salinity had no significant effect on root Ca^{2+} of all five species. *Panicum turgidum* had significantly greater concentrations of root Ca^{2+} compared with the other species at 8 and 16 dSm^{-1}.

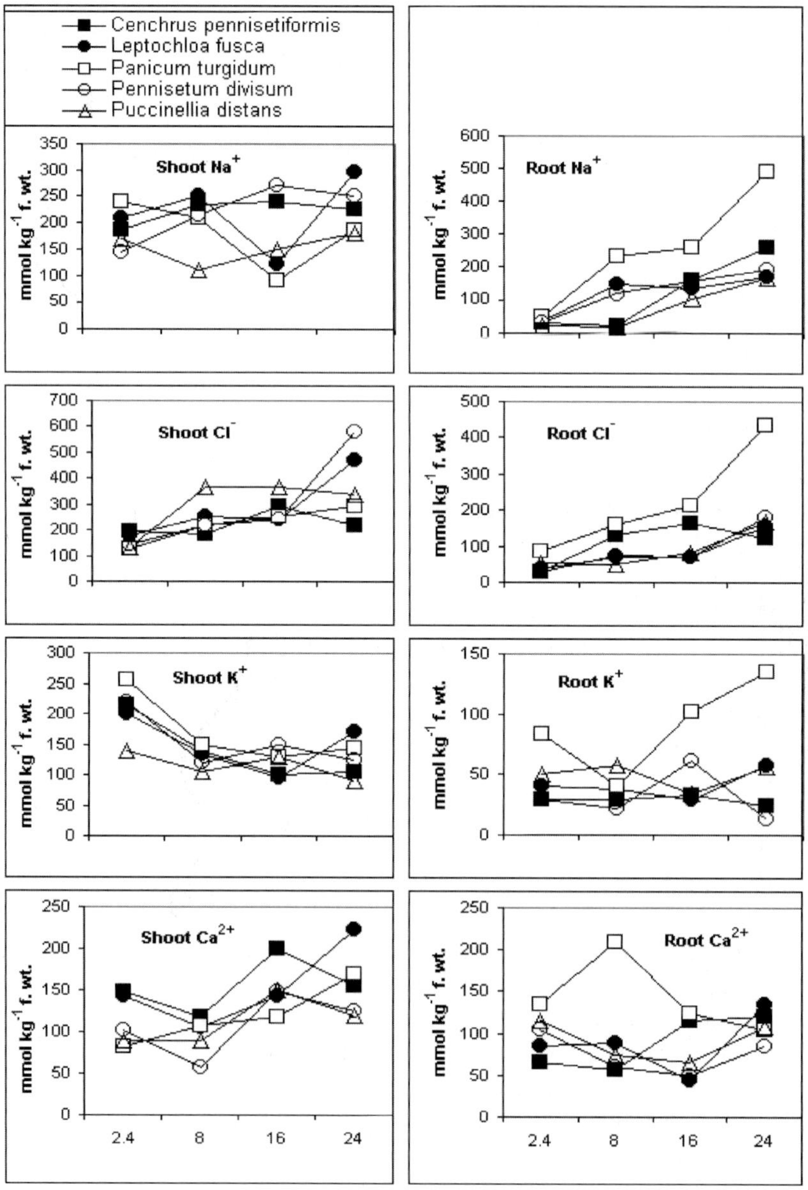

Figure 3. *Na^+, Cl^-, K^+ and Ca^{2+} concentrations (mmol kg^{-1} fresh wt) of shoots and roots of five grass species after six weeks growth at varying salinity levels in sand culture.*

The analyses of different components of soluble carbohydrates (monosaccharides, disaccharides and trisaccharides) in the leaves of three grass species are presented in (Table 7). *Cenchrus pennesetiformis* and *Panicum turgidum* had similar amounts of monosaccharides in the normal and stressed treatments. *Leptochloa fusca* had the lowest accumulation of monosaccharides among the three species. Disaccharide concentration in *Cenchrus pennesetiformis* decreased markedly in the salt treatment, whereas in the remaining two species it increased slightly compared with the control. Trisaccharides were very low in all three species in comparison with monosaccharides and disaccharides. However, in the salt treatment their concentration remained almost unaffected in *Cenchrus pennesetiformis* and *Panicum turgidum*, whereas trisaccharides were detected in *Leptochloa fusca*.

Table 7. *Soluble carbohydrates (mg g^{-1} dry leaf weight) in three grass species grown for six weeks under salt stress.*

		Grass species		
Characteristics	*Salt treatments*	*Cenchrus pennesetiformis*	*Leptochloa fusca*	*Panicum turgidum*
	Control	14.9±1.34	2.4±0.14	14.4±1.46
Mono saccharides	Stressed (16 dS m^{-1})	10.8±0.87	4.7±0.21	8.6±0.37
	Control	104.0±6.32	29.4±2.47	99.4±3.87
Di saccharides	Stressed (16 dS m^{-1})	51.0±3.21	36.6±2.89	117.0±6.67
	Control	0.30±0.02	1.93±0.16	4.03±0.13
Tri saccharides	Stressed (16 dS m^{-1})	0.40±0.03	0.00±0.00	3.30±0.14

3.3. *Experiment 3: Assessment of salt tolerance in germplasm in some potential native grasses.*

Plant height was adversely affected by all levels of salinity (Table 8). The maximum plant height was recorded in *Lasiurus scindicus* accession LS 3/6 at all salinity treatments. The least affected accession due to salinity was BJ 1/2 of *Panicum antidotale*, showing only 9.99% decrease compared to control (Figure 4). It was followed by LS 3/6 of *L. scindicus* (11.58% decrease) and KS 1/1 of *P. antidotale* (13.11% decreases).

Increase in salinity level caused a significant reduction in shoot fresh weight of all accessions (Table 9). The maximum decrease was noted in accession Local-16 of *Cenchrus ciliaris*, showing a decrease of 68.85% under the highest salinity level. The least affected accessions were KS 1/2 and Local-2 of *Cenchrus ciliaris* and LS 3/6 of *Lasiurus scindicus*, showing a decrease of 23.99, 27.46 and 25.33 %, respectively at the highest salinity level. The reduction in plant dry weight was almost similar to that of plant fresh weight (Table 10). The most affected accession was Local-16 of *C. ciliaris* and the least affected accessions were Local-2 and KS 1/2 of *C. ciliaris* and KS 3/6 of *L. scindicus* that decreased by 19.77, 34.27 and 27.83%, respectively. Accession KS 1/1 of *P. antidotale* was also least affected due to salinity in terms of fresh or dry weights of shoots.

Table 8. *Effect of salinity on plant height of different accessions of three grasses from Cholistan desert.*

			Plant height (cm)		
			Conductivity dS m^{-1}		
Acc. No.	*1.25 (Control)*	*10*	*15*	*20*	*25*
Panicum antidotale					
MW 1/1	77.40 cd	73.30 e	72.10 ef	68.10 efg	61.30 ef
BJ1/2	78.10 cd	77.30 de	74.20 cde	72.10 def	70.30 cd
RD 1/1	73.50 d	69.60 e	65. 40 f	61.30 g	58.40 f
RD 1/2	79.50 cd	77.30 de	70.10 ef	68.30 efg	65.00 def
KH 1/6	55.60 e	50.10 f	46.70 g	45.30 h	40.70 g
KS 1/1	76.30 cd	73.20 e	70.70 ef	68.40 efg	66.30 de
Cenchrus ciliaris					
Local-2	91.33 b	84.60 c	81.20 bc	77.30 bcd	74.30 bc
Local-4	88.30 b	83.70 cd	79.70 bcd	74.80 cde	71.70 cd
Local-8	92.10 b	90.10 bc	85.50 ab	83.20 ab	70.20 cd
Local-10	90.10 b	85.40 bc	80.40 bcd	76.70 bcd	74.20 bc
Local-14	80.50 cd	76.30 e	73.20 de	68.20 efg	66.70 de
Local-15	81.40 c	75.50 e	70.30ef	63.50 g	60.40 ef
Local-16	76.40 cd	74.27 e	70.40ef	65.33 fg	45.50 g
KS 1/2	95.30 ab	92.20 ab	86.30ab	81.13 bc	79.30 b
Lasiurus scindicus					
LS 3/6	99.30 a	98.20 a	91.53a	89.33 a	87.80 a

Means with the same letters in each column do not differ significantly at the 5% level

A decrease due to salinity was again recorded in flag leaf area (Table 11), but all the accessions were similarly affected. However, *Cenchrus ciliaris* accession KS 1/2 was the best of all the accessions examined, showing a decrease of 23.86 % under the highest salinity level.

Enhanced accumulation of shoot Na^+ under salinity was found in all the accessions (Figure 5, Table 12), and the highest increase was recorded in *Cenchrus ciliaris* accession Local-14 (650.76%) followed by accession KH 1/6 (228.45 increase). Accession Local-8 of *C. ciliaris* showed the lowest accumulation of Na^+ in the shoots at all salinity levels. A decrease in shoot K^+ was recorded in all the accessions except LS 3/6 of *Lasiurus scindicus* (Table 13), in which 44.84% increase was noted at 20 dSm^{-1}. Accessions Local-4 and Local-2 of *Cenchrus ciliaris* and RD 1/1 of *Panicum antidotale* were the least affected

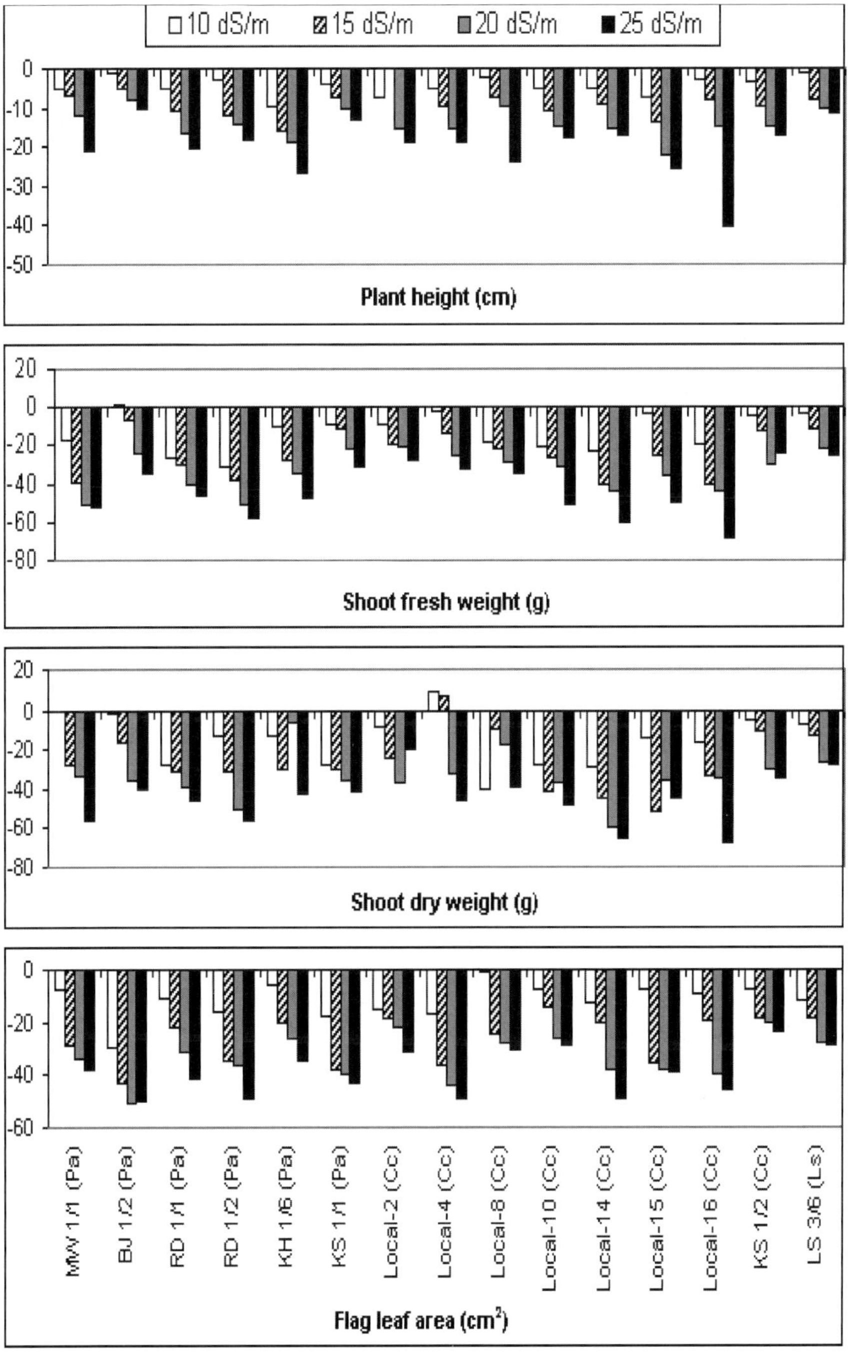

Figure 4. *Percent increase or decrease with regard to control in some growth attributes in different accessions of* Panicum antidotale *(Pa),* Cenchrus ciliaris *(Cc) and* Lasiurus scindicus *(Ls) from Cholistan subjected to salinity stress.*

showing a decrease of 24.85, 31.25 and 23.40% at the highest salinity treatment, respectively. Salt stress caused a reduction in shoot Ca^{2+} in all accessions (Table 14). Increased Ca^{2+} concentration was recorded in only BJ 1/2 of *P. antidotale* at 10, 15 and 20 dS m^{-1}. A stability in this attribute was recorded in accession LS 3/6 of *L. scindicus*, Local-8 and KS 1/2 of *C. ciliaris* and BJ 1/2 of *P. antidotale*. The best among them was LS 3/6, showing a decrease of only 4.58% at the highest salinity level.

On the basis of 50% reduction of shoot dry weight under salinity levels, *Leptochloa fusca* and *Cynodon dactylon* can be regarded as highly tolerant species (Table 15). *Panicum antidotale* is of moderate tolerance while *Pennisetum divisum*, *Echinochloa colona* and *Desmostachya bipinnata* are relatively less tolerant.

Table 9. *Effect of salinity on shoot fresh weights of different accessions of three grasses from Cholistan desert.*

	Shoot fresh weight (g)				
	Conductivity dSm^{-1}				
Acc. No.	*1.25 Control*	*10*	*15*	*20*	*25*
Panicum antidotale					
MW 1/1	6.43 cdef	5.34 cd	3.94 cd	3.12 ef	3.06 de
BJ1/2	7.02 abcd	7.12 ab	6.54 ab	5.32 abc	4.58 bc
RD 1/1	5.84 efg	4.32 de	4.11 cd	3.50 ef	3.11 de
RD 1/2	4.89 g	3.37 e	3.01 d	2.41 f	2.08 e
KH 1/6	5.99 efg	5.41 c	4.32 c	3.91 de	3.12 de
KS 1/1	7.87 ab	7.12 ab	7.01 a	6.12 ab	5.43 ab
Cenchrus ciliaris					
Local-2	8.12 a	7.41 ab	6.53 ab	6.42 a	5.89 a
Local-4	7.93 ab	7.77 a	6.82 ab	5.93 abc	5.34 ab
Local-8	7.84 ab	6.39 bc	6.13 ab	5.59 abc	5.12 ab
Local-10	5.42 fg	4.32 de	3.98 cd	3.74 e	2.68 e
Local-14	5.34 fg	4.10 e	3.20 cd	2.99 ef	2.14 e
Local-15	7.56 abc	7.32 ab	5.67 b	4.87 cd	3.84 cd
Local-16	6.87 bcde	5.54 c	4.12 cd	3.87 de	2.14 e
KS 1/2	7.42 abc	7.12 ab	6.53 ab	5.23 bc	5.64 ab
Lasiurus scindicus					
LS 3/6	6.87 bcde	6.67 ab	6.12 ab	5.34 abc	5.13 ab

Means with the same letters in each column do not differ significantly at the 5% level

4. DISCUSSION

4.1. Experiment 1: Biodiversity assessment of Greater and Lesser Cholistan.

Vegetation of Cholistan is specific at different soil types. Only *Ochthochloa compressa* was found in all the three types, i.e., sand dunes, interdunal flats and saline patches. Vegetation at saline / sodic soils was very specific and very few

Table 10. *Effect of salinity on shoot dry weights of different accessions of three grasses from Cholistan desert.*

		Shoot dry weight (g)			
		Conductivity dS m^{-1}			
Acc. No.	*1.25 Control*	*10*	*15*	*20*	*25*
Panicum antidotale					
MW 1/1	2.34 bc	2.32 ab	1.68 bcd	1.56 a	1.02 cde
BJ 1/2	2.51 ab	2.48 ab	2.11 abc	1.62 a	1.49 abcd
RD 1/1	2.00 bc	1.45 cd	1.38 cd	1.21 ab	1.09 bcde
RD 1/2	1.63 c	1.42 cd	1.12 d	0.81 b	0.71 de
KH 1/6	2.01 bc	1.76 bcd	1.41 cd	1.89 a	1.16 bcde
KS 1/1	3.11 a	2.26 ab	2.16 abc	1.99 a	1.82 ab
Cenchrus ciliaris					
Local-2	2.63 ab	2.41 ab	1.99 abc	1.67 a	2.11 a
Local-4	2.48 ab	2.71 a	2.68 a	1.67 a	1.34 abcde
Local-8	2.34 bc	1.41 cd	2.11 abc	1.93 a	1.42 abcde
Local-10	1.92 bc	1.38 d	1.12 d	1.21 ab	0.99 cde
Local-14	1.94 bc	1.38 d	1.08 d	0.79 b	0.68 e
Local-15	2.52 ab	2.16 abc	1.21 d	1.63 a	1.38 abcde
Local-16	2.23 bc	1.86 bcd	1.48 bcd	1.45 ab	0.72 de
KS1/2	2.48 ab	2.37 ab	2.22 ab	1.74 a	1.63 abc
Lasiurus scindicus					
LS 3/6	2.30 bc	2.12 abcd	2.01 abc	1.68 a	1.66 abc

Means with the same letters in each column do not differ significantly at the 5% level

Table 11. *Effect of salinity on flag leaf area of different accessions of three grasses from Cholistan desert.*

		Flag leaf area (cm^2)			
		Conductivity dS m^{-1}			
Acc. No.	*1.25 Control*	*10*	*15*	*20*	*25*
Panicum antidotale					
MW 1/1	13.53 d	12.52 ef	9.63 c	9.00 bc	8.34 bc
BJ 1/2	12.78 d	9.04 f	7.22 c	6.34 c	6.45 bc
RD 1/1	12.71 d	11.31 ef	9.87 c	8.73 bc	7.45 bc
RD 1/2	12.73 d	10.72 f	8.34 c	8.11 bc	6.54 bc
KH 1/6	11.82 d	11.12 ef	9.43 c	8.76 bc	7.72 bc
KS 1/1	18.13 c	14.94 de	11.28 c	11.00 b	10.30 b
Cenchrus ciliaris					
Local-2	22.48 b	19.01 bc	18.31 ab	17.61 a	15.40 a
Local-4	28.30 a	23.40 a	18.06 ab	15.91 a	14.37 a
Local-8	24.40 b	24.15 a	18.42 ab	17.65 a	17.00 a
Local-10	23.70 b	21.80 ab	20.30 a	17.40 a	16.80 a
Local-14	13.40 d	11.70 ef	10.70 c	8.31 bc	6.80 bc
Local-15	11.34 d	10.47 f	7.36 c	7.01 bc	6.93 bc
Local-16	10.42 d	9.48 f	8.42 c	6.32 c	5.64 c
KS 1/2	18.78 c	17.33 cd	15.27 b	14.98 a	14.30 a
Lasiurus scindicus					
LS 3/6	10.25 d	9.01 f	8.37 c	7.42 bc	7.31 bc

Means with the same letters in each column do not differ significantly at the 5% level

Table 12. *Effect of salinity on concentration of shoot Na^+ of different accessions of three grasses from Cholistan desert.*

		Shoot Na^+ (mg g^{-1} d. wt.)			
			Conductivity dS m^{-1}		
Acc. No.	1.25Control	10	15	20	25
Panicum antidotale					
MW 1/1	3.88 bcd	7.16 ab	7.92 abc	8.41 abc	8.24 bcde
BJ 1/2	3.15 bcd	5.35 bc	6.78 abc	8.00 abc	9.42 abcd
RD 1/1	2.83 bcd	4.76 bc	4.93 c	6.42 bcd	7.34 cde
RD 1/2	3.41 bcd	6.78 ab	7.47 abc	8.93 abc	10.21 abc
KH 1/6	3.48 bcd	7.56 ab	8.91 a	9.32 ab	11.43 a
KS 1/1	5.06 bc	5.15 bc	6.73 abc	7.81 abc	7.93 bcde
Cenchrus ciliaris					
Local-2	5.71 b	6.83 ab	7.23 abc	8.14 abc	9.14 abcd
Local-4	4.31 bcd	6.81 ab	7.71 abc	7.89 abc	8.23 bcde
Local-8	8.51 a	9.48 a	9.50 a	9.52 a	9.63 abcd
Local-10	4.27 bcd	4.83 bc	5.76 bc	7.23 abcd	10.41 ab
Local-14	1.32 d	3.42 c	5.42 c	7.82 abc	9.91 abc
Local-15	2.64 cd	3.24 c	5.88 bc	6.93 abcd	7.87 bcde
Local-16	4.13 bcd	7.32 ab	8.45 ab	9.43 ab	10.56 ab
KS 1/2	3.62 bcd	4.82 bc	5.67 bc	5.94 cd	6.81 de
Lasiurus scindicus					
LS 3/6	2.87 bcd	3.54 c	8.63 ab	4.72 d	5.34 e

Means with the same letters in each column do not differ significantly at the 5% level

Table 13. *Effect of salinity on concentration of shoot K^+ of different accessions of three grasses from Cholistan desert.*

		Shoot K^+ (mg g^{-1} d. wt.)			
			Conductivity dS m^{-1}		
Acc. No.	1.25 Control	10	15	20	25
Panicum antidotale					
MW 1/1	6.82 a	5.78 ab	4.52 abcde	3.46 bcd	2.31 defg
BJ 1/2	5.08 bc	4.89 bc	4.13 bcdef	3.86 abc	3.14 bcde
RD 1/1	4.23 cde	4.12 cd	3.99 cdefg	3.68 bcd	3.24 bcd
RD 1/2	6.38 a	5.67 ab	5.43 a	4.69 ab	4.28 ab
KH 1/6	3.58 ef	2.93 de	2.41 i	1.93 e	1.85 fg
KS 1/1	6.82 a	5.78 ab	4.52 abcde	3.46 bcd	2.31 defg
Cenchrus ciliaris					
Local-2	6.88 a	6.72 a	5.12 abc	5.02 a	4.73 a
Local-4	5.03 bcd	4.96 bc	4.78 abcd	4.26 ab	3.78 abc
Local-8	4.35 cde	3.81 cde	3.71 defgh	2.82 cde	2.72 cdef
Local-10	3.94 cdef	3.11 de	2.78 ghi	2.13 e	1.56 fg
Local-14	4.17 cde	3.13 de	2.98 fghi	2.12 e	1.33 g
Local-15	4.83 cde	4.03 cd	3.37 efghi	2.48 de	1.94 efg
Local-16	3.99 cdef	2.68 e	2.41 i	1.96 e	1.37 g
KS 1/2	6.12 ab	5.63 ab	5.31 ab	4.72 ab	3.84 abc
Lasiurus scindicus					
LS 3/6	2.81 f	2.94 de	2.63 i	4.07 abc	1.81 fg

Means with the same letters in each column do not differ significantly at the 5% level

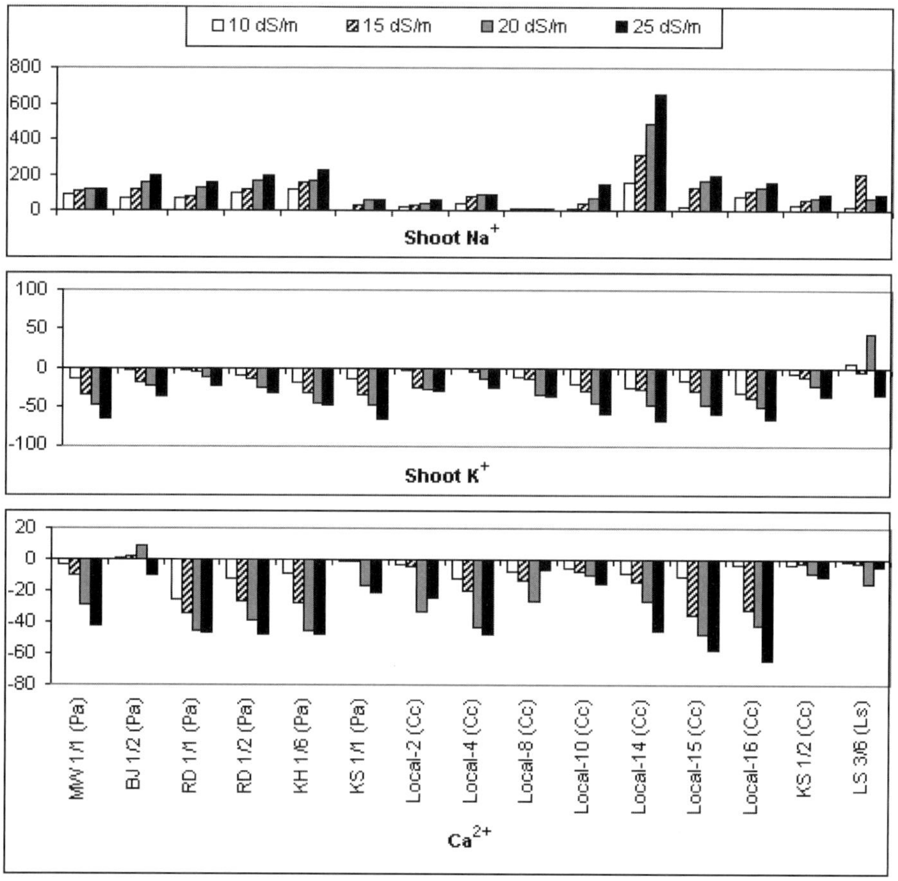

Figure 5. *Percent increase or decrease with regard to control in concentrations of different ions in the shoots of different accessions of* Panicum antidotale *(Pa),* Cenchrus ciliaris *(Cc) and* Lasiurus scindicus *(Ls) from Cholistan subjected to salinity stress.*

species were recorded there. Jalal-ud-Din and Farooq (1975) reported *Salsola baryosma, Suaeda fruticosa, Tamarix aphylla* and *Prosopis glandulosa* at Lalsohanra, but in the present investigation in addition to *Salsola baryosma* and *Suaeda fratiasa, Aeluropus lagopoides, Ochthochloa compressa* and *Trianthema triquetra* were also recorded quite frequently.

Vegetation at interdunal flats varied greatly from the other habitats and also the maximum number of species was recorded there. Rao et al. (1989) reported two dominant grasses, *Cymbopogon jwarancusa* and *Lasiurus scindicus*, whereas in our survey *Aristida adscensionis* dominated all the species. Other dominant species were *Cenchrus biflorus* and *Sporobolus ioclados, Tamarix aphylla, Acacia nilotica* and *Prosopis cineraria* were also occasionally reported which were dominant species according to Rao et al. (1989).

Table 14. *Effect of salinity on concentration of shoot Ca^{2+} of different accessions of three grasses from Cholistan desert.*

	Shoot Ca^{2+} (mg g^{-1} d. wt.)				
	Conductivity dS m^{-1}				
Acc. No.	1.25Control	10	15	20	25
Panicum antidotale					
MW 1/1	1.58 ef	1.52 fg	1.43 ef	1.12 cde	0.91 d
BJ 1/2	1.80 ef	1.83 efg	1.84 de	1.96 be	1.63 bcd
RD 1/1	4.78 a	3.57 ab	3.12 ab	2.61 ab	2.57 a
RD 1/2	4.12 ab	3.62 ab	3.01 ab	2.54 ab	2.13 abc
KH 1/6	1.68 ef	1.54 fg	1.22 ef	0.91 de	0.87 d
KS 1/1	3.76 bc	3.71 a	3.70 a	3.15 a	2.96 a
Cenchrus ciliaris					
Local-2	2.90 d	2.81 bcd	2.76 bc	1.93 bc	2.20 abc
Local-4	1.78 ef	1.56 fg	1.42 ef	1.01 de	0.93 d
Local-8	1.52 cf	1.41 fg	1.32 ef	1.12 cde	1.42 cd
Local-10	0.98 f	0.93 g	0.90 f	0.88 e	0.83 d
Local-14	1.68 ef	1.53 fg	1.43 ef	1.24 cde	0.91 d
Local-15	2.37 de	2.11 def	1.53 ef	1.25 cde	1.01 d
Local-16	3.12 cd	3.00 abc	2.12 cde	1.81 bcd	1.12 d
KS 1/2	2.71 d	2.63 cde	2.64 bcd	2.48 ab	2.41 ab
Lasiurus scindicus					
LS 3/6	1.53 ef	1.51 f	1.50 ef	1.30 cde	1.46 cd

Means with the same letters in each column do not differ significantly at the 5% level

Table 15. *Salt tolerance of some forage grasses (salt level at which 50% dry matter reduction takes place) in some grass species from Cholistan desert.*

Grass species	Average root zone salinity (dS m^{-1})
Cenchrus ciliaris	18.0
Cenchrus pennesetiformis	17.5
Cynodon dactylon	13.2-21.0
Desmostachya bipinnata	9.0
Echinochloa colona	11.2
Leptochloa fusca	14.6-22.0
Panicum antidotale	16.0
Panicum turgidum	17.0
Pennisetum divisum	11.5

Many grasses of the Cholistan desert are of great economic importance. *Aristida adscensionis* can be eaten by cattle when young (Bor, 1941), but it is a grass of low stature giving little foliage and poor feeding. Nothing is definite about its economic value (Cope, 1982). *Cenchrus* spp., i.e, *C. ciliaris, C. pennesetiformis, C. biflorus* and *C. prieurii*, all are excellent fodder species. *Cenchrus pennesetiformis* remains green during extreme dry conditions, while

C. biflorus is only eaten at young stages because it develops spiny bristly inflorescence at the adult stage.

Cynodon dactylon is one of the best fodder grasses and is highly crush-proof. *Cymbopogon jwarancusa* is not a fodder species, but *Desmostachya bipinnata* and *Imperata cylindrical* are only consumed by buffaloes when young. *Dichanthium annulatum* and *Echinochloa colona*, although susceptible to environmental stresses like salinity and drought, are the finest fodder species, eagerly eaten by livestock at all stages.

Eragrostis ciliaris affords good grazing. *Lasiurus scindicus*, a drought tolerant species, is a valuable fodder species of desert areas and is relished by camel, goat and cattle. *Ochthochloa compressa* is also a good fodder grass for cattle and horses.

Panicum turgidum although with fewer leaves, its stems are very palatable and frequently heavily grazed, being particularly relished by cattle and camel. *Panicum antidotale* is of doubtful value as fodder but is grazed when quite young as it afterwards acquires a bitter or saltish taste. *Phragmites karka* is said to be poisonous to cattle, but in any case it is far too coarse for fodder.

Saccharum spontaneum gives a favourite buffalo fodder. *Saccharum bengalense* is of little value as a fodder and palatable at only very young stages. *Stipagrostis plumosus* is of great value as a fodder, but its production is quite low in the area. However, it provides dense cover in other Middle East countries.

4.2. Experiment 2: Response of some arid zone grasses to brackish water.

The results for the biomass production clearly shows that *Leptochloa fusca* and *Puccinellia distans* were highly tolerant to varying salinity levels of the growth medium compared with the other three species. The better performance of these two species is expected as they were already found to be highly salt tolerant *Leptochloa fusca* (Malik et al., 1986), and *Puccinellia distans* (Ashraf et al., 1986). *Pennisetum divisum* was the poorest of all species, whereas *Cenchrus pennesetiformis* and *Panicum turgidum* were intermediate in performance in response to salinity.

In the present study the salinity treatments were prepared by mixing different salts in ratio, which correspond to the composition of subsoil saline water from the Cholistan desert. Therefore, considerable interaction of different ions in all species has been observed at each salinity level. Each species has used its own specific selective ion transport mechanism in response to varying salinity treatments. For instance, the highly tolerant *Leptochloa fusca* accumulated relatively greater concentrations of Na^+ and Cl^- in the shoots at the highest salinity level, whereas these concentrations were low in the roots. Thus the species used a typical halophytic mechanism (Flowers et al., 1977). Since the species possesses characteristic salt glands (Malik et al., 1986) it is possible that Na^+ and Cl^- absorbed by roots are rapidly translocated to leaves for onward excretion through salt glands. *Leptochloa fusca* also accumulated high concentrations of both K^+ and

Ca^{2+} for maintaining low Na^+/K^+ and Na^+/Ca^{2+} ratios in the shoots. High Na^+/K^+ and Na^+/Ca^{2+} ratios have already been found responsible for increasing membrane permeability in plants (Greenway & Munns, 1980; Muhammad et al., 1987).

In contrast, the second highly salt tolerant species, *Puccinellia distans* partially included Na^+ in both shoots and roots, but accumulated high concentration of Cl^- in the shoots. The low concentrations of both Na^+ and K^+ in the shoots of *Puccinellia distans* show that it does not have selectivity to both K^+ and Na^+ as was suggested by Greenway and Munns (1980) that some mesophytes are selective to K^+ while others are selective to both K^+ and Na^+.

The relatively most salt sensitive species, *Pennisetum divisum* showed a clear relationship between its poor growth and patterns of ion accumulation under saline conditions. Its high accumulation of both Na^+ and Cl^- in the shoots can be related to the early findings of Wyn Jones et al. (1984) in which it was demonstrated that *Agropyron intermedium* was salt sensitive compared with *Agropyron junceum* because it efficiently accumulated both Na^+ and Cl^- in its leaves. In addition, low Na^+/K^+ and Na^+/Ca^{2+} ratios were not maintained by this species at varying salinity treatments.

The other two species, *Cenchrus pennesetiformis* and *Panicum turgidum* were relatively intermediate in salinity tolerance. *Cenchrus pennesetiformis* absorbed large amount of Na^+ in the shoots, but at the same time it accumulated high concentrations of Ca^{2+} in shoot so as to maintain Na^+/Ca^{2+} ratio low. In contrast, *Panicum turgidum* maintained low concentrations of both Na^+ and Cl^- in the shoots, although the concentrations of these ions were high in its roots. This type of mechanism is very common in many salt tolerant mesophytes (Maas & Nieman, 1978; Greenway & Munns, 1980). The same authors advocated that salt excluders have the ability to restrict the uptake of salts into the shoot. This might be due to the phenomenon that toxic ions such as Na^+ are absorbed in considerable amount, but are reabsorbed from the root or the shoot and is either stored in the roots or retranslocated to the soil.

Of the free sugars, monosaccharides (glucose and fructose) and disaccharides (sucrose) are known to play important role in osmotic adjustment in plants subjected to salinity stress (Gorham et al., 1981; Weimberg et al., 1984). In the present study, neither change in water content nor other osmotically active solutes was measured. Therefore, the contribution of monosaccharides and disaccharides to osmotic adjustment is not clear. However, it seems reasonable to assume a substantial contribution by glucose, fructose, and sucrose to osmotic adjustment due to their levels in the three grass species under salt stress. It is evident that the contribution of trisaccharides is likely to be almost negligible in the three species, except for a marginal role in *Panicum turgidum*. A significant reduction in the sucrose level of *Cenchrus pennesetiformis* in the salt treatment is in agreement with the results of Downton (1977) who found a decrease in the sucrose level in grapevines under saline conditions, but is in contrast to the report by Weimberg et al., (1984) who found a considerable increase in this carbohydrate in the leaves of *Sorghum bicolor* in Cl^- and SO_4^{2}- salinities.

A lack of any difference in the levels of monosaccharides and disaccharides in *Leptochloa fusca* and *Panicum turgidum* is quite similar to what was found in *Capsicum frutescens* and *Atriplex nummularia* (El-Shourbag & Koshk, 1975), but is also in contrast to *Sorghum bicolor* under salinity (Weimberg et al., 1984).

4.3. *Experiment 3: Assessment of salt tolerance in germplasm in some potential native grasses.*

Although salt stress caused a significant reduction in growth attributes of all accessions examined, a great amount of variation was observed in this set of germplasm. For example, accession KS 1/2 and Local-2 of *Cenchrus ciliaris*, LS 3/6 of *Lasiurus scindicus* and KS 1/1 of *Panicum antidotale* were superior to the other accessions in all growth parameters measured. However, the better growth performance of these accessions can be easily related to their low uptake of Na^+ and relatively high uptake of K^+ in the shoots under saline conditions, a phenomenon which has been commonly found in a number of mesophytes (Greenway & Munns, 1980). In conclusion, the salt tolerant accessions of three grass species found in this set of germplasm could be exploited for economic utilization of Cholistan land, particularly those where subsoil water is brackish.

Taken overall, it is not difficult to say that *Cenchrus pennesetiformis* and *Panicum turgidum* intermediate in salt tolerance can be grown in those areas of the Cholistan desert having moderately saline subsoil water for irrigation. These species have high forage value for all types of livestock and already well adapted to the prevailing environmental conditions of the area (Ashraf & Bokhari, 1987). Thus these two species could be of great value for economic utilization of the desert area. The other two highly salt tolerant species *Leptochloa fusca* and *Puccinellia distans* may not be suitable for the area as they both are adapted to entirely different environmental conditions than those of Cholistan and also highly sensitive to drought conditions. With the proper management at Cholistan, the production of these grasses can be enhanced which can bring socio-economic gain for the farmers rehabilitated in these areas. Our present study also confirms that these grasses can be introduced on salt affected and drought prone areas where livestock is facing severe problems of fodder shortage.

5. LITERATURE CITED

Ashraf, M. & Bokhari, M.H. 1987. Biological approach for economic utilization of the Cholistan desert. Biologia 33: 27-34.

Ashraf, M., McNeilly, T. & Bradshaw, A.D. 1986. The potential for evaluation of salt (NaCl) tolerance in seven grass species. New Phytologist 103: 299-309.

Bor, N.L. 1941. Indian Forest Records. New Series Botany, vol. II Common grasses of the United Provinces. New Dehli: Government of India Press. 220 pp.

Cope, T.A. 1982. Poaceae, vol.143. In: E. Nasir & S.I. Ali, (Eds.), Flora of Pakistan. National Herbarium (Steward Collection). Islamabad, Pakistan: Pakistan Agricultural Research Council. 680 pp.

Downton, W.J.S. 1977. Photosynthesis in salt stressed grapevines. Australian Journal Plant Physiology 4: 183-192.

El-Shourbagy, M.N. & Koshk, H.T. 1975. Sodium Cl^- effects on the sugar metabolism of several plants. Phytochemistry 17: 101-108.
Flowers, T.J., Troke, P.F. & Yeo, A.R. 1977. The mechanism of salt tolerance in halophytes. Annual Review Plant Physiology 28: 89-121.
Gorham, J., Hughes, L. & Wyn Jones, R.G. 1981. Low molecular weight carbohydrates in some salt stressed plants. Physiology Plant 31: 149-190.
Greenway, H. & Munns, R.A. 1980. Mechanism of salt tolerance in non-Halophytes. Annual Review Plant Physiology 31: 149-190.
Hussain, F. 1983. Field and Laboratory Manual of Plant Ecology. Islamabad, Pakistan National Academy of Higher Education. University Grants Commission. 422 pp.
Hussain, S.S. 1992. Pakistan Manual of plant Ecology. Islamabad, Pakistan: National Book Foundation. 376 pp.
Jalal-ud-Din & Farooq, M. 1975. Soil variation in relation to forest management in Lalsohanra irrigated plantation. Pakistan Journal Forestry 25: 5-13.
Leopold, A.S. 1963. The Desert. New York, New York: Time-life International. 160 pp.
Maas, E.V. & Nieman, R.H. 1978. Physiology of plant tolerance to salinity. In: G.A. Jung (Ed.), Crop tolerance to suboptimal land conditions. CSSA & SSA (Eds.), Crop tolerance to suboptimal land conditions. Madison, Wisconsin:Soil Science Society of America. 277-299 pp.
Malik, K.A., Aslam, Z. & Naqvi, M. 1986. Kallar Grass-A Plant for Saline Land. 93 pp. Faisalabad, Pakistan: The Nuclear Institute for Agriculture and Biology.
Muhammad, S., Akbar, M. & Neue, H.U. 1987. Effect of Na/Ca and N/K ratios in saline culture solution on the growth and mineral nutrition of rice (*Oryza sativa* L.). Plant Soil 104: 57-62.
Muzaffar, M. 1997. Atlas of Pakistan Survey of Pakistan. Murree Road Rawalpindi, Pakistan, Director Map Publication.143 pp.
Rao, A.R., Arshad, M. & Shafiq, M. 1989. Perenial Grass Germplasm of Cholistan Desert and Its Phytosociology. Bahawalpur, Pakistan: CHIDS, Islamia University. 84 pp.
Weinburg, R., Lerner, H.R. & Poljakoff-Mayber, A. 1984. Changes in growth and water soluble solute concentrations in *Sorghum bicolor* stressed with sodium and potassium salts. Physiology Plant 62: 472-480.
Wyn Jones, R.G., Gorham, J. & McDonnell, E. 1984. Organic and Nonorganic Solute Contgents as Selection Criteria for Salt Tolerance in Triticeae. In: R.C. Staples & G.H. Toenniessen (Eds.), Salinity Tolerance in Plants: Strategies for Crop Improvement. New York: Wiley Interscience. 189-203 pp.

CHAPTER 4

VARIABILITY OF FRUIT AND SEED-OIL CHARACTERISTICS IN TUNISIAN ACCESSIONS OF THE HALOPHYTE CAKILE MARITIMA (BRASSICACEAE)

MOHAMED ALI GHARS[1], ALMED DEBEZ[1], ABDERRAZZAK SMAOUI[2], MOKTAR ZARROUK[2], CLAUDE GRIGNON[3] AND CHEDLY ABDELLY[1, 3]

[1]*INRST, Laboratoire d'Adaptation des Plantes aux Contraintes Abiotiques, BP 95, Hammam-Lif 2050, Tunisia.*
[2]*INRST, Laboratoire de Caractérisation et de la Qualité de l'Huile d'Olive, BP 95, Hammam-Lif 2050, Tunisia.*
[3]*B&PMP, Agro-M INRA, 34060 Montpellier CEDEX 1, France.*
[3]*corresponding author: chedly.abdelly@inrst.rnrt.tn*

Abstract. The annual coastal halophyte *Cakile maritima* shows considerable ecological and economic importance. This study describes the variability of the fruit characteristics in twelve accessions of this species, harvested along the Tunisian littoral. Depending on the plant-sampling site, seed-oil contents ranged from 25% to 39% on the seed fresh weight basis. Reserve lipids (Triacylglycerols) represented 80% to 97% of the total lipids, and erucic acid, was the major fatty acid (25% to 35% of the total fatty acids). Fruit morphology and ion status within stems, siliques, and seeds exhibited great fluctuation. *Cakile maritima* seems to protect its reproductive organs from the salt harmful accumulation.

Abbreviations: 16:0: palmitic acid, 16:1: palmitic acid, 17:0: heptadecanoic acid, 18:0: stearic acid, 18:1: oleic acid, 18:2: linoleic acid, 18:3: linolenic, 20:0: arachidic acid, 20:1: gadoleic acid, 22:0: behenic acid, 22:1: erucic acid, DAG: diacylglycerols, DW: dry weight, FFA: free fatty acids, HEA: high-erucic acid, MAG: monoacylglycerols, MAR: Mean annual rainfall, PL: polar lipids, TAG: triacylglycerols, TLC: thin layer chromatography.

1. INTRODUCTION

Coastal areas worldwide are among the most vulnerable ecosystems threatening progressive salinization (Van Zandt et al., 2003). The negative impact of this phenomenon is particularly exacerbated in the irrigated semi-arid and arid regions, characterized by unfavorable environmental-soil features like high water consumption for irrigation, salt drainage, greater evapo-transpiration rates and secondary root-zone salinization (Smedema & Shiati, 2002). In Tunisia, the (semi)

M.A. Khan and D.J. Weber (eds.), Ecophysiology of High Salinity Tolerant Plants, 55-67.

arid Mediterranean bio-climatic regions (rainfall ranging from 250 to 700 mm) are frequently irrigated with salinized water. Consequently, about 10% of the whole territory and 20% of the cultivated lands are salinized (Hachicha et al., 1994).

Halophytes are exposed to high salt levels in their native habitats, have evolved several mechanisms to deal successfully with these stressful conditions (Zhang, 1996; Donovan et al., 1997). Several among them are recognized as new crop-species (Grieve & Suarez, 1997; Glenn et al., 1998, Khan & Duke, 2001), either like fodder (Khan et al., 2000) or for oil production (Puppala & Fowler, 2002; Bashan et al., 2000; Pita Villamil et al., 2002; Dierig et al., 2003). However, exploiting halophytes requires first the identification of species and their habitats, the ecophysiology and their eventual economical utilization (Khan & Duke, 2001; Kefu et al., 2002).

The present work is a part of a Tunisian project, which aimed at identifying local plant genetic resources of saline ecosystems, presenting potential ecological and/or economical interest. The exploration of the Tunisian coastal areas allowed us to recognize three oilseed halophytes: *Zygophyllum album* (Zygophyllaceae), *Crithmum maritimum* (Umbelliferous), and *Cakile maritima* (Brassicaceae). Oil extraction from *C. maritima* seeds revealed appreciable oil content (ca. 40% on the fresh weight basis). In addition, this oil was rich in erucic acid (Zarrouk et al., 2004), thus potentially appropriate for industrial applications. Alone, the US market demand for this product was estimated at about 18000 metric tons with the majority being imported (Bhardwaj & Hamama, 2003), which gives an indication of the extent of the need for erucic acid.

The purpose of this investigation was to evaluate the variability of seed-oil composition and of fruit and seed mass. Salt distribution among the harvested organs was also determined.

2. MATERIALS AND METHODS

2.1. Description of the species

The mediterranean annual halophyte *C. maritima* achieves its whole life cycle exclusively on coastal sandy dunes and is naturally absent from the inland localities or rocky coasts (Clausing et al., 2000). Nitrogen is required for the development of this succulent plant (Pakeman & Lee, 1991), which shows optimal growth at moderate salt levels (50-100 mM NaCl) and survives even at salinity close to seawater (500 mM NaCl), despite accumulating large amounts of sodium in the leaves (Debez et al., 2004). The fruit consisting of a two-segmented silique (Rodman, 1974), contributes to the plant dispersal when transported away by the wind and waves, owing to its long-term floating ability (Barbour & Rodman, 1970; Maun & Payne, 1989).

2.2. Sampling sites and plant sampling methodology

Though characterized as mediterranean, Tunisian climate displays multiple bioclimatic stages, ranging from the humid (northern) to the saharan (southern). The semi-arid stage prevails in the main part of the country. Twelve harvest sites were selected along the coast, from the north of the country to the sub-saharan region (Figure 1). Here are the different regions and the corresponding bioclimatic stage characterized by the mean annual rainfall (MAR): Tabaraka (humid, MAR > 1500 mm) for, Bizerte (sub-humid, MAR 1000 mm - 1500 mm), Raoued, La Marsa, Hammamet, Enfidha, Sousse, Monastir and Mahdia (semi-arid with temperate winter, MAR 200 mm - 700 mm), Sfax, Jerba and Zarzis

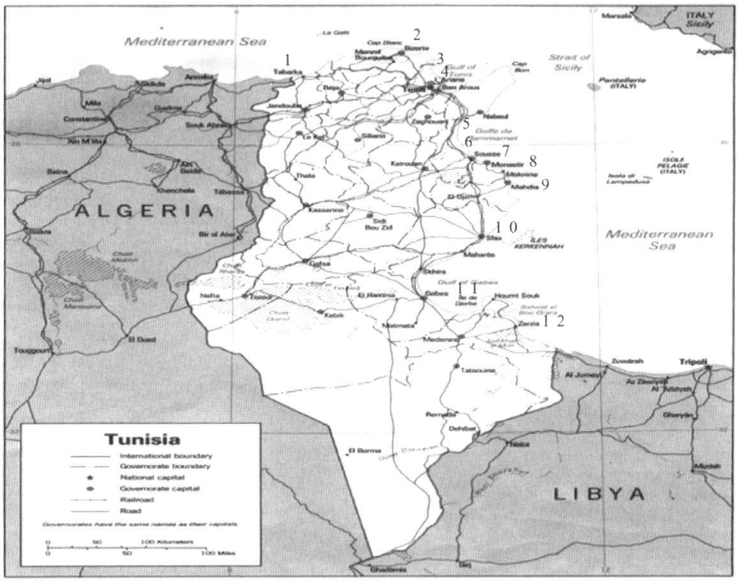

Figure 1. *General map of Tunisia showing the different sampling sites of* C. maritima *seeds. The numbers correspond to the following accessions: 1- Tabarka, 2- Bizerte, 3- Raoued, 4- La Marsa, 5- Hammamet, 6- Enfidha, 7- Sousse, 8- Monastir, 9- Mahdia, 10- Sfax, 11- Jerba, and 12- Zarzis.*

(Arid with temperate winter, MAR 100 mm - 200 mm). Mature siliques and desiccated twigs of C. maritima were harvested between August and October 2001.

2.3. Fruit and seed mass

Samples (n = 300) from each accession were used to estimate fruit and seed mass variability. Masses of each silique and of the corresponding seed manually extracted from the fruit were determined with a Mettler microbalance.

2.4. Ion relationships

Na^+ contents were determined by flame emission photometry on 0.5 % HNO_3 extracts (Corning, UK). Chloride was assayed by coulometry (Büchler chloridometer) on the same extract. Five replicates of the different plant organs (twigs, siliques, and seeds) were used.

2.5. Lipid analysis

Seed total lipids were extracted as described by Allen & Good (1971), with methanol/chloroform/water (1:1:1, vol/vol/vol). According to Mangold (1964), triacylglycerols (TAG) were separated by thin layer chromatography (TLC), using silica gel plates (Merck G 60). Fatty acid methyl esters were prepared (Metcalfe et al., 1966) and analyzed by Flame Ionization Detector-Gas Chromatography using a Hewlett-Packard model 4890 D with a HP innowax (30 m x 0.25 mm) capillary column. The column was maintained isothermally at 210°C . Fatty acid methyl esters were quantified by adding heptadecanoic acid (17:0) as an internal standard. Results are the means of three samples.

2.6. Statistical analysis

Data were statistically analyzed (one-way ANOVA) using the SPSS. Duncan post-hoc test was carried-out if significant differences at $P < 0.05$ occurred among the accessions.

3. RESULTS

3.1. Seed-oil composition

Significant differences among the accessions were found for the seed-oil content, with values ranging from ca. 25% to 39% on the fresh weight basis (Figure 2A), thus confirming the oleaginous potential of this halophyte. Seeds harvested from the humid and sub-humid regions (respectively Tabarka and Bizerte) exhibited the highest seed-oil contents (38.8%), whereas the lowest level was found in La Marsa accession (25.4%). Seed-oil percentage in the remaining provenances was important, ranging from ca. 32% for Raoued and Zarzis to 37% for Monastir.

Saturated, monounsaturated and polyunsaturated fatty acids were found in *C. maritima* seed-oil (Table 1). Regardless of the sampling site, erucic acid (22:1) was the most abundant fatty acid, accounting for 25% to 32% of the total fatty acids. Mahdia (semi-arid region) and Zarzis (arid region) showed the highest erucic acid contents (up to 31.9 %), while the lowest contents (ca. 25 %) were recorded in Hammamet and Raoued (the both from the semi-arid region). Appreciable levels of C_{18} unsaturated fatty acids were found: oleic (18:1), linoleic (18:2), and linolenic (18:3) acid percentages reached up to ca. 25 %, 18 %, and 16 % respectively (Table 2).

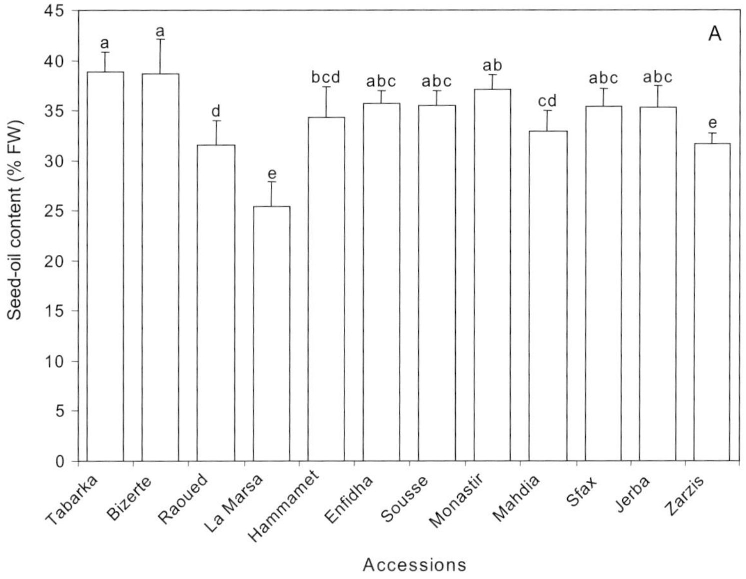

Figure 2a. *Variability of seed-oil content and erucic acid proportion in seed*
Tag. A: Oil content in seeds of C. maritima *accessions.*
Means of 3 replicates ± standard error. Means followed by at least one same letter were not significantly different between the accessions at P < 5 %.

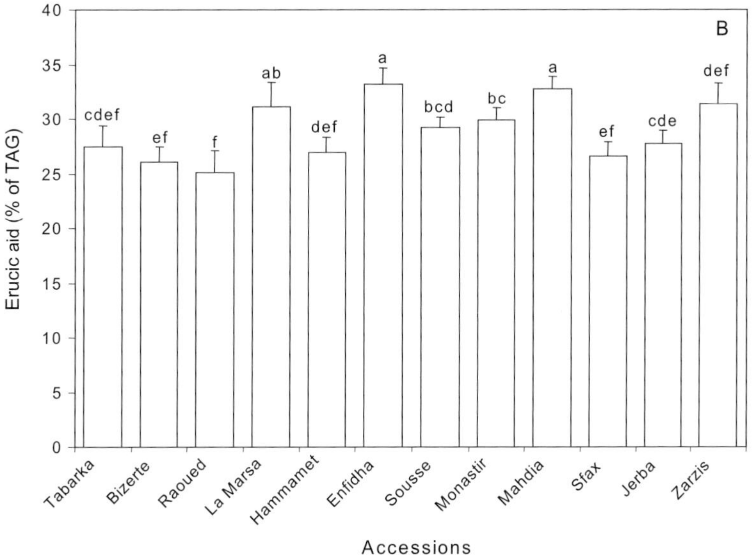

Figure 2b. *Variability of seed-oil content and erucic acid proportion in seed*
Tag. B: Erucic acid percentage in seed of C. maritima accessions. Means of 3 replicates ± standard error. Means followed by at least one same letter were not significantly different between the accessions at P < 5%.

The lipid categories identified were triacylglycerols (TAG), diacylglycerols (DAG), monoacylglycerols (MAG), free fatty acids (FFA), and polar lipids (PL) (Table II). For all the accessions, TAG representing the main form of reserve lipids in oil-seeds, were the major lipid class, contributing for up to 97% of the total lipids. Significant differences were found depending on the sampling site, Jerba being characterized by the highest level (96.9%), unlike Raoued, which showed the lowest one (82.6%). Similarly to what was described for the seed-oil, TAG fatty acid pattern analysis revealed once more the prevalence of erucic acid, which accounted for 25.1% (Raoued) to 33.2% (Enfidha) of this lipid category (Figure 2B).

3.2. Variability of fruit mass and ion relationships

One way-ANOVA statistical analysis revealed significant differences in mean individual masses of fruits, teguments and seeds. Thc both accessions harvested from the humid and sub-humid bioclimatic regions (Tabarka and Bizerte) were characterized by high seed masses (ca. 7.0 mg), while Sfax accession displayed the heaviest seeds (8.8 mg) (Figure 3). Jerba and Zarzis accessions, belonging to the same bioclimatic stage, but closer to the sub-saharan coast, showed the lowest seed mass (3.5-4.5 mg). This tendency was also true when considering the tegument and silique mass.

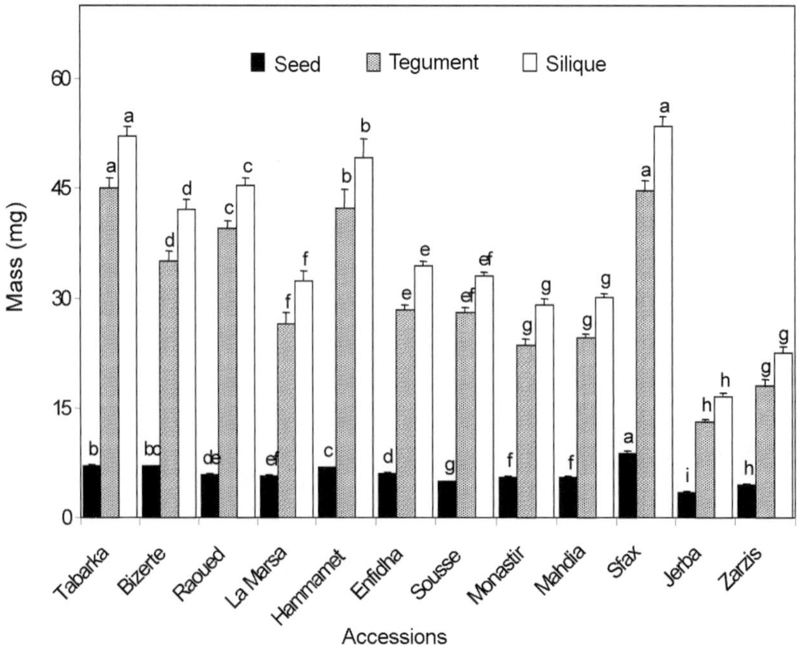

Figure 3. *Variability of fruit and seed morphological characteristics. Mean individual mass of* Cmaritima *silique, and its corresponding tegument and seed. Means of 300 replicates ± standard error. For each organ, means followed by at least one same letter were not significantly different between the accessions at P < 5%.*

Table 1. Variability of fatty acid composition in seed-oil among C, maritima accessions. Means of 3 replicates ± standard error. Means followed by at least one same letter were not significantly different between the accessions at P < 5%.

Accession	16:0	16:1	18:0	18:1	18:2	18:3	20:0	20:1	22:0	22:1
Tabarka	5.71	1.41	1.97	19.70	16.70	16.95	1.44	8.98	1.24	25.90
	± 1.09ab	± 0.03a	± 0.33ef	± 3.83bed	± 2.85ab	± 2.11de	± 0.27de	± 1.00ab	± 0.17de	± 1.89bcd
Bizerte	5.00	0.72	2.34	21.22	14.04	17.25	1.56	8.36	1.66	27.84
	± 0.74abcd	± 0.05c	± 0.44cde	± 3.57b	± 2.73cd	± 2.75de	± 0.47bcde	± 1.32abc	± 0.27c	± 1.19b
Raoued	5.86	0.89	2.28	19.95	16.51	18.05	1.60	8.88	0.60	25.38
	± 0.80a	± 0.80b	± 0.24cdef	± 1.32bcd	± 1.11ab	± 1.09cd	± 0.12abcd	± 0.21ab	± 0.24g	± 1.10cd
La Marsa	4.98	0.24	2.43	20.19	16.47	15.70	1.90	8.20	2.54	27.35
	± 0.34abcd	± 0.06fg	± 0.23cde	± 0.74bc	± 0.81ab	± 0.58c	± 0.41abcd	± 1.13bc	± 0.22a	± 0.53bc
Hammamet	5.92	0.40	2.21	15.97	17.60	22.29	1.41	8.32	1.06	24.82
	± 0.07a	± 0.07d	± 0.05cdef	± 0.13ef	± 0.11a	± 0.16a	± 0.08c	± 0.10bc	± 0.10ef	± 1.86d
Enfidha	4.88	0.21	2.98	16.00	16.88	21.63	1.62	8.30	0.74	26.77
	± 0.16bcde	± 0.05fg	± 0.08ab	± 0.08ef	± 0.06ab	± `0.17ab	± 0.18abcde	± 0.18bc	± 0.11fg	± 1.19bcd
Sousse	5.51	0.26	2.07	16.41	17.68	19.81	1.48	7.80	1.67	27.30
	± 0.35abc	± 0.08ef	± 0.20def	± 0.55ef	± 1.2la	± 1.31bc	± 0.41cde	± 0.93bc	± 0.46c	± 1.49bc
Monastir	5.38	0.27	2.46	18.72	17.95	16.83	1.93	8.71	0.75	27.01
	± 0.65abc	± 0.12ef	± 0.25cde	± 0.38bcde	± 0.68a	± 1.19de	± 0.27abc	± 0.34ab	± 0.10fg	± 1.52bcd
Mahdia	4.71	0.24	1.86	15.69	14.98	20.18	1.56	7.32	1.58	31.89
	± 0.26cde	± 0.05fg	± 0.21f	± 1.34f	± 1.13bcd	± 1.45ab	± 0.18bcde	± 0.58c	± 0.24cd	± 1.23a
Sfax	4.11	0.15	2.44	17.73	12.69	21.71	1.49	8.61	0.98	30.10
	± 0.43de	± 0.02g	± 0.36cde	± 1.05cdef	± 1.01cd	± 1.46ab	± 0.27cde	± 0.09abc	± 0.10efg	± 1.35a
Jerba	3.97	0.43	3.09	24.55	16.57	12.80	2.05	9.68	2.22	24.64
	± 0.45c	± 0.04d	± 0.39a	± 1.94a	± 1.63ab	± 1.20f	± 0.24a	± 0.82a	± 0.46a	± 1.40d
Zarzis	5.23	0.34	2.61	17.04	14.66	16.97	2.10	8.11	2.17	31.21
	± 0.62abc	± 0.04de	± 0.30bc	± 1.13def	± 1.27bcd	± 0.65de	± 0.19ab	± 0.69bc	± 0.34ab	± 1.73a

Table 2. Variability of lipid categories proportion in seed-oil among C, maritima accessions. Means of 3 replicates ± standard error. Means followed by at least one same letter were not significantly different between the accessions at P < 5%.

Accession	TAG	DAG	MAG	FFA	PL	Total
Tabarka	91.63 ± 1.27b	1.73 ± 0.25e	1.41 ± 0.35d	0.55 ± 0.12f	4.67 ± 0.7ed	100
Bizerte	88.90 ± 1.61ed	3.23 ± 0.33c	4.08 ± 0.41a	2.82 ± 0.31d	0.97 ± 0.40f	100
Raoued	82.62 ± 2.72e	2.73 ± 0.58d	3.47 ± 0.20a	4.50 ± 0.71e	6.42 ± 1.70b	100
La Marsa	95.65 ± 1.71a	1.39 ± 0.06ef	0.54 ± 0.06ef	0.91 ± 0.08f	1.51 ± 0.23f	100
Hammamet	91.32 ± 2.09bc	1.45 ± 0.07ef	1.74 ± 0.09cd	1.69 ± 0.13e	3.80 ± 0.11de	100
Enfidha	91.27 ± 1.55bc	2.89 ± 0.17cd	0.86 ± 0.10e	1.92 ± 0.11e	3.07 ± 0.13e	100
Sousse	84.17 ± 1.78e	2.44 ± 0.31d	0.88 ± 0.38e	4.21 ± 0.42c	7.68 ± 0.58a	100
Monastir	92.01 ± 1.55b	1.76 ± 0.31e	0.70 ± 0.11e	1.56 ± 0.23e	3.98 ± 0.19cd	100
Mahdia	88.13 ± 1.19d	4.73 ± 0.51b	2.18 ± 0.33b	1.91 ± 0.20e	3.06 ± 0.32e	100
Sfax	84.30 ± 1.36e	5.22 ± 0.51a	0.91 ± 0.09e	3.41 ± 0.68cd	6.15 ± 0.60b	100
Jerba	96.86 ± 1.64a	0.53 ± 0.07g	0.21 ± 0.05f	0.61 ± 0.09f	1.80 ± 0.11f	100
Zarzis	84.63 ± 1.45e	1.19 ± 0.16f	2.11 ± 0.39bc	7.15 ± 0.54a	4.92 ± 0.51e	100

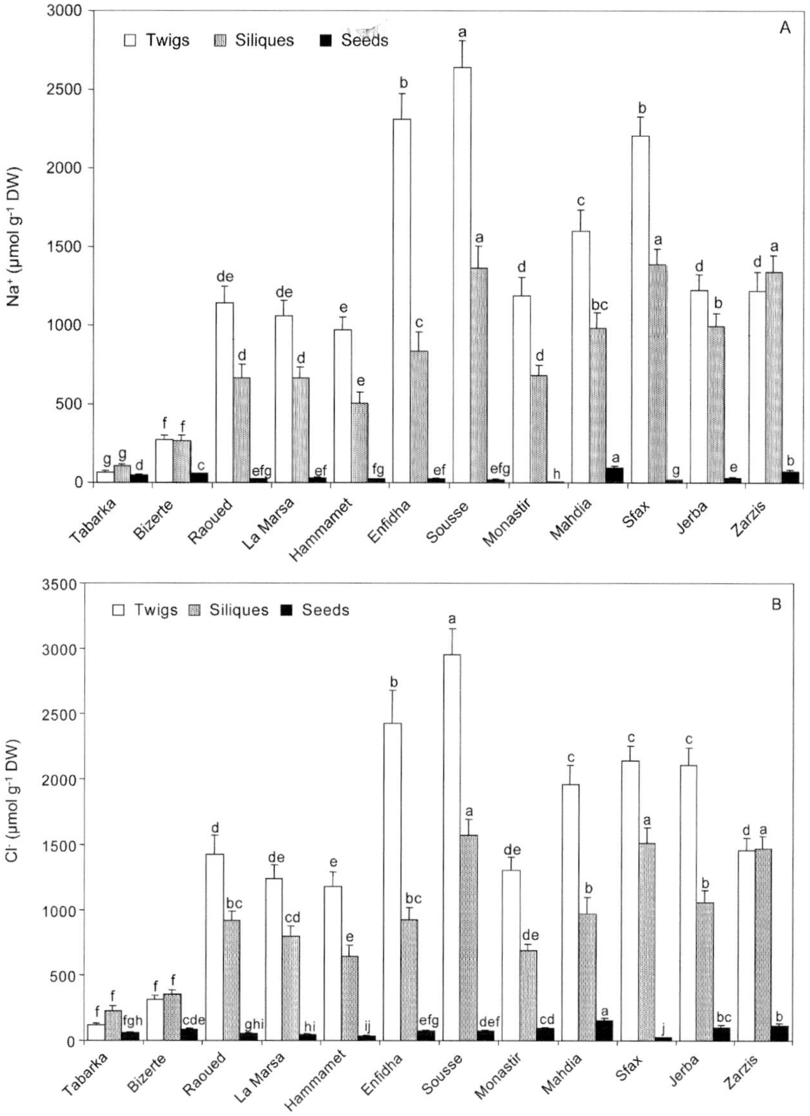

Figure 4. *Na^+ and Cl^- distribution pattern in C. maritima organs.*
A: Sodium accumulation in desiccated twigs, siliques and seeds of C. maritima *accessions. Means of 5 replicates ± standard error. For each parameter, means followed by at least one same letter were not significantly different between the accessions at P < 5%.*
B: Chloride accumulation in desiccated twigs, siliques and seeds of C. maritima accessions. Means of 5 replicates ± standard error. For each organ, means followed by at least one same letter were not significantly different between the accessions at P < 5%.

Significant variation in Na^+ and Cl^- accumulation pattern within C. maritima organs occurred between the twelve accessions (Figures. 4A & 4B). In addition, irrespective of the accession, the lowest sodium or chloride contents were registered in the seeds, in comparison with the levels accumulated in the shoots or in the silique teguments. This constitutes a strong evidence for the ability of *C. maritima* to protect the reproductive organs from the harmful salt accumulation. As shown by figure 4A, Na^+ amounts reached 64.8 and 118.3 µM g^{-1} DW in the desiccated twigs and siliques from Tabarka respectively, unlike the arid and semi-arid area accessions, especially Sousse which exhibited the highest Na^+ amounts in the twigs and siliques (ca. 2645 and 2955 µM g^{-1} DW respectively). Seeds collected from Mahdia were the most salt-charged (93.1 µM g^{-1} DW). Similar trends were found for chloride levels (Figure 4B).

4. DISCUSSIONS AND CONCLUSION

Our data show that *C. maritima* is characterized by great intra-specific heterogeneity, which seems to be accession-dependent. Furthermore, this halophyte could be considered as a promising crop with regard to its seed-oil content. Indeed, its seeds contained up to 39% of oil, a level close to that of several conventional oleaginous species like *Carthamus tintorius*, *Sesamum indicum* (Karleskind, 1996), and some rapeseed genotypes (Bhardwaj & Hamama, 2003). This Brassicaceae showed higher seed-oil contents when compared with other halophytes, among which *Mesembryanthemum crystallinum* (ca. 8%) (Pasternak et al., 1985), Salicornia *bigelovii* Torr. (ca. 30%) (Glenn et al., 1991; Bashan et al., 2000), but was lower than other Brassicaceae such as *Cakile edentula* (O'Leary et al., 1985) or *Crambe abysinnica* (Mandal et al., 2002) (50% and 60%, respectively).

Previous studies on *C. maritima* ecotypes from Morocco and Australia showed the existence of such a variability in seed-oil content, with up to 49 % for the former accession (Pasternak et al., 1985) versus ca. 35% for the latter (Hocking, 1982). This intra-specific variability in the amount and the quality of the extracted oil could be related to environmental factors, such as precipitation (Rana & Ahmed, 1981; Pannelli et al., 1994), temperature (Champolivier & Merrien, 1996), and salinity (Zarrouk & Cherif, 1983). Seed-oil of *Cakile maritima* contained high proportions of erucic acid (22:1) (25% to 32%), a compound considered as anti-nutritional for both humans and animals (Rajcan et al., 1999), but widely used in the industry (Domergue et al., 1999;Kaushik. & Agnihotri, 2000).

Seed size was close to that of conventional crops like mustard, winter colza and sunflower (Karleskind, 1996). The smallest fruits and seeds were produced by the arid zone accessions (Jerba and Zarzis), particularly contrasting with those from the humid and sub-humid areas (Tabarka and Bizerte). This corroborates previous studies, which showed such diversity within *C. maritima* populations harvested from central Europe or from Santa Cruz (California) (Barbour, 1970;

Maun & Payne, 1989). It was suggested that seed and fruit variability might improve the ability of halophytes to establish and cope with fluctuating conditions in their native environments (Mandák, 2001). In this way, Maun & Payne (1989) showed the presence of a morphological variability for *Cakile maritima* silique and seed mass. This variability was correlated with the dispersal ability, seedling emergence, and more generally, the plant response. A further dimorphism was reported for seed size in relation with their position on the fruit (Barbour, 1970).

Na^+ and Cl^- ions were not uniformly accumulated between the vegetative organs (twigs), silique teguments and seeds of *C. maritima*, their repartition within the plants organs showing the existence of a decreasing gradient from the twigs to the seeds. Indeed, the twig:silique:seed Na^+ accumulation ratios reached up to 176:99:1 for Monastir accession, while those of Cl^- establishing at 85:60:1 for Sfax accession. Barbour (1970) found a similar pattern of Na^+ and Cl^- repartition among the above cited organs of *C. maritima*, with values reaching 359:74:1 for sodium and 200:30:1 for chloride. Thus, ion relationship investigation confirmed the aptitude of *C. maritima* to control salt repartition within its different tissues, avoiding the harmful salt accumulation in the reproductive organs. This is of high ecological adaptive significance, since it determines the seed viability and thus the species establishment and distribution in saline biotopes. However, the amplitude of the shoot-silique-seed Na^+ or Cl^- accumulation ratios would traduce variability in the efficiency of this high ecological significance feature. This suggests also that the studied populations cope with different salinity levels in their original ecosystems.

The silique salt charge has been reported to impair the germination of this Brassicaceae and could partially explain the salt-induced seed dormancy (Hocking, 1982), which would also require salt leaching from the soil to germinate in its natural habitats (Ungar, 1978). More recently, we showed that seed germination of *C. maritima* was inhibited by salinities higher than 200 mM NaCl. Nevertheless, seeds remained viable and fully recovered their germination capacity once transferred to distilled water (Debez et al., 2004). On the other hand, avoiding salt accumulation in the fruits seems decisive for the use of crop plants in saline conditions. Indeed, Brassica plants overexpressing AtNHX1, a gene enabling salt accumulation in leaves and not in the seeds, were able to grow with unaffected seed yield and seed-oil quality up to 200 mM NaCl (Zhang et al., 2001).

Finally, *C. maritima* is potentially a HEA-crop though high intra-specific variability occurred. The preliminary field yields estimations (ca. 2 tons ha^{-1} for Raoued accession), the salt tolerance capacity of this species, and its ability to limit salt accumulation in the seeds in particular, are encouraging elements for the future exploitation of this halophyte. Several parameters (the financial profitability of the farming systems, the efficiency of salt uptake by the plants, and long-term salt accumulation in the root-zone and the watertable) still need to be taken in account for a reliable evaluation of the sustainability of the introduction of this halophyte as an alternative crop (Barrett-Lennard, 2002; Bañuelos et al., 2003). In this way, large-scale field trials with the most interesting accessions and the impact of the irrigation with salty water on the yield, the seed viability, and the seed-oil quality of *C. maritima* are currently being investigated.

5. ACKNOWLEDGEMENTS

This work was achieved in the framework of a national project financially supported by the "Secrétariat d'Etat à la Recherche Scientifique et à la Technologie" in Tunisia, and also by the Tunisian-French "Comité Mixte de Coopération Universitaire" (CMCU), network #02F0924. We thank Mr. Jamel Bouraoui from the National Institute of Meteorology in Tunis for providing us with the climatic data, PD Dr. Hans-Werner Koyro from the Institute of Plant Ecology of the Justus-Liebig University (Giessen, Germany), and PD Dr. Bernhard Huchzermeyer from the Botany Institute (University of Hannover, Germany) for their constructive suggestions.

6. REFERENCES

Allen, C. & Good, P. 1971. Acyl lipids in photosynthetic systems. Methods Enzymology 23: 523-547.

Bañuelos, G.S., Sharmarsakar, S., Cone, D. & Stuhr, G. 2003. Vegetative approach for improving the quality of water produced from soils in the westside of central California. Plant and Soil 249: 229-236.

Barbour, M.G. 1970. Germination and early growth of the strand plant *Cakile maritima*. Bulletin of the Torrey Botanical Club 97: 13-22.

Barbour, M.G. & Rodman, J.E. 1970. Saga of the west coast sea roket, *Cakile edentula* ssp. *California* and *Cakile maritime*. Rhodora 72: 370-386.

Barrett-Lennard, E.G. 2002. Restoration of saline land through revegetation. Agriculture Water Management 53: 213-226.

Bashan, Y., Moreno, M. & Troyo, E. 2000. Growth promotion of the seawater irrigated oilseed halophyte *Salicornia bigelovii* inoculated with mangrove rhizosphere bacteria and halotolerant *Azospirillum* spp. Biology and Fertility of Soils 32: 265-272.

Bhardwaj, H.L. & Hamama, A.A. 2003. Accumulation of glucosinolate, oil, and erucic acid in developing *Brassica* seeds. Industrial Crops and Products 17: 47-51.

Champolivier, L. & Merrien, A. 1996. Evolution de la teneur en huile et de sa composition en acides gras chez deux variétés de tournesol (oléique ou non) sous l'effet de températures différentes pendant la maturation des graines. Oléagineux, Corps Gras, Lipides 3: 140-144.

Clausing, G., Vickers, K. & Kadereit, J.W. 2000. Historical biogeography in linear system: Genetic variation of Sea Rocket (*Cakile maritima*) and Sea Holly (*Eryngium maritimum*) along European coasts. Molecular Ecology 9: 1823-1833.

Debez, A., Ben Hamed, K., Grignon, C. & Abdelly, C. 2004. Salinity effects on germination, growth, and seed production of the halophyte *Cakile maritima*. Plant and Soil (in press).

Dierig, D.A., Grieve, C.M. & Shannon, M.C. 2003. Selection for salt tolerance in *Lesquerella fendleri* (Gray) S. Wats. Industrial Crops and Products 17: 15-22.

Domergue, F., Cassagne, C. & Lessire, R. 1999. Les Acyl-CoA élongases des graines : l'autre système de synthèse d'acides gras. Oléagineux, Corps Gras, Lipides 6: 101-110.

Donovan, L.A., Richards, J.H. & Schaber, E.J. 1997. Nutrient relations of the halophytic shrub, *Sarcobatus vermiculatus*, along a soil salinity gradient. Plant and Soil 190: 105-117.

Glenn, E.P., Brown, J.J. & O'Leary, J.W. 1998. Irrigating crops with seawater. Scientific American 279: 76-81.

Glenn, E.P., O'Leary, J.W., Watson, M.C., Thompson, T.L. & Kuehl, R.O. 1991. *Salicornia bigelovii* Torr.: an oilseed halophyte for seawater irrigation. Science 251: 1065-1067.

Grieve, C.M. & Suarez, D.L. 1997. Purslane (*Portulaca oleracea* L.): A halophytic crop for drainage water reuse systems. Plant and Soil 192: 277-283.

Hachicha, M., Job, J.O. & Mtimet, A. 1994. Les sols salés et la salinisation enTunisie. Sols de Tunisie 5: 271-341.

Hocking, P.J. 1982. Salt and mineral nutrient levels in fruits of two strand species, *Cakile maritima* and *Arctotheca populifolia*, with special reference to the effect of salt on the germination of *Cakile*. Annals of Botany 50: 335-343.

Karleskind, A. 1996. Oil and fats manual: a comprehensive treatise. Vol.1. Paris, France: Lavoisier Publishing. 1-168 pp.

Kaushik, N. & Agnihotri, A. 2000. GLC analysis of Indian rapeseed-mustard to study the variability of fatty acid composition. Biochemical Society Transactions 28: 581-583.

Kefu, Z., Hai, F. & Ungar, I.A. 2002. Survey of halophyte species in China. Plant Science 163: 491-498.

Khan, M.A. & Duke, N.C. 2001. Halophytes - A resource for the future Wetlands Ecology Management 6: 455-456.

Khan, M.A., Ungar, I.A. & Showalter, A.M. 2000. Effects of salinity on growth, water relations and ion accumulation of the subtropical perennial halophyte, *Atriplex griffithii* var. *stocksii*. Annals of Botany 85: 225-232.

Mandák, B. & Pylek, P. 2001. The effects of light quality, nitrate concentration and presence of bracteoles on germination of different fruit types in the heterocarpous *Atriplex sagittata*. Journal of Ecology 89: 149-158.

Mandal, S., Yadav, S., Singh, R., Begum, G., Suneja, P. & Singh, M. 2002. Correlation studies on oil content and fatty acid profile of some Cruciferous species. Genetic Resources and Crop Evolution 49: 551-556.

Mangold, H.K. 1964. Thin layer chromatography of lipids. Journal of the American Oil Chemistry Society 47: 726-773.

Maun, M.A. & Payne, A.M. 1989. Fruit and seed polymorphism and its relation to seedling growth in the genus *Cakile*. Canadian Journal of Botany 67: 2743-2750.

Metcalfe, D., Schmitz, A. & Pelka, J.R. 1966. Rapid preparation of fatty acid esters from lipids for gas chromatographic analysis. Analytic Chemistry 38: 524-535.

O'Leary, J.W., Glenn, E.P. & Watson, M.C. 1985. Agricultural production of halophytes irrigated with seawater. Plant and Soil 89: 311-321.

Pakeman, R.J. & Lee, J.A. 1991. The ecology of the strand line annuals *Cakile maritima* and *Salsola kali*. 2. The role of nitrogen in controlling plant performance. Journal of Ecology 79: 155-165.

Pannelli, G., Servili, M., Selvaggini, R., Baldioli, M. & Montedoro, G.F. 1994. Effect of agronomic and seasonal factors on olive (*Olea europaea* L.) production and on the qualitative characteristics of the oil. Acta Horticulturae 356: 239-243.

Pasternack, D., Danon, A., Aronson, J.A. & Benjamin, R.W. 1985. Developing the seawater agriculture concept. Plant and Soil 89: 337-348.

Pita Villamil, J.M., Perez-Garcıa, F. & Martınez-Laborde, J.B. 2002. Time of seed collection and germination in rocket, *Eruca vesicaria* (L.) Cav. (Brassicaceae). Genetic Resources and Crop Evolution 45: 47-51.

Puppala, N. & Fowler, J.L. 2002. *Lesquerella* seed pretreatment to improve germination. Industrial Crops and Products 17: 61-69.

Rajcan, I., Kasha, K.J., Kott, L.S. & Beversdorf, W.D. 1999. Detection of molecular markers associated with linolenic and erucic acid levels in spring rapeseed (*Brassica napus* L.). Euphytica 105: 173-181.

Rana, M.S. & Ahmed, A.A. 1981. Characteristics and composition of Libyan olive oil. Journal of the American Oil Chemistry Society 58: 630-631.

Rodman, J.E. 1974. Systematics and evolution of the genus *Cakile*(Cruciferae). Contributions from the Asa Gray Herbarium of Harvard University 205: 3-146.

Smedema, L.K. & Shiati, K. 2002. Irrigation and salinity: a perspective review of the salinity hazards of irrigation development in the arid zone. Irrigation and Drainage Systems 16: 161-174.

Ungar, I.A. (1978). Halophyte seed germination. Botanical Review, 44, 233- 264.

van Zandt, P.A., Tobler, M.A., Mouton, E., Hasenstein, K.H. & Mopper, S. 2003. Positive and negative consequences of salinity stress for the growth and reproduction of the clonal plant, *Iris hexagona*. Journal of Ecology 91: 837-846.

Zarrouk, M. & Cherif, A. 1983. Teneur en lipides des halophytes et résistance au sel. Zeitschrift für Pflanzenphysiolgie 112 : 373-380.

Zarrouk, M., El Almi, H., Ben Youssef, N., Sleimi, N., Smaoui, A., Ben Miled, D. & Abdelly, C. 2004. Lipid composition of local halophytes seeds: *Cakile maritima, Zygophyllum album* and *Crithmum maritimum*. In: H. Lieth, & M. Moschenko. (Eds.), Cash Crop Halophytes Recent Studies. Ten Years after the Al Ain Meeting. Dordrecht, Netherlands: Kluwer Academic Publishers. 121-126 pp.

Zhang, J. 1996. Interactive effects of soil nutrients, moisture and sand burial ion the development, physiology, biomass and fitness of *Cakile edentula*. Annals of Botany 78: 591-598.

Zhang, X.X., Hodson, J.N., Williams, J.P. & Blumwald, E. 2001. Engineering salt-tolerant *Brassica* plants: Characterization of yield and oil quality in transgenic plants with increased vacuolar sodium accumulation. Proceedings of the National Academy of Sciences 98: 12832-12836.

CHAPTER 5

SALT TOLERANT PLANTS FROM THE GREAT BASIN REGION OF THE UNITED STATES

DARRELL J. WEBER AND JOSEPH HANKS

Department of Integrated Biology, Brigham Young University
Provo, Utah 84602.USA

Abstract. The Great Basin Region (305,710 square kilometers) includes Nevada, half of Utah and some parts of Oregon, Idaho and California. It is so named because it does not have any rivers that empty into the ocean. The results of this drainage are saline valley soils, salt playa and saline lakes. Halophytes are the best-adapted plants for growing in the region. A large number of native and introduced halophyte plants are present in the Great Basin. Many saline tolerant plants have been used as forage for animals, to reduce soil erosion, for saline tolerant grasses, and for ornamental purposes.

1. INTRODUCTION

An early explorer of the Western United States during 1843-45, Captain John C. Fremont, observed that rivers in the area that are now Nevada and part of Utah did not flow to the ocean. Instead the waters flowed to shallow lakes between the mountain ranges and the water evaporated over time resulting in saline soil and saline lakes. Some of the persistent saline lakes include Great Salt Lake in Utah and Pyramid Lake in Nevada. Fremont coined the term ‘Great Basin’ to describe the area that is not drained by rivers to the sea.

The Great Basin region is the largest U.S. desert and covers an area of 190,000 square miles. The west border is the Sierra Nevada Range and the east border is the Rocky Mountains. The north border is the Columbia Plateau and the south border is the Mojave and Sonoran deserts (see figure 1)(MacMahon, 1979). There are many mountain ranges with broad saline soil valleys between them in the Great Basin region.

The Great Basin is considered to be a cool or cold desert due to its more northern latitude as well as it higher elevation (normally from 1312 m to 2132 m). The precipitation is around 17 to 30 cm per year in the valley bottoms. As one moves up in elevation on the mountains, the amount precipitation increases. Salt

M.A. Khan and D.J. Weber (eds.), Ecophysiology of High Salinity Tolerant Plants, 69-106.

playas are a conspicuous part of the valley floors. Because of the long history of saline soils in the valleys, natural selection for salt tolerance has produced a large number of salt tolerant plants largely in the Chenopodiaceae (West, 1983). Introduced salt tolerant plants (both intentional and accidental introductions) do well in this saline enviroment and there are a huge number of introduced species present.

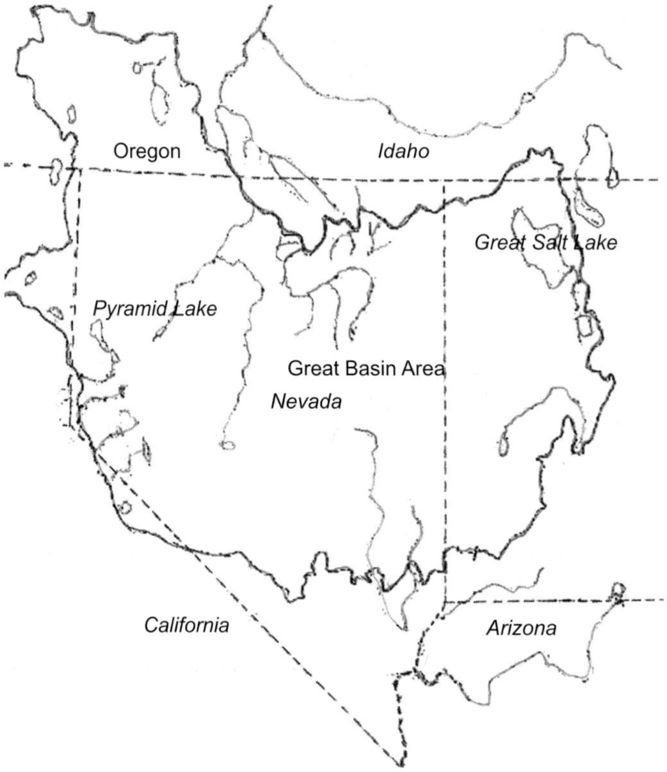

Figure 1. *Map showing the boundaries of the Great Basin Area.*

A vertical vegetational zonation exists from the bottom of the valley floors to the top of the mountains. There are three shrub-steppe zones (a) Salt-desert shrub, (b) Sagebrush semidesert, (c) Sagebrush steppe. The salt-desert shrub zone is dominanted by perennials that are mainly halophytes. At the highest saline conditions (121 mmho cm^{-1}) typically with free NaCl on the soil surface, the species found are normally *Allenrolfea occidentalis, Salicornia utahensis, Salicornia rubra, Distichlis spicata* and some *Atriplex gardneri*. When the soil salinity is lower (96 mmho cm^{-1}) *Atriplex gardneri, Sarcobatus vermiculatus, Atriplex falcata, Kochia americana* and *Suaeda torreyana* are commonly present.

At higher elevations the soils are well formed but in some cases made by still have high levels of Na^+. Under these conditions the most conspicuous shrub species is *Atriplex confertifolia* (Trimble, 1989). Assocated perennial species include *Ceratoides lanata, Kochia americana, Artemisia spinescens* and several species of *Atriplex*. Perennial grasses are present in the salt gradient such as *Sporobolus airoides, Elymus elymoides* and *Elymus cinereus*. At the lowest levels of soil salinity, various species of *Atriplex* and *Ceratoides* thrive.

The semidesert sagebrush zone and the sagebrush steppe are dominanted by *Artemisia tridentata*. It is subdivided into three subspecies: *tridentata* in mesic low elevation areas; *wyomingensis* across a wide range of elevations and at intermedicate moisture levels and *vaseyana* at intermediate moisture levels and higher elevations (West, 1983).

Perennials are dominant in the Great Basin and native annuals are extremely scarce. However exotic annual species are common and make up the majority of the annuals in the Great Basin. Some of the most common are *Bromus tectorumn, Descurainia pinnata, Halogeton glomeratus, Lepidium perfoliatum, Malcomia africana,* and *Salsola pestifer*. These are typically disturbance species and develop best when the perennials have died out or been removed by events such as fire. The interaction between exotic annuals and native perennials vary but normally healthy stands of native perennials tend to outcompete the exotic annuals. But when disturbances occur such as fire or overgrazing by cattle, the exotic annuals are more successful (Booth, 2000). The list of salt tolerant plants in the Great Basin will include native and introduced salt tolerant plants.

2. CHARACTERISTICS OF SELECTED GREAT BASIN HALOPLHYTES

2.1. Achillea millefolium L. Milfoil yarrow.

Achillea millefolium can grow in saline soils (Szabolcs, 1994).

2.2. Agropyron species. Grasses.

Agropyron species represent a range of grasses that grow in the Great Basin region. There has been some rearrangement in the classification of *Agropyron* and many of the previous *Agropyron* species are now listed in the genus *Elymus*. Only the grasses with some salt tolerance will be mentioned.

2.3. Agropyron cristatum (L.) Gaerin. Crested wheatgrass.

Agropyron cristatum was introduced from Eastern Europe and has been used extensively in reseeding sagebrush areas in the Great Basin. It produces good forage for livestock and wildlife. It has some tolerance to saline soils (Sokolov et al., 1979) but it cannot tolerate high salinity (Butler et al., 1974).

2.4. Agrostis stolonifera (L.) Linn. Carpet bentgrass. Perennial.
Agrostis stolonifera grows in saline meadows and salt grass communities. It can tolerate 36% salinity (Rozema, 1975).

2.5. Alhagi marourum Medicus. Camel-thorn. Perennial. Introduced from Eurasia.
Alhagi marourum provides spring forage even though it has thorns (Le Houerou, 1994). It grows in saline meadows. Mainly present in the southern areas of the Great Basin. (Welsh et al., 1993).

2.6. Allenrolfea occidentalis (Wats.) Kuntze. Iodine bush (pickleweed).
Allenrolfea occidentalis commonly called iodine bush (sometime it is called pickleweed) is one of the most salt tolerant stem succulent shrub species in the Great Basin region. It grows in the salt playas where the majority of the salt is NaCl with concentrations up to 1027 mM (Hansen & Weber, 1975; Weber et al., 2002). Seed production is abundant even in very dry years (Young et al., 1995). Flowering occurs in late July and plants produce seeds in late September. Most of the seeds are dispersed around the parent plant, but wind and water can cause long distance seed dispersal. The seeds are uniform in size and color with no evidence of polymorphism (Gul & Weber, 1999: Gul & Weber, 2000). Seeds of *A. occidentalis* persist in the soil of the upper zone in large densities. The large seed bank is important in maintaining the population of *A. occidentalis* (Gul & Weber, 2001). While germination decreases with increased salinity, seeds can germinate in 800 mM NaCl. Inhibition of seed germination was greater in the dark than in the light (Gul & Weber, 2000c). The best seed germination occurred at 25°C night and 35°C day. Cooler temperatures decreased the germination rate. Seeds that has been in high saline solutions for 20 days and then transferred to distilled water germinated readily, indicating tolerance to exposure to high salt conditions (Gul et al., 2000c). Fusicoccin completely reversed the inhibitory effects of salinity on seed germination. Ethephon significantly promoted germination at all salinities tested (Gul & Weber, 1998). When *A. occidentalis* was grown at three different densities, the highest dry mass of roots occurred at 600 mM NaCl at low density (2000 plants m^{-2}). Water potential of the plants became more negative with increasing salinity due to the accumulation of NaCl in the leaves (Gul et al., 2000b). Na^+ and Cl^- concentrations in shoots and roots increased when NaCl level was increased, while other cations (K^+, Ca^{++}, Mg^{++}) decreased (Gul et al., 2001). Photosynthesis functioned reasonably well at high salinities, but extremely high salinity did decrease dry mass of roots and shoots (Gul et al., 2000b). Metabolic heat rate and respiration rate were determined. High metabolism, respiration and growth were found to be the highest during May and June and the lowest during the dry month of August (Harris et al., 2001). The salt playa where *A. occidentalis* grows has extreme osmotic stress and the precipitation is very low. Predawn xylem water potentials in May for 1991 and 1992 were –2.2 and 3.3 MPa for *A. occidentalis* (Trent et al., 1997).

It is common in many salt playas for saline mounds to form around individual *A. occidentalis* plants. These mounds average 0.3 m in height. The mounds appear to be favorable for plant recruitment and survivorship. This value is lost after a number of years because of the accumulation of salts (Blank et al., 1998).

The oil content of the seeds of *A. occidentalis* was 14%. In terms of practical use, lipid analysis of the seed oil indicated that it was 88% unsaturated fatty acids and 11% saturated fatty acids, which would make it a good cooking, oil (Weber et al., 2001; Hess & Weber, 1995). Kovalev and Krylova (1992) evaluated *A. occidentalis* as a plant to improve plant cover and forage production on saline lands in arid regions. It is a possible feed for animals if mixed with other forage plants.

2.7. Ambrosia acanthicarpa Hook. Bur ragweed. Annual.
Ambrosia acanthicarpa can grow in the salt desert environment. It is not a good forage plant and its pollen may be associated with allergic reactions.

2.8. Anagallis arvensis L. Scarlet pimpernel. Annual. Native to Europe,
Anagallis arvensis is a salt tolerant weedy plant with some value as fodder (Pasternak, 1990).

2.9. Anemopsis californica (Nutt.) H. & A.Yerba mansa.
Anemopsis californica is found growing in wet saline areas (Mason, 1957).

2.10. Aristida arizonica Vasey. Arizona threeawn. Perennial.
Aristida arizonica is found in some of the salt desert shrub community, which would indicate some salt tolerance. It is not a good forage plant (Welsh et al., 1993).

2.11. Aristida purpurea Nutt. Syn Aristida longiseta. Purple threeawn. Perennial.
Aristida purpurea is found in some salt desert community suggesting that it has some salt tolerant. It is not a good forage plant and it even has a nickname of 'no-eatum' (Welsh et al., 1993).

2.12. Armeria maritima (Mullo.) Willd. Sea-pink.
Armeria maritima is an ornamental plant that escaped into the native environment. It has some salt tolerance (Cooper, 1982).

2.13. Arundo donax L. Giant reed. Perennial. Native to Eurasia.
Arundo donax grows in disturbed sites and in saline seeps. It has been evaluated as a fodder (Pasternak, 1990).

2.14. Asparagus officinalis L. Asparagus. Periennal. Native to Eurasia.
Asparagus officinalis is a food plant that often escapes and grows in the native environment. It has some salt tolerance.

2.15. Astragalus species.
The genius of *Astragalus* has at least 111 species and numerous varieties. It is probably one of the largest genus of flowering plants in the Great Basin. The plant has attractive flowers in a range of colors. Some of the *Astragalus* species are poisonous. The toxin is a combination of pyrrolizidine alkaloids and selenium. However, large amounts of the plant need to be consumed before it is lethal. It is not a good forage plant but some animals become addicted to eating certain species of *Astragalus*. The odd behavoir patterns of these animals that eat *Astragulus* have resulted in the condition being calling 'Loco disease'. The genius has a wide range of habitat. Only those *Astragalus* species associated with the salt desert area will be listed. The species listed below grow in the salt brush community and are considered to have some salt tolerance.

2.16. Astragalus ampullarius Wats. Gumbo milkvetch. Perennial.
It grows on the saline shales of the Chinle Formation.

2.17. Astragalus asclepiadoides Jones. Milkweed milkvetch. Perennial.
It grows on saline shales.

2.18. Astragalus coltonii var. coltonii (Jones) Rydb. Colton milkvetch. Perennial.

2.19. Astragalus cymboides Jones. Canoe milkvetch. Perennial.

2.20. Astragalus diversifolius Gray. Mesic milkvetch. Perennial.

2.21. Astragalus ensiformis Jones. Pagumpa milkvetch. Perennial.

2.22. Astragalus flacus var. argillosus (Jones) Barneby.Clay milkvetch. Perennial.

2.23. Astragalus flacus var. flavus (Nutt.) Rydb. Yellow milkvetch. Perennial.

2.24. Astragalus flexuosus var diehlii (Jones) Barneby. Diehl's milkvetch. Perennial.

2.25. Astragalus geyeri Gray. Geyer's milkvetch. Annual.

2.26. Astragalus lonchocarpus Torr. Great rushy milkvetch. Perennial.

2.27. Astragalus lentiginosus Dougl. Ex Hook. Freckled Milkvetch. Perennial.
There are 10 varieties to this species.

2.28. Astragalus megacarpus (Nutt.) Gray. Great bladdery milkvetch. Perennial.

2.29. Astragalus moencoppensis Jones. Moenkopi milkvetch. Perennial.

2.30. Astragalus musiniensis Jones. Ferron milkvetch. Perennial.

2.31. Astragalus mollissimus Torr. Woolly locoweed. Perennial.

2.32. Astragalus nelsonianus Barneby. Nelson's milkvetch. Perennial.
It grows in saline soils.

2.33. Astragalus newberryi var castoreus Jones. Beaver Dam Milkvetch. Perennial. There are two varieties of this specie.

2.34. Astragalus praelongus var praelongus Sheldon. Stinking milkvetch. Perennial.

2.35. Astragalus racemosus Pursh. Alkali milkvetch. Perennial.

2.36. Astragalus rafaelensis Jones. San Rafael milkvetch. Perennial.

2.37. Astragalus sabulonum Gray. Gravel milkvetch. Annual.

2.38. Astsragalus sabulosus Jones. Cisco milkvetch. Perennial.

2.39. Astragalus saurinus Barneby. Dinosaur milkvetch. Perennial.

2.40. Astragalus toanus Jones. Toana milkvetch. Perennial.

2.41. Astragalus uncialis Barneby. Currant Milkvetch. Perennial.

2.42. Astragalus wingatanus Wats. Fort Wingate milkvetch. Perennial.

2.43. Astragalus zionis var zionis Jones. Zion milkvetch. Perennial.

2.44. Atriplex species.
Atriplex is a complex genius with at least 23 species and at least 9 varieties. In addition, the genius forms hybrids with native and introduced species (Welsh et al., 1993). The result of this genetic plasticity is the development of a range of plants that can occupy numerous habitats. More detailed descriptions of some of the *Atriplexs* will follow with the focus on those species that occupy the salt desert environment.

2.45. Atriplex canescen (Pursh) Nutt. Four wing saltbush. Perennial.
Atriplex canescen (Four wing saltbush) is wide spread in the Great Basin region and has value as a forage species for wildlife and livestock. While it is found in the salt desert shrub zone, it is not as salt tolerant as *A. confertifolia* (Hodgkinson, 1987). It forms hybrids with *A. confertifolia* and *A. gardneri* (Welsh et al., 1993).

It is often used in reclamation projects. Temperature and salinity have a variable effect on seed germination (Mikhiel et al., 1992).

2.46. Atriplex confertifolia (Torr. & Frem.) Wats. Shadscale.
This salt-tolerant shrub commonly called shadscale is wide spread in the Great basin region. It is a valuable browse plant for wildlife and livestock, especially sheep even though it has spikes. *Atriplex confertifolia* is resistant to overgrazing. It forms hybrids with *A. canescens, A. garrettii, A. carrugata and A. gardneri* (Welsh et al., 1993). *Atriplex confertifolia* is a C_4 photosynthsis plant. Carbon isotope discrimination (Delta value) of the water in *A. confertifolia* was measured along a salinity gradient. The Delta values correlated with the salinity gradient. It was suggested that salinity induces an increase in the bundle sheath leakiness (Sandquist & Ehleringer, 1995). Salt bladders in the plant appear to function for salt removal in *A. confertifolia* (Sen & Rajpurohit, 1982; Schirmer & Breckle, 1982) Analysis of the seeds by EDAX revealed that the highest concentration of sodium, chlorine, potassium and calcium was in the seed coat. The endosperm was low in all elements whereas phosphorous was the highest in the embryo (Khan et al., 1987).

A comparison of six *Atriplex* species was made in relation to Na adsorption ration (SAR), electrical conductivity (EC) and alkalinity. The ranking with these three factors from highest to lowest be: *A, corrugata, A. obovata, A. cuneata, A. falcate, A. confertifolia* and *A. canescens* (Hodgkinson, 1987). Evapotranspiration rates for *A. confertifolia* was from 2 mm per day in June to a low of 0.3 mm per day in August. The soil-water potential associated with *A. confortifolia* was –10 bars in April and went to –70 bars in August (Evens et al., 1981). Seedlings of A. *confertifolia* were grown in soil irrigated with different salt solutions. The best yields were obtained with the 50 meq/liter NaCl treatment (Kleinkopf et al., 1975). Kenagy (1972) indicated that the peripheral tissues of the leaves were hypersaline and the inner tissue of *A. confertifolia* was not. When *A. confertifolia* was grown in three levels of sodium, there was a steep sodium gradient from the leaves to the roots (Wallace et al., 1973). In evaluating the forage value of *A. confertifolia,* the best yield was obtained with pure stands of *A. confertifolia*. Mixed stands with other plants such as native grass reduced the yield (Goodman, 1973).

2.47. Atriplex corrugata Watts. Mat-saltbush.
Atriplex corrugata grows well on fine textured saline soils often with considerable clay. It was the most salt tolerant plant in a comparison of *A. corrugata, A. obovata, A. cuneata, A. falcate, A. confertifolia* and *A. canescens* (Hodgkinson, 1987).

2.48. Atriplex elegans (Moq) D. Dietr. Wheelscale orach. Annual.
Atriplex elegans grows in disturbed sites with salinity (Richardson & McKell, 1980).

2.49. Atriplex gardneri (Moq.) D. Dietr. (Syn Atriplex nuttallii). Gardner saltbush.
Atriplex gardneri is a widely distributed complex of genotypes of great plasticity. There are six varieties listed namely, var. *bonnevillensis*, var. *cuneata*, var. *gardneri*, var. *falcate,* var. *welshii*, and var. *tridentate* (Welsh et al., 1993). The complex commonly grows in fine-textered saline soils. Polypoidy is common and diploids, triploids, tetraploids, hexaploids and higher polyploids have been found in the complex (Welsh et al., 1993). Hybrids between *A. gardneri* and *A. canescens*, *A. garrettii, A. carrugata* and *A. confertifolia* are common (Welsh et al., 1993). Salo et al. (1999) evaluated the ability of several halophytes to grow in a saline waste marsh from a power generating plant and *A. gardneri* was one of the most successful shrubs grown. *Atriplex gardneri* is able to survive in high salinity levels (121 mmho cm^{-1}) but Goodman, (1973) found it could be grown over an eighty-fold range of salinity. The amount of biomass produced reflected the salinity gradient. It is a good forage plant for livestock and wildlife and can withstand heavy grazing.

2.50. Atriplex heterosperma Bunge. Two-seeded orach. Annual.
Atriplex heterosperma grows in greasewood and salt grass communities. It likes saline lowlands (Osmund et al., 1980).

2.51. Atriplex hortensis L. Garden orach. Annual. Introduced from Europe.
Atriplex hortensis grows in disturbed sites with saline soils (Osmund et al., 1980).
2.52. Atriplex lentiformis (Torr.) Wats. Big saltbush. Perennial.
Atriplex lentiformis grows in the warm desert shrub communities and can tolerate 46% salinity (Pasternak, 1990). It has been used for fodder in some areas.

2.53. Atriplex obovata Moq. New Mexico saltbush. Perennial.
Atriplex obovata is a shrub type that grows in the salt desert. It was ranked 2nd in salt tolerant in the comparison of six Atriplex species (Hodgkinson, 1987). It has been evaluated as a animal fodder (Pasternak, 1990).

2.54. Atriplex patula L. Fathen saltplant. Annual.
Atriplex patula has two varieties listed *A. triangularis* and *A. patula* (Welsh et al., 1993). They are annual herbs and normally are associated with saline mucky soils. It is reported that *A. triangularis* will tolerate 10% salinity (Gallagher, 1986) and that *A. patula* can still grown in 49% salinity (Glenn & O'Leary, 1984). The germination of the dimorphic seeds was studied by Katembe et al., (1998) and they concluded that the impact of NaCl on seed germination was a combination of an osmotic effect and a specific ion effect.

2.55. Atriplex semibaccata R. Br. Australia saltbush.
Atriplex semibaccata is a desert shrub that was introduced from Australia. It grows mainly in disturbed sites and can tolerate 10% salinity (Pasternak, 1990).

2.56. Atriplex rosea L. Tumbling orach.
Atriplex rosea is an annual herb that grows in salt marshes throughout the Great Basin region. It has dimorphic seeds (black and brown). Both types of seeds will germinate in 1000 mM NaCl and the best temperature conditions for germination was 20-30°C. Variation in temperature affects germination of the black seeds more than brown seeds. The brown seeds germinated better in saline solutions than did the black seeds. The brown seeds can germinate shortly after seed production whereas the black seeds appear to be capable of surviving harsher conditions and can germinate at later time periods (Khan et al., 2004b). The oil content of seeds of *A. rosea* was 13%. Lipid analysis of the seed oil indicated that it was 83% unsaturated fatty acids and 17% saturated, which would make it a good cooking oil (Weber et al., 2001). A number of studies had been made on the halophytes that colonizes the potash mine dumps in Germany and France, *Atriplex rosea* was a common halophyte under those conditions (Garve & Garve, 2000),

2.57. Atriplex torreyi Wats. Torrey's saltbush.
Atriplex torreyi grows in shadscale and creosote bush communities. The photosynthesis pathways have been studied by Welkie & Caldwell, (1970).

2.58. Atriplex truncata (Torr.) Gray. Wedge orach. Annual.
Atriplex truncata grows in the saline saltgrass-greasewood communities. It has been used as a fodderplant. (Welsh et al., 1993).

2.59. Other Atriplex species.
Below are listed other *Atriplex* species associated with the salt desert community which indicates some salt tolerance but there is little additional information available on their characteristics.

2.60. Atriplex argentea Nutt. Silver orach. Annual.

2.61. Atriplex garrettii Rydb. Garrett's saltbush. Perennial.

2.62. Atriplex pleiantha W. A. Weber. Four-corners orach. Annual,

2.63. Atriplex powellii Wats. Powell's orach. Annual.

2.64. Atriplex saccaria Wats. Stalked orach. Annual.

2.65. Atriplex graciliflora Jones. Blue valley orach. Annual.

2.66. Atriplex wolfii Wats. Slender orach. Annual.

2.67. Bassia hyssopifolia (Pallas) Kuntze. Bassia. Annual.
Bassia hyssopifolia is an annual that grows in saline soils and toleratetes up to 35% (Glenn & O'Leary, 1984). It also forms hybrids with *Kochia scoparia* (Welsh et al., 1993).

2.68. Beta vulgaris L. Sugar beet. Introduced from Europe.
Beta vulgaris is a commercial crop that can tolerate some salinity. It has escaped into the native environment but there are no significant populations (Welsh et al., 1993).

2.69. Beckmannia syzigachne (Steudel) Fern. American sloughgrass.
Beckmannia syzigachne grows in saline meadows (Mason, 1957).

2.70. Buchloe dactyloides (Nutt.) Engeim. Buffalograss. Perennial.
Buchloe dactyloides is the dominent species of the shortgrass prairie of the Western Great Plains. The Great Basin has a limited amount of *B. dactyloides*. It is an excellent feed for livestock and can be a winter feed when not covered by snow. Overall *B. dactyloides* is not considered to be very salt tolerant. Wu and Lin (1994) found considerable genetic variation for salt tolerance and felt that salt tolerant lines could be selected. They collected diploid, tetraploid, and hexaploid *B. dactyloides* from Mexico to Nebraska. A salt exclusion mechanism was found to be present and that the shoots had a preferential exclusion of sodium uptake. Marcum (1999) exposed six grasses in the subfamily Chloridoideae to salinities up to 600 mM NaCl in solution culture. He concluded that salinity tolerance involves saline ion exclusion, leaf salt gland ion secretion and the accumulation of the compatible solute such as glycinebetaine. Lin and Lin (1996) studied the plasma membranes of *B. dactyloides* under salt stress conditions. They found unsaturated fatty acids ($C_{14:1}$, $C_{18:1}$ and $C_{18:3}$) were present in the salt tolerant clone of *B. dactyloides* but not in the salt sensitive clone. They concluded that the chemical properties of the plasma memberane were partially responsible for the salt-tolerance. Wu et al., (1988) studied the uptake of selenium and NaCl in *B. dactyloides*. Selenium uptake was increased as the salinity increased. In contrast NaCl uptake was not increased as the concentration of selenium increased.

2.71. Calystegia sepium (L.) R. Br. Hedge bindweed. Perennial.
Calystegia sepium grows in marshes and can tolerate salinity (Mason, 1957).

2.72. Calamovilfa gigantean (Nutt.) Scribn. & Merr. Big sandreed. Perennial.
Calamovilfa gigantean is found in sandy sites. It is too coarse for good fodder (Welsh et al., 1993).

2.73. Castilleja exilis A. Nels. Annual paintbrush. Annual.
Castilleja exilis is found in saline meadows, seeps bogs and with salt grass (Mason, 1957).

2.74. Ceratoides lanata (Pursh) J. T. Howell. Winterfat. Perennial.
Ceratoides lanata is a salt tolerant perennial that grows throughout the Great Basin. It is widely distributed in brackish-water playas but it can also grow in non-saline areas (Romney & Wallace, 1980). Seeds of *C. lanata* have no dormancy and 90% of the seeds germinated in non-saline conditions. Seed germination decreased with the increase in salinity, but more than 10% of the seeds of C. lanata germinated at 900 mmol/L NaCl. Almost all of the seeds germinated in less than 24 hr (Khan et al., 2004a). Gibberellic acid had no effect in alleviating salinity effects, however, kinetin and fusicoccin substantially alleviated the effect of salinity on seed germination. Ethephon almost completely reverted the effect of salinity (Khan et al., 2004a). The evapotranspiration of *C. lanata* was determined to have a high of 2 mm per day in June and dropped to a low of 0.3 mm per day in August. The soil-water potential during this time dropped nearly linearly from less than –10 bars in April to about –70 bars in September (Evans et al., 1981). The forage value is considered good for sheep, pronghorn sheep, elk and mule deer and fair for livestock. It is most valuable as forage in the winter time (Stubbendieck et al., 1991).

2.75. Chenopodium represents a group of annual herbs.
The genus is complex and includes some 15 species and a number of varieties are listed.

2.76. Chenopodium album L. Lambs quarters. Annual. Introduced from Eurasia.
Chenopodium album is reported to tolerate 8% salinity (Hoffman & Shannon, 1986).

2.77. Chenopodium ambrosoides L. Annual. Mexican tea. Introduced from Mexico.
Chenopodium ambrosoides is present in disturbed sites.

2.78. Chenopodium fremontii Wats. Fremont goosefoot. Annual.
Chenopodium fremontii is found in the salt desert zone in disturbed area.

2.79. Chenopodium glaucum L. Oakleaf goosefoot. Annual.
Seeds of *Chenopodium glaucum* were treated with a range of concentration of NaCl, mixed salts and PEG-6000. Seed germination percentage of *C. glaucum* decreased with increased salinity and osmotic potential. The optimal germination was in a salt solution at 2.9 mmol L^{-1}. When the non-germinated seeds from the treatments were transferred to distilled water more than 90% of the seeds germinated. It was concluded that certain degrees of salt stress and water stress were a reversible inhibition to seed germination (Duan et al., 2004). Tanaka and Tanaka, (1980) determined that *C. glaucum* was diploid with n=9, 2n=18.

2.80. Chenopodium murale L. Nettlefleaf goosefoot. Introduced from Eurasia. *Chenopodium murale* is considered to be a weed but can tolerate 11% salinity. (Glenn & O'Leary, 1984)

2.81. Chenopodium rubrum L.Red goosefoot. Annual.
Seeds of *C. rubrum* germinated best in the light. Stratification improved the seed germination percentages. Germination percentages were highest in alternating day/night temperatures of 15/25°C and 20/30°C. Germination was the highest when the seeds were placed on the top of the soil, but the percentage declined sharply when the seeds were covered. Germination was reduced when the level of salinity was 2000 mg/l (Galinato et al., 1987). *Chenopodium rubrum* and *Chenopodium salinum* were analysed for mineral ion composition. The dominant ions were Na^+ and Cl^- in *C. fremontii* where as *C. salinum* had relatively high K^+, Mg^{++}, Ca^{++} and low Cl^- contents (Redmann & Fedec, 1987). *Chenopodium rubrum* grew on soil containing 1.56% NaCl (Simon, 1977).

2.82. Chloris virgata Swartz. Feather fingergrass. Annual.
Chloris virgata is a grass that grows along waterways and irrigation sites. It does have some salt tolerance (Shainberg & Shalhevet, 1984).

2.83. Chrysothamus is a genus with 10 species.
Those species associated with salt desert areas are:

2.84. Chrysothamus albidus (Jones) Greeene.
Alkali rabbitbrush. Perennial shrub. In a study relating to control of Alkali Rabbitbrush, it was found that the plant was very tolerant to herbicides (Roundy et al., 1983).

2.85. Chrysothamus depressus Nutt. Dwarf rabbitbrush. Perennial shrub. *Chrysothamnus nauseosus* (Pallas) Britt. Rubber rabbitbrush. There are 14 varieties of this species of which the following are associated with the salt desert (Weber et al., 1993).

2.86. Chrysothamnus nauseosus var. albicaulis (Nutt.) Rydb. Whitestem rabbitbrush.
Davis et al., (1985) found a high rate of photosynthesis in this plant.

2.87. Chrysothamus nauseosus var. comsimilis (Greene) Hall. Greenish rabbitbrush.
Donovan et al., (1996) determined the roots grew down 3.4 to 5.0 m to ground water. The predawn xylem pressure potential was –1.0 MPa and a midday xylem pressure of –2.2 MPa. The seasonal maxima content was 0.4% Na and 2.4% K of the leaf dry wt. Xylem ionic contents indicated the some Na was being excluded at the root (Donovan et al., 1996). Seed germination fell below 10% at –1.64 MPa

(NaCl and PEG). Seedlings grew satisfactory at –0.82 MPa (100 mmol/l, NaCl) but declined substantially at –1.3 MPa (200 mmol/l, NaCl) (Donovan et al., 1996)

2.88. Chrysothamus nauseosus var. turbinatus (Jones) Blake. Dune rabbitbrush.

2.89. Cordylanthus parviflorus (Ferris) Wiggins. Small flower birdbeak.
Cordylanthus parviflorus grows in saline meadows with salt grass (Mason, 1957).

2.90. Cressa truxillensis H. B. K.
Cressa truxillensis has been grown with 4% seawater in a saline habitat (Glenn & O'Leary, 1985).

2.91. Croton californicus Muell. Arg.Mohave croton. Perennial.
Croton californicus grows in sandy sites and sand-dropseed (Shreve & Wiggins, 1964).

2.92. Crypsis schoenoides (L.) Lam. Common pricklegrass. Annual. Native to Eurasia.
Crypsis schoenoides grows on the margins of ponds and has some salt tolerance (Welsh et al., 1993).

2.93. Cuscuta salina Engelm in Gray. Salt dodder. Parasitic plant.
Cuscuta salina is parasitc on other plants but is reported to be saline tolerant (Mason, 1957).

2.94. Cynodon dactylon (L.) Pers. Bermuda grass. Perennial grass.
Cynodon dactylon was introduced in the US as early as 1751. It is good forage for cattle. It grows commonly in saline soils. Because of its high saline tolerance it has been tested and grown in a large number of countries in the world. In the Great Basin area it is considered more of a weed than a forage plant. Several salt tolerant grasses were exposed to salinities up to 600 mmol NaCl in solution culture and *C. dactylon* was among the most tolerant of the grasses (Marcum, 1999). *Cynodon dactylon* has been grown as a turf grass and irrigated with saline water with satisfactory results if the level of salinity is not too high in the water (Schaan et al., 2003). When lambs were fed diets with 30% *C. dactylon* hay, the carcass merit of the lambs was excellent and was not affected by the inclusion of the halophyte forage in the diet (Swingle et al., 1996). Marcum and Murdock (1994) concluded that glycinebetaine and proline might make a significatnt contribution to cytoplasmic osmotic adjustment under salinity in C_4 grass such as *C. dactylon*. Marcum and Murdock (1994) observered that *C. dactylon* maintained low levels of Na^+ and Cl^- under high salinity, which appears to indicate ion regulation due in, part to efficient leaf salt glands. Under non-irrgated conditions, *C. dactylon* produced 13.8 t/ha of forage (Gonzalex & Heilman, 1977)

2.95. Dactyloctenium aegyptium (L.) Beauv.
Dactyloctenium aegyptium has been reported to be present in the Great Basin but no significant populations have been found.

2.96. Digitaria sasnguinalis (L.) Scop. Hairy crabgrass.
Introduced from Eurasia. *Digitaria sasnguinalis* has some salt tolerance but is considered a weed.

2.97. Distichlis spicata (L.) Greene. Saltgrass. Native.
Distichlis spicata is very salt tolerant and is widely distributed in North America. Not an excerptionally palatable forages but is often grazed for lack of other forage, particularily in the winter. Yensen, (2002) has developed a more palatable saltgrass (*Distichlis spicata* var. *yensen-4a*) that appears to have good potental for forage grown on saline soils. Saltgrass (*D. spicata)* provides excellent protection againt erosion by wind and water (Welsh et al., 1993). The highest metabolism, respiration, efficiency and growth occur during May and June and the lowest was during the hot, dry month of August (Harris et al., 2001). Six halophyte grasses were grown in salinities up to 600 mM NaCl in solution culture and *Distichlis spicata* var *stricta* (Torr.) Beetle was the most salt tolerant (Marcum, 1999). In certain areas *D. spicata* and some other halophytes are associated with mound formation (0.3 m in height) and the mounds appear to favor plant recruitment and survivorhip (Blank et al., 1998). Six salt tolerant forage plants were irrigated with brackish water and *D. spicata* was the most salt tolerant (Pasternak et al., 1993). Ketchem et al., (1991) found that proline accumulation occurred rapidly in *D. spicata* when NaCl-induced stress was started. The threshold salinity levels for *D. spicata* was around 0.5 M NaCl and the accumulated proline was 27.4 mu.mol/g fresh wt. (Cavalier & Huang, 1979).

2.98. Echinochloa crus-galli (L.) Beauv. Barnyard grass. Annual.
Echinochloa crus-galli grows in moist area and can tolerate some salinity.

2.99. Elaegnus angustifolia L. Russian Olive. Perennial.
Elaegnus angustifoli is an introduced tree from Europe that has spread through out the Great Basin (Welsh et al., 1993). It grows in moist areas and has some salt tolerance.

2.100. Eleocharis palustris (L.) R. & S. Common spikerush. Perennial.
Eleocharis palustris grows in swampy areas, saline meadows and on the margins of ponds (Welsh et al., 1993).

2.101. Eleochari parvula (R. & S.) Dwarf spikerush. Perennial.
Eleochari parvula grows in margins of ponds and on flood plans and has some salt tolerance (Welsh et al., 1993).

2.102. Eleusine indica (L.) gaertner. Goosegrass. Annual. Introduced from Europe.

Eleusine indica can just tolerate 55 dS/m of NaCl but most of the plants die at that level (Wiecko, 2003). It was introduced from Euroasia and is considered a weed.

2.103. Elymus L.

Elymus L. is an important grass genius with some 20 species. Many species of *Agropyron* are now included in the genius *Elymus*. The species hybridizes with each other making a complex genius. Many of the species are subject to infection by the fungus *Claviceps purpurea* (Fr.) Tul, commonly called Ergot. The sclerotia of *C. purpurea* contain alkaloids that can cause abnormal behavior and death in animals that consume it (Welsh et al., 1993).

2.104. Elymus canadensis L. Canada wildrye. Perennial. Native.

Elymus canadensis can grow in saline meadows. It is good forage for cattle and horses, fair for sheep and wildlife when the plants are young (Stubbendieck et al., 1991).

2.105. Elymus cinereus (Scribn. & Merr.) A. Love syn Leymus cinereus.

Great Basin wildrye. Native. Perennial. *Elymus cinereus* can grow in moderately saline soils (EC 7 dSm^{-1}). For successful establishment, *E. cinereus* seedlings needed frequent precipitation in April through June (Roundy, 1985). *Elymus cinereus* var *magnar* grew well at reduced osmotic potentials and also had high germinability (Roundy et al., 1989). It is good forage for cattle and is fair for sheep and wildlife.

2.106. Elymus elongates (Host) Runem (syn Agropyron elongatum).

Tall Wheatgrass. *Elymus elongates* is native to Euroasia and was introduced into the Great Basin region in the 1930s (Welsh et al., 1993). It is an excellent source of forage and it also cures well as winter forage when left standing. However, when cured it is a coarse grass. *Elymus elongates* is more salt tolerant than Basin wildrye and is more like to be established on saline soils (Roundy, 1985). Belligno et al., (2002) grew *E. elongates* with 0, 20, 40, and 60% seawater. Its fresh weight, dry weight and organic matter decreased as salinity increased, but plants were still growing at the highest salinity level. Organic components responded to increased salinity and proline may be a saline stress indicator. Berdahl and Barker (1981) used a salt solution of 225 meq/l to screen seeds of *E. elongates* for salt tolerance. Then they grew the seedlings in a gradient of saline solutions (15 meq/l to 540 meg/l to screen the seedlings. They found that salt tolerance in seed germination was not closely associated with salt tolerance in later stages of plant development. In an evaluation of six grasses, *E. elongates* was among the most salt tolerant (McElgunn & Lawrence, 1973).

2.107. Elymus elymoides (Raf.) (syn Sitanion hystrix (Nutt.) J.G. Smith Swezey. Squirreltail. Perennial. Native.
Elymus elymoides is a native invader of disturbed and saline sites. Because of the bristlelike awns, it is not good forage feed unless it is grazed before the awns form (Welsh et al., 1993).

2.108. Elymus hispidus (Opiz.) Meld. (syn Agropyron intermedium).
Intermediate wheat grass. Perennial. Introduced from Eurasia. *Elymus hispidus* was introducted from the USSR in the 1930s and has been used for forage. Not very salt tolerant (Welsh et al., 1993)

2.109. Elymus junceus Fisch. Russian wildrye. Native to the USSR.
Introduced into the US in 1927. *Elymus junceus* grows in the salt desert shrub zone. It is good forage for livestock and remains green into August. The yield on saline soil was 800-1800 lb/ac (Glenn et al., 1974). It has relatively low seedling vigor and poor seed production (Welsh et al., 1993).

2.110. Elymus lanceolatus (Scribn. & Sm.) Gould. Thickspike wheatgrass. Perennial. Elymus lanceolatus is moderately salt tolerance and has low forage value. It does have sod–holding ability and is often used in revegetating mine spoils (Welsh et al., 1993).

2.111. Elymus salinus Jones. Salina wildrye. Perennial.
Elymus salinus grows in the salt desert shrub zone. In the spring it provides fair forage but when the plant is mature it is unpalatable to livestock.

2.112. Elymus smithii (rydb.) Gould (syn *Agropyron smithii*).
Western wheatgrass. Perennial. *Elymus smithii* is an octoploid species. It is a domant grass in the Great Basin area and is considered a valuable range grass. It grows in the salt desert shrub zone (Scheetz et al., 1981).

2.113. Elymus triticoides Buckley Beardless or creeping wildrye.
Perennial. *Elymus triticoides* is salt tolerant and grows in saline meadows and salt desert shrub zone. Huges et al., (1975) found it could tolerate up to 20,000 ppm of NaCl.

2.114. Ephedra viridis Cov. Mormon tea.
Ephedra viridis grows in the salt desert but is not considered to be very salt tolerant. Seed germination of *E. viridis* was highest at 82% at 15/25°C. In the presence of –8 bar osmotic potential obtained with ethylene glycol or NaCl, the germination was reduced to 14% (Young et al., 1977). It is poor forage for livestock but is browsed some by bighorn sheep and jackrabbits (Stubbendieck et al., 1991). There is some use of *E. viridis* as a health food drink.

2.115. Eragrostis curvula (Schrader) Nees. Weeping lovegrass. Perennial.
Eragrostis curvula grows in blackbrush communities and has some salt tolerance (Ryan et al., 1975).

2.116. Festuca arundinacea Schreber. Tall Fescue. Introduced from Europe.
Festuca arundinacea is salt tolerant and does well in heavy alkaline soils. It is a valuable forage plant, but excess ingestion of Tall Fescue may lead to lamemess in the hind feet of susceptible cattle (Kingsbury, 1964).

2.117. Festuca myuros L. Myur fescue. Annual. Introduced from Eurasia.
Festuca myuros grows in saline meadows and is salt tolerant. It is used to control erosion in some areas.

2.118. Festuca ovina L. Sheep fescue. Perennial.
Festuca ovina is an excellent forage grass and normally grows best in non-saline sites. Indications are that it does have some salt toleance.

2.119. Festuca rubra L. Red fescue. Perennial.
Festuca rubra has some salt tolerance and has been evaluated as forage (Pasternak, 1990).

2.120. Fimbristylis spadicea (L.) Vahl. Fimbristylis. Perennial.
Fimbristylis spadicea grows in wet meadows and is salt tolerant (Welsh et al., 1993).

2.121. Flaveria compestris J. B. Johnston. Marshweed.
Flaveria compestris is found in moist areas and saline seeps (Powell, 1978).

2.122. Frankenia pulverulenta L. Wisp-weed. Annual herb.
Introduced from Europe. *Frankenia pulverulenta* grows in saline soils.and has some value as fodder (Pasternak, 1990).

2.123. Glaux maritima L. Sea milkwort. Perennial.
Glaux maritima is a succulent perennial herb that grows in the saline seeps (Mason, 1957).

2.124. Glycyrrhiza glabra L. licorice. Perennial.
Glycyrrhiza glabra grows in moist areas.

2.125.Grayia spinosa (Hook.) Moq. Hopsage. Perennial.
Grayia spinosa can grow in the salt brush zone (Romney & Wallace, 1980). It is a valuable browse plant for livestock especially sheep. The ranchers call it applebush because of its palatability (Welsh et al., 1993).

2.126. Halimodendron halodendron (Pallas) Schneid.
Salt plant. Introduced from Asia. *Halimodendron halodendron* grows on moist saline soils (Welsh et al., 1993).

2.127. Halogeton glomeratus (Bieb.) C. A. Mey. Halogeton. Annual herb. Introduced from Eurasia. *Halogeton glomeratus* was introduced into Nevada in the early 1930s and has spread through the Great Basin area. It establishes itself in disturbed sites and can grow in salt grass areas. The plant is rich in oxalates and if a large amount is ingested it can change the blood pH of the animal and cause death. In 1970s around 1250 sheep died from *Halogeton* poisoning (Welsh et al., 1993). Thc planting of more favorable salt tolerant plants has reduced the populations of *H. glomeratus* (Young, 2002). Increased salinity decreases the seed germination of *H. glomerstus* but some seeds (10%) germinated at 800 mM NaCl. The rate of germination was the highest at 25/35 °C and the lowest was at 5/15 °C. Inhibition of seeds of *H. glomeratus* by NaCl was higher in the dark as compared to the light. Seed transferred from being in salt solutions for 20 days germinated quickly at a high percentage level (Khan et al., 2001b). The oil content of the seed of *H. glomeratus* was 24%. Lipid analysis of the seed oil indicated that it was 84% unsaturated fatty acids and 16% saturated, which would make it a good cooking oil (Weber et al., 2001).

2.128. Holosteum umbellatum L. Holosteum. Annual herb. Introduced from Europe.
Holosteum umbellatum was present in a population of *Allenrolfea occidentalis*. Some *H. umbellatum* plants were growing under *A. occidentalis*. The soil contained seeds of *H. umbellatrum* (Gul et al., 2000a).

2.129. Heliotropium curassavicum L. Salt heliotrope.
Annual or short-lived perennial. *Heliotropium curassavicum* has two varieties. *Heliotropium curassvicum* var. *obovatum* DC is common in saline seeps and in saltbrush zone. *Heliotropium curassvicum* var. *oculatum* Johston is called pretty heliotrope. It is also found in saline seeps, saltbush and saltgrass zones. *Heliotropium curassvicum* grows in soil salinity levels up to 16 dS/m. Seed germination of *H. curassvicum* was reduced by increased salinity (Dhankhar et al., 2002). Gulzar and Khan (1998) comparied water relations between inland and coastal halophytis and concluded that coastal populations are more stressed than inland populations.

2.130. Hordeum brachyantherum Nevski. Meadow barley.
Hordeum brachyantherum grows in moist and marshy sites. It has some salt tolerance (Mason, 1957).

2.131. Hordeum jubatum L. Foxtail barley. Perennial.
Hordeum jubatum grows in saline meadows and in the salt desert shrub zone. It has fair forage prior to formation of seed heads (Dodd & Coupland, 1966).

2.132. Hordeum marinum Hudson. Mediterranean barley. Introduced from Europe.
Hordeum marinum is found in the salt desert shrub zone, which would indicate salt tolerance (Szaboles, 1994). In Europe it is considered an indicator plant of saline conditions (Ferrari & Gerdol, 1987). Seeds of *H. marinum* appear to have some dormancy (Onnis & Bellettato, 1972).

2.133. Hordeum murinum L. Rabbit barley.
Hordeum murinum can grow in saline soils (Welsh et al., 1993).

2.134. Hordeum pusillum Nutt. Little barley.
Hordeum pusillum is found in the salt desert shrub zone, which would indicate some salt tolerance (Welsh et al., 1993).

2.135. Hordeum vulgare L. Common barley.
Hordeum vulgare is the common commercial barley and it sometimes escapes into the native habitate. It has some salt tolerance.

2.136. Iva axillaris L. Poverty weed. Perennial.
Iva axillaries grow in the salt shrub community (Welsh et. al., 1993).

2.137. Juncus arcticus Willd. Wiregrass. Perennial.
Juncus arcticus is found in saline meadows and marshes (Cooper, 1982).

2.138. Juncus gerardii Lois. Black grass. Perennial.
Juncus gerardii grows in saline mud flats (32% salinity) and has been used as a fodder plant (Rozema, 1975).

2.139. Juncus torreyi Cov. Torrey's rush. Perennial.
Juncus tooreyi is found in saline meadows and marshes (Welsh et al., 1993).

2.140. Juncus tweedyi *Rydb. Tweedy's rush. Perennial.*
Juncus tweedyi grows in wet places and can tolerate some salinity (Cooper, 1982).

2.141. Kochia americana Wats. Native. Perennial.
Kochia americana is excellent forage for sheep, cattle and deer. It is high in protein during the fall (Stubbendieck et al., 1991).

2.142. Kochia prostrata (L.) Schrader. Prostrate Kochia. Annual. Introduced from Eurasia.

Kochia prostrata seeds do not mature until late October. The best germination (97%) was obtained after seeds were air dried and stored for 3 months at 4°C (Waller et al., 1983). The optimum germination occurred between 15-25°C. *K. prostrata* can grow on highly saline soils (25-38 dSm^{-1}) (Rao et al., 1995). Plants grown in saline soils (17 mmho/cm) accumunlated Na^{+} and Cl^{-} contents of 50 and 85 meq/100 g dry matter respectively but no salt injury symptoms were visible (Francois, 1976). It is considered to be good forage.

2.143. Kochia scoparia (L.) Schrader. Summer–cypress. Annual. Introduced from Eurasia.

Kochia scoparia is a salt tolerant species. Optimum germination occurs at 25/35 °C. Seeds germinated in a range of salinity up to 1000 mM NaCl (Khan et al., 2001a). Germination was high for seeds transferred from saline solutions which indicated that high concentrations of NaCl did not inhibit germination permanently (Khan et al., 2001a). Seed production is high and a single plant may have 15,000 seeds. The oil content of *K. scoparia* was 10%. Lipid analysis of the seed oil indicated that 81% was unsaturated fatty acids and 19% saturated (Weber et al., 2001). It can be grown using seawater as the irrigation water (Shamsutdinov et al., 1996). It is considered to be good forage. Cattle can be feed up to 40% *Kochia* in their diet (Lieth & Masoom, 1993).

2.144. Lactuca tatarica (L.) C. A. Mey. Blue lettuce. Perennial.

Lactuca tatarica is found in saline marshes (Welsh et al., 1993).

2.145. Lepidium latifolium L. Broadleaf pepperplant. Perennial.

Lepidium latifolium is found in moist area and can tolerate some salinity (Welsh et al., 1993).

2.146. Lepidium montanumm Nutt. Mountain pepperplant. Perennial.

There are 9 varieties listed for the species. The two listed below are known to grow in the salt desert shrub community.

2.147. Lepidium montanum var jonesii (Rydb.) C.L. Hitche Jones epperplant

2.148. Lepidium montanum var montanum Nutt.

2.149. Leptochloa filiformis (Lam.) Beauv. Red spranagletop. Annual Leptochloa filiformis is rarely found in the Great Basin. It favors marshes (Mason, 1957).

2.150. Leptochloa uninervia (Presl) Hitche. & Chase Mexican sprangetop. Leptochloa uninervia grows in moist areas.and can tolerate some salinity. It has been evaluated as a fodder (Pasternak, 1990).

2.151. Limonium sinuatum (L.) Mill. Winged limonium. Perennial. Native to southern Europe.
Limonium sinuatum is an escaped ornamental that will tolerate 39% salinity (Pasternak, 1990).

2.152. Limonium vulgare Mill.Common Limonium.Perennial.
Limonium vulgare is also an escaped ornamental that will tolerate 23% salinity (Pasternak, 1990).

2.153. Limosella aquatica L. Mudwort. Perennial.
Limosella aquatica grows in mudflats and saline seeps.

2.154. Lobularia maritime (L.) Desv. Sweet alyssum.
Lobularia maritime is an ornamental plant with some salt tolerance that often escapes into the native environment (Welsh et al., 1993).

2.155. Lolium perenne L. Ryegrass.Native to Europe.
Lolium perenne grows in waste places and is wide spread in the world.

2.156. Lycium L.
Lycium L. has 5 species in the genera but only two species appear to have some salt tolerance.

2.157. Lycium barbarum L. Matrimony vine. Native to Eurasia.
Lycium barbarum grows in disturbed sites. Hu et al., (2003) used hydroxyproline to screen for salt tolerance in *L. barbarum*. Wang et al., (1995) used ethyl methanesulfonate to select a salt tolerant mutant of *L. barbarum*. It grew well in 8% PEG.

2.158. Lycium Torreyi Gray Torrey's lycium.
Lycium Torreyi grows in the Tamarix and greasewood communites.

2.159. Malvella leprosa (Ortega) Krap.
Alkali mallow. Perennial. *Malvella leprosa* grows in saline meadows and seeps (Welsh et al., 1993).

2.160. Medicago falcate L.Yellow alfalfa.Perennial.Native to Europe.
Medicago falcate is an introduced forage plant that escaped into the native environment (Welsh et al., 1993).

2.161. Medicago polymorpha L. Bur clover. Introduced from Europe. Perennial.
Medicago falcate is a weedy plant that grows widely on saline soils in different environments (Szaboles, 1994).

2.162. Medicago sativa L. Lucern, alfalfa.Perennial.Introduced from Europe.
Medicago sativa is a major forage crop in the Great Basin area that is grown with irrigation. Many plants have escaped and are part of the non-cultivated population in the Great Basin. Alfalfa has some salt tolerance. Ashraf et al., (1994) found *M. sativa* retained a large proportion of Na^+ and Cl^- in it roots with little transport of these ions to the shoots.

2.163. Melilotus L.
There are three species in this genus and all of them have some salt tolerance. Rogers and Evans (1996) evaluated the three species of *Melilotus* in salt solutions up to 180 mM of NaCl.While all were salt tolerant, *M. alba* was the most salt tolerant.

2.164. Melilotus alba Desr. ex Lam.White sweet-clover.
An evaluation of 11 species of grasses and 10 species of leguminous plants indicated that *M. alba* showed a high level of salt tolerance (Zawadzka, 1976).

2.165. Melilotus indica (L.) All. India sweet clover.
Ashraf et al., (1994) found higer amounts of Na^+, K^+, and Ca^{++} in the shoots of *Mellilotus indica* than *Medicago Sativa. Melilotus indica* was grown in sand culture treated with salt solutions of 0, 80, 160, 240 mol m^{-3} NaCl and the plants maintained low leaf osmotic potential and high turgor potential (Ashraf, 1993).

2.166. Melilotus officinalis (L) Pallas.Yellow sweet clover.
On saline areas *M. alba* and *M. officinalis* increased the productivity of native pastures 3 –5 times over the unimproved patures (Zhambakin et al., 1974). In salt tolerant evaluation trials in Siberia, *M. officinalis* was the most salt tolerant (Baitkanov, 1977). The production of *M. officinalis* grown on saline alkaline soil was 36.7 ton fresh forage/ha (Baitkanov, 1977).

2.167.Monolepis nuttalliana (Schultes) Greene. Poverty-Weed. Annual. Monolepis nuttalliana grows in the saltbush community.

2.168. Nitrophila occidentalis (Moq.) Wats. Niterwort.
Nitrophila occidentalis grows in the saline clay with the pickleweed population.

2.169. Parkinsonia aculeata L. Retama.
Parkinsonia aculeata is a small ornamental tree that escaped into the native environment and has some salt tolerance (Welsh et al., 1993).

2.170. Paspalum distichum L. syn Paspalum vaginatum. Knotgrass. Perennial.
Paspalum distichum grows in saline meadows and salt shrub communities (Glenn, 1987).

2.171. Petunia parviflora Juss. Streamside petunia.
Petunia parviflora grows along stream in the gravel areas and has some salt tolerance.

2.172. Phalaris arundinacea L. Reed canary grass. Perennial.
Phalaris arundinacea grows in wet meadows including saline areas. It is good forage prior to maturity (Welsh et al., 1993).

2.173. Phragmites australis (Cav.) Trin. ex Steudel Common reed.
Phragmites australis grows in saline marshes. Good growth of *P. australis* was obtained at 1.5% permill salinity but all plants died at 35% permill salinity (Hartzendorf & Rolletschek, 2001). Lissner and Hans-Henrik, (1997) suggested that *P. australis* adapts to saline conditions by adjusting the level of osmotically active solutes in its leaves. When *P. australis* was grown in 300 mM NaCl, the sucrose, Cl^- and Na^+ concentrations increased in the shoots. K^+ contributed the most to the leaf osmotic potenial (Matoh et al., 1988). The Na^+ exclusion from the shoot depended on the biological activity of the shoot base (Matsshita & Matoh, 1992). Farnsworth and Meyerson (2003) found *P. australis* could displace other plant species in brackish marshes.

2.174. Phyla nodiflora (L.) Greene. Matted frog-fruit.
Phyla nodiflora grows in moist area and along ditchbanks. It can tolerate 10% salinity (Pasternak, 1990).

2.175. Plantago elongata Pursh Longleaf plantain. Annual.
Plantago elongata grows in the saline sites in the salt desert shrub community.

2.176. Plantago eriopoda Torr. Woollybase plantain. Perennial.
Plantago eriopoda grows in wet meadows and rabbitbrush communities. It can tolerate some salinity (Welsh et al., 1993).

2.177. Plantago insularis Eastw. Insular plantago. Annual.
Plantago insularis is found in the shadscale community and has some salt tolerance (Welsh et al., 1993).

2.178. Plantago lanceolata L. English plantain. Perennial. Native to Eurasia.
Plantago lanceolata us considered to be a weedy plant but it does have some salt tolerance (Welsh et al., 1993).

2.179. Plantago major L Broadleaf plantain. Native to Europe.
Plantago major grows in disturbed sites and is considered a weedy plant. It does have some salt tolerance.

2.180. Pluchea camphorata (L.) DC. Camphorweed.
Pluchea camphorata is present in saline seeps and moist area (Welsh et al., 1993)

2.181. Pluchea sericea (Nutt.) Cov. Perennial. Arrowweed.
Pluchea sericea grows in bottomland and saline seeps. Glenn et al., (1998) evaluated the salt tolerance of *P. sericea* and found a 3-5% reduction in relative growth when one gram of salt per liter was present in the soil. At high salinity levels *P. sericea* can still obtain sufficient water from the soil (Vandersande et al., 2001).

2.182. Poa bulbosa L. Bulbous bluegrass. Perennial. Native to Eurasia.
Polypogon monspeliensis grows in waste areas and is salt tolerant (Szaboles, 1994)

2.183. Polypogon monspeliensis (L.) Desf. Rabbitfoot grass. Annual. Native to Eurasia.
Polypogon monspeliensis grows in moist sites including saline areas (Welsh et al., 1993).

2.184. Polygonum avivulare L. Knotweed. Annual.
Polygonum avivulare is a weedy plant with some salt tolerance (Ungar, 1974).

2.185. Polygonum ramosissimum Michx. Bushy knotweed. Annual.
Polygonum ramosissimum grows in saltmarsh meadows.

2.186. Portulaca halimoides L. Dwarf Purslane. Annual.
Portulaca halimoides is found in sandy sites of fourwing saltbush and is salt tolerant (O'Leary & Glenn, 1994).

2.187. Portulaca oleracea L. Purslane. Annual.
Portulaca oleracea grows in disturbed sites and is salt tolerant (O'Leary & Glenn, 1994).

2.188. Prosopis odorata (Torr. & Frem.) Torr. & Frem. Emend Welsh. Screwbean.
Prosopis odorat is present in the salt shrub community (Welsh et al., 1993).

2.189. Prosopis glandulosa Torr. Honey mesquite. Perennial.
Prosopis glandulosa grows in the desert shrub community and is salt tolerant (Felker et al., 1981).

2.190. Potentilla anserina L. Common silverweed. Perennial.
Potentilla anserina grows in meadows and moist areas and has some salt tolerance.

2.191. Puccinellia Parl.
Puccinellia Parl. has five species that grow in the Great Basin. Three of the species are salt tolerant and grow in the salt desert shrub communities.

2.192. Puccinellia distans (L.) Parl. Weeping alkaligrass. Perennial. Native to Eurasis.
Puccinellia distans is a grazing–tolerant grass (Pehrsson & Thorssell, 1990). It has been used as a turf grass (Torello & Rice, 1986; Torello & Symington, 1984)). Accumulation of soluble carbohydrates in the plant make it possible for *P. distans* to grow on saline soils having low water potential without having extensive salt accumulation in the plant (Albert & Popp, 1978). *Puccinelli distans* grew in salt solutions up to 24 dS m^{-1} and the shoots had a high Cl^- and low Na^+ and K^+ content (Ashraf & Yasmin, 1997).

2.193. Puccinellia fasciculate (Torr.L) Bicknell.Torrey's alkali grass. Perennial.
Puccinellia faciculate has been reported to grow in hypersaline conditions of up to 7.5 % NaCl (Partridge & Wilson 1987).

2.194. Puccinellia nuttalliana (Schult.) A. S. Hitchc. Nuttall's alkaligrass. Perennial.
Puccinellia nuttalliana is very salt tolerant and it is more salt tolerant than the mycorrhizal fungi that colonize its roots at lower saline conditions. (Johnson-Green et al., 2001). Increased soil salinity did reduce the germination rate and early growth of *P. nuttalliana* (Macke & Ungar, 1971).

2.195. Ranunculus cymbalaria Pursh. Marsh buttercup. Annual.
Ranunculus cymbalaria is found in marsh and moist areas. It has some salt tolerance (Welsh et al., 1993).

2.196. Raphanus raphanistrum L. Wild radish. Annual.
Raphanus raphanistrum is a weedy plant with some salinity tolerance. (Welsh et al., 1993).

2.197. Ruppia maritima L. ditchgrass. Perennial.
Ruppia maritima grows in marshes and around ponds of brackish water (Mason, 1957).

2.198. Salicornia L.
Salicornia L. has annual and perennial species. They are very salt tolerant plants.

2.199. Salicornia europaea var rubra L. syn (Salicornia rubra A. Nels). Samphire Annual.
Salicornia europaea var. *rubra* is a succulent annual that is classified as an obligate halophyte. It benefits from certain levels of NaCl in fact it does not grow

well in the absence of salt. It grows near the bottom of saltpan where the salinity is the highest (Harris et al. 2001). The optimal growth of *S. rubra* is at 200 mM NaCl (Khan et al., 2001e). While the fresh and dry weight decreases with increased salinity, some plants were able to grow in 1000 mM NaCl. Seeds of *S. rubra* are capable of germinating at 1000 mM NaCl at 25/35°C alternating temperature regimes. Kinetin, GA_3, and ethephon substantially alleviated the effects of salinity (Khan et al., 2000; Khan et al., 2002a).

2.200. Salicornia utahensis Tidestr. syn Salicornia pacifica var utahensis (Tidestr.) Munz. Utah samphire.

Salicornia utahensis is a succulent perennial that grows well in high saline conditions (salt playa) (Keller et al., 2002). Reproduction is mainly from the roots although a seed bank exists in the salt playa (Gul & Weber, 2001). The outer cells (palisade) of the shoots of *S. utahensis* contain chloroplast, which photosynthesize. The inner region (cortex) lacks chloroplasts (Hansen et al., 1972). The osmotic potential values increase from the base of the stem shoot (75 to 90 atms range) to the top (110 to 170 atms range (Hansen & Weber, 1975). Over the growing season, the concentration of Na^+ gradually increased to 16.1% whereas potassium ion concentration decreased slightly through the growing season (Hansen & Weber, 1975). Weber et al., (1979) found that the carbon fixing enzyme, RuBPCase from *S. utahensis* and *S. rubra* were almost as salt sensitive as RuBPCase from Tomato (Chong-Kyun & Weber, 1980). Using wavelength dispersive X-ray microanalysis to analyse the ion distribution in a cross section of S. utahensis, Weber et al., (1977) found that Na^+, K^+, and Cl^- concentrations were very low in the palisade region but high in the cortex area. This suggested compartmentalization of the ions in the nonphotosynthetic cortex cells. Similar ion distribution results were obtained using Energy Dispersive X-Ray Microanlysis (Khan et al., 1985; Khan et al., 1986). Using silver Cl^- precipitation to locate ions in the cell, Hess et al., (1975) found that the Cl^- ions were concentrated in the vacuoles in the palisade and cortex cells but the Cl^- ions were low in organelles such as chloroplasts (Hess et al., 1975; Hansen et al., 1975). Using lead phosphate as a marker for ATPase, it was determined that the ATPase was located along the plasma memberanes in the palisade and cortex cells of shoots (Weber et al., 1980). Chong-Kyun and Weber (1980) isolated ATPase from *S. utahensis* and found it was salt tolerant and could function in 3M NaCl. Over time sections of the shoot may accumulate so much salt that it kills the cells, but the vascular tissue in the center of the shoot continues to function as water transport to the sections above the dead section (Weber, 1982). Analysis of the oil from seeds of *S. utahensis* indicate that it has potential as a cooking oil. Poulin (1981) found the ash content of *S. europea* was 8.5 g per 100 g fresh wt and the energy content was 279 calories per 100 grams.

2.201. Salsola collina Pall. Pallas tumbleweed.

Plants found in disturbed sites

2.202. Salsola paulsenni Litv.Barbwire Russian thistle.
Plants found in disturbed sites.

2.203. Salsola pestifer A. Nels. syn Salsola liberica Sennen & Pau and Salsola kali ds is commonly known as tumble weed or Russian thistle. Introduced from Russia.
Seeds of *Salsola pestifer* will germinate at different concentrations of salt and a few will even germinate at 1000 mM NaCl. Increased salinity decreased the germination percent. Cooler temperatures significantly inhibited germination, while warmer nights (25°C) and warmer days (35°C) showed higher germination. Seeds transferred from salt solutions to distilled water after 20 days recovered quickly at cooler temperatures but the recovery germination percentation decreased with increase in salinity and temperature (Khan et al., 2002b). It is a C_4 photosynthesis halophyte (Glagoleva & Chulanovskaya, 1996). It has fair forage value for cattle and sheep in the early spring when the plant is tender. Biomass yield was increased by exposure to saline conditions (Fowler et al., 1988).

2.204. Sarcobatus vermiculatus (Hook.) Torr. Greasewood. Native. Perennial.
Sarcobatus vermiculatus is usually considered an indicator plant for saline soils. It is a fair forage plant for livestock, big game, small mammals and birds. Soluble oxalates particularly in young twigs can cause mortality in sheep and in rare cases in cattle. It has some value as firewood. *Sarcobatus vermiculatus* is able to maintain high leaf nitrogen in spite of the high sodium (Drenovsky & Richards, 2003). It has deep roots (nearly 13 meters deep). *Sarcobatus vermiculatus* has two chromosome races (n=18 and n=36) with the dipoloid being the common form (Sanderson et al., 1999). Mound formation (0.3 m in height) is common with S. vermiculatus. Donovan et al., (1997) found that *S. vermiculatus* could accumulate large amounts of leaf Na and still maintain adequate uptake of N, P, K, Ca and Mg over an extreme salinity gradient (non-saline to highly saline). Eddleman and Romo (1987) concluded that rapid uptake of Na by the germinating embryos and seedlings were an adaptive mechanism for developing and maintaining a favorable water balance in soils with low osmotic potentials. Dobb and Donovan (1999) suggested that the Na uptake of *Sarcobatus* seedlings enhanced it ability to deal with declining Psi(s) and to become established in more saline areas. Some seeds of S. vermiculatus could germinate in salt solution of 800 mM NaCl. The optimal germination was obtained at a temperature regime of 20/30 °C at all salinity concentrations. But even seeds placed in 1000 mM NaCl for 20 days germinated readily when transferred to distilled water. This indicated that high salinity is not detrimental to the seeds of *S. vermiculatus* (Khan et al., 2001d). The oil content of seeds of *S. vermiculatus* was 18%. Lipid analysis of the seed oil indicated that 79% was unsaturated fatty acids and 21% saturated (Weber et al., 2001).

2.205. Scripus acutus Muhl. Ex Bigelow. Hardstem bulrush. Perennial.
Scripus acutus grows in the margins of ponds, in marshes and saline seeps (Mason, 1957).

2.206. Scirpus americanus Pers. Perennial Olney's threesquare.
Scirpus americanus is found in saline meadows and in saltgrass communities.

2.207. Scirpus maritimus L. Alkali bulrush. Perennial.
Scirpus maritimus is found in saline meadows and seeps. It can tolerate 21% salinity (Mepham & Mepham, 1985).

2.208. Scirpus nevadensis Wats. Nevada bulrush. Perennial.
Scirpus nevadensis is found in mosist saline meadows and is salt tolerant (Welsh et al., 1993).

2.209. Scirpus pungens Vahl. Common threesquare. Perennial.
Scirpus pungens is found in saline seeps and is saline tolerant (Welsh et al., 1993).

2.210. Scirpus validus Vahl. Softstem bulrush. Perennial.
Scirpus validus grows in marshland and has some salt tolerance (Welsh et al., 1993).

2.211. Sesuvium verrucosum Raf. Seapurslane. Perennial.
Sesuvium verrucosum grows in saline areas and tolerate up to 46% saline (O'Leary & Glenn, 1994).

2.212. Setaria viridis (L.) Beauv. Green bristlegrass. Native to Eurasis.
Setaria viridis grows in dry and moist areas and is considered to be a weed. But does have some salt tolerance.

2.213. Sidalcea neomexicana Gray. New Mexico checker. Perennial.
Sidalcea neomexicana is found in wet meadows and stream banks and has some salt tolerance (Welsh et al., 1993).

2.214. Spartina gracilis Trin. Alkali cordgrass. Perennial.
Spartina gracilis grows in saline sites (Welsh et al., 1993).

2.215. Spartina pectinata Link Prairie cordgrass. Perennial.
Spartina pectinata grows in moist saline sites (Welsh et al., 1993).

2.216.Spergularia marina (L) Griseb. syn Spergularia salina (Presl.) J & C Presl.
Salt sandspurry. Annual. Introduced from Europe. *Spergularia marina* grows in saline areas. On salt brine soils, autumn seeded *S. marina* produced the greatest yields (Keiffer & Ungar, 2002). In a density study of halophytes, the survival of

Eddleman, L.E. & Romo, J.T. 1987. Sodium relations in seeds and seedlings of *Sarcobatus vermiculatus*. Soil Science 143: 120-123.

Evans, D.D., Sammis, T.W. & Cable, D.R. 1981. Actual evapotranspiration under desert conditions, In D.D. Evans & J.L Thames (Eds.), Water in Desert Ecosystems (US/IBP Synthesis Series No.11). pp. 195-218. Stroudsburg, Pennsylvania: Dowden, Hutchinson and Ross, Inc.

Farnsworth, E.J. & Meyerson, L.A. 2003. Comparative ecophysiology of four wetland plant species along a continuum of invasiveness. Wetlands 23: 750-762.

Felker, P., Clark, P.R., Lagg, A.E. & Pratt, P.F. 1981. Salinity tolerance of the treelegumes Mesquite (*Prosopis glandulosa* var *torreyana*, *P. velutina* and *P. articulata*). Algarrobo (*P. chilensis*), Kaiwe (*P. pallida*) and Tamarugo (*P. tamarugo*) grown in sand cultures on nitrogen free media. Plant and soil 61: 311-317.

Ferrari, C. & Gerdol, R. 1987. Numerical syntaxonomy of badland vegetation in Apennines Italy. Phytocoenologia 15: 21-38.

Fisher, D.D., Schenk, H.J., Thorsch, J.A. & Ferren Jr, W.R. (1997). Leaf anatomy and subgeneric affiliations of C_3 and C_4 species of *Suaeda* (Chenopodiaceae) in North America. American Jouranl of Botany 84: 1198-1210.

Fowler, J.L., Hageman, J.H., Suzukida, M. & Assadian, H. 1988. Evaluation of the salinity tolerance of Russian thistle, a potential forage forage crop. Agronomy Journal 80: 250-258.

Francois, L.E. 1976. Salt tolerance of prostrate summer cypress *Kochia prostrata*. Agronomy Journal 68: 455-456.

Galinato, M.I. & van der Valk, AG. 1987. Seed germination traits of annuals and emergents recruited during drawdowns in the Delta Marsh, Manitoba, Canada. Aquatic Botany 26: 89-102.

Gallagher, J.L. 1986. Halophytic crops for cultivation at seawater salinity. Plant and soil 89: 323- 336.

Garve, E. & Garve, V. 2000. Halophytes at potash-mine dumps in Germany and France (Alsace). Tuexenia 20: 375-417.

Glagoleva, T.A. & Chulanovskaya, M.V. 1996. Photosynthetic metabolism and assimilate translocation in C_4 halophytes inhabiting the Ararat Valley. Russian Journal of Plant Physiology 43: 349-357.

Glenn, E., Tanner, R., Mendez, S., Kehret, T., Moore, J.D. & Hull Jr., A.C. 1974. Species for seeding arid rangeland in southern Idaho. Journal of RangeManagement 27: 216-218.

Glenn, E.P. & O'Leary, J.W. 1984. Relationship between salt accumulation and water content of dicotyledonous halophytes. Plant, Cell and Environment 7:253-261.

Glenn, E.P. & O'Leary, J.W. 1985. Productivity and irrigation requirements of halophytes grown with seawater in the Sonoran Desert. Journal of Arid Environments 9: 81-91.

Glenn, E.P. 1987. Relationship between cation accumulation and water contents of salt tolerant grasses and sedge. Plant, Cell and Enivironment 10: 205-212.

Gonzalez, C.L. & Heilman, M.D. 1977. Yield and chemical composition of coastal bermudagrass, rhodesgrass, and volunteer species grown on saline and nonsaline soils. Journal of Range Management 30: 227-230.

Goodman, P.J. 1973. Physiological and ecotypic adaptations of plants to salt desert condtions in Utah. Journal of Ecology 61: 473-494.

Gul, B. & Weber, D.J. 1998. Effect of dormancy relieving compounds on the seed germination of *Allenrolfea occidentalis* under salinity stress. Annals of Botany 82: 555-560.

Gul, B. & Weber, D.J. 1999. Morphological characteristics of seeds of five halophytes. Indian Journal of Botany 78: 331-338.

Gul, B. & Weber, D.J. 2000. Effect of salinity, light and temperature on the seed germination of *Allenrolfea occidentalis*. Canadian Journal of Botany 77: 240-246.

Gul, B. & Weber, D.J. 2001. Seed bank dynamics in a Great Basin salt playa. Journal of Arid Environments 49: 785-794.

Gul, B., Khan, M.A. & Weber, D.J. 2000c. Alleviation of salinity and dark-enforced dormancy in *Allenrolfea occidentalis* seeds under various thermoperiods. Australian Jounal of Botany 48: 745- 752.

Gul, B., Weber, D.J. & Khan, M.A. 2000a. Population dynamics of a perennial halophyte, *Allenrolfea occidentalis*. In E. Durant McArther, W. Kent Ostler, Carl L. Wambolt (Eds.), Proceedings: Shrubland Ecotones, 1998. Proceedings RMRS-P-000. pp. 124-130. Ogden, Utah: USDA, Forest Service, and Rocky Mountain Research Station.

Gul, B., Weber, D.J. & Khan, M.A. 2000b. Effect of salinity and planting density on the physiogical responses of *Allenrolfea occidentalis*. Western North American Naturalist 60: 188-197.

Gul, B., Weber, D.J. & Khan, M.A. 2001a. Growth, ionic and osmotic Relations of an *Allenrolfea occidentalis* population in an inland salt playa of the Great Basin Desert. Jouranl of Arid Environment 48: 445-460.

Gulzar, S. & Khan, M.A. 1998. Diurnal water relations of inland and coastal halophytic populations from Pakistan. Jounal of Arid Environment 40: 295-305.

Hansen, D.J. & Weber, D.J. 1975. Environmental factors in relation to the salt content of *Salicornia pacifica* var. utahensis (Tidestrom) Munz. Great Basin Naturalist 35: 86-96.

Hansen, D.J., Hess, W.M., Weber, D.J. & Andersen, W.R. 1972. Ultrastructure and physiological investigations of the halophyte *Salicornia utahensis*. Journal of Cell Biology 55: 105A.

Harris, L.C., Gul, B., Khan, M.A., Hansen, L.D. & Smith, B.N. 2001. Seasonal changes in respiration of halophytes in salt playas in the Great Basin, U.S.A. Wetlands Ecology and Management 9: 463-468.

Hartzendorf, R. 2001. Effects of $NaCl^-$ salinity on amino acid and carbohyudrate contents of *Phragmites australis*. Aquatic Botany 69: 195-208.

Hess, W.M. & Weber, D.J. 1995. Morphology of epicuticular wax and comparison of lipids of *Allenrolfea occidentalis*. In: M.A. Khan and I.A. Ungar (Eds.), Biology of Salt Tolerant Plants. pp 107-117. Karachi, Pakistan: University of Karachi Press.

Hess, W.M., Hansen, D.J. & Weber, D.J. 1975. Light and electron microscopy localization of Cl^- ions in cells of *Salicornia pacifica* var. *utahensis*. Canadian Journal of Botany 53: 1176-1187.

Hodgkinson, H.S.1987. Relationship of saltbush species to soil chemical properties. Journal of Range Management 40: 23-26.

Hoffman, G.J. & Shannon, M.C. 1986. Relating plant performance and salinity. Rech. and Revegetation Researh 5: 211-225.

Hopkins, C.O. & Blackwell Jr., W.H. 1977. Synopsis of Suaeda chenopodiaceae in North American. SIDA Contributions to Botany 7: 147-173.

Hu, Bo-ran, Xu, Wen-biao, Zhang, Shuang-ling, & Ma, Feng-wang. 2003. Screening and characterizing of hydroxyproline-resistant variants in *Lycium barbarum* and analysis of its salt tolerance. Xibei Zhiwu Xuebao 23: 422-427.

Hughes, T.D., Butler, J.D. & Sanks, G.D.1975. Salt tolerance and suitability of various grasses for saline roadsides. Journal of Environmental Quality 4: 65-68.

Jain, B.L. & Muthana, K.D. (1982). Performance of different tree species under saline irrigation at nursery stage. Myforest 1: 175-180.

Johnson-Green, P., Kenkel, N.C. & Booth, T. 2001. Soil salinity and arbuscular mycorrhizal colonization of *Puccinellia nuttalliana*. Mycological Research 105: 1094-1110.

Katembe, W.J., Ungar, I.A. & Mitchell, J.P. 1998. Effect of salinity on germination and seedling growth of two *Atriplex* species (Chenopodiaceae). Annuals of Botany 82: 167-175.

Keiffer, C.H. & Ungar, I.A. 1997a. The effect of extended exposure to hypersaline conditions on the germination of five inland halophyte species. American Jouranl of Botany 84: 104-111.

Keiffer, C.H. & Ungar, I.A. 1997b. The effects of density and salinity on shoot biomass and ion accumulation in five inland halophytic species. Canadian Journal of Botany 75: 96-107.

Keiffer, C.H. & Ungar, I.A. 2001. The effect of competition and edaphic conditions on the establishment of halophytes on brine effected soils. Wetlands Ecology and Management 9: 469- 481.

Keiffer, C.H. & Ungar, I.A. 2002. Germination and establishment of halophytes on brine-affected soils. Journal of Applied Ecology 39: 402-415.

Keller, E.A., Harris, L.C., Khan, A., Zou, Jiping, Smith, B.N. & Hansen, L.D. 2002. *Salicornia utahensis* grows at high salt concentrations but only at high temperatures. In: International Conference of the Society for Ecological Restoration abstracts, pp. 176. Tucson, Arizona: Society for Ecological Restoration.

Kenagy, G.J. 1972. Saltbush leaves: excision of hypersaline tissue by a kangaroo rat. Science 178: 1094-1906.

Ketchum, R.E.B., Warren, R.S., Klima, L.J., Lopex-Gutierrez, F. & Nabors, M.W. 1991. The mechanism and regulation of proline regulation of proline accumulation in suspension cell cultures of the halophytic grass Distichilis spicata. L. Journal of Plant Physiology 137: 368-374.

Khan, M.A., Gul, B. & Weber, D.J. 2000. Germination responses of *Salicornia rubra* to temperature and salinity. Journal of Arid Environments 45: 207-214.

Khan, M.A., Gul, B. & Weber, D.J. 2001a. Influence of salinity and temperature on the germination of *Kochia scoparia*. Wetlands Ecology and Management 9: 483-489.

Khan, M.A., Gul, B. & Weber, D.J. 2001b. Seed germination characteristics of *Halogeton glomeratus*. Canadian Journal of Botany 79: 1189-1194.

Khan, M.A., Gul, B. & Weber, D.J. 2001d. Seed germination in relation to salinity and temperature in *Sarcobatus vermiculatus*. Biologia Plantarum 45: 133-135.

Khan, M.A., Gul, B. & Weber, D.J. 2001e. Effect of salinity on the growth and ion content of *Salicornia rubra*. Communication in Soil Science and Plant Analysis 32: 2965-2977.

Khan, M.A., Gul, B. & Weber, D.J. 2002a. Improving seed germination of *Salicornia rubra* (Chenopodiaceae) under saline conditions using germination regulating chemicals. Western North American Naturalist 62: 101-105.

Khan, M.A., Gul, B. & Weber, D.J. 2002b. Seed germination in the Great Basin halophyte *Salsola iberica*. Canadian Journal of Botany 80: 650-655.

Khan, M.A., Gul, B. & Weber, D.J. 2004a. Action of plant growth regulators and salinity on seed germination of *Ceratoides lanata* Canandian Journal of Botany 82: 37-42.

Khan, M.A., Gul, B. & Weber, D.J. 2004b. Temperature and high salinityn effects in germinating dimorphic seeds of *Atriplex rosea*. Western North American Naturalist 64: 193-201.

Khan, M.A., Gul, B., & Weber, D.J. 2001c. Germination of dimorphic seeds of *Suaeda moquinii* (Torrey) Greene under high salinity stress. Australian Journal of Botany 49: 185-192.

Khan, M.A., Weber, D.J. & Hess, W.M. 1987. Elemental Compartmentalization in seeds of *Atriplex triangularis* and *Atriplex confertifolia*. Great Basin Naturalist 47: 91-95.

Khan, M.A., Weber, D.J. & Hess, W.M. 1985. Elemental distribution in seeds of the halophytes *Salicornia pacifica* var. *utahensis* and *Atriplex canescens*. American Journal of Botany 72: 1672- 1675.

Khan, M.A., Weber, D.J. & Hess, W.M. 1986. Elemental distribution in shoots of *Salicornia pacifica* var. *utahensis* as determined by energy dispersive X-ray microanalysis using a cryochamber. Botanical Gazette 147: 16-19.

Kingsbury, J.M. 1964. Poisonous plants of the United States and Canada Englewood Cliffs. New Jersey: Prentce-Hall Inc.

Kleinkopf, G.E., Wallace, A. & Cha, J.W. 1975. Sodium relations in desert plants: 4. Some physiological responses of *Atriplex confertifolia* to different levels of sodium Cl^-. Soil Science, 120, 45-49.

Kovalev, V.M. & Krylova, N.P. 1992. Use of halophytes to improve arid pastures. Sel'skokhozyaistvennaya Biologiyha 4: 135-141.

Lieth, H. & Al Masoom, A.A. 1993. Kochia: A new alternative for forage under high salinity conditions in Mexico. Towards the rational use of high salinity tolerant plants. In: H. Lieth & A. A. Al Masoom (Eds.), Vol 1: Deliberations about high salinity tolerant plants and ecosystems. pp. 459-464. Dordrecht, Netherlands: Kluwer Academic Publishers

Lieth, H., Moschenko, M., Lohmann, M., Koyro, H.W. & Hamdy, A. (Eds.), 1999. Halophytes uses in different climates vol. 13. Leiden, Netherlands: Backhus Publishers,

Lin, Hong, & Wu, Lin, 1996. Effects of salt stress on root plasma membrane characteristics of salt- tolerant and salt-sensitive buffalograss clones. Environmental and Experimental Botany 36: 239- 254.

Lissner, J. & Hans-Henrik, S. 1997. Effects of salinity on the growth of *Phragmites australis*. Aquatic Botany 55: 247-260.

Macke, A.J. & Ungar, I.A. 1971. The effects of salinity on germination and early growth of *Puccinellia nuttalliana*. M. Canadian Journal of Botany 49: 515-520.

MacMahon, J.A. 1979. North American deserts: Their floral and faunal components. In: D.W. Goodall & R.A. Perry. (Eds.), Arid-land Ecosytems: Structure Functioning and Management. pp. 21-82, Cambridge UK: Cambridge University Press.

Marcum, K.B. & Murdoch, C.L. 1994. Salinity tolerance mechanisms of six C_4 Turfgrasses. Journal of the American Society for Horticultural Science 119: 779-784.

Marcum, K.B. 1999. Salinity tolerance mechanisms of grasses in the subfamily Chloridoideae. Crop Science 39: 1153-1160.

Mason, H.C. 1957. A flora of the marshes of California. Berkely.California: University of California Press.

Matoh, T., Matsushita, N. & Takahashi, E. 1988. Salt tolerance of the reed plant *Phragmites communis*. Physiologia Plantarum 72: 8-14.

Matsushita, N. & Matoh, T. 1992. Function of the shoot base of salt-tolerant reed (*Phragmites communis trinius*) plants for Na exclusion from the shoots. Soil Science and Plant Nutrition 38: 565-571.

McElgunn, J.D. & Lawrence, T. 1973. Salinity tolerance of Altai wild ryegrass and other forage grasses. Canadian Journal of Plant Science 53: 303-307.

Mepham, R.H. & Mepham, J.S. 1985. The flora of tidal forests-a rationalization of the use of the term "mangrove". South African Journal of Botany 51: 75-99.

Mikhiel, G.S., Meyer, S.E. & Pendleton, R.L. 1992. Variation in germination response to temperature and salinity in shrubby *Atriplex* species. Journal of Arid Environments 22: 39-49.

O'Leary, J.W. & Glenn, E.P. 1994. Global distribution and potential for halophytes. In: V.R. Squires and A.T Ayoub. (Eds.), Tasks for Vegetation Science 32, pp. 7-17. Dordrecht, Netherlands: Kluwer Academic Publishers.

Onnis, A. & Bellettato, R. 1972. Dormancy and salt-tolerance in two spontaneous species of *Hordeum* (*Hordeum murinum* L. and *Hordeum marinum* Huds.) Nuovo Gior Bot Ital 106: 101- 113.

Osmund, C.B., Bjorkman, O. & Anderson, D. 1980. Physiological processes in plant ecology: Towards a synthesis with Atriplex. In Ecological Studies 36, pp. 357 New York, New York: Springer.

Partridge, T.R. & Wilson, J.B. 1987. Salt tolerance of salt marsh plants of Otago, New Zealand. New Zealand Journal of Botany 25: 559-566.

Pasternak, D. 1990. Fodder production with saline water. Project Report. pp. 173 Ben-Gurion University of the Negev, Israel: The institute for applied research.

Pasternak, D., Nerd, A. & De Malach, Y. 1993. Irrigation with brackish water under desert conditions: IX. The salt tolerance of six forage crops. Agricultural Water Management, 24, 321- 334.

Pehrsson, O. & Thorssell, S. 1990. Reconstuction of grazed inland water habitats. Fauna och Flora Naturhistoriska Riksmuseet 85: 225-233.

Poulin, G. 1981. The commercial production of a salt marsh native plant, *Salicornia europea* (Chenopodiaceae). Association for Arid Lands Studies, Proceedings, 4, 23-25.

Powell, A.M. 1978. Systematics of *Flaveria* (Flaveriiaeo-Asteraceae). Annuals of the Missouri Botanical Garden 65: 590-636.

Rao, G.G., Ravender, S. & Bhargava, G.P. 1995. Species diversity on salt affected soils under canal command areas and in Bhal region in Gujarat. State Indian Forester 121: 1143-1150.

Redmann, R.E. & Fedec, P. 1987. Mineral ion compostion of halophytes and associated soils in Western Canada. Communications in Soil Science and Plant Analysis 18: 559-580.

Richardson, S.G. & McKell, C.M. 1980. Salt tolerance of two saltbush species grown in processed oil shale. Journal of Range Management 33: 460-463.

Rogers, M.E., Noble, C.L., & Pederick, R.J. 1996. Identifying suitable grassspecies for saline areas. Australian Journal of Experimental Agriculture 36: 197-202.

Romney, E.M. & Wallace, A. 1980. Ecotonal distribution of salt tolerant shrubs in the Northen Mojave desert Nevada USA. Great Basin Naturalist Memoirs 4: 134-139.

Roundy, B.A. 1985. Emergence and establishment of basin wild rye *Elymus cinereu* and tall wheat grass *Agropyron elongatum* in relation to moisture and salinity. Journal of Range Management 38: 126-131.

Roundy, B.A., Cluff, G.J., Young, J.A. & R.A. Evans. 1983. Treatment of inland saltgrass and greasewood sites to improve forage production. In: S.B. Monsen & N. Shaw, (Eds.), Managing intermountain rangelands – improvement of range and wildlife habitats. pp. 54-61. Ogden, Utah: USDA Forest Service.

Roundy, B.A., Young, J.A. & Evans, R.A. 1989. Seedling growth of three Great Basin wildrye collections at reduced osmotic potential. Agriculture, Ecosystems and Environment 25: 245-251.

Rozema, J. 1975. Biology of halophytes. In: R. Choukr-Allah, D.V. Malcolm, & A. Hamdy. (Eds.), Halophytes and biosaline agriculture. pp. 17-31. New York, New York: Marcel Dekker Inc.

Ryan, J., Miyamoto, S. & Stroehlein, J.L. 1975, Salt and specific ion effects on germination of four grasses. Journal of Range Management 28: 61-64.

Salo, L.F., Artiola, J.F. & Goodrich-Mahoney, J.W. 1999. Evaluation of revegetation techniques of a saline flue gas desulfurization sludge pond. Journal of Environmetal Quality 28: 218-225.

Sanderson, S.C., Stutz, H.C., Stutzm, M. & Roos, R.C. 1999. Chromosome races in *Sarcobatus* (Sarcobataceae, Caryophyllales). Great Basin Naturalist 59: 301-314.

Sandquist, D.R. & Ehleringer, J.R.1995. Carbon-isotope discrimination in the C_4 shrub *Atriplex confertifolia* along a salinity gradient. Great Basin Naturalist 55: 135-141.

Scheetz, J.G., Majerus, M.E. & Carlson, J.R. 1981. Improved plant materials and their establishment to reclaim saline seeps in Montana. In Agronomy Abstracts. 73rd annual meeting, p.96. Madison, Wisconsin: American Society of Agronomy. Schirmer, U. & Breckle, W. of Nebraska.

Swingle, R.S., Glenn, E.P. & Squires, V. 1996.Growth performance of lambs fed mixed diets containing halophyte ingredients. Animal Feed Science and Technology 63: 137-148.

Szaboles, I. 1994. Salt affected soils as the ecosystem for halophytes. In: V.R. Squires, & A.T. Ayoub (Eds.), Tasks for vegetation science vol 32. pp 19-24. Dordrecht, Netherlands: Kluwer Academic Publisher.

Tanaka, R. & Tanaka, A. 1980. Karyo morphological studies on halophytic plants 1. Some taxa of *chenopodium*. Cytologia (Tokyo) 45: 257-270.

Telenikus, A. & Torstensson, P. 1989. The seed dimorphism of *Spergularia marina* in relation to dispersal by wind and water. Oecologia (Berlin) 80: 206-210.

Torello, W.A. & Rice, L.A. 1986. Effects of sodium Cl^- stress on proline and cation accumulation in salt sensitive and tolerant turfgrasses. Plant and Soil 93: 241-248.

Torello, W.A. & Symington, A.G. 1984. Screening of turfgrass species and cultuivars for Cl^- tolerance. Plant and Soil 82: 155-162.

Trent, J.D., Blank, R.R. & Young, J.A. 1997. Ecophysiology of the temperate desert halopyte *Allenrolfea occidentalis* and *Sarcobatus vermiculatus*. Great Basin Naturalist 57: 57-65.

Trimble, S. 1989. The sagebrush ocean-a natural history of the Great Basin. 248 pp. Reno Nevada: University of Nevada Press,

Ungar, I.A. 1974. Halophyte communities of Park County, Colorado. Torrey Botany Club 101: 145- 152.

Ungar, I.A. 1988. A significant seed bank for *Spergularia marina* Caryophyllaceae. Ohio Journal of Science 88: 200-202.

Vandersande, M.W., Glenn, E. & Walworth, J.L. 2001. Tolerance, E. M. 1973. Sodium relations in desert plants. 2. Distribution of cations in plant parts of three different species of *Atriplex*. Soil Science 115: 390-394.

Waller, S.S., Britton, C.M., Schmidt, D.K., Stubbendikeck, J. & Sneva, F.A. 1983. Germination characteristics of 2 varieties of *Kochia prostrata*. Journal of Range Management 36: 242-245.

Wang, LunShan, Lu, Wei, Sun, Tong, Li, HuiJuan, & Wang, YaFu. 1995. In vitro selection of sodium Cl-tolerant variants of the tea tree and differentiation of regenerated plants. Hereditas (Beijing) 17: 7-11.

Weber, D.J. 1982. Mechanisms of salt tolerance in *Salicornia pacifica* var *utahensis*. In: A. San Pietro (Ed.), Bio-Saline Research: A Look at the Future. pp. 555-558. New York, New York: Plenum Press.

Weber, D.J. 1995. Mechanisms and reactions of halophytes to water and salt stress. In: M.A. Khan and I.A. Ungar (Eds.), Biology of Salt Tolerant Plants. pp. 170-180. Karachi, Pakistan: University of Karachi Press.Welsh, S.L., Atwood, N.D., Goodrich, S. & Higgins, L.C. 1993. A Utah Flora. 986 pp. Provo, Utah: Brigham Young University Press.

Weber, D.J., Gul, B. & Khan, M.A. 2002. Halophytic characteristics and potentiall uses of *Allenrolfea occidentalis*. In: R. Ahmad & K.A. Malik, (Eds.), Prospects for saline agriculture, pp. 333-352. Dordrecht, Netherlands: Kluwer Academic Publishers,

Weber, D.J., Gul, B., Khan, M.A., Williams, T., Wayman, P. & Warner, S. 2001. Composition of vegetable oil from seeds of six native halophytic shrubs. In: E. Durant McArthur, Daniel J. Fairbanks, (Eds.), Proceedings: Shrubland Ecosystem Genetics and Biodiversity; 2000 pp. 237- 290.Proceedings RMRS-P-21. Ogden, Utah: U.S. Department of Agriculture, Forestry Service Rocky Mountain Research Station.

Weber, D.J., Hegerhorst, D.J., Bhat, R.B. & McArthur, E.C. 1993. Rubber Rabbitbrush (*Chrysothamnus nauseosus*), a multi-use desert shrub. In: Tasks for Vegetation Science vol 28. pp. 343-350. Dordrecht, etherlands : Kluwer Academic Publishers.

Weber, D.J., Hess, W.M. & Kim, Chong-Kyun. 1980. Distribution of ATPase in cells of *Salicornia pacifica* var. *utahensis* as determined by lead phosphate precipitation and x-ray microanalysis. New Phytologist 84: 285-291.

Weber, D.J., Rasmussen, H.P. & Hess, W.M. 1977. Electron microprobe analyses of salt distribution in the halophyte *Salicornia pacifica* var. *utahensis*. Canadian Journal of Botany 55: 1516-1523.

Welkie, G.W. & Caldwell, M. 1970. Leaf anatomy of species in some dicotyledonous families as related to the C_3 and C_4 pathways of carbon fixtaion. Canadian Journal of Botany 48: 2135-2146.

West, N.E. 1983. Great Basin-Colorado Plateau sagebruch semi-desert. pp. 331-349. In: N.E. West (Eds.), Ecosystems of the World 5/Temperate Deserts and Semi Deserts. Amsterdam. Netherlands: Elsevier.

Wiecko, G. 2003. Ocean water as a substitute for postemergence herbicides in tropical turf. Weed Technology 17: 788-791.

Wu, L. & Lin, H. 1994. Salt tolerance and salt uptake in diploid and polyploidy buffalograsses (*Buchloe dactyloides*). Journal of Plant Nutrition 17: 1905-1928.

Wu, L., Huang, Z.Z. & Burau, R.G. 1988. Selenium accumulation and selenium salt co-tolerance in five grass species. Crop Science 28: 517-522.

Yensen, N.P. 2002, Prospects for saline agriculture. In: R. Ahmad & K A. Malik, (Eds.), pp. 321-332. Dordrecht, Netherlands: Kluwer Academic Publishers.

Young, J.A. 2002. Halogeton grazing management: historical perspective. Journal of Range Management 55: 309-311.

Young, J.A., Blank, R.R., Palmquist, D.E. & Trent, J.T. 1995. Allenrolfea deserts in Western North America. Journal of Arid Land Studies 4: 197-205.

Young, J.A., Evans, R.A. & Kay, B.L. 1977. Ephedra seed germination. Agronomy Journal 69: 209-211.

Zawadzka, M. 1976. Salt tolerance of grasses and leguminous plants. Acta Agrobotanica 29: 85-98.

Zhambakin, Z.A., Pryanishnikov, S.N., Baitkanov, K.A. & Salukov, P.A. 1974. Improvement and rational management of native grasslands in Kazakhstan. In: proceedings of the 12th International Grassland Congress. Improvement of natural and production of seeded meadows and pastures. pp. 434-441. Moscow, Russia: 12th International Grassland Congress.

CHAPTER 6

ROLE OF CALCIUM IN ALLEVIATING SALINITY EFFECTS IN COASTAL HALOPHYTES

BILQUEES GUL AND M. AJMAL KHAN[1]

Department of Botany, University of Karachi, Karachi-75270, Pakistan
[1]corresponding author:halophyte_ajmal@yahoo.com

Abstract: The purpose of the current investigation was to study the effect of Ca^{2+} on the seed germination of coastal species like *Arthrocnemum indicum, A. macrostachyum, Desmostachya bipinnata, Halopyrum mucronatum* and *Urochondra setulosa* under NaCl salinity. Seed germination was inhibited with the increase in NaCl concentration and species differ in their range of NaCl tolerance. Inclusion of $CaCl_2$ partially or completely alleviated NaCl effects on the germination of all species. In the case of *A. macrostachyum* calcium application completely alleviated salinity effects at all concentrations of NaCl, while in the case of *A. indicum* and *D. bipinnata* calcium completely alleviated NaCl effects except at 800 mM NaCl. While in the case of *H. mucronatum* and *U. setulosa* Ca^{2+} completely alleviated the NaCl effects at low concentration and partially alleviated at higher concentrations.

1. INTRODUCTION

The success of bio-saline agriculture is greatly dependent on the germination response of seeds of un-conventional halophytic crops (Khan, 2003). The soils where halophytes normally grow becomes more saline due to rapid evaporation of water particularly during summer, therefore, surface of the soil tend to have higher soil salinity and higher water potential (Khan & Gul, 2002). Seed germination in arid and semi-arid regions usually occurs after the rains, which help in reducing soil surface salinity (Khan, 2003). The germination of halophytes are inhibited by salinity for the following reasons: a) causing a complete inhibition of germination process at salinities beyond the tolerance limit of species, b) delaying the germination of seeds at salinities that cause some stress to seeds but do not prevent germination, c) causing the loss of viability of seeds due to high salinity and temperature and d) upsetting growth regulator balance in the embryo to prevent successful initiation of germination process. There is a great deal of variability in the response of halophytes to increasing salinity, moisture, light, and temperature stresses and their interactions (Khan & Ungar, 2000, 2001). Seeds of halophytes

M.A. Khan and D.J. Weber (eds.), Ecophysiology of High Salinity Tolerant Plants, 107-114.

often germinate best under non-saline conditions and their germination decreases in salinity (Khan, 2003; Ungar, 1995). Halophytic species that dominate the region have shown variable response to NaCl tolerance during germination (Khan & Gulzar, 2003). *Halopyrum mucronatum* failed to germinate at or above 300 mM NaCl (Khan & Ungar, 1995) while *Aeluropus lagopoides* (Linn.) Trin. Ex Thw., *Sporobolus ioclados* (Nees ex Trin.) Nees and *Urochondra setulosa* (Trin.) C.E. Hubbard could germinate in up to 500 mM NaCl approaching seawater salinity (Khan & Gulzar, 2003). Stem succulent halophyte like *A. macrostachyum* could germinate in up to 800 mM NaCl (Khan & Gul, 1998).

It is well known that Ca^{2+} alleviates the adverse effects of salinity on many plant species (Rengle, 1992; Marschner, 1995; Agboola et al., 1998; Munns, 2002; Ebert et al., 2002). Adding Ca^{2+} to root media with NaCl favors plant growth in both halophytic (Colmer et al., 1996) and non-halophytic species (LaHaye & Epstein, 1969; Cramer et al., 1986; Kurth et al., 1986; Suhayda et al., 1992; Kinrade, 1999). Calcium alleviated the toxic effects of Na^+ and Mg^{2+} on the germination of *Kalidum caspicum* (Tobe et al., 1999, 2001) and *Hordeum vulgare* (Bliss et al., 1986). Tobe et al. (2002) showed that Ca^{2+} successfully alleviated the toxicity of various chloride and sulfate salts on the germination of *Kalidium capsicum*. Plants occupying the coastal area around Karachi suffer high temperature, drought and salinity stresses. It may be that the establishment of these halophytes is facilitated by the alleviation of salt toxicity to its radicles by the Ca^{2+} present in the alkaline soils of this region.

The present study was done to investigate the effect of NaCl on the seed germination of coastal halophytes and to determine whether $CaCl_2$ application alleviate the salinity effect.

2. MATERIALS AND METHODS

Seeds of *Arthrocnemum indicum, A. macrostachyum, Desmostachya bipinnata, Halopyrum mucronatum* and *Urochondra setulosa* were collected in 2004 from a salt flats located at coastal regions from Karachi to Gadani. Tetrazolium test showed 100% viability in seeds. Seeds were surface sterilized with 1% sodium hypochlorite and germination experiments were started immediately. Germination was carried out in 50 mm by 9 mm (Gelman no.7232) tight-fitting plastic petri dishes with 5 ml of test solution (0 to 1000 mM NaCl depending on the range of tolerance of a given species). Calcium (10.0 mM $CaCl_2$) was also applied in addition to NaCl to study whether it would promote germination under saline conditions. Four replicates of 25 seeds were each used for each treatment. The germination experiments were carried out at a temperature of 20-30°C. A 24 hr cycle was used; where the higher temperature was 30°C coincided with the 12-light period (Sylvania cool white light, 110 μmol photons. m^{-2}. s^{-1}) and the lower temperature 20°C coincided with the 12-dark period. Germination was recorded at every alternate day. The results were analyzed using a three way ANOVA.

A Bonferroni test was carried to determine the significance among individual means (SPSS, 1999).

3. RESULTS

A two-way ANOVA indicated there was a significant ($F = 217.8$; $P < 0.001$) effect of calcium in alleviating germination-inhibiting effects of NaCl ($F = 46.8$; $P < 0.01$), and all significant interaction ($P < 0.001$) on the germination of *Arthrocnemum indicum* seeds. *Arthrocnemum indicum* showed less than 20% germination under non-saline conditions and the seed germination decreased with the increase in salinity (Figure 1). Inclusion of calcium significantly alleviated salinity effects at all concentrations (Figure 1).

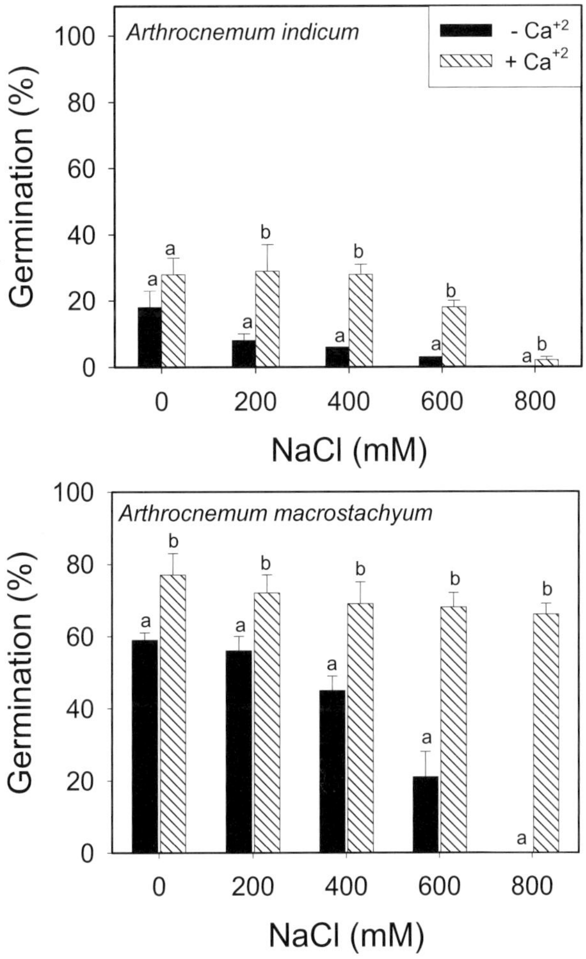

Figure 1. *Effect of calcium chloride (10 mM) on the seed germination of* Arthrocnemum indicum *and* A. macrostachyum *under saline conditions. (Mean with in each having different letters are significantly different from one another (P < 0.05), Bonferroni test).*

Germination in *A. macrostachyum* seeds were about 60% in control and germination was inhibited with the increase in salinity ($F = 50.7$; $P< 0.001$), however, application of calcium ($F= 351.7$; $P < 0.001$) almost completely alleviated salinity effects on germination at all salinity concentrations (Figure 1). A two-way ANOVA indicated significant ($F= 220.8$; $P< 0.001$) effect of calcium in alleviating germination-inhibiting effects of various salts ($F = 56.8$; $P < 0.001$) and all significant interaction ($P < 0.001$) on the germination of *Desmostachya bipinnata* seeds. All seeds of *D. bipinnata* were germinated under controlled conditions and, however, few seeds germinated at 400 mM NaCl. Inclusion of calcium completely alleviated salinity effects in all concentrations except at 800 mM NaCl where 63% seeds germinated in comparison to none in non-treated control (Figure 2). A two-way ANOVA indicated significant ($F= 198.4$; $P < 0.001$) effect of calcium in alleviating germination-inhibiting effects of various salts ($F = 144.6$; $P < 0.001$), and all significant interaction ($P < 0.001$) on the germination of *H. mucronatum* seeds. In *H. mucronatum*, calcium completely alleviated salinity effects at low concentrations and partially alleviated at higher levels (Figure 2). A two-way ANOVA indicated significant ($F = 179.8$; $P < 0.001$) effect of calcium in alleviating germination-inhibiting effects of various salts ($F = 78.8$; $P < 0.001$) and all significant interaction ($P < 0.001$) on the germination of *U. setulosa* seeds. Maximum seed germination was obtained in non-saline control (Figure 3). Addition of NaCl inhibited seed germination and few seeds germinated at 400 mM NaCl (Figure 3) and calcium ($CaCl_2$) alleviated seed germination when it was included in the growth medium with NaCl (Figure 3).

4. DISCUSSION

The diversity of coastal vegetation of Pakistan is very low and the vegetation on the entire coast is dominated by *Avicennia marina, Arthrocenmum macrostachyum, Suaeda fruticosa, Cressa cretica* and *Halopyrum mucronatum* other species are occasional in appearance (Khan & Gul, 2002). The soil samples analyzed from the adjacent areas showed high concentrations of sodium chloride and the base rock of the region is calcareous in nature and the amount of $CaCO_3$ in soil of this region is high.

Halophytes of the area are reported to tolerate variable concentrations of NaCl which includes *Arthrocnemum macrostachyum, A. indicum* (1000 mM, Khan & Gul, 1998; Khan, unpublished), *Salsola imbricata* (800 mM, Mehrunnisa et al., 2005), *Cressa cretica* (800 mM, Khan, 1991), *Limonium stocksii, Aeluropus lagopoides, Sporobolus ioclados* and *Urochondra setulosa* (500 mM, Zia & Khan, 2004; Khan & Gulzar, 2003), *Halopyrum mucronatum* (300 mM, Khan & Ungar, 2001). Our results with *A. indicum, A. macrostachyum, D. bipinnata, H. mucronatum* and *U. setulosa* showed a partial to complete reversal of the injurious effects of sodium chloride with the application of $CaCl_2$.

Tobe et al. (2002) also showed that calcium alleviated the effects of NaCl, Na_2SO_4, $MgCl_2$, and PEG on the germination of *Kalidium capsicum* indicating that the toxicity of different salts to radicles originates from a common mechanism. They attributed this inhibition due to osmotic effect which prevents

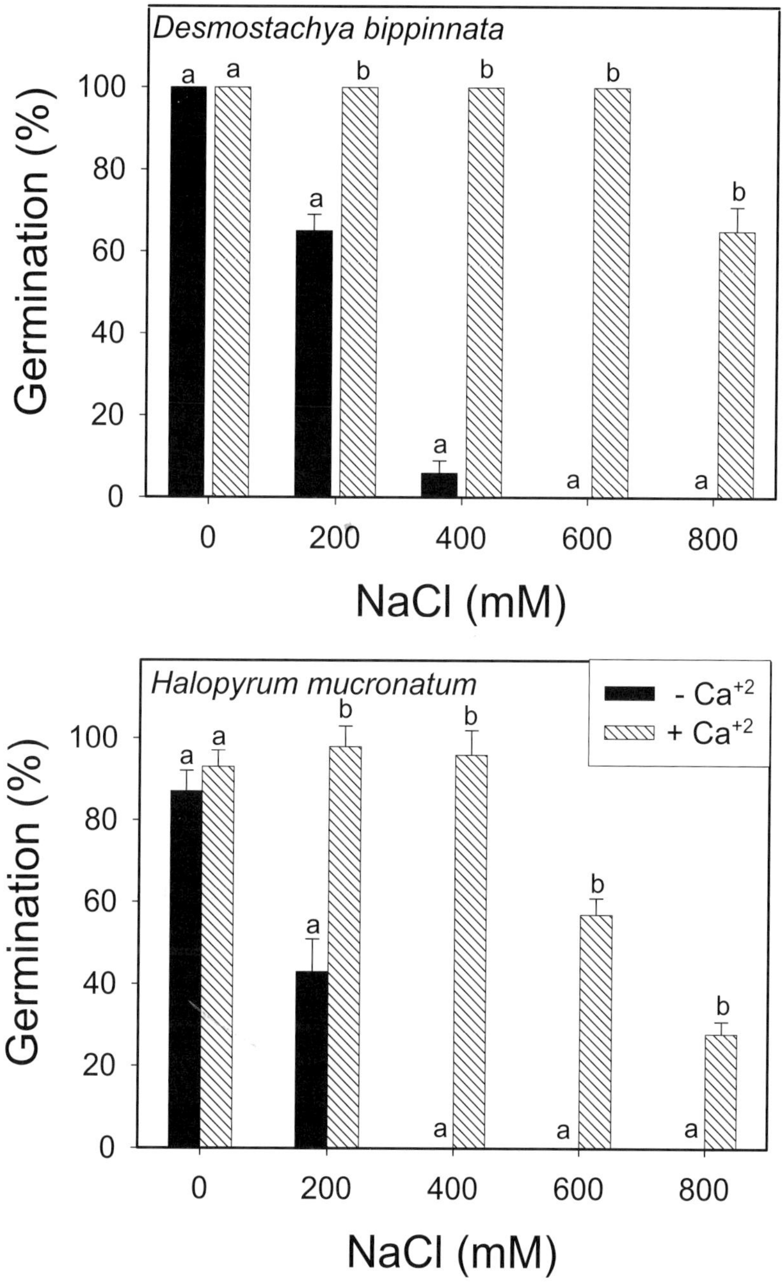

Figure 2. *Effect of calcium chloride (10 mM) on the seed germination of* Desmostachya bippinnata *and* Halopyrum mucronatum *under saline conditions. (Mean with in each having different letters are significantly different from one another (P < 0.05), Bonferroni test).*

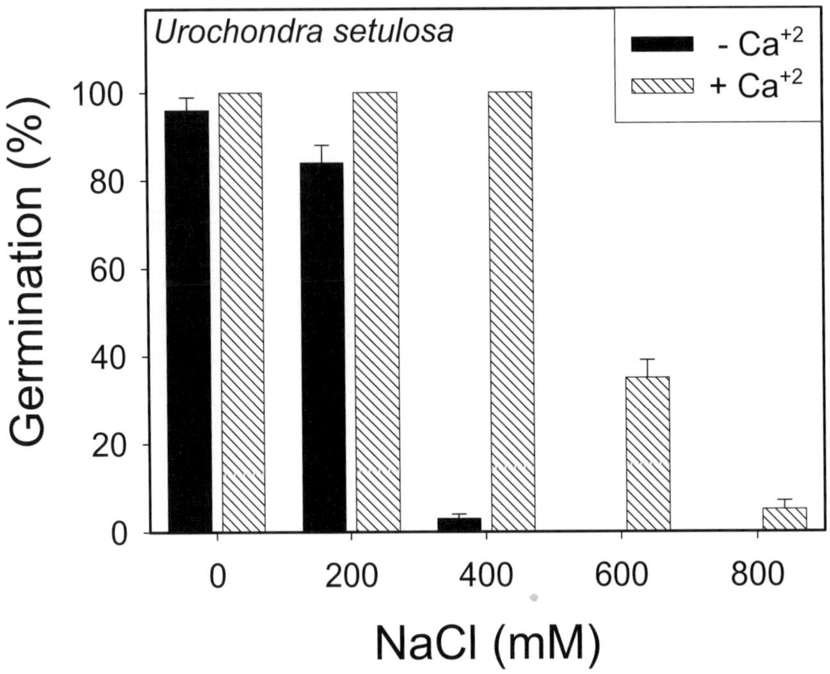

Figure 3. *Effect of calcium chloride (10 mM) on the seed germination of* Urochondra setulosa *under saline conditions. (Mean with in each having different letters are significantly different from one another ($P < 0.05$), Bonferroni test).*

radicle elongation. Other studies have mixed results some showed little ameliorative effects (Leidi et al., 1991) while other showed a significant calcium effect in alleviating salinity stress on germination (Redmann, 1974; Bliss et al., 1986; Katembe et al., 1998; Rengel, 1992; Tobe et al., 1999, 2001, 2002). The role of calcium to alleviate the adverse effect of NaCl and other salts on plant species (Rengel, 1992; Marshner, 1995; Tobe et al., 2002; Girija et al., 2002; Munns, 2002) is well known. Experimental evidence implicates Ca^{2+} function in salt adaptation (Parida & Das, 2005). Externally supplied Ca^{2+} reduces the toxic effects of NaCl, presumably by facilitating higher K^+/Na^+ selectivity (Liu & Zhu, 1998; Lauchli & Schubert, 1989). High salinity also results in increased cytosolic Ca^{2+} that is transported from the apoplast and intracellular compartments (Knight et al., 1997). The resultant transient Ca^{2+} increase potential stress signal transduction and leads to salt adaptation (Mendoza et al., 1994; Knight et al., 1997). A prolonged elevated Ca^{2+} level may, however, also pose a stress; if so, reestablishment of Ca^{2+} homeostasis is a requisite (Parida & Das, 2005). The changes in $[Ca^{2+}]^{Cyt}$ perturbations following combinations of oxidative stress and hyper-osmotic stress correlated well with the expression of Ca^{2+}-regulated osmotic stress induced genes, and the acquisition of osmotic stress tolerance (Knight et al., 1998). More research on Ca^{2+} in plants has been focused at the cellular level due to realization that $[Ca^{2+}]^{Cyt}$ was an obligate intracellular messenger coordinating the cellular responses to numerous developmental cues and environmental challenges (White & Broadley, 2003).

Arthrocnemum indicum, A. macrostachyum, D. bipinnata, H. mucronatum and *U. setulosa* are among the dominant vegetation along the Pakistani coast. Presence of Ca^{2+} salt in natural and artificial conditions may increase the emergence and the annual productivity of these species that may be a good reason for their survival and dominance in such harsh environmental conditions.

5. REFERENCES

Agboola, D.A. 1998. Effect of saline solutions and salt stress on seed germination of some tropical forest tree species. Tropical Biology 45: 324-331.

Bliss, R.D., Platt-Aloia, K.A. & Thompson, W.W. 1986. The inhibitory effect of NaCl on barley germination. Plant Cellular Environment 9: 727-733.

Colmer, T.D., Fan, T.W-M., Highashi, R.M. & Lauchli, A. 1996. Interactive effects of Ca^{2+} and NaCl salinity on the ionic relations and proline accumulation in the primary root tip of *Sorghum bicolor*. Physiology Plant 97: 421-424.

Cramer, G.R., Läuchli, A. & Epstein, E. 1986. Effect of NaCl and $CaCl_2$ on ion activities in complex nutrient solutions and root growth of cotton. Plant Physiology 81: 792-797.

Ebert, G., Eberele, J., Ali-dinar, H. & Ludders, P. 2002. Ameliorating effects of Ca $(NO_3)_2$ on growth, mineral uptake and photosynthesis of NaCl-stresses guava seedlings (*Psidium guajava* L.). Science Horticulture 93: 125-135.

Girija, C., Smith, B.N. & Swamy, P.M. 2002. Interactive effects of sodium chloride and calcium chloride on the accumulation of proline and glycinebetaine in peanut (*Arachis hypogaea* L.). Environmental & Experimental Botany 47: 1-10.

Katembe, W.J., Ungar, I.A. & Mitchell, J.P. 1998. Effect of salinity on germination and seedling growth of two *Atriplex* species (Chenopodiaceae). Annuals Botany 82: 167-175.

Khan, M.A. 1991. Studies on germination of *Cressa cretica* L. seeds. Pakistan Journal Weed Science Research 4: 89-98.

Khan, M.A. 2003. Halophyte seed germination: Success and Pitfalls. In: A.M. Hegazi, H. M. El-Shaer, S. El-Demerdashe, R. A. Guirgis, A. Abdel Salam Metwally, F. A. Hasan, & H. E. Khashaba (Eds.), International symposium on optimum resource utilization in salt affected ecosystems in arid and semi arid regions. Cairo, Egypt: Desert Research Centre. 346-358 pp.

Khan, M.A. & Gul, B. 1998. High salt tolerance in the germinating dimorphic seeds of *Arthrocnemum macrostachyum*. International Journal Plant Science 159: 826-832.

Khan, M. A. and B. Gul. 2002. Salt tolerant plants of coastal sabkhas of Pakistan. In H. Barth. and B. Boer. Sabkha Ecosystems: Volume 1: The Arabian Peninsula and Adjacent Countries. pp. 123-140. Kluwer Academic Press, Netherlands.

Khan, M.A., & Gulzar, S. 2003. Light, salinity and temperature effects on the seed germination of perennial grasses. American Journal Botany 90: 131-134.

Khan, M.A. and I.A. Ungar. 2000. Alleviation of salinity-enforced dormancy in *Atriplex griffithii* Moq. var. *stocksii* Boiss. Seed Science & Technology. 28: 29-37.

Khan, M.A. & Ungar, I.A. 2001. Alleviation of salinity stress and the response to temperature in two seed morphs of *Halopyrum mucronatum* (Poaceae). Australian Journal Botany 49: 777-783.

Kinraide, T.B. 1999. Interactions among Ca^{2+}, Na^+ and K^+ in salinity toxicity: quantitative resolution of multiple toxic and ameliorative effects. Journal Experimental Botany 50: 1495-1505.

Kurth, E., Cramer, G.R., Läuchli, A. & Epstein, E. 1986. Effects of NaCl and $CaCl_2$ on cell enlargement and cell production in cotton roots. Plant Physiology 82: 1102-1106.

Knight, H., Trewavas, A.J. & Knight, M.R. 1997. Calcium signaling in *Arabidopsis thaliana* responding to drought and salinity. Plant Journal 12: 1067-1078.

Knight, H., Brandt, S. & Knight, M.R. 1998. A history of stress alters drought calcium signaling pathways in *Arabidopsis*. Plant Journal 16: 681-687.

LaHaye, P.A. & Epstein, E. 1969. Salt toleration by plants: enhancement with calcium. Science 166: 395-396.

Lauchli, A. & Schubert, R. 1989. The role of calcium in the regulation of membrane and cellular growth processes under salt stress. In: J. H. Cherry. (Ed.), Environmental Stress in Plants. NATO ASI Series, Springer-Verlag, Berlin: NATO. 131-138 pp.

Leidi, E.O., Nogales, R. & Lips, S.H. 1991. Effect of salinity on cotton plants grown under nitrate or ammonium nutrition at different calcium levels. Field Crop Research 26: 35-44.

Liu, J. & Zhu, J.K. 1998. A calcium sensor homolog required for plant salt tolerance. Science 280: 1943-1945.
Marshner, H. 1995. Mineral nutrition of higher plants. Academic Press. London.
Mehrunnisa, Weber, D.J. & Khan, M.A. 2005. Dormancy, germination and viability of *Salsola imbricata* seeds in relation to light, temperature and salinity. Journal of Arid Environment, (Submitted).
Mendoza, I., Rubio, F., Rodriguez-Navarro, F. & Pardo, J.M., 1994. The protein phosphatase calcineurin is essential for NaCl tolerance of *Saccharomyces cerevisiae*. Journal Biological Chemistry 269: 8792-8796.
Mohammad, S. & Sen, D.N. 1990. Germination behaviour of some halophytes in Indian desert. Indian Journal of Experimental Biology 28: 546-549.
Munns, R. 2002. Comparative physiology of salt and water stress. Plant Cell & Environment 25: 239-250.
Parida, A.K. & DAS, A.B. 2005. Salt tolerance and salinity effects on plants: a review. Ecotoxicology Environmental Safety 60: 324-349.
Redmann, R.E. 1974. Osmotic and specific ion effects on the germination of alfalfa. Canadian Journal of Botany 52: 803-808.
Rengel, Z. 1992. The role of calcium in salt toxicity. Plant Cell & Environment 15: 625-632.
SPSS. 1999. SPSS 9.0 for Windows Update. Chicago, Illinois:SPSS Inc.
Suhayda, C.G., Redmann, R.E., Harvey, B.L. & Cipywnyk, A.L. 1992. Comparative response of cultivated and wild barley species to salinity stress and calcium supply. Crop Science 32: 154-163.
Tobe, K., Zhang, L. & Omasa, K. 1999. Effects of NaCl on seed germination of five non-halophytes species from a Chinese desert environment. Seed Science and Technology 27: 851-863.
Tobe, K., Li, X. & Omasa, K. 2002. Effect of sodium magnesium and calcium salts on seed germination and radicle survival of a halophyte, *Kalidium caspicum* (Chenopodiaceae). Australian Journal of Botany 50: 163-169.
Tobe, K., Zhang, L., Qui, G.Y., Shimizu, H. & Omasa, K. 2001. Characteristics of seed germination in five non-halophytic Chinese desert shrub species. Journal of Arid Environment 47: 191-201.
Ungar, I.A., 1995. Seed germination and seed-bank ecology in halophytes. In: Kigel, J. & Galili, G. (Eds.), Seed development and germination. NewYork,New York: Marcel Dekker. 599-628 pp.
White, P.J. & Broadley, M.R. 2003. Calcium in plants. Annals of Botany 92: 487-511.
Zia, S. & Khan, M.A. 2004. Effect of light, salinity and temperature on the seed germination of *Limonium stocksii.* Canadian Journal of Botany 82: 151-157.

CHAPTER 7

CALORESPIROMETRIC METABOLISM AND GROWTH IN RESPONSE TO SEASONAL CHANGES OF TEMPERATURE AND SALT

BRUCE N. SMITH[1,4], LYNEEN C. HARRIS[2], EMILY A. KELLER[1], BILQUEES GUL[3], M. AJMAL KHAN[3] AND LEE D. HANSEN[2]

[1]*Department of Plant and Animal Science, Brigham Young University, Provo, UT 84602, U.S.A.*

[2]*Department of Chemistry and Biochemistry, Brigham Young University, Provo, Utah 84602, U.S.A.*

[3]*Department of Botany, University of Karachi, Karachi-75270, Pakistan*

[4] *Corresponding author:bruce_smith@byu.edu*

Abstract. Heat rate (Rq) and respiration rate (RCO_2) determined by calorimetric measurements on plants adapted to high salt environments were used to define upper and lower limits of temperature and salt concentrations in both laboratory and field grown plants. Species-specific responses to seasonal differences in temperature and salinity determine plant survival in a cold desert, salt-playa environment where most of the moisture is received as winter snow. Increased soil salinity in playas is in parallel with increased environmental temperatures, thus exposing plants to two stresses simultaneously.

1. INTRODUCTION

1.1. Temperature

A biotic and biotic stresses can determine the limits to plant growth by altering both metabolic rate and efficiency. For a given population of plants, there is a range of temperatures for optimal growth and extremes leading to stress and even to death. C. Hart Merriam (1894) observed that different "life zones" could be characterized with increases in altitude from desert to mountaintop on the San Francisco peaks in Arizona. The presumed abiotic stresses were temperature and rainfall.

Over the last several years calorimetry has been employed to determine the limits to plant growth in response to environmental stress. Temperature limits for

M.A. Khan and D.J. Weber (eds.), Ecophysiology of High Salinity Tolerant Plants, 115-125.

growth have been shown to differ for eleven populations of *Bromus tectorum* L. in the Great Basin (Hemming et al., 1999; McCarlie et al., 2003). Similar responses to differences in temperature were found among cultivars of soybean (Hemming et al., 2000), and also among cultivars of corn (*Zea mays* L.) (Taylor et al., 1998). Sagebrush (*Artemisia tridentata* Nutt.) populations on a single hillside with a total vertical range of 85 meters could be distinguished on the basis of adaptation to growth for small differences in temperature (Smith et al., 2002).

Recently laboratory growth rates as a function of temperature for seedlings of three species of *Eucalyptus* (Figure 1) were compared with the environmental temperature range in which the native species are found (Criddle et al., 2005).

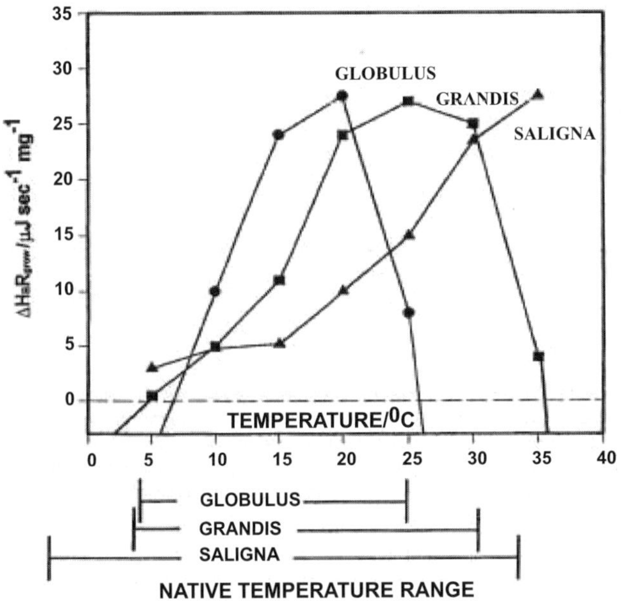

Figure 1. *Normalized rates of growth (as $\Delta H_B R_{grow}$) for three* eucalyptus species *(*Eucalyptus globulus *Labill,* E. grandis *W. Hill ex Maiden,* E. saligna *Sm.). Values were calculated from measurements of respiratory heat and CO2 production rates as a function of temperature. Growth rate values as a function of temperature are compared with the environmental temperature range for growth of the species at the bottom of the figure. (After Criddle et al., 2005).*

1.2. Salinity

Growth of most species is inhibited by high salt concentrations, i.e. greater than about 0.5 M NaCl, the concentration of salt in seawater (Ungar, 1991). Some C_4 plants have an absolute requirement for low levels of NaCl. However growth of halophytes is often stimulated by salinity (Flowers et al., 1977). Previous studies on growth of desert species (*Atriplex griffithii, Halopyrum mucronatum, Haloxylon recurvum* and *Suaeda fruticosa*) from Pakistan showed that low salinity levels promoted growth (Khan et al., 1998). Increasing salt to 0.425 M promoted growth of *Cressa cretica*, but growth in 0.85 M salt was not significantly different from controls grown without added salt (Khan & Aziz, 1998).

Halophytes from the Great Basin in western North America (*Salicornia rubra, Salicornia utahensis, Suaeda torreyana, Allenrolfea occidentalis*) show a similar pattern of growth promotion at moderate salinities (0.4 to 0.6 M NaCl) but declined with further increases in salinity (Harris et al., 2001; Khan et al., 2000). Thus it seems that many salt-tolerant plants have a physiological growth response to both minimum and maximum levels of salinity. What is the mechanism for such responses?

Ethylene is produced in response to plant stress. Mung bean hypocotyl sections exposed to salinity stress (chlorides of Ca, K, Mg, and Na) induced CO_2 production and ethylene production early on, followed by ethane (Chrominski et al., 1986a). In *Allenrolfea occidentalis*, salinity promotes conversion of ACC to ethylene (Chrominski et al., 1986b; Chrominski et al., 1988). In turn there is evidence for the production of dimethylsulfonium propionate as an osmoprotectant for terrestrial glycophytes (Chrominski et al., 1989). Response to salinity often involves production of proline and/or glycine-betaine (Girija et al., 2002; Khan et al., 1998) as well as polyamines and carbohydrates (Jouve et al., 2004). Salinity as well as drought can increase production of abscisic acid (Swamy & Smith, 1999).

In studies where the effects of temperature were also examined, seed germination of *Triglochin maritima* from the Great Basin was most inhibited by exposure to high salinities at suboptimal thermoperiods (Khan & Ungar, 1999). Halophytes growing in the field showed the highest respiration rate, efficiency, and growth during May and June with the lowest metabolic rates during the hot, dry month of August (Harris et al., 2001).

1.3. Photosynthesis

Photosynthesis is the process which traps light energy to reduce carbon dioxide (a low energy molecule) to carbohydrate and other high energy molecules which can then be used to support life processes.
Photosynthesis:

$$CO_2 + H_2O + \text{sunlight} \text{ ----> } [CH_2O] + O_2 \qquad (1)$$

Green plant tissues differ remarkably in photosynthetic rate. However, despite much effort to correlate variation in CO_2 uptake with growth rates, no consistent results have been obtained (Nelson, 1988). Insufficient carbon assimilation does not explain why alpine plants are so small and why biomass accumulation per unit land area is so low (Korner & Larcher, 1988). Many investigators are now convinced that respiration is a better predictor for plant growth than is photosynthesis (Hay & Walker, 1989).

1.4. Respiration

Respiration consists of two integrated parts - catabolism and anabolism (Figure 2). Oxidative release of energy (heat and ATP) by breaking carbon and hydrogen bonds from photosynthetic products (carbohydrate, lipids, protein, etc.) with the release of carbon dioxide is catabolism. Anabolism uses photosynthate plus energy released by catabolism (ATP) to synthesize compounds needed for growth, defense, reproduction, etc. Heat is also produced in anabolism due to the second law of thermodynamics. Respiration is usually measured as the rate of oxygen uptake or carbon dioxide evolution. However, gas exchange is not sufficient to predict growth or the ability of the plant to handle stress from abiotic or biotic factors. Metabolic heat must be measured from both catabolism and anabolism to predict the rate and efficiency of growth.

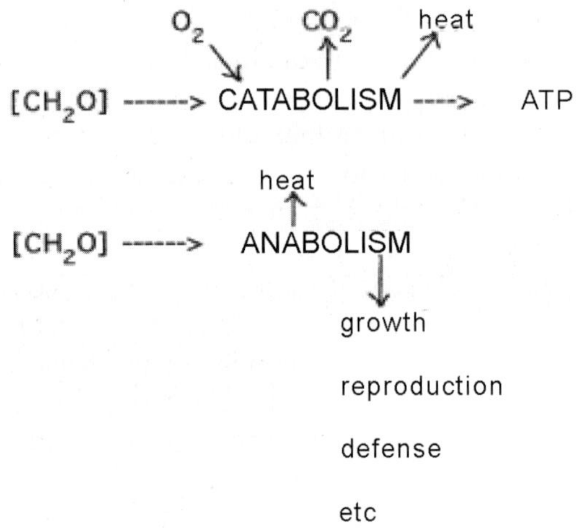

Figure 2. *A model for respiration showing interaction between catabolism and anabolism.*

1.5. Calorimetry

Using modern calorimeters, it is possible to make rapid, isothermal measurements of metabolic heat loss (R_q) from both catabolism and anabolism (Criddle & Hansen, 1999; Hansen et al., 1994). The catabolic respiration rate (RCO_2) is determined for small samples (about 100 mg fresh weight) of plant tissues. Knowing the heat rate (R_q) and the respiration rate (RCO_2), the relative specific growth rate (R_{SG}) or anabolic rate can be predicted:

$$R_{SG}\Delta H_B = (455RCO_2 - R_q) \qquad (2)$$

Where R_{SG} is the specific growth rate in terms of moles of carbon incorporated per gram of biomass, Rq is the specific heat rate in μW/mg, RCO_2 is the rate of CO_2 evolution in the dark at pmol mg^{-1} sec^{-1}, and ΔH_B is the enthalpy change for structural biomass formation (as kJ/mol carbon). If photosynthate is stored as starch or sugars (which have chemical oxidation states of zero), and assuming that ΔH_B is constant with temperature, Thornton's constant (-455 kJ mole^{-1}) may be introduced (Thornton, 1917).

Since the method measures energy changes (R_q) as well as gas exchange rates (RCO_2), equation (2) can also be expressed as:

$$R_{SG} = RCO_2[/(1-)] \qquad (3)$$

where is the substrate carbon conversion efficiency. Thus growth rate is directly proportional to both respiration rate and efficiency.

2. METHODS AND MATERIALS

Plants on the playas of the Great Basin must survive cold winters, hot summers, and large temperature changes during the growing season (common diurnal variation = 20 to 30 °C). During the period of high summer temperatures (up to 45 °C), which is also the time of least rainfall, water evaporates from the shallow playas, increasing salt concentration in the soil (Table 1). The center of the playa is often white with salt, with no vegetation growing there by the end of the summer. Vegetation occurs in concentric rings around the central area of salt. The most highly salt tolerant species, such as the annual forbs, *Salicornia rubra* A. Nels., grow adjacent to the center. The next outward concentric ring is characterized by perennial forbs, *Salicornia utahensis* Tidestr. Farther away from the center, in slightly less saline soil, is found the grass, *Distichlis spicata* (L.) Greene. At a still greater distance from the center of the playa can be found the shrub, *Allenrolfea occidentalis* (Wats.) Kuntze (Welsh et al., 1987). Salt content in the playa was mostly sodium chloride (Table 1). Concentration of salts changed with the season. However salt distribution within a zone also was not constant in concentration. Plants were taken from each concentric zone from areas of relatively higher and lower salinity (Harris et al., 2001). Stem and leaf tissue was collected from each of the species in high and low salt soil (admittedly a very subjective and inaccurate judgment) in the months of May, June, and August. Seedlings of *Salicornia utahensis* were grown hydroponically in the laboratory at several temperatures and NaCl concentrations. Stem and leaf tissue were placed in ampules and run in triplicate in the Calorimetry Sciences Corporation model 4100 MC or the Hart Scientific model 7707 calorimeters. Metabolic heat rates (R_q) were measured on 10 to 30 mg dry weight of tissue. Respiration rates (RCO_2) were measured with addition of a NaOH trap and consequent heat of carbonate formation (Criddle & Hansen, 1999; Hansen et al., 1994). From these measurements, metabolic efficiency and growth rate can be predicted.

Table 1. *Mean concentration (mmol kg^{-1}) of soil ions on the salt playa near Goshen, Utah during 1997 (Harris et al., 2001).*

Months	Na^+	Cl^-	K^+	Ca^2+	NO_3^{2-}
May	7030a	10175a	8.3a	439a	7.5a
June	11470b	12523b	13.4a	497a	2.9a
August	15870c	12989b	19a	345b	3.3a

Values in each column having the same letter are not significantly different at p<0.05, Bonferroni test.

3. RESULTS

Salicornia rubra and *S. utahensis* (Figure 3) indicated that June and August had the most efficient metabolism while predicted growth, though never large, was best in May and June. Negative R_{SG} values does not mean that the tissue was dead, only dormant. We find that summer dormancy is very common in desert plants. In comparing these graphs, please note that the scales are sometimes different. In Figure 4, *Distichlis spicata* and *Allenrolfea occidentalis* had more efficient metabolism in May and June. Growth rate was predicted to be best in May and June with growth close to zero in August.

Salicornia utahensis and other species on salt playas are adapted to low NaCl contentrations at low temperatures and higher salt concentrations at higher temperatures. The mechanism for adapting to the sum of two apparently deleterious stresses is certainly not clear as yet.

Salicornia utahensis seedlings in the laboratory grew best in 300 mM NaCl at 32°C. Plants held at 25/15°C (day/night) grew best at 600 mM salt. In growth chambers set at 10, 20, and 32°C, plants survived and grew (but not much) at 1,500 mM NaCl.

Salicornia growth predicted by respiration ($\Delta H_B R_{SG}$) as measured in the calorimeter generally increased with increasing salt concentrations (Figure 5) as indicated by the curve of a third-order polynominial fit by least squares to all of the data without regard to temperature. The dashed line indicates zero growth. Growth is predicted at moderate salinity (200 to 800 mM) and moderate temperatures.

Plotting the same data another way (Figure 6), and fitting the data to a second order polynomial without regard to salt shows growth generally decreases with increasing temperature. The data points at 1,000 mM NaCl at 40 and 45°C are anomalous in both (Figures 5 & 6). These two points at high values of $\Delta H_B R_{SG}$ do not indicate high growth rates (the plants remain very small). Instead these values are a result of imposed stress and indicate the presence of a significant amount of anaerobic respiration.

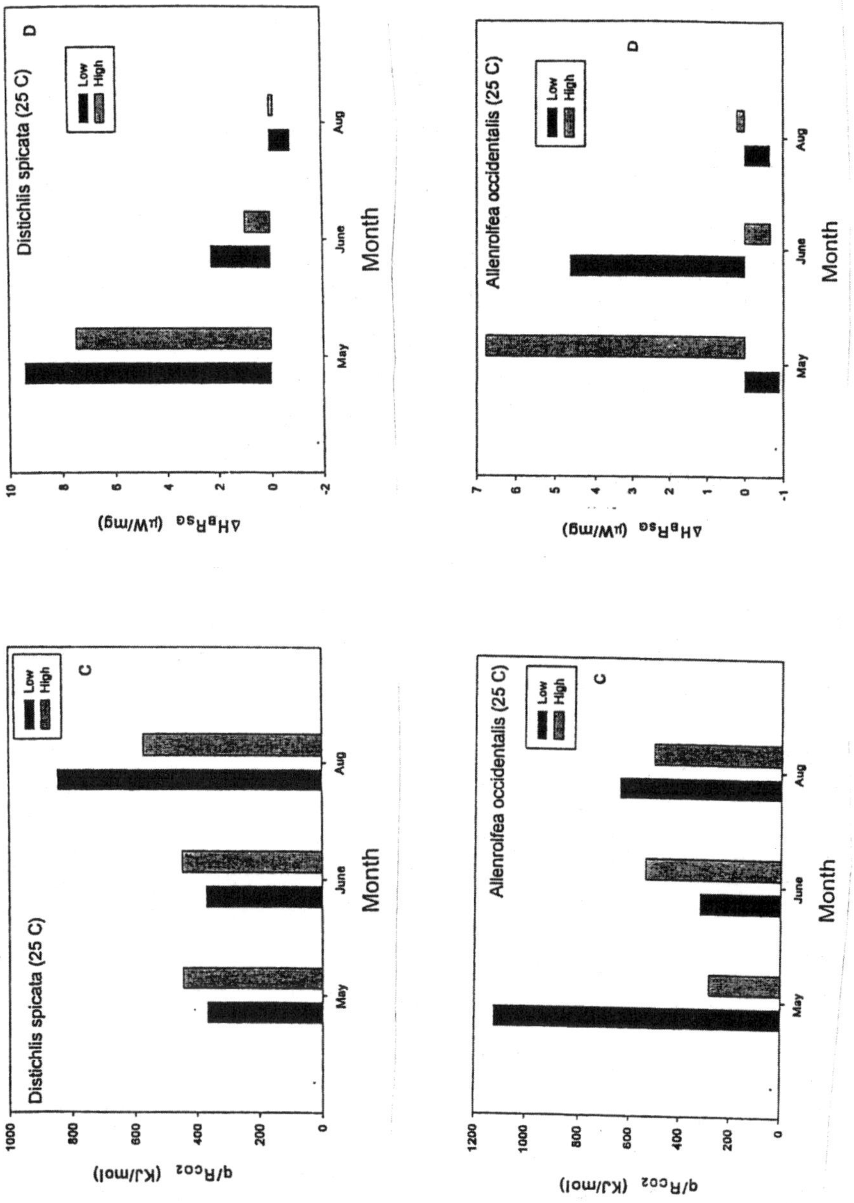

Figure 3. Salicornia rubra *and* S. utahensis *stem and leaf tissue was collected in the field from soil relatively high and low in NaCl in May, June, and August of 1997. Isothermal calorimetric measurements were made at 25°C: C. The ratio of metabolic heat rate to respiration rate (q/R_{CO2}) or efficiency. Smaller numbers indicate greater efficiency.* D. Predicted *specific growth rate $\Delta H_B R_{SG}$ (After Harris et al., 2001).*

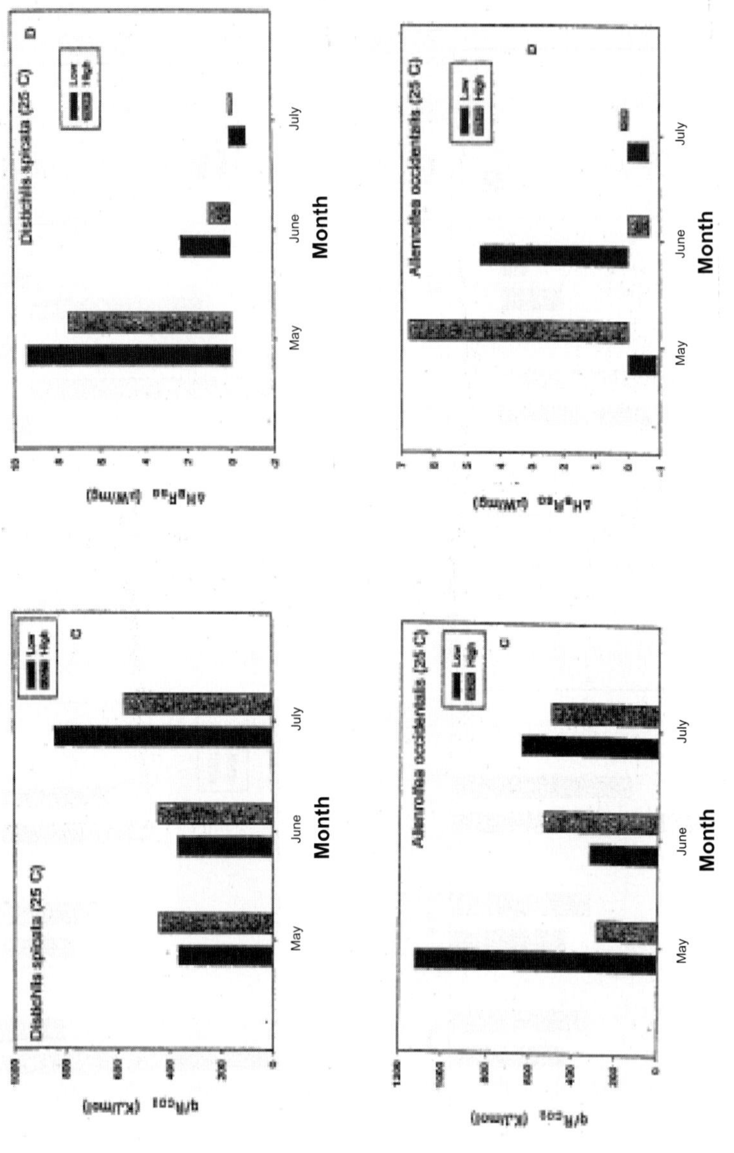

Figure 4. Distichlis spicata *and* Allenrolfea occidentalis *stem and leaf tissue was collected and measured as described in Figure 3. (After Harris et al., 2001).*

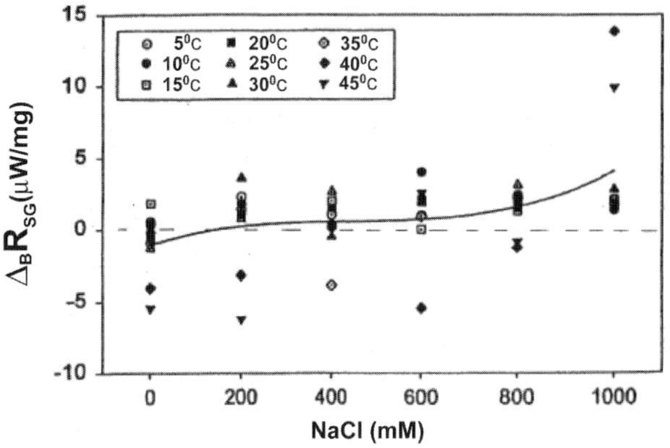

Figure 5. *Specific growth rate, $R_{SG}\Delta H_B$, calculated from metabolic heat and CO_2 rates as a function of the salt concentration in the growth medium.*

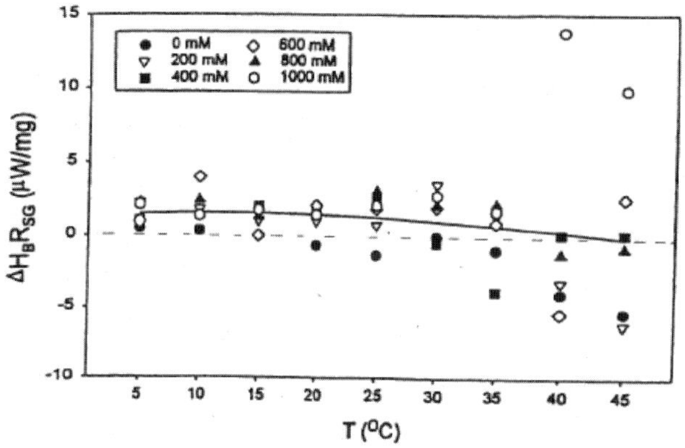

Figure 6. *Specific growth rate, $R_{SG}\Delta H_B$, calculated from metabolic heat and CO_2 rates as a function of the measurement temperature.*

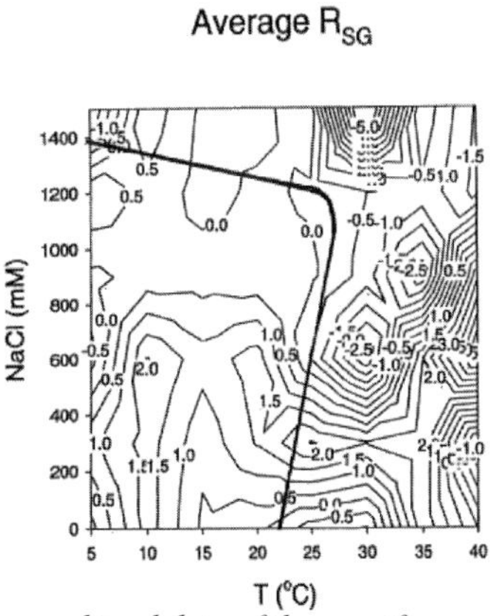

Figure 7. *Summary plot of the combined data of the specific growth rates, $R_{SG}\Delta H_B$, as a funjction of the measurement temperature and salt concentration in the growth medium.*

4. DISCUSSION

Some glycophytes and even some halophytes (Girija et al., 2002) have the capacity to synthesize osmoticants such as proline or glycine-betaine to prevent damage due to salinity. The relatively low concentrations of glycine-betaine in *S. utahensis* may help during germination in the spring with low salt concentrations. Unlike some plants (Khan et al., 1998), *Salicornia* does not show a linear increase in glycine-betaine with salt exposure. Other osmoticants exist (Flowers et al., 1977) and may be operative but were not measured. Weber et al., (1977) did show that salt tolerance in *S. utahensis* was based, in part, on exclusion of salt from the photosynthetic cells and on the ability of the succulent stems to function even though sections were dead owing to high salt. In addition to salt and temperature stresses, there is often a low partial pressure of oxygen in the roots due to standing water or a high water table (Ungar, 1991).

The ability to grow crop plants in saline soils is highly desirable but difficult to achieve. Indeed, after 10 years of research using transgenic plants to alter salt tolerance, significant improvement in growth has not been shown (Flowers, 2004). *Chenopodium quinoa* has been shown to have greater salt tolerance than wheat apparently due to a variety of mechanisms (Wilson et al., 2002). Young poplar trees grown in 150 mM NaCl produced both osmoprotectants and antioxidants (Jouve et al., 2004).

Combined effects of temperature and salt need to be understood. Presumably the imposition of one kind of stress reduces tolerance for other simultaneous stresses. To test this hypothesis, we examined the interaction of both low and high salt and low and high temperature stresses. Reduction of biomass accumulation, as well as characteristics of energy metabolism, are used as indicators of stress.

5. REFERENCES

Chrominski, A., Bhat, R.B., Weber, D.J. & Smith, B.N. 1988. Osmotic stress dependent conversion of aminocyclopropane-1-carboxylic acid (ACC) to ethylene in the halophyte, *Allenrolfea occidentalis*. Environmental and Experimental Botany 28: 171-174.

Chrominski, A., Khan, M.A., Weber, D.J. & Smith, B.N. 1986a. Ethylene and ethane production in response to salinity stress. Plant, Cell and Environment 9: 687-691.

Chrominski, A., Weber, D.J., Smith, B.N. & Hegerhorst, D.F. 1989. Is dimethylsulfonium propionate an osmoprotectant of terrestrial glycophytes? Die Naturwissenschaften 76: 473-475.

Chrominski, A., Weber, D.J., Smith, B.N. & Khan, A.M. 1986b. NaCl salinity dependent conversion of ACC to ethylene in the halophyte, *Allenrolfea occidentalis*. Die Naturwissenschaften 73: 274-278.

Criddle, R.S. & Hansen, L.D. 1999. Calorimetric methods for analysis of plant metabolism. In: R.B. Kemp, (Ed.), Handbook of thermal analysis and calorimetry. Vol. 4. Amsterdam, Netherlands: Elsevier. 711-763 pp.

Criddle, R.S., Hansen, L.D., Smith, B.N., Macfarlane, C., Church, J.N., Thygerson, T., Jovanovich, R.T. & Booth, B. 2005. A thermodynamic law of adaptation of plants to environmental temperatures. Pure and Applied Chemistry, in press.

Flowers, T.J. 2004. Improving crop salt tolerance. Journal of Experimental Botany 55: 307-319.

Flowers, T.J., Troke, P.F. & Yeo, A.R. 1977. The mechanism of salt tolerance in halophytes. Annual Review of Plant Physiology 28: 89-121.

Girija, C., Smith, B.N. & Swamy, P.M. 2002. Interactive effects of sodium chloride and calcium chloride on the accumulation of proline and glycinebetaine in peanut (*Arachis hypogaea* L.). Environmental and Experimental Botany 47: 1-10.

Hansen, L.D., Hopkin, M.S., Rank, E.R., Anekonda, T.S., Breidenbach, R.W. & Criddle, R.S. 1994. The relation between plant growth and respiration: A thermodynamic model. Planta 194: 77-85.

Harris, L.C., Gul, B., Khan, M.A., Hansen, L.D. & Smith, B.N. 2001. Seasonal changes in respiration of halophytes in salt playas in the Great Basin, U. S. A. Wetlands Ecology and Management 9: 463-468.

Hay, R.K.M. & Walker, A.J. 1989. An introduction to the physiology of crop yield. Essex. Great Britian: Longman Scientific and Technical. 87-106 pp.

Hemming, D.J.B., Meyer, S.E., Smith, B.N. & Hansen, L.D. 1999. Temperature dependence of respiration differs among cheat grass (*Bromus tectorum* L.) populations. Great Basin Naturalist 59: 355-360.

Hemming, D.J.B., Monaco, T.A., Hansen, L.D. & Smith, B.N. 2000. Respiration as measured by scanning calorimetry reflects the temperature dependence of different soybean cultivars. Thermochimica Acta 349: 131-134.

Jouve, L., Hoffmann, L. & Hausman, J.F. 2004. Polyamine, carbohydrate, and proline content changes during salt stress exposure of aspen (*Populus tremula* L.): Involvement of oxidation and osmoregulation metabolism. Plant Biology 6: 74-80.

Khan, M.A. & Aziz, S. 1998. Some aspects of salinity, density, and nutrient effects on *Cressa cretica*. Journal of Plant Nutrition 21: 769-784.

Khan, M.A., Gul, B. & Weber, D.J. 2000. Germination responses of *Salicornia rubra* to temperature and salinity. Journal of Arid Environments 45: 207-214.

Khan, M.A. & Ungar, I.A.1999. Effect of salinity on seed germination of *Triglochin maritima* under various temperature regimes. Great Basin Naturalist 59: 144-150.

Khan, M.A., Ungar, I.A., Showalter, A.M. & Dewalt, H.D. 1998. NaCl-induced accumulation of glycinebetaine in four subtropical halophytes from Pakistan. Physiologia Plantarum 102: 487-492.

Korner, C. & Larcher, W. 1988. Plant life in cold climates. Symposium of the Society of Experimental Biology 42: 25-57.

McCarlie, V.W., Hansen, L.D., Smith, B.N., Monsen, S.B. & Ellingson, D.J. 2003. Anabolic rates measured by calorespirometry for eleven subpopulations of *Bromus tectorum* match temperature profiles of local microcliimates. Russian Journal of Plant Physiology 50: 183-191.

Merriam, C.H. 1894. Laws of temperature control of the geographic distribution of terrestrial animals and plants. National Geographic Magazine 6: 229-238.

Nelson, C.J. 1988. Genetic associations between photosynthetic characteristics and yield: review of the evidence. Plant Physiological Biochemistry 26: 543-556.

Smith, B.N., Monaco, T.A., Jones, C., Holmes, R.A., Hansen, L.D., McArthur, E.D. & Freeman, D.C. 2002. Stress-induced metabolic differences between populations and subspecies of *Artemisia tridentata* from a single hillside. Thermochimica Acta 394: 205-210.

Swamy, P.M. & Smith, B.N. 1999. Role of abscisic acid in plant stress tolerance. Current Science 76: 1220-1227.

Taylor, D.K., Rank, D.R., Keiser, D.R., Smith, B.N., Criddle, R.S. & Hansen, L.D. 1998. Modeling temperature effects on growth-respiration relations of maize. Plant, Cell and Environment 21: 1143-1151.

Thornton, W.M. 1917. Philosophy Magazine 33: 196-203.

Ungar, I.A. 1991. Ecophysiology of vascular halophytes. Boca Raton, FL: CRC Press. 209 pp.

Weber, D.J., Rasmussen, H.P. & Hess, W.M. 1977. Electron microprobe analyses of salt distribution in the halophyte *Salicornia pacifica* var. *utahensis*. Canadian Journal of Botany 55: 1516-1523.

Welsh, S.L., Atwood, N.D., Higgins, L.C. & Goodrich, S. 1987. A Utah Flora. Great Basin Naturalist Memoirs, No. 9. Provo, Utah: Brigham Young University Press. 894 pp.

Wilson, C., Read, J.J. & Abvo-Kassen, E. 2002. Effect of mixed-salt salinity on growth and ion relations of a quinoa and a wheat variety. Journal of Plant Nutrition 25: 2689-2704.

CHAPTER 8

EVALUATION OF ANTHOCYANIN CONTENTS UNDER SALINITY (NACL) STRESS IN BELLIS PERENNIS L.

R. A. KHAVARI-NEJAD, M. BUJAR AND E. ATTARAN

Biology Department, Teacher Training University, Tehran, Iran

Abstract. The scavenging of reactive oxygen under stressed environments is one of the suggested roles of anthocyanins in leaves of plants. The present work was designed to evaluate the alteration of anthocyanin concentration in leaves of daisy under high NaCl concentration. Because, high NaCl concentration is a common stress factor in natural and agricultural system, which has detrimental effect on crop productivity, it is critical to recognize the mechanism of protection in plants against NaCl stress. The data of this survey suggest that the increase of the anthocyanin content in some levels of high NaCl salinities is a kind of defense response to this abiotic environmental stress.

1. INTRODUCTION

NaCl Stress is a major factor limiting crop production, because it affects almost all plant functions (Bohnert & Jensen, 1996). Therefore it is important to understand how plants respond and adapt to this stress. Adaptation of plant cells to high NaCl salinity involves osmotic adjustment and the compartmentation of toxic ions, whereas an increasing body of evidence suggests that high salinity also includes oxidative stress. One of the most important and ubiquitous compound which acts both as osmotic adjusters (Chalker-Scott, 1999) and scavenger of active oxygen, is anthocyanin (Gould, 2000). Anthocyanins are water-soluble pigments derived from flavonoids via shikimic acid pathway. The best known function of this colorful pigment is that they do not only account for beautiful color in the petals of flowering plants, but also serve as factors in plant reproduction by recruiting pollinators and seed disperses (Winkel-Shirly, 2002). However, the role of anthocyanin in plant foliage has long been the subject of study and speculation. Foliar anthocyanins arise in a great diversity of plant species across a broad range of environments, often occurring in response to environment stress as nutrient deficiencies, salinity stress and low temperature (Steyn et al., 2002).

M.A. Khan and D.J. Weber (eds.), Ecophysiology of High Salinity Tolerant Plants, 127-134.

The main purpose in the present work was to survey the evaluation of anthocyanin concentration in the leaves of one of the medicinal plants, *Bellis perrenis.* L., under different levels of salt stress, and to understand the possible role of anthocyanin in protection of plants from NaCl stress.

2. MATERIALS AND METHODS

2.1. Plant material and growth condition

The seeds of daisy plants (Bellis *perennis* L.) were sown in the special plate and sprayed with fresh water each day to avoid drying. After 10 days seedlings were transferred into separated pots and were watered with Hoagland nutrition solution. From the 28th day they were treated by different levels of salt stress, 15 mM NaCl, 25 mM NaCl, 50 mM NaCl, 75 mM NaCl and 100 mM NaCl. After 40 days the treated plants along with control plants were harvested and the chemical analysis on the alteration of the total anthocyanin and protein in the leaves of daisy plants were began.

2.2. Chemical Analysis

Fresh leaves (0.05 g) of daisy plants were homogenized in HCl 15%: ethanol: water (85:15) in 4°C. Extracts were centrifuged, and then the absorbances of the supernatants were measured with spectrophotometer and compared with pure anthocyanin control. Anthocyanin content of each sample was estimated at 530 nm and 653 nm. The anthocyanins absorbed maximally at 530 nm, subtraction of 0.24 A653 compensated for the small overlap in absorbance at 530 nm by the chlorophylls (Gould, 2000).

2.3. HPLC Analysis

To confirm the conclusions of spectrophotometric studies, the extract was examined by HPLC. The supernatant was purified by chromatography column; the analysis was conducted by using two reverse phases. The dimension of column was 500 x 12 mm as a stable phase. The solvent was $CHCl_{3:}$ methanol: water (20: 10: 15). Samples were injected at a flow rate of 1 mL min^{-1} into HPLC (Crystal 200), under certain condition of 25 min linear gradients of methanol in 0.005% H_3PO_4 using 4.8 x 250 mm, $C_{48.}$ Anthocyanin was detected at 535 nm (Murray, 1991).

2.4. Other Assays

Protein was assayed according to the method of Bradford (1976) using bovine serum albumin as the standard.

3. RESULTS

3.1. Result of high NaCl salinity on leaf development.

NaCl salinity stress (100 mM NaCl) had fatal effect on plants and all of them died after two treatments. In general, NaCl salinity causes cholorosis and reduces leaf growth and shoot development especially at the level of 50 mM NaCl and 75 mM NaCl concentrations (Figure 1). Figures 1 and 2 illustrate the effect of NaCl salinity on growth of aerial parts of *Bellis perennis* L.

Figure 1.*The Effect of NaCl Salinity Stress at the level of 25 Mm.*

Figure 2. *The reduction of leaf growth in treated plants with 75 mM NaCl in comparison with controls.*

3.2. Effect of salt stress on anthocyanin concentration

Induction of salt stress on the accumulation of anthocyanins in the leaves was assayed by a spectrophotometer. The result was recorded, which shows that the accumulation of anthocyanin increased significantly in 15 and 25 mM NaCl in comparison with control plants (Tables 1, 2 & 3, Figure 1), whereas the concentration of anthocyanin in plants treated with 50 and 75 mM NaCl concentration were significantly reduced. This observation was further confirmed by HPLC analysis.

Since, many or most, environmental stresses induce oxidative stress and protein denaturation, so diverse stresses induce similar cellular adaptive response, such as the production of stress proteins, up-regulation of oxidative stress protectors, and accumulation of protective solutes. In this case, salt stress like intensive light, may affect the activity of enzymes, which produce anthocyanins. Matsumoto reported that activity of phenylalanine ammonia-lyase (PAL), which is a key enzyme in anthocyanin biosynthesis, is enhanced as plants are exposed to stress. As a result, this enzyme may show activity in low salt concentration (Table 1).

Table 1. *Effect of NaCl treatment on the accumulation of anthocyanin in leaves of* Bellis perennis *L.*

Anthocyanin Concentration (μgg^{-1} fw)	*Parameter NaCl mM*
3.3±5.77	Control
1.9±75.122	15
93.8±25. 124	25
54.5±5.68	50
29.4±5.78	75

Table 2. *Deference mean Student's t Test, p < 0.05: *Student's t Test p< 0.001: ***

75	*50*	*25*	*15*	*Contorl*	*NaCl (mM)*
1	9	**75.48	**25.45	X_1	Control
**25.44	**25.54	5.1	X_2		15
**75.45	**75.55	X_3			25
10	X_4				50
X_5					75

Table 3. *Effects of NaCl treatments on the accumulation of total protein in leaves of* Bellis perennis *L. (Mean ±SE).*

Protein Concentration mg. g^{-1}.dw	*Parameter NaCl mM*
01.29±385	Control
78.12±5.124	15
18.11±5.173	25
11.4±5.149	50
7.4±5.257	75

3.3. Effects of NaCl salt stress on the concentration of total protein in leaves of Bellis perrenis *L.*

Reduction of total protein in leaves is a common response of plant to salinity stress, because this environmental stress induces protein denaturation. As a result this reduction was observed in this assay too (Table 3 and Figure 6). The reason for the reduction of the concentration of total protein in leaves is that, in the stress condition metabolic pathway may be stopped and shifted to anthocyanin production, stopping the cell cycle leads to the accumulation of phenylalanine which results not only in an increase in proteins but also anthocyanins (Sato, 1996).

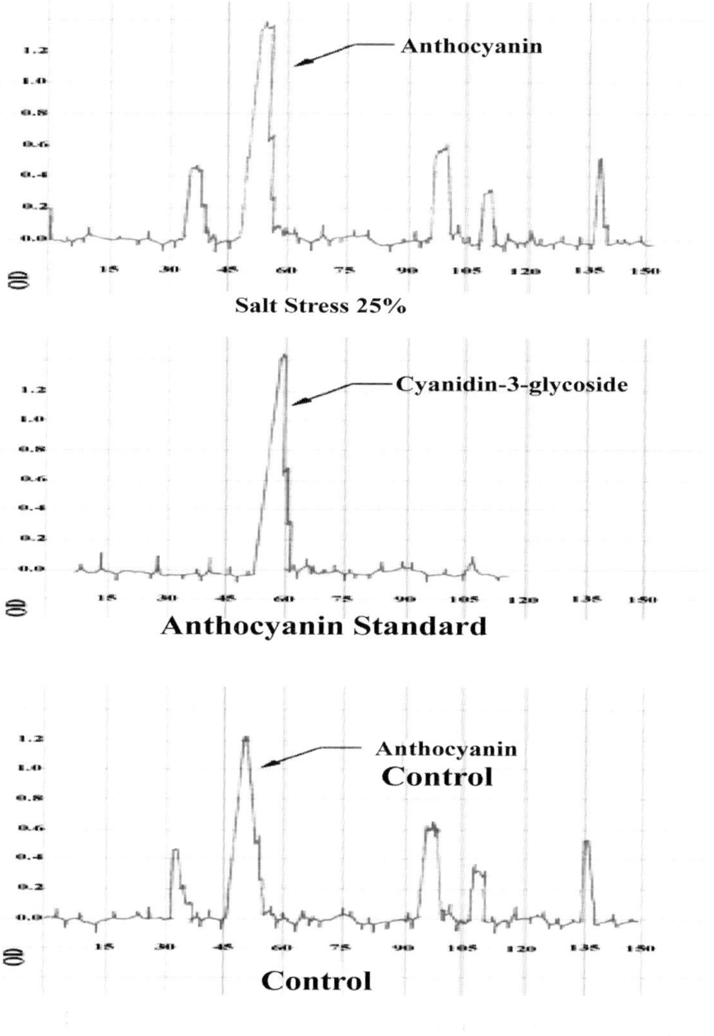

Figure 3. *HPLC analysis of anthocyanin.separation by HPLC revealed a major peak that eluted at 60 min.*

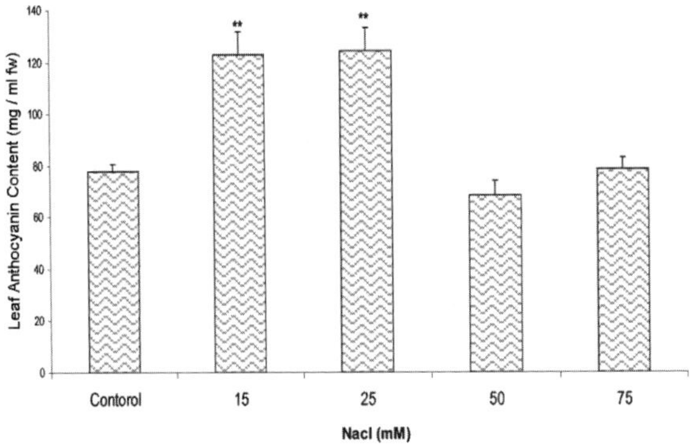

Figure 4. Effect of NaCl salinity on the concentration of anthocyanin (Spectrophotometer).

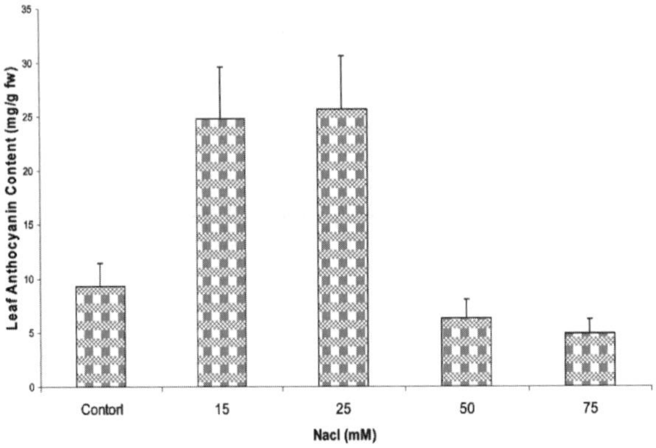

***Figure 5.** The effects of NaCl salinity on the concentration of anthocyanin (HPLC).*

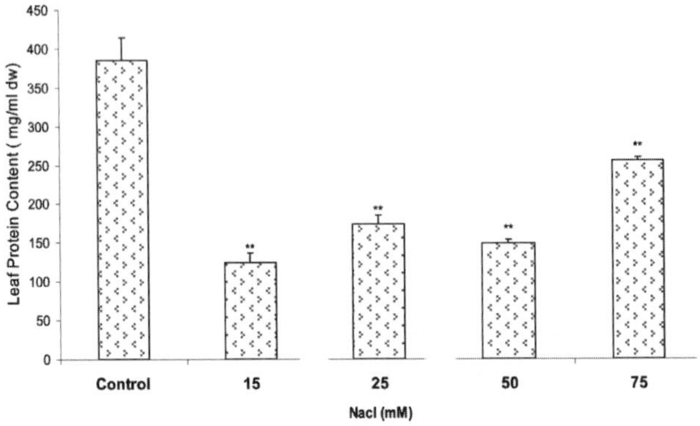

***Figure 6.** Effects of NaCl salinity stress on the concentration of total protein in leaves (Spectrophotometer).*

High NaCl stress is a major environmental factor that limits plant growth and productivity (Boyer, 1982). The detrimental effect of high concentration of salt stress such as the death of plants and/or decrease in photosynthetic activities has been understood and, the evidence for the effect of salt stress-inducing changes in plant metabolism is also well documented (Greenway & Muns, 1980). NaCl stress in addition to induction of the known compounds of osmotic stress and ion toxicity is also manifest as an oxidative stress, and all of these contribute to its deleterious effect (Gueta-Dahan et al., 1997). Given that, all the plants have some mechanism to protect themselves from environmental stress, it is essential to recognize these mechanisms. One of the most remarkable possible ways of resistances, which is suggested currently, is the induction of anthocyanins in leaves as a response to stress.

The present work showed that, the accumulation of anthocyanin in the plants grown under NaCl comparison with control plants was different. The increase of anthocyanins in plants at 15 and 25 mM NaCl concentrations demonstrate that anthocyanins serve as a protection from NaCl Stress.

Firstly, the severe consequence of NaCl stress is the loss of protoplasmic water leading to the increase concentration of ions such as Cl^- (Mundree, 2002). Therefore the osmotic potential of leaves will be reduced, and perhaps it will contribute to the decrease of stomatal conductance (Chalker-Scott, 1999). To avoid the lack of water in cells, the production of anthocyanin increase in specific tissue of plants under NaCl stress. Because anthocyanins in leaf tissues have function as osmotic adjusters. Secondly, there is a break down of chlorophyll and a release of reactive oxygen intermediates such as O^-_2 and H_2O_2 cause by NaCl stress in plants. Probably, owing to antioxidant capabilities of anthocyanins, this water–soluble pigments acts as a scavenger and scavenges oxygen radicals and inhibits lipid peroxidation (Chalker-Scott, 1999).

Thus, anthocyanins in leaf tissue protect leaves from photo oxidative and serve as a scavenger (Cooper, 2001). So, it is convincing that the ability to alter anthocyanin production under stressful condition could help plants to protect themselves from harmful effects.

4. REFERENCES

Aveto, P. 1995. Acetylence and terpenoids of *Bellis perennins*.L Phytochemistry 1: 141-147.

Boyer, J.S. 1982. Plant productivity and environment. Science 218: 443-448.

Chalker-Scott, L. 1999. Environmental significance of anthocyanin in plant stress responses. Photochemistry and Photobiology 70: 1-17.

Cooper, D.G. 2001. Contribution of jeffry halborn and co-workers to study of anthocyanin. Phytochemistry 56: 229-236.

Gilbert, G. 1998. Amino acid accumulation in sinks and source tissues of *Coleus blumei* benth during salinity stress. Journal of Experimental Botany 9: 107-114.

Gould, K. 2000. Functional roles of anthocyanin in leaves of *Quinita serreta* A.cunn. Journal of Experimental Botany 51: 1107-1115.

Greenway, H. & Muns R. 1980. Mechanisms of salt tolerance in nonhalophytes. Annual Review Plant Physiology 31: 149-190.

Gueta-Dahan, Y. Yaniv, Z. Zilinskas, B.A. & Ben-Hayyim, G. 1997. Salt and oxidative stress: similar and specific responses and their relation to salt tolerance in Citrus. Planta 203: 460-469.

Hamiltton, W. 2001. Mitochondrial adaptation to NaCl. Complex I is protected by anti-oxidants and small heat shock protein, whereas complex II is protected by praline and betain. Plant Physiology 126: 1266-1274.

Hoch, W. 2003. Resorption Protection. Anthocyanins Facilitate Nutrient Recovery in Autumn by Shielding Leaves from Potentially Damaging Light Levels. Plant Physiology 133: 1296-1305.

Mundree, G. 2002. Physiological and molecular insights into drought tolerance. African Journal of Biotechnology 1: 28-38.

Murrym, R. 1991. Dihydroflavonol reductase activities in relation to differential anthocyanin accumulation in juvenile and mature phase Hedera helix L. Plant Physiology 197: 343-351.

Sato, K. 1996. Culturing condition affecting the production of anthocyanin in suspended cell culture of strawberry. Plant Science 98: 91-113.

Steyn, W. 2002. Anthocyanins in vegetative tissues: a proposed unified function in photoprotection. New Phytologist 155: 349 - 361.

Toki, K. 1991. Three cyaniding 3-glucuronyl glucosides from red flowers of *Bellis perrenis* L. Phytochemistry 30: 3769-3771.

Winkel-Shirly, B. 2001. Flavonoid Biosynthesis. Plant physiology 127: 1619-1928.

Yamakashi, H.A 1997. Function of color. Trend in Plant science 2: 7-8.

CHAPTER 9

A COMPARATIVE STUDY ON RESPONSES OF GROWTH AND SOLUTE COMPOSITION IN HALOPHYTES SUAEDA SALSA AND LIMONIUM BICOLOR TO SALINITY

XIAOJING LIU[1], DEYU DUAN[1], WEIQIANG LI[1], T. TADANO[2] AND M. AJMAL KHAN[3]

[1]Research Center for Agricultural Resources, Institute of Genetics and Developmental Biology, Chinese Academy of Sciences, 286 Huaizhong Road, Shijiazhuang, Hebei 050021, P. R. China. [2]Faculty of Applied Biology, Tokyo University of Agriculture, 1-1-1 Sakuraoka, Setagaya-ku, Tokyo, 156-8502, Japan. [3]Department of Botany, University of Karachi, Karachi-75270, Pakistan

Abstract. *Suaeda salsa* (leaf succulence) and *Limonium bicolor* (secreting) are common halophytic species grown in coastal saline soil area in China. They possess different physiological adaptations which help them to avoid salt stress. Their mechanism of salt tolerance is varied and not properly understood. Therefore, the proposed plan to grow them in highly saline conditions could be hampered. The present study was designed to study the effect of salinity on growth and various solute compositions. Growth of *S. salsa* showed a 94% and 48% increases in comparison to control in 50 and 100 mM NaCl respectively in both shoot and root while at high salinity (400 mM NaCl) shoot and root dry weight were not significantly different from control. However, in *L. bicolor* root showed little promotion of shoot growth at 150 and 100 mM NaCl respectively and growth was substantially inhibited at 400 mM NaCl. *Suaeda salsa* accumulated more Na^+ and Cl^- ions in comparison to *L. bicolor*. These ions accumulated more in shoots of *S. salsa* whereas distributions of ions were similar in both shoots and roots of *L. bicolor*. Shoot soluble sugar decreased and proline increased with increase of external salinities of both species but shoots of *L. bicolor* contained relatively higher amount of sugar and proline at high salinity levels.

1. INTRODUCTION

Salinisation of soils and groundwater is a serious land-degradation problem in arid and semi-arid areas, and is increasing steadily in many parts of the world due to poor irrigation and drainage practices, which cause a great reduction for crop productivity (Lambers, 2003). As an alternative method to restore saline land, the utilization of halophytes attracted more attention due to their salt tolerance characteristics and potential economic values (Flowers, 1977; Glenn et al., 1999; Lieth, 1999; Barrett-Lennard, 2002; Zhao et al., 2002).

M.A. Khan and D.J. Weber (eds.), Ecophysiology of High Salinity Tolerant Plants, 135-143.

The principal mechanism of halophytic adaptation may be the high Na^+ and Cl^- absorption. The halophytes which are able to grow at high salinity can generate a high turgor in their cells by the high internal Na^+ and Cl^- concentrations. They also have some adaptive features like salt secretion (salt glands & bladders) or increase succulence to deal with high concentration of ions in the cell (Greenway & Munns, 1980). Difference of growth and physiological response between closed related plants to salinity is particularly interesting and may result to the identification of a number of factors that influence salt tolerance (Tester & Davenport, 2003). These differences can be used for the management practices of different halophytic species even for screen and selection of salt tolerant crops.

Suaeda salsa (Chenopodiaceae) a leaf succulent annual and *L. bicolor* (Plumbaginaceae) a perennial secreting halophyte has been reported to be widely distributed in saline areas of China. Both species could be used cash crops in high saline soils of China. Young seedling of *S. salsa* can be used as vegetable and the seeds contain approximately 25% of high quality oil. *Limonium bicolor* is a traditional Chinese herb, which has a haemostatic function (Zhao & Li, 1999; Zhao et al., 2002). The objective of this study was to compare the salt tolerance and mechanism of osmoregulation of the halophytes with different physiological adaptations.

2. MATERIALS AND METHOD

2.1. Plant materials and growth conditions

Seeds of *S. salsa* and *L. bicolor* were collected from coastal saline soils area in Hebei Province, China and were sown in plastic containers filled with clean sands and kept in a greenhouse at Nanpi Eco-Agricultural Experimental Station, Chinese Academy of Sciences under an approximately 28°C maximum and 23°C minimum temperature. After seed emergence the plants were irrigated daily with a half-strength Hoagland solution (Hoagland & Arnon, 1950). One week later, seedlings with a uniform size were selected and transferred to 8 L plastic tubs containing aerated Hoagland solution. Five plants were transplanted in each tub. Salinity treatments included 0 (CK), 50, 100, 200 and 500 mM NaCl. The NaCl concentrations were gradually increased by 100 mM NaCl increments at one day interval until to reach the maximum level of each treatment. For each treatment there were 3 replications, and the solutions were changed weekly to avoid ion accumulation. The duration of treatments was 30 days.

2.2. Growth measurements

At the end of experiments, plants of each tub were harvested and divided into shoots and roots. The shoots and roots were washed with cold deionized water, blotted dry, and fresh weights were measured. Then they were oven dried at 70°C for 48 hours and the dry weights were measured.

2.3. Chemical analysis

The dried shoot and root samples were ground separately and were wet ashed with HNO_3 digestion. The Na^+, K^+, Ca^{2+} and Mg^{2+} concentrations were analyzed by Hitachi 170-10 Atomic Absorption Spectrophotometer. Shoot soluble sugar was determined by the phenol-sulfuric acid method (Dubois et al., 1956) and proline by the Ninhydrin method of Bates et al., 1973.

2.4. Statistical analysis

Data were analyzed by using the SPSS 11.0 software and means were tested using Duncan's multiple range tests ($p < 0.05$).

3. RESULTS

3.1. Growth

Shoot and root dry weight of *S. salsa* was highest in 50 mM NaCl, and no significant growth inhibition was recorded at highest salinity treatments (Figure 1). There was small but significant root growth promotion with increase in salinity but root growth at highest salinity was not significantly different from control. Shoot growth in *L. bicolor* showed a little promotion at low salinities (Figure 1) and some inhibition at higher salinities. There was no effect of salinity on the roots growth of *L. bicolor* except at higher concentrations.

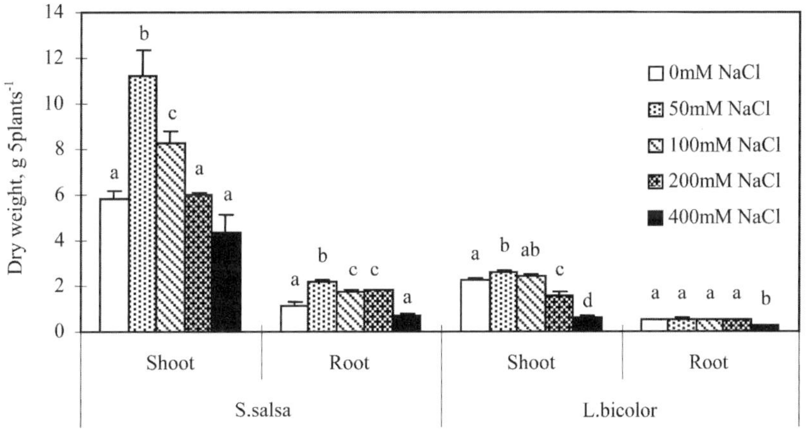

Figure 1. *Effect of salinity on shoot and root dry weight of two plant species. Same letters within each species indicate no significant difference ($p < 0.05$) according to Duncan's multiple range tests. Bars indicate s.e. of means (n=3).*

3.2. Ions

Sodium content of *Suaeda salsa* shoot increased rapidly with the increase in salinity reaching to about 19000 μmol.g.dw^{-1} (Table 1). However, in comparison root ion concentration could only reach about 2000 μmol.g.dw^{-1}. While Na concentration in *Limonium bicolor* was even less than 1500 μmol.g.dw^{-1} and there was not significant difference between root and shoot. Potassium concentration decreased with increase in salinity (Table 1). The decrease was more pronounced in the leaf of both plants particularly in the case of *S. salsa* (Table 1). Calcium concentration decreased with the increase in salinity and the decrease was higher in the roots of both plants in comparison to shoots (Table 1). Magnesium concentration of shoots of *S. salsa* decreased substantially with increase in salinity while small reduction is recorded in *Suaeda salsa* roots and for *L. bicolor* for both roots and shoots (Table 1).

Table 1. Effect of salinity on Cl, Na, K, Ca, and Mg (μmol/g DW) accumulation in the shoots and roots of S. salsa *and* L. bicolor.

			NaCl concentration, mM				
			0	*50*	*100*	*200*	*400*
Cl	*S. salsa*	Shoot	1939.4a*	2858.3b	3120.7b	3608.4c	5798.5d
		Root	229.3a	1524.1b	1538.1b	2039.1c	3192.4d
	L. bicolor	Shoot	970.4a	1651.0b	2263.4c	2334.4c	2703.9d
		Root	367.9a	630.4b	1281.0c	1730.3d	3334.7e
Na	*S. salsa*	Shoot	383.6a	3558.2b	9139.9c	17638.7d	19315.7e
		Root	347.3a	1267.2b	1444.6b	1730.0b	2740.4c
	L. bicolor	Shoot	27.4a	698.9b	1010.4c	1374.2d	2170.4e
		Root	103.9a	501.4b	1107.4c	1455.1d	1504.8d
K	*S. salsa*	Shoot	2180.9a	1652.1a	719.1b	603.7b	320.4c
		Root	1286.6a	1103.3a	920.8ab	887.3ab	797.2b
	L. bicolor	Shoot	585.8a	581.1a	496.6a	338.9b	355.3b
		Root	361a	341.4a	290.2a	249.6ab	103.9b
Ca	*S. salsa*	Shoot	46.7a	26.6b	21.3c	14.7d	12e
		Root	53.1a	46.6a	35.2ab	25.8ab	14.4b
	L. bicolor	Shoot	38.1a	34.5a	21.1b	16.3b	11.7b
		Root	26.2a	21.8a	20.1a	19.7a	16.7a
Mg	*S. salsa*	Shoot	273.4a	220.9a	138.1b	82.9b	56.2b
		Root	160.5a	149.9a	143.6a	133.8a	88.7a
	L. bicolor	Shoot	140.0a	103.7b	101.9b	84.9bc	78.2c
		Root	139.2a	131.7a	102.6b	100.4b	87.4b

* *Means in a row followed by the same letter are not significantly different ($p<0.05$).*

The shoots of *S. salsa* accumulated higher Na^+ than that of roots which was increased with the increase in NaCl concentrations until 200 mM then decreased (Figure 2). In *L. bicolor*, shoots Na^+ and roots Na^+ was almost at the same level. Shoots K^+ of *S. salsa* were higher than that of roots K^+ in 0 and 50 mM NaCl and lower in the higher NaCl concentrations. However in *L. bicolor* shoots K^+ were higher than roots K^+ in all the treatments. *S. salsa* shoots accumulated lower Ca^{2+} than that of roots in all treatments, and *L. bicolor* shoots accumulated lower Ca^{2+} only in higher concentrations. With the present of 100 mM and 50 mM or more NaCl, shoot Mg^{2+} was lower than roots Mg^{2+}.

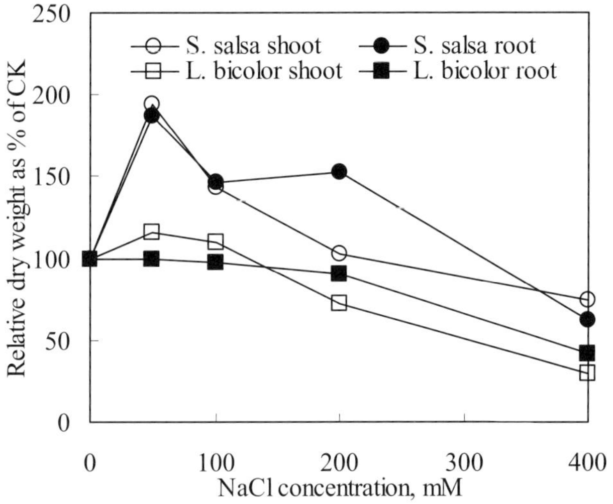

Figure 2. *Comparison of salinity effect on relative shoot and root dry weight of* S. salsa *and* L. bicolor.

3.3. Soluble sugar and Proline

In both species, shoot soluble sugar contents were decreased with the increase of NaCl concentrations (Figure 3).

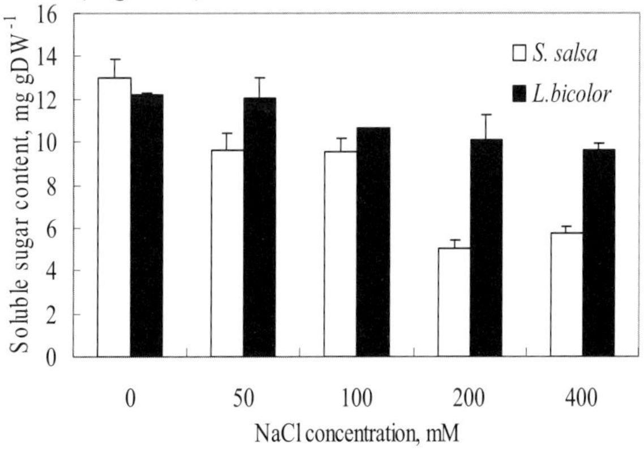

Figure 3. *Effect of salinity on leaf soluble sugar content of* S. salsa *and L.* bicolor. *Bars indicate s.e. of means (n=3).*

The shoot soluble sugar declined in 400 mM NaCl to 39% and 78% in comparison to control in *S. salsa* and *L. bicolor* respectively. The total shoot soluble sugar contents of *S. salsa* were lower than that of *L. bicolor* under saline conditions. Shoot proline contents were increased with the increase in NaCl concentrations in both species, and the increase of proline in *L. bicolor* was greater at higher salinity levels (Figure 4).

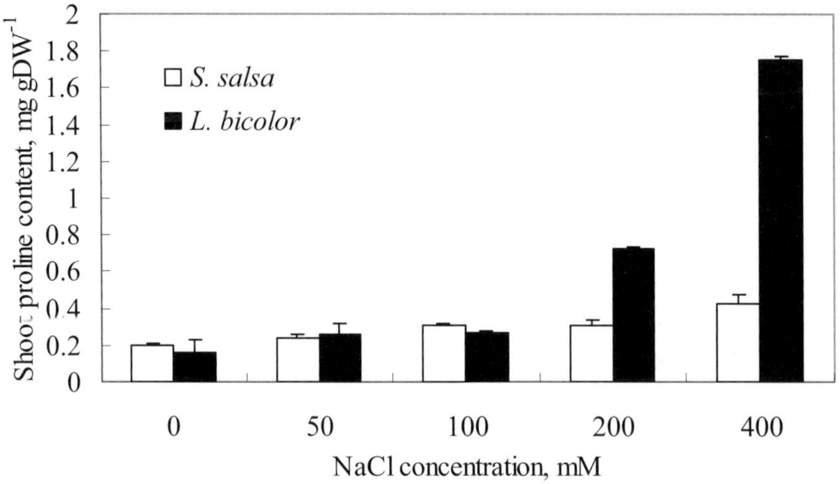

Figure 4. *Effect of salinity on leaf proline content of* S. salsa *and* L. bicolor. *Bars indicate s.e. of means (n=3).*

The proline content of *S. salsa* and *L. bicolor* shoots grown in 400 mM was increased 1-fold and 9-fold respectively.

4. DISCUSSION

Suaeda salsa showed optimal growth at 100 mol m^{-3} NaCl. Comparative results have been reported for *S. depressa* (Williams & Ungar, 1972), *S. maritima* var. *flexils* and var. *marcrocarpa* (Boucaud & Ungar, 1976), *S. australis* (Robinson & Downtown, 1985), *S. maritima* (Yeo & Flowers, 1980), *S. ussuriensis* (Kef-Fu et al., 1986), *S. salsa* (Ke-Fu et al., 1995). In fact, many dicotyledonous halophytes show optimal growth in the presence of salt (Naidoo & Raghunan, 1990; Khan et al., 2001). Secreting halophyte, *L. bicolor,* shoot growth was significantly inhibited above 100 mol m^{-3} NaCl but leaf succulent, *S. salsa*, showed no effect of salinity. Though the lower concentration of NaCl increased root growth in *S. salsa*, however, in comparison root growth of *L. bicolor* showed some inhibition at 400 mM NaCl. At moderate NaCl concentrations, the increase in weight was greater in *S. salsa* in comparison to *L. bicolor*. This may be due to the rise in inorganic ion content in *S. salsa* (Flowers et al., 1977) to absorb more water to maintain high succulence. It was considered that the increase in growth was the possible compensatory mechanisms for regulating salt concentration (Albert, 1975).

Suaeda salsa and *L. bicolor*, as most of halophytes can accumulate higher Na^+

and the accumulation was increased with the increase of NaCl levels. However comparing the two species, there are some differences in the ion accumulation and partitioning. The Na^+ accumulation in *S. salsa* plant was higher than that of *L. bicolor*. *Suaeda salsa*, as leaf succulent halophyte, accumulated most of Na^+ in shoots which was ranged from 74% to 94% of total Na^+ through all NaCl treatments, similar to the result that more than 90% of Na^+ in halophytes is in the shoot (Flowers, 1975). However *L. bicolor* as halophyte with salt glands, Na^+ accumulations was almost similar between shoots and roots. The difference between two species in the internal ion regulation may be due to the different mechanisms of salt tolerance, especially with the leaf succulence and salt glands. Though both species had the highest shoot water content in 100 mM NaCl (for *S. salsa* 93%, for *L. bicolor* 87.5%), *S. salsa* maintained the higher water content (>92%) in all the NaCl concentrations and the water content in shoots of *L. bicolor* varied greatly (from 87.5% in 100 mM NaCl to 81% in 400 mM NaCl) (Figure 5).

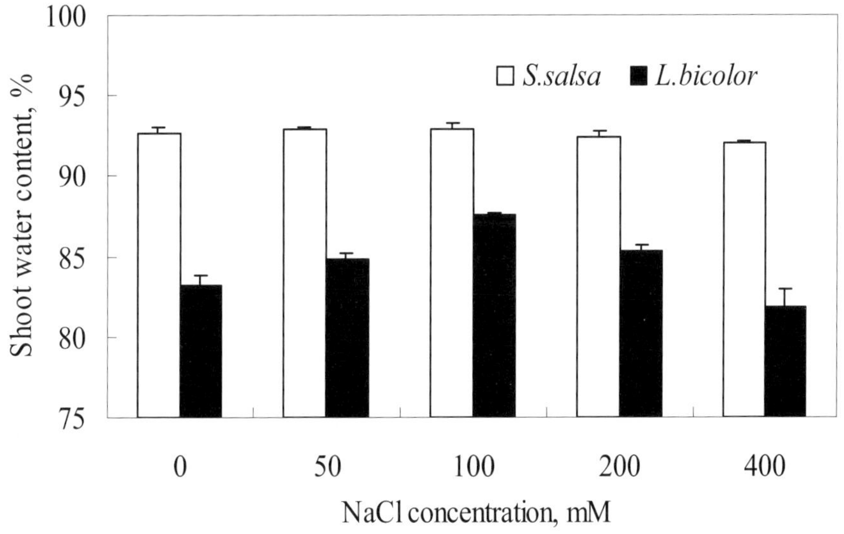

Figure 5. *Effect of salinity on shoot water content of* S. salsa *and* L. bicolor. *Bars indicate s.e. of means (n=3).*

The higher shoot water content avoids the higher ion accumulation in *S. salsa*. Nevertheless, in *L. bicolor*, salt glands removed salt from leaf surface that has reached to the shoot, so salt content in shoots and roots was similar, and also the glands remove water from leaf which results in lower shoot water content (Tester & Davenport, 2003).

The accumulations of K^+, Ca^{2+} and Mg^{2+} in shoots and roots of both species were decreased with the increase of NaCl in solutions. Similar results have been reported for *Salvadora persica* (Maggio et al., 2000) and *Atriplex canescens* (Richardson & McKell, 1980). However, if comparing the species, K^+, Ca^{2+} and

Mg^{2+} in *L. bicolor* was more stable than in *S. salsa*, this may be due to removal of salt through salt glands to keep the ion balance between shoots and roots. *Suaeda salsa* accumulated more K^+ in roots than in shoots under higher salinity levels. The higher K^+ in roots may continuously transport K^+ to shoots to avoid K^+ deficiency in shoots (Zhang, 2002).

Sugars, proline and other organic solutes are considered to improve salt tolerance by contributing to osmotic balance and preserving enzyme activity in the presence of toxin ions (Shannon, 1997; Greenway & Munns, 1980). In this study it was shown that in both species the shoot soluble sugar content was declined and shoot proline content was increased with the increase of NaCl concentration. However comparing the two species, the shoots of *L. bicolor* contained relatively higher soluble sugar and proline than that of *S. salsa* especially at higher salinity levels. Rathert (1984) reported that salinity causes greater sucrose increases in salt-sensitive species than in tolerant species. Greenway and Munns (1980) indicated that adaptive role of proline is related to survival rather than to maintenance of growth. Based on these reports, it is assumed that the higher soluble sugar and proline contents in shoots of *L. bicolor* than that of *S. salsa* indicate that *S. salsa* is more salt tolerance than *L. bicolor*.

The results of present study indicated that leaf succulent halophyte *S. salsa*, and secreting halophyte *L. bicolor*, showed increase in growth at moderate salinities. The higher salinity inhibited the growth of *L. bicolor* more than *S. salsa*. The ion accumulations and partitioning were different. *Suaeda salsa* accumulated a large amount of ions than *L. bicolor* and most of the ions were in the shoots, but for *L. bicolor* the ions in shoots and roots were approximately similar. Under higher salinity conditions, the shoot soluble sugar was decreased and proline increased. The shoot of *L. bicolor* contained relatively higher soluble sugar and proline than that of *S. salsa* especially at higher salinity levels. Based on the context, it was considered that *S. salsa* was more salt tolerant than *L. bicolor*. The abovementioned observations suggest that *S. salsa* is more salt tolerant, however, exact mechanism of other salt tolerance needs to studied.

5. ACKNOWLEDGEMENT

This work was carried out under the JSPS Research Fellowship granted to Xiaojing Liu, and was supported in part by China-Pakistan Joint Research Project (16-413), Innovation Project of Chinese Academy of Science (KWCX-SW-317-02) and Hebei "10th 5-year" Research Project (03220169D).

6. REFERENCES

Albert, R. 1975 Salt regulation in halophytes. Oecologia 21: 57-71.

Barrett-Lennard, E.G. 2002 Restoration of saline land through revegetation. Agri. Water Management 53: 213-226.

Bates, L.S., Waldren, R.P. & Teare, J.D. 1973. Rapid determination of free proline for water stress studies. Plant Soil 39: 205-207.

Boucaud, J. & Ungar, I.A. 1976. Influence of hormonal treatments on the growth of two halophytic species of *Suaeda*. American Journal of Botany 63: 694-699.

Dubois, M., Guilles, K.A., Hamilton, J.K., Rebers, P.A. & Smith, F. 1956. Colorimetric method for determination of sugars and related substances. Annals of Chemistry 28: 350-356.

Flowers, T.J. 1975. Halophytes. In Ion Transport in Plant Cells and Tissues. D.A. Baker & J.L. Hall. (Eds.), Amsterdam: North Holland. 309-334 pp.

Flowers, T.J. 1977. The mechanisms of salt tolerance in halophytes. Annual Review of Plant Physiology 28: 89-121.

Glenn, E.P., Brown, J.J. & Blumwald, E. 1999. Salt tolerance and crop potential of halophytes. Critical Review of Plant Science 18: 227-255.

Greenway, H. & Munns, R. 1980. Mechanisms of salt tolerance in nonhalophytes. Annual Review of Plant Physiology 31: 149-190.

Hoagland, D.R. & Arnon, D.I. 1950. The water-culture method for growing plants without soil. California Agriculture Experimental.Station Circular No. 347.

Jennings, D.H. 1976. The effect of sodium chloride on higher plants. Biological Review 51: 453-486.

Lambers, H. 2003. Dry land salinity: A key environmental issue in southern Australia. Plant Soil 257: □-□

Lieth, H. 1999. Development of crops and other useful plants from halophytes. In: Halophyte Uses In Different Climates □. H. Lieth et al., (Ed.), Backhus Publishers, Leiden. 1-17 pp.

Maggio, A., Reddy, M.P. & Joly, R.J. 2000 Leaf gas exchange and solute accumulation in the halophyte *Salvadora persica* grown at moderate salinity. Environtment Expexperimental Botany 44: 31-38.

Naidoo, G.R. & Rughunanan, R. 1990. Salt tolerance in the succulent halophyte, *Sarcocornia natalensis*. Journal of Experimental Botany 41: 497-502.

Rathert, G. 1984 Sucrose and starch content of plant parts as possible indicator for salt tolerance of crops. Australian Journal of Plant Physiology 11: 491-495.

Richardson, S.G. & McKell, C.M. 1980 Water relations of *Atriplex canescens* as affected by the salinity and moisture percentage of processed oil shale. Agronomy Journal 72: 946-950.

Robinson, S.P. & Downtown, J.S. 1980. Potassium, sodium and chloride ion concentrations in leaves and isolated chloroplasts of the halophyte *Suaeda australis* R. Br. Australian Journal of Plant Physiology 12: 471-479.

Shannon, M.C. 1997. Adaptation of plants to salinity. Advance Agronomy 60: 76-120.

SPSS Inc 2002 SPSS: SPSS 11.0. SPSS Inc. USA.

Tester, M. & Davenport, R. 2003 Na^+ tolerance and Na^+ transport in high plants. Annals of Botany 91: 503-527.

Williams, M.D. & Ungar, I.A. 1972. The effect of environmental parameters on the germination, growth, and development of *Suaeda depressa* (Pursh) Wats. American Journal of Botany 59: 912-918.

Yeo, A.R. & Flowers, T. J. 1980. Salt tolerance in the halophyte *Suaeda maritima* L. Dum.: Evaluation of the effect of salinity upon growth. Journal of Experimental Botany 31: 1171-1183.

Zhang, H. 2002. A study on the characters of content of inorganic ions in salt-stressed *Suaeda salsa*. *Acta Bot. Boreal.-Occident. Sin.* 22: 129-135.

Zhao, K.F & Li, F.Z 1999. Halophytes in China. Science Press, Beijing.

Zhao, K.F., Li, Ming-liang. & Jia-yao, L. 1986. Reduction by GA_3 of NaCl-induced inhibition of growth and development in *Suaeda ussuriensis*. Australian Journal of Plant Physiology 13: 547-551.

Zhao, K.F., Hai, F., & Harris, P.J.C. 1995. The physiological basis of growth inhibition of halophytes by potassium. In: Biology of Salt Tolerant Plants. M.A. Khan and I.A. Ungar (Eds.), Department of Botany, University of Karachi, Pakistan. 221-227 pp.

Zhao, K.F., Fan, H. & Ungar, I.A. 2002. Survey of halophyte species in China. Plant Science 163: 491-498.

Zhao, K.F., Fan, H., Z, San. & S, Jie. 2003. Study on the salt and drought tolerance of *Suaeda salsa* and *Kalanchoe claigremontiana* under iso-osmotic salt

CHAPTER 10

ALLEVIATION OF SALINITY STRESS IN THE SEEDS OF SOME BRASSICA SPECIES

M. ÖZTURK[1], S. BASLAR[2], Y. DOGAN[2] AND M. S. SAKCALI[3]

[1]Ege University, Bornova-Izmir, Turkey
[2]Dokuzeylul University, Buca-Izmir, Turkey
[3]Marmara University, Institute of Sciences, Istanbul, Turkey

Abstract. Uptake of various salt and sugar solutions by *Brassica* varies with change in species and temperature. The seeds of *B. nigra* show a dormancy period of 6 months, whereas other species germinated immediately. The seeds generally not sensitive to photoperiod, however, exposure to extended dark period reduced seed germination. *Brassica nigra* and *B. juncea* show medium tolerance up to 1 percent salinity level but at 1.5 percent level tolerance decreases by 32%. An application of IAA and GA_3 alleviates the salinity stress up to 1.5% level and to some extent at 2% level. Alleviation of salt stress occurred more in *B. juncea* than *B. nigra*. In *B. oleracea* var. *botrytis* and *B. oleracea* var. oleracea the application of growth regulators improves germination and seedling growth.

1. INTRODUCTION

Soil salinity greatly decreases the productivity of economically important plants (Sheikh et al., 1976). However, many genetic variations in the tolerance of seedlings to saline soils have been observed which are evaluated for crop improvement on such soils. Seed germination and seedling emergence are regarded as the most sensitive and critical stages in plant growth and development on such habitats (Ozturk et al., 1993, 1995). In the studies carried out on the alleviation of limiting effects of salt stress on the germination, plant growth regulators like IAA, KIN and GA_3 applied externally have been shown to improve physical and metabolic conditions of seed germination (Braun et al., 1976; öztürk et al., 1994). Lately much attention has been paid towards this subject (Khan & Ungar, 1998; Khan et al., 2002) but little information is available on the performance of economically important oilseed and vegetable crops on saline soils. One of these is Brassicaceae family. Most of the plant species from this family are being used for centuries as vegetables, fodder, medicinal purposes, etc. The family members generally show distribution in the temperate and cold parts

M.A. Khan and D.J. Weber (eds.), Ecophysiology of High Salinity Tolerant Plants, 145-156.

with 380 genera and 3000 species. One of these is the genus *Brassica* with generally herbaceous annuals, biennials or perennials, leaves are sessile, amplexicaule and petiolate, simple or lobed, flowers yellow or yellowish-white, fruits are linear and oblong siliqua. In Turkey 6 species of *Brassica* are distributed naturally namely; *B. campestris* L. (Field Mustard), *B. cretica* Lam. (Cretan Mustard), *B. elongata* Ehrh. (Elongated Mustard) *B. nigra* (L.) Koch.(Mustard), *B tournefortii* Gouan. (Wild turnip-rape) and *B. deflexa* Boiss. (Özturk et al.,1983). In addition to these, many taxa are cultivated, two of these being *Brassica oleracea* L.var. *botrytis* and *Brassica oleracea* L. var. *oleracea*. This investigation has been undertaken with the aim of evaluating the factors effective in the water uptake and determination of salt-hormone interactions during the germination and seedling emergence of four economically important taxa from the genus *Brassica* under different light, temperature and salinity conditions.

2. MATERIALS AND METHODS

Seeds of *Brassica oleracea* var. *oleracea*, *Brassica oleracea* var. *botrytis* and *B. juncea* were purchased from local vegetable seed market and *B. nigra* was collected from its natural populations in Adana and Izmir. These were surface sterilized with one percent sodium hypochlorite for one minute. Water uptake in four replicates of 25 seeds each was recorded in 50 ml of distilled water in seven centimeter diameter disposable petri dishes at 5, 15, 25, 35°C for 8, 18, 24, 48 hours for *B. nigra*, *B. juncea* and 10, 20, 30°C for 24, 48, 96 hours for *B. oleracea* var. *oleracea*, *B. oleracea* var. *botrytis*. The dry weight was recorded with the help of Chyo JL–200 balance. In germinated seeds weight recorded together with radicle. For water uptake from salt solution five salinity concentrations (0.1, 0.5, 1, 2, 3 molar NaCl) were prepared and placed in petri dishes with 10 ml. of the solution added to each and covered by parafin film. These were left at 20°C for 8, 18, 24 ,48 hrs. for *B. nigra*, *B. juncea* and 24, 48, 96 hours for *B. oleracea* var. *oleracea*, *B. oleracea* var. *botrytis*. For sucrose solution five concentrations (0.1, 0.5, 1, 2, and 3 molar) were prepared and 10 ml from each left in petri dishes containing 25 seeds per replicate at 20°C for 8, 18, 24, 48 hrs. for *B. nigra*, *B. juncea* and 24, 48, 96 hours for *B. oleracea* var. *oleracea*, *B. oleracea* var. *botrytis*. Seeds were dried with tissue paper and weight recorded.

In the following experiments four replicates with 50 seeds per replicate were used. The effects of temperature (10, 15, 20, 25, 30°C) were followed with the seeds in the preset incubators. The light treatment was given as 12/12 -h light/dark and 24-h dark in the germination cabinets (with 700 nm cool-white fluorescent lamps-GE) at 20°C. Salinity levels used were 0.5, 1, 1.5, 2 and 3% NaCl (20°C) and hypocotyl-radicle length measured after 96 hours. The growth regulators (10, 100 ppm of GA_3, IAA) were applied for two hours; then the seeds were left in 10 ml of 0.5, 1, 1.5, 2, and 3 percent NaCl solutions placed at 20°C. After 96 hours the length of radicle-hypocotyls of germinated seeds were measured. All studies on the germination were carried out on two folds of Whatman number 1 filter paper placed in 7 cm diameter disposable petri dishes covered by parafin film. The emergence of the radicle was accepted as the criterion for germination.

3. RESULTS AND DISCUSSION

3.1. Imbibition in relation to the temperature

The water plays a constructive role for starting the biochemical and physiological processes in seeds and its uptake is regarded as the most important factor in the chain of events triggerring the germination. It is mainly controlled by the structural features of the seeds, seed coat permeability, temperature and salinity. An increase in the volume, respiration, enzyme activity, transportation of food, cell growth and division followed by differentiation of tissues and organs (although not fully developed) due to water uptake during seed germination is controlled by the critical hydration level varying from species to species. The role of temperature and moisture in soil for the germination of seeds has been well stressed by several investigators because a proper combination of critical temperature and humidity for the seeds to germinate in the soil is of great importance (Khan & Khan, 1978; Khan, 1980; Mayer & Poljakoff-Mayber, 1982; Baskin & Baskin, 1998;). During seed germination as soon as water uptake starts, membrane systems are reorganised and this process is mainly related to the temperature and the plant species (Uygunlar et al., 1985). An assessment of the water uptake of *B. nigra*, *B. juncea*, *B. oleracea* var. *botrytis* and *B. oleracea* var. *oleracea* at different temperatures and time intervals revealed that imbibition of water by the seeds from the soil was closely related with the temperature (Figure 1).

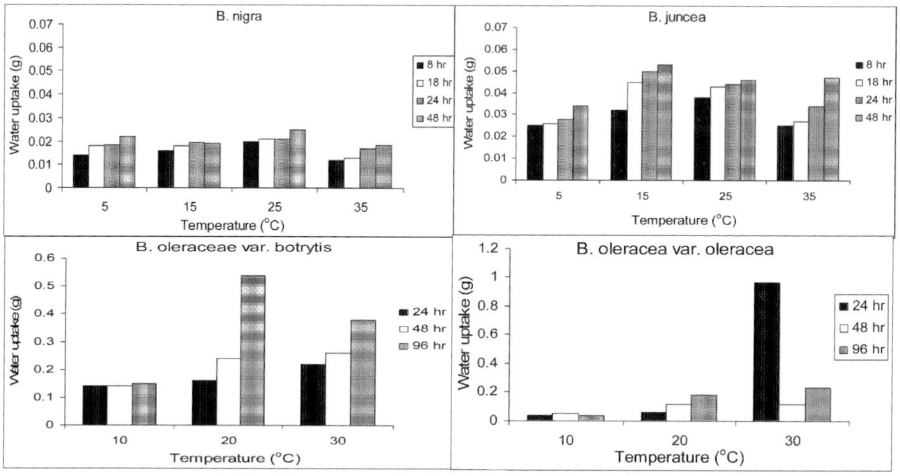

Figure 1. *Water uptake in* Brassica *species at different temperatures.*

The water uptake of *B. nigra* seeds during 8 to 48 hr at the temperatures ranging between 5 to 35 °C as well as the total amount of water imbibed differed from each other. The water absorbed was maximum at 25° C in *B. nigra*, but at 15 ° C in *B. juncea*, however, minimum water uptake took place at 5 ° C in *B. juncea* and 35° C in *B. nigra*. In *B. oleracea* var. *botrytis* water uptake by the seeds at 10, 20 and 30°C shows that it is maximum at 20° C and at 30° C in *B. oleracea* var. *oleracea*, but very little at 10 ° C in both taxa. In general a continuous increase has been observed in the water uptake of seeds at different temperatures. These results are in general agreement with those of Ozturk & Mert (1983).

3.2. Imbibition in relation to different osmotic solutions

In order to ascertain the effects of different osmotic solutions on the imbibition, the seeds of *B. nigra* and *B. juncea* were placed in 5 different salt concentrations and water uptake recorded (Figure 2).

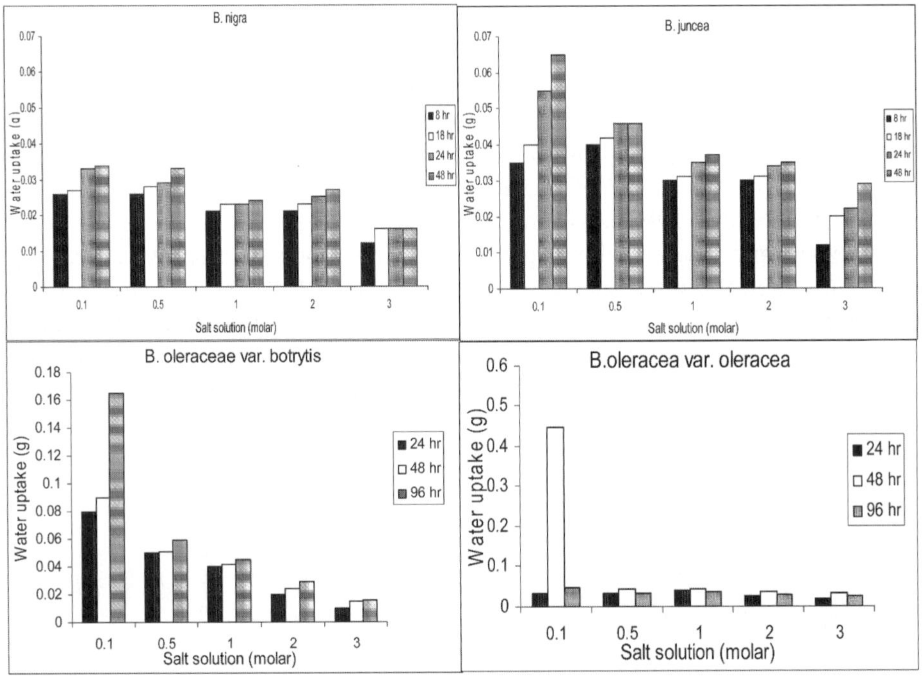

Figure 2. *Water uptake in* Brassica *species from different salt solutions.*

The seeds in general absorb water from all the osmotic solutions, however, the uptake increases with time. On the other hand there is a decrease in the imbibition as the osmotic pressure of the solution increases. These results coincide with those of Vardar & Ahmet (1971). A maximum water uptake was observed at 0.1 molar NaCl solutions and the least at 3 molar in *B. nigra*, *B. juncea* and *B. oleracea* var. *botrytis* whereas in the seeds of *B. oleracea* var. *oleracea* the uptake decreases abruptly from 0.5 molar onwards, even at 0.1 percent level water uptake is low after 48 hours. The seeds left in different sugar solutions showed maximum water uptake in 0.1 followed by 0.5 molar solution, minimum in 3 molar solutions (Figure 3). In *B. oleracea* var. *botrytis* water uptake occurs in all salt and sugar solutions but is maximum (0.165 g and 0.131 g) at the lowest salt (0.1% NaCl) and sugar (0.1%) concentrations. Similarly in *B. oleracea* var. *oleracea* water uptake occurs in all salt and sugar solutions, however, the maximum water uptake (0.45 g and 0.34 g) occurs at the lowest concentrations (0.1 molar). Higher concentrations of salt and sugar inhibit the water uptake in the seeds resulting in lower germination ratio. This coincides with the findings of Ozdemir et al., (1994). The arrangement of membrane systems in relation to temperature, osmotic principles and water uptake in seeds is regulated by the osmotic principles, the reason for this is inhibition of water uptake due to the toxicity of high Cl^- ions or osmotic pressure because of high salt and sugar concentrations (Ozturk et al., 1993).

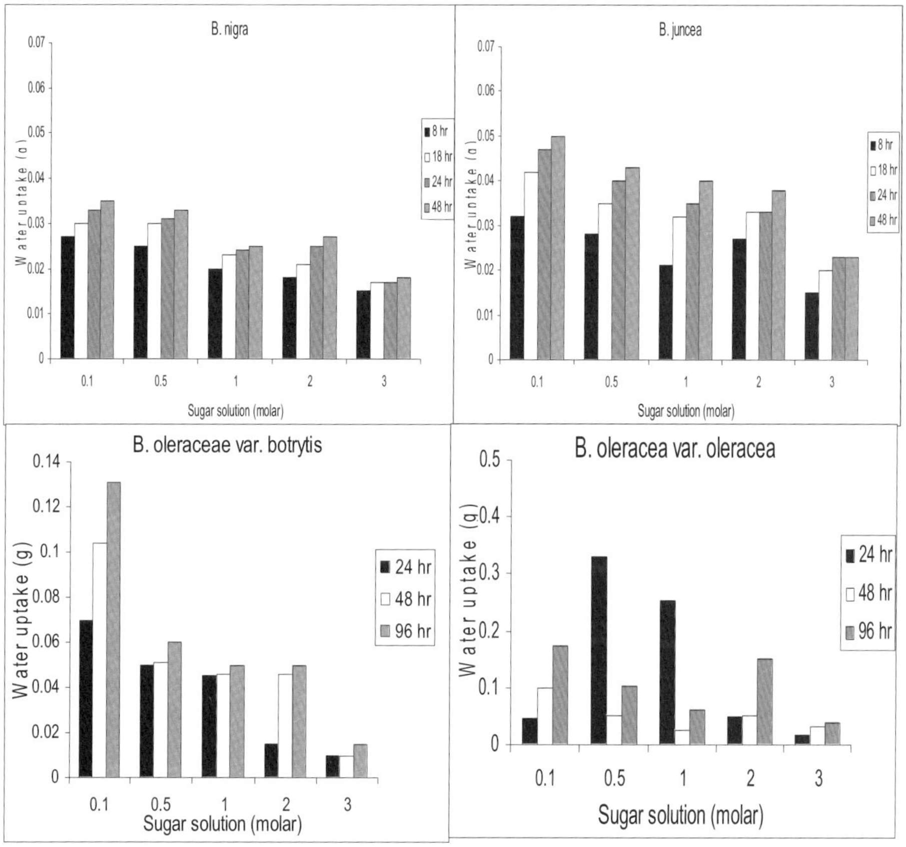

Figure 3. *Water uptake in* Brassica *species from different sugar solutions.*

3.3. Germination in relation to different light and temperature conditions

The complex changes during seed germination involve different metabolic events. It is not thus surprising to find a close dependance of this phase on temperature and light, even neolithic farmers knew about the influence of light on the seed germination (Vidaver, 1980). Seeds of different species show different germination behaviour in response to different light and temperature conditions (Ozturk et al., 1983). In the case of non- dormant seeds, the effect is reflected in the rate and speed of germination and germination rate increases with an increase in temperature up to an optimum. In some seeds even a single shift in temperature proves highly effective, others achieve an optimal state from 0-5°C, but range widens between 10-40 °C in some cases (Ozturk & Pirdal, 1986). Similarly, light may be inhibitory to germination as in *Ranunculus arvensis* and *R. muricatus* (Ahmet, 1968, 1969) or promotive as in *Ranunculus laetus* (Ahmet, 1970). Certain seeds germinate after a short exposure, whereas others need long- or short-day photoperiod, many however appear indifferent to light. Studies on the temperature and light interactions in the germination behaviour of the *Brassica* taxa investigated by us revealed that the seeds of *B. nigra* show a dormancy period of 6 months, whereas other species germinated immediately. Both *B. nigra* and *B. juncea* attain 77% germination under 12/12 hrs light/dark condition at 20-25 °C in 5 days, but germination decreases at 10 and 30 °C, however, an alternate temperature of 20/15 °C under 12/12 hrs light/dark condition proves more effective by increasing the percentage up to 89%. The seeds are indifferent to photoperiod but in continuous dark the percentage germination decreases up to 51%. In the seeds of *B. oleracea* var. *botrytis* optimum germination occurs at 15°C in continuous light and at 25 °C in darkness, whereas in *B. oleracea* var. *oleracea* germination generally occurs at 20-25 °C under 12/12 hrs light/dark condition. Seeds of these taxa appear to be indifferent to the light condition. This is in agreement with the findings of Ozdemir et al. (1994) whose studies on some members of Brassicaceae reveal that maximum germination takes place at 25-30° C irrespective of the light condition.

3.4. Germination in relation to different osmotic solutions

Many investigators have evaluated the germination and growth behaviour of seeds of several crop species in relation to different salinity levels (Ansari, 1970, 1972; Khan & Gulzar, 2003). As reported in these studies high soil salinity seriously affects the emergence of seedlings and growth as well. However, 100% germination has been recorded in water soaked non-chilled seeds up to 100 mM NaCl, whereas germination occurred in up to 200 mM NaCl in chilling treated seeds (Sharma & Kumar, 1999). In our studies germination and seedling emergence of *B. nigra* and *B. juncea* was followed. The results revealed that these species show medium tolerance in up to 1 percent salinity level as compared to the control but at 1.5 percent level tolerance decreases by 32%. This decrease can be attributed to inhibition in the germination and slow germination rate. The results depict that these taxa are sensitive to salinity at the stage of seedling emergence. An application of IAA and GA_3 alleviates the salinity stress up to 1.5% level and to some extent at 2% level in *B. juncea*, but was not effective at 3% level (Figure 4). GA_3 proves more effective than IAA and alleviation of salt stress was higher in *B. juncea* than *B. nigra*.

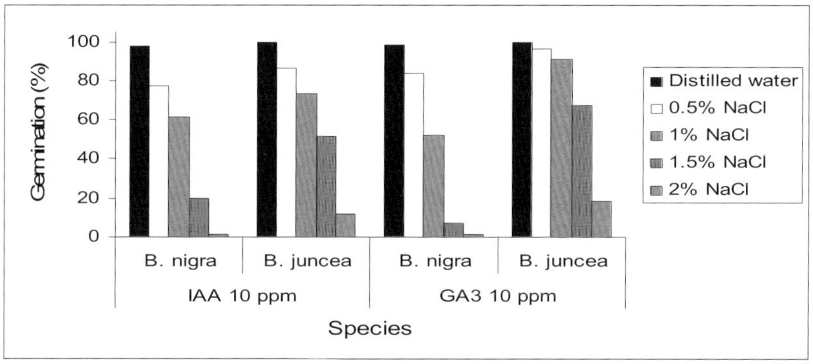

Figure 4. *Salt-Hormone interactions in the germination of* Brassica *species.*

In *B. oleracea* var. botrytis 98% germination was observed in the series where 0.5% NaCl was applied. This was very close to the control (100 %). However at 1% NaCl germination was 84% and at 2 and 3% germination was completly inhibited. When seeds were treated with 10 and 100 ppm GA_3 and IAA growth regulators for two hours and then left in 0.5, 1 and 2% NaCl solutions, maximum germination was observed in 0.5% NaCl +10 ppm IAA (96%) and 0.5% NaCl + 10 ppm GA_3 (96%). An improvement was observed at 10 ppm GA_3 in 1 percent NaCl solution (Figures. 5 & 6).

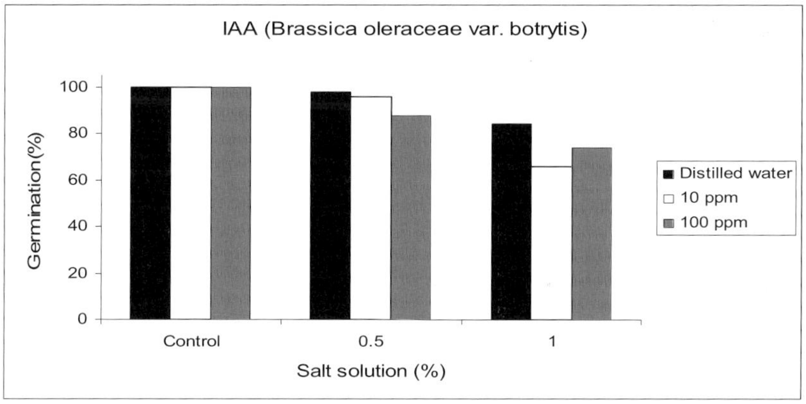

Figure 5. *Salt-IAA interactions in the germination of* B. oleraceae *var.* botrytis.

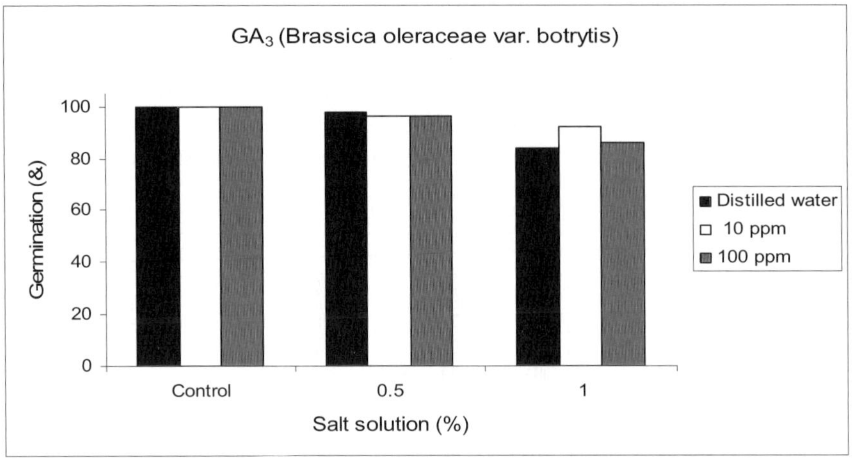

Figure 6. *Salt-GA_3 interactions in the germination of* B. oleraceae *var.* botrytis.

Radicle and hypocotyle lengths were also measured to follow the impact on seedling growth (Table 1). The length of radicle was between 32 to 44 mm and that of hypocotyle 13 to 15 mm in GA_3 and IAA. These results are very close to the control without salt. In 0.5% and 1% salt concentrations the length of radicle as well as hypocotyle decreased by 50 to 27%. But application of these in combination with the GA_3 and IAA (2 hours applications) alleviated the stress as compared to the salt alone. The maximum length of radicle (33.8 mm) and hypocotyle (15.7 mm) was observed in 0.5% NaCl + 100 ppm GA_3. In the case of IAA 0.5% NaCl + 10 ppm IAA proved more effective as compared to other concentrations.

Table 1. *The lengths of radicle (Rd) and hypocotyl (Hp) in different salt concentrations in* B. oleracea *var.* botrytis *after pretreatment with GA_3 and IAA for 2 hours.*

		Control (mm)		*% 0.5 NaCl (mm)*		*% 1 NaCl (mm)*	
		Rd	*Hp*	*Rd*	*Hp*	*Rd*	*Hp*
D/W		42.6±1	15.3±.3	20.1±1	10.6±.6	10.9±2	4.1±.5
GA_3 ppm	10	44.3±1	15.1±.5	32.4±1	13.7±.4	17.2±.7	8.0±.4
	100	40.2±1	14.8±.5	33.8±1	15.7±.8	17.8±.9	8.1±.7
IAA	10	37.5±1	13.6±.5	20.4±.3	13.4±.6	4.9±.1	5.1±.6
ppm	100	32.3±1	13.0±.6	10.6±1	9.1±.6	3.3±.1	3.4±.4

In *B. oleracea* var. *oleracea* growth regulators were observed to alleviate the inhibitory effect of higher salt concentrations. In the controls, 0.5, 1 and 1.5% NaCl solutions germination was higher but decreased by 60-78% at 2% NaCl. However application of 10 ppm IAA and 10 ppm GA_3 resulted in 82-84% germination even at 2% NaCl, however, 100 ppm IAA and 100 ppm GA_3 resulted in a decrease in the germination at this level. 10 ppm of IAA and GA_3 proved more effective in alleviating the salt stress (Figures. 7 & 8).

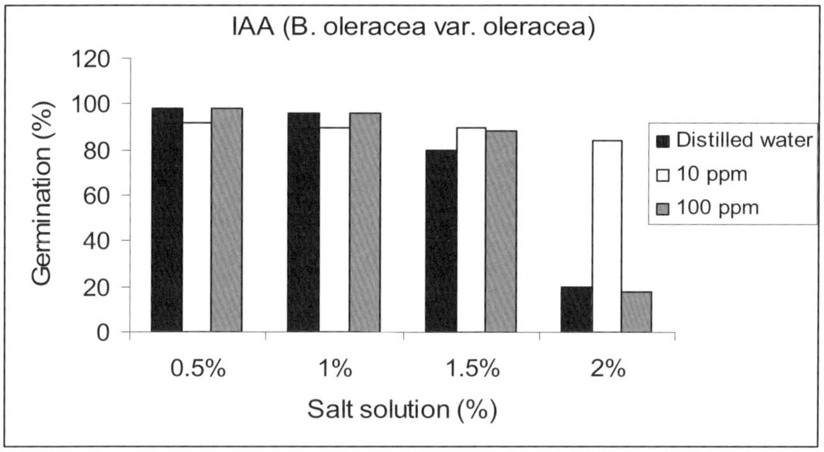

Figure 7. *Salt-IAA interactions in the germination of* B. oleraceae *var.* oleracea.

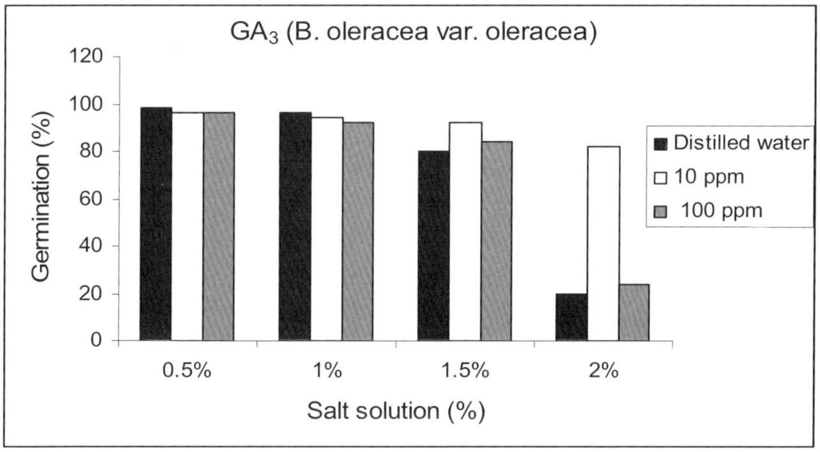

Figure 8. *Salt-GA_3 interactions in the germination of* B. oleraceae *var.* Oleracea.

These results agree with our earlier findings on *Brassica* species (Özturk et al.,1993; Ozdemir et al.,1994). The application of growth regulators improves the length of hypocotyl and radicle as compared to those in distilled water. The slight increase in lengths observed at 0.5% NaCl as compared to the control could be attributed to the increase in the potential osmotic pressure of root cells. However, in 1 and 1.5% NaCl solutions the length of radicle and hypocotyl was greatly decreased. This is due to their sensitiveness to the NaCl solution. An application of growth regulators like GA_3 and IAA depicted that 10 and 100 ppm of GA_3 alleviate the inhibitory effect of higher salt concentrations to some extent (Table 2).

Table 2. *The lengths of radicle (Rd) and hypocotyl (Hp) in different salt concentrations in* B. oleracea *var.* oleracea *after pretreatment with GA_3 and IAA for 2 hours.*

		Control		% 0.5 NaCl		% 1 NaCl		% 1.5 NaCl	
		Rd (mm)	Hp (mm)	Rd (mm)	Hp (mm)	Rd (mm)	Hp (mm)	Rd (mm)	Hp (mm)
D.Water		40±1	15±1	41±1	25±2	11±2	10±1	7±1	4±1
GA_3	10 ppm	41±7	29±2	35±2	22±2	20±1	17±1	17±1	15±2
	100 ppm	49±6	33±1	37±2	29±2	22±2	21±1	19±1	17±1
IAA	10 ppm	42±3	22±1	29±1	10±1	4±1	8±1	2±1	3±1
	100 ppm	47±2	13±1	23±2	11±1	1±2	4±1	1±1	1±1

4. SUMMARY

The water uptake and germination requirements of *Brassica nigra*, *Brassica juncea*, *B. oleracea* var. *botrytis*, and *B. oleracea* var. *oleracea*, were studied under control conditions in the laboratory. The water uptake treatments included 5, 15, 25, 35°C temperatures for 8, 18, 24, 48 hours with *B. nigra*, *B. juncea* and 10, 20, 30 °C for 24, 48, 96 hours with *B. oleracea* var. *oleracea*, *B. oleracea* var. *botrytis*. In the case of water uptake from salt and sugar solutions five concentrations (0.1, 0.5, 1, 2, 3 molar) were used and experiments followed at 20° C for 8, 18, 24 ,48 hrs. with *B. nigra*, *B. juncea* and 24, 48, 96 hours with *B. oleracea* var. *oleracea*, *B. oleracea* var. *botrytis*. Seeds were dried with tissue paper and weight recorded. The water uptake of *B. nigra* seeds during 8 to 48 hr at the temperatures ranging between 5 to 35°C as well as the total amount of water imbibed differed from each other. The water absorbed was maximum at 25°C in *B. nigra*, but at 15°C in *B. juncea*, however, minimum water uptake took place at 5 °C in *B. juncea* and 35 °C in *B. nigra*. In *B. oleracea* var. *botrytis* water uptake by the seeds at 10, 20 and 30 °C shows that it is maximum at 20°C and at 30°C in *B. oleracea* var. *oleracea*, but very little at 10°C in both taxa. A maximum water uptake was observed at 0.1 molar NaCl solutions and the least at 3 molar in *B. nigra*, *B. juncea* and *B. oleracea* var. *botrytis* whereas in the seeds of *B. oleracea* var. *oleracea* the uptake decreases abruptly from 0.5 molar onwards, even at 0.1 percent level water uptake is low after 48 hours. The seeds left in different sugar solutions showed maximum water uptake in 0.1 followed by 0.5 molar solution, minimum in 3 molar solutions. In *B. oleracea* var. *botrytis* water uptake occurs in all salt and sugar solutions but was maximum (0.165 gr and 0.131 gr) at the lowest salt (0.1% NaCl) and sugar (0.1%) concentrations. Similarly in *B. oleracea* var. oleracea water uptake occurs in all salt and sugar solutions, however, the maximum water uptake (0.45 gr and 0.34 gr) occurs at the lowest concentrations (0.1 molar). The effects of temperature (10, 15, 20, 25, 30° C) and light (12/12-h light/dark, 24-h dark) were followed in the germination cabinets (with 700 nm cool-white fluorescent lamps-GE). *Brassica* taxa investigated by us revealed that the seeds of *B. nigra* show a dormancy period of 6 months, whereas other species germinated immediately. Both *B. nigra* and *B. juncea* reach 77% germination under 12/12 hrs light/dark condition at 20-25 °C in 5 days, but germination decreases at 10 and 30 °C, however, an alternate temperature of 20/15 °C under 12/12 hrs light/dark condition proves more effective by increasing the percentage up to 89 %. The seeds are indifferent to photoperiod but in continuous dark the percentage germination decreases up to 51 %. In the seeds of *B. oleracea* var. *botrytis* optimum germination occurs at 15 °C in continuous light and at 25° C in darkness, whereas in *B. oleracea* var. *oleracea* germination generally occurs at 20-25 °C under 12/12 hrs light/dark condition. Seeds of these taxa appear to be indifferent to the light conditions. In the salinity (0.5, 1, 1.5, 2 and 3% NaCl at 20 °C) studies and their interactions with growth regulators (10, 100 ppm of GA_3, IAA) applied for two hours revealed that *B. nigra* and *B. juncea* show medium tolerance up to 1 percent salinity level as compared to the control but at 1.5 percent level tolerance decreases by 32 %. An application of IAA and GA_3 alleviates the salinity stress upto 1.5% level and to some extent at 2% level in *B. juncea*, but is not effective at 3% level. GA_3 proves more effective than IAA and alleviation of salt stress is more in *B. juncea* than *B. nigra*. In *B. oleracea* var. *botrytis* 98% germination was observed in the series where 0.5% NaCl was applied. However at 1% NaCl germination was 84% and at 2 and 3% germination was completly inhibited. When seeds were treated with 10 and 100 ppm GA_3 and IAA growth regulators for two hours and then left in NaCl solutions, maximum germination was observed in 0.5% NaCl +10 ppm IAA (96 %) and 0.5 %NaCl + 10 ppm GA_3 (96%). An improvement was observed at 10 ppm GA_3 in 1 percent NaCl solution. The length of radicle was between 32 to 44 mm and that of hypocotyle 13 to 15 mm in GA_3 and IAA. In 0.5% and 1% salt concentrations the length of radicle as well as hypocotyle decreased by 50 to 27%. But application of these in combination with the GA_3 and IAA alleviated the stress as compared to the salt alone. The maximum length of radicle (33.8 mm) and hypocotyle (15.7 mm) was observed in 0.5% NaCl + 100 ppm GA_3. In the case of IAA 0.5 % NaCl + 10 ppm IAA proved more effective as compared to other concentrations. In *B. oleracea* var. *oleracea* growth regulators were observed to alleviate the inhibitory effect of higher salt concentrations. In the control, 0.5, 1 and 1.5% NaCl solution germination was higher but decreased by 60-78% at 2% NaCl. However application of 10 ppm IAA and 10 ppm GA_3 resulted in 82-84% germination even at 2% NaCl, however, 100 ppm IAA and 100 ppm GA_3 resulted in a decrease in the germination at this level. 10 ppm of IAA and GA_3 proved more effective in alleviating the salt stres. The application of growth regulators improves the length of hypocotyl and radicle as compared to those in distill water. The slight increase in lengths observed at 0.5% NaCl as compared to the control. However, in 1 and 1.5% NaCl solutions the length of radicle and hypocotyl get highly decreased. This is due to their sensitiveness to the NaCl solution. An application of growth regulators like GA_3 and IAA depicted that 10 and 100 ppm of GA_3 alleviate the inhibitory effect of higher salt concentrations to some extent.

5. REFERENCES

Ahmet, M. 1968. Some aspects of the autecology of *Ranunculus arvensis*. Scientific Reports of the Faculty of Science, Ege University 62: 1-19.

Ahmet, M. 1969. Some autecological studies of *Ranunculus muricatus*. Scientific Reports of the Faculty of Science, Ege University 63: 1-13.

Ahmet, M. 1970. Ecology of *Ranunculus laetus*. Phyton (Austria)14: 1-8.

Ansari, R. 1972. Effect of salinity on some *Brassica* oilseed varieties. Pakistan Journal of Botany 4: 55-63.

Baskin, C.C. & Baskin, J.M. 1998. Seeds: ecology, biogeography, and evolution of dormancy and germination. San Diego, California: Academic Press. 666 pp.

Braun, J.W. & Khan, A.A. 1976. Alleviation of salinity and high temperature stress by plant growth regulators permeated into Lettuce seeds via acetone. Journal of American Society for Horticultural Science 101: 716-721.

Khan, A.A. 1980. The physiology and biochemistry of seed dormancy and germination. New York, New York: North Holland. 447 pp.

Khan, M.A. & M. I.Khan. 1978. Effect of light and temperature on seedlings raised under sodium chloride salinity. Pakistan Journal of Botany 10: 167-172.

Khan, M.A. & Ungar, I.A. 1998. Seed germination and dormancy of *Polygonum aviculare* L. As influenced by salinity, temperature, and gibberellic acid. Seed Science & Technology 26: 107-117.

Khan, M.A., Gul, B. & Weber, D.J. 2002. Improving seed germination of *Salicornia rubra*. (Chenopodiaceae) under saline conditions using germination regulating chemicals. Western North American Naturalist 62: 101-105.

Khan, M.A. & Gulzar, S. 2003. Light, salinity and temperature effects on the seed germination of perennial grasses. American Journal of Botany 90: 131-134.

Mayer, A.M. & Poljakoff-Mayber, A. 1982.The Germination of Seeds. (3rd Edn). New York: Pergamon Press. 211 pp.

Ozdemir, F., Ozturk, M., Guvensen, A. & Gemici, M. 1994. Investigations on the seed germination of some Brassicaceae members. National Biology Congress Proceedings, Edirne,Turkey 12: 49-52.

Ozturk, M. & Mert, H.H. 1983. Water relations and germination of seeds of *Inula graveolens* (L). Desf. Biotronics 12: 11-17.

Ozturk, M., Secmen, O. & Segawa, M. 1983. Ecological aspects of seed germination in *Myrtus communis* L. Memoirs of Faculty of Integrated Arts & Science, Hiroshima, Japan 8: 63-68.

Ozturk, M., Hinata, K., Tsunoda, S. & Gomez, C. 1983. A general account of the distribution of the cruciferous plants in Turkey. Ege University Science Faculty Journal 4: 87-98.

Ozturk, M. & Pirdal, M. 1986. Studies on the germination of *Asphodelus aestivus* Brot. Biotronics 15:55-60.

Ozturk, O., Gemici, M., Yilmazer, C. & Ozdemir, F. 1993. Alleviation of salinity stress by GA_3, KIN and IAA on seed germination of *Brassica campestris* L. Doga. Turkish Journal of Botany 17: 47-52.

Ozturk, M., Esiyok, D., Ozdemir, F., Olcay, G. & Oner, M. 1994. Studies on the effects of growth substances on the germination and seedling growth of *Brassica oleracea* L.var. *acephala* (Kara lahana). Ege University Science Faculty Journal 16: 63-70.

Ozturk, M., Ozdemir, F., Eser, B., Adiyahsi, O.I. & Ilbi, H. 1995. Studies on the salt-hormone interactions in the germination and seedling growth of some vegetable species. In: M.A. Khan.& I.A. Ungar (Eds.), Biology of Salt Tolerant Plants, Michigan: Book Crafters. 59-64 pp.

Sharma, P.C. & Kumar, P. 1999. Alleviation of salinity stress during germination in *Brassica juncea* by pre-sowing chilling treatments to seeds. Biologia Plantarum (Prague) 42: 451-455.

Sheikh, K.H., Ozturk, M. & Zeybek, N. 1976. Performance of *Inula graveolens* (L). Desf., in saline soils. In: Plant production under saline conditions, CENTO Symposium, Adana,Turkey. 154-164 pp.

Uygunlar, S., Yazgan, M. & Ozturk, M. 1985. Role of water in seed germination. Turkish Journal of Botany 9/3: 620-630.

Vardar, Y. & Ozturk, M. 1971. Water relations of *Myrtus communis* seeds. Verhandlungen der Schweizerischen Naturforschenden Gesellschaft, Switzerland. 70-75 pp.

Vidaver, W. 1980. Light and seed germination. In: A.A. Khan (Ed.), The physiology and biochemistry of seed dormancy and germination. Amsterdam, Netherlands: North Holland 181-192. pp.

CHAPTER 11

SALINE TOLERANCE PHYSIOLOGY IN GRASSES

KENNETH B. MARCUM

Department of Applied Biological Sciences
Arizona State University

1. INTRODUCTION

Salinization of agricultural lands is accelerating, with over 1 Mha of irrigated lands deteriorating to non-productivity each year (Hamdy, 1996; Choukr-Allah, 1996). Currently from 100 Mha to 1000 Mha of irrigated land is salt-affected due to human activity (Szabolcs, 1989; Oldeman et al., 1991). Though much of this land is currently too saline for conventional agriculture, it has the potential for growing salt tolerant forages, grasses (Poaceae) playing a dominant role (Ghassemi & Jakeman, 1995).

With over 7,500 species, the Poaceae inhabit the earth in greater numbers, and have a greater range of Chlorideimatic adaptation than any other plant family (Hitchcock, 1971; Gould & Shaw, 1983). Therefore, it is not surprising that grasses show an extreme range in salinity tolerance, from salt-sensitive (ex. meadow foxtail *Alopecurus pratensis* L.), to salt-tolerant halophytic (ex. saltgrass *Distichlis spicata* L.) (Richards, 1954; Maas, 1986; Aronson, 1989).

In this paper growth responses and physiological adaptations to salinity of eight C_4 grass species studied in my lab will be discussed, representing an extreme range of tolerance. Physiological mechanisms of salt tolerance will be discussed, and cross-referenced to salinity studies involving other grass species. The grasses, listed in (Table 1), will be indicated in this paper by genus, except for *Sporobolus*, where genus abbreviation is followed by species names.

1.1. Growth responses to salinity and relative salinity tolerance

1.1.1. Shoot Growth Responses

Plant salinity tolerance depends not only on genotype, but also on environmental and

M.A. Khan and D.J. Weber (eds.), Ecophysiology of High Salinity Tolerant Plants, 157-172.

cultural conditions. Therefore, absolute salinity tolerance cannot be determined with certainty, but rather on a relative basis (to other genotypes), given uniform growing conditions (Maas & Hoffman, 1977; Maas, 1986). Growth indicators used in these studies (shoot weight, % canopy leaf firing, rooting depth, and root weight) were highly correlated with one another (r^2 ranging from 0.65 to 0.8), indicating their mutual effectiveness in predicting relative salinity tolerance.

Table 1. *Grasses studied.*

Scientific Name	*Common Name*
Bouteloua curtipendula (Michx.) Torr.	Sideoats grama
Buchlon dactyloides (Nutt.) Engelm.	Buffalograss
Cynodon dactylon (L.) Pers.	Bermudagrass
Distichlis spicata var. *stricta* (Torr.) Beetle	Desert saltgrass
Sporobolus airoides (Torr.) Torr.	Alkali sacaton
Sporobolus cryptandrus (Torr.) Torr.	Sand dropseed
Sporobolus virginicus (L.) kunth	Seashore dropseed
Zoysia japonica Steud.	Japanese lawngrass

Relative salinity tolerance is often quantified as the salt level resulting in a 50% reduction in shoot growth (yield), or alternatively, the threshold salinity, i.e. salinity level where yield begins to decline, followed by the rate, or slope, of yield reduction (Maas & Hoffman, 1977; Carrow & Duncan, 1998). Fifty percent shoot growth reduction occurred at media salinities ranging from 140 mM (approximately 11 dS m^{-1}) for *Bouteloua*, to >600 mM (>46 dS m^{-1}) for *Distichlis* (Figure 1). Using this as criteria, salinity tolerance decreased in the order: *Distichlis* = S. v*irginicus* > *S. airoides* > *Cynodon* > *Zoysia* > *S. cryptandrus* > *Bouteloua* = *Buchlon*. Reid et al. (1993) also reported 50% shoot growth decline at 12 dS m^{-1} for three *Buchlon* cultivars. Data for *Zoysia* reveals a high genetic diversity, with 50% shoot growth reduction occurring from 170 to 375 mM Na Chloride, depending on cultivar or accession (Marcum & Murdoch, 1994; Marcum et al., 1998). Genetic diversity is also seen within the *Cynodon* genus (de Wet & Harlan, 1970). Fifty percent shoot growth reductions for bermudagrass cultivars and/or accessions has been reported as 24 and 33 dS m^{-1} (Dudeck & Peacock, 1993),2 4 and 31 dS m^{-1} (Francois, 1988), and 17 to 22 dS m^{-1} (Dudeck et al., 1983). The halophytic nature of *S. airoides*, *S. virginicus*, and *Distichlis* has been reported (Butler et al., 1974; Maas & Hoffman, 1977; Aronson, 1989; Marcum & Murdoch, 1992). In several studies, shoot growth of *Distichlis* was not affected by salinities up to 40 dS m^{-1} (Parrondo, 1978; Kemp et al., 1981).

Salt-sensitive plants (glycophytes) and moderately salt-tolerant plants (mesophytes) generally have a flat yield response to salinity prior to a threshold salinity level, beyond which shoot growth declines. In contrast, highly salt-tolerant plants often display stimulated shoot, and root growth at moderate salinity levels, followed by yield de cline (Maas & Hoffman, 1977; Maas, 1986; Carrow et al., 1998). Increased shoot growth (relative to control) under moderate salinity (100 mM

Na Chloride, or 8 dSm^{-1}) was evident in *Distichlis*, *S. airoides* and *S. virginicus* (Figure 1). All other grasses displayed progressive shoot growth reductions at all salinity levels. Salt-stimulated shoot growth has been observed in other salt tolerant and halophytic grasses. Shoot growth peaked at 90 mM Na Chloride (8 dSm^{-1}), then declined in *Halopyrum mucronatum* (L.) Stapf, a perennial grass found on coastal dunes of Pakistan (Khan et al., 1999). Shoot growth was stimulated with increasing salinity up to 25 mM Na Chloride (2.5 dSm^{-1}), then declined, in 2 of 6 *Sporobolus* species studied (*S. stapfianus* and *S. pellucidus*) (Wood & Gaff, 1989).

1.1.2. Root Growth Responses

Root growth stimulation (increased root mass, rooting depth, or both) in salt tolerant grasses is typically a more common, accentuated response to moderate salinity stress than shoot growth stimulation (Maas & Hoffman, 1977). The net result is generally an increase in root/shoot ratios, which may be a salinity tolerance mechanism to counter low external water potential by increasing plant absorptive area (Bernstein & Hayward, 1958; Donovan & Gallagher, 1985). Increased rooting depth, relative to control plants, was observed in *Distichlis*, *S. airoides*, *S. virginicus,* and *Cynodon* under salinity stress (Figure 1). However, relative rooting depth de Chlorideined at high salinity for *Cynodon*, but not in the halophytic grasses. In contrast, rooting depth of *Buchlon*, *Bouteloua*, and *S. cryptandrus* progressively declined with increasing salinity stress.

Root stimulation has been observed in a number of salt tolerant and halophytic grasses. Root dry weights linearly increased with increasing salinity up to 450 mM Na Chloride (35 $dS\ m^{-1}$) in *S. virginicus*, resulting in a root/shoot ratio of 2.2, relative to 0.5 (control) (Marcum, 1992). Blits and Gallagher (1991) reported a doubling in root mass of *S. virginicus* grown in seawater, relative to fresh water. Though root growth (length) increased under moderate salinity stress, relative to control, shoot growth de clined in *Chloris gayana* L. (Waisel, 1985), *Cynodon* (Ackerson & Youngner, 1975), and *Zoysia* (*Zoysia japonica* Steud. and *Z. matrella* [L.] Merr.) (Marcum & Murdoch, 1990). Rooting decline under even mild salinity stress has been previously reported in *Buchlon* (Wu & Lin, 1993), and in other moderate to salt-sensitive grasses, such as *Poa pratensis* L. (Torello & Symington, 1984), *Paspalum notatum* Flugge (Dudeck & Peacock, 1993), and *Festuca rubra* L. (Khan & Marshall, 1981). Total root dry weight (data not shown) was highly correlated with rooting depth (r=0.83).

1.2. Physiological adaptations to salinity

1.2.1. Ion Exclusion

It has long been accepted that the major causes of plant growth inhibition under salinity stress are osmotic stress (osmotic inhibition of plant water absorption), and specific ion effects, including toxicities and imbalances (Bernstein & Hayward, 1958; Greenway et al., 1966; O'Leary, 1971). In comparison to salt tolerant, or halophytic dicotyledonous

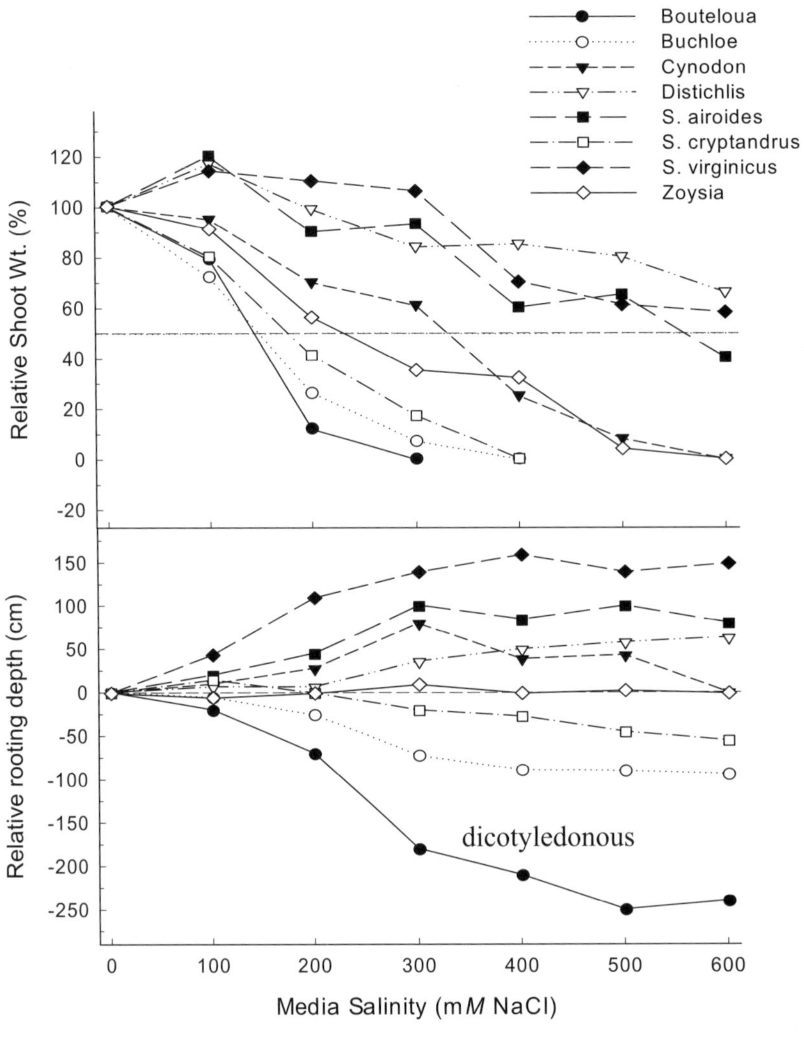

Figure 1. *Relative shoot dry weight [(treatment wt./control wt.) X 100] and relative rooting depth (salinity treatment length minus control length) of grasses exposed to increasing salinity levels in solution culture. Vertical bars represent LSD ($P < 0.05$) values for mean comparison at each salinity level.*

plants, monocots (including Poaceae) tend to exclusion saline ions from shoots, thereby minimizing toxic effects (Albert & Popp, 1977; Gorham et al., 1985, 1993). Saline ion exclusion from shoots was strongly associated with salinity tolerance among these eight grasses representing the range of salinity tolerance present in the Poaceae (Figure 2). Sodium shoot ion content mirrored that of Chloride, and is not shown. Chloride and Na^+ accumulated to extremely high levels in *Bouteloua* shoots, and high levels in *Buchlon* and *S. cryptandrus* shoots, but was maintained at concentrations similar to the growth media in *Cynodon* and halophytic *Distichlis*, *S. virginicus*, and *S. airoides* shoots, particularly at high salinity. Salinity tolerance of other grasses has been related to saline ion exclusion. Salinity tolerance in *Sorghum halepense* (L.) Pers., relative to *Sorghum bicolor* (L.) Moench was associated with shoot Cl^- concentration (Yang et al., 1990). Similarly, salt-tolerant *Agropyron elongatum* (Host) Palisot de Beauvois accessions ex Chlorideuded Na^+ and Cl^- from shoots (while maintaining fairly high K^+ contents) to a greater extent than salt-sensitive *Agropyron desertorum* (Fisch. ex Link) Schult. accessions (Johnson, 1991). In contrast, salt-tolerant *Puccinellia distans* (L.) Parl and *P. lemmoni* (Vasey) Scribn. were found to accumulate more Na^+ and Cl^- in shoots than did moderately salt-tolerant *Agrostis stolonifera* L. (Harivandi & Butler, 1992).

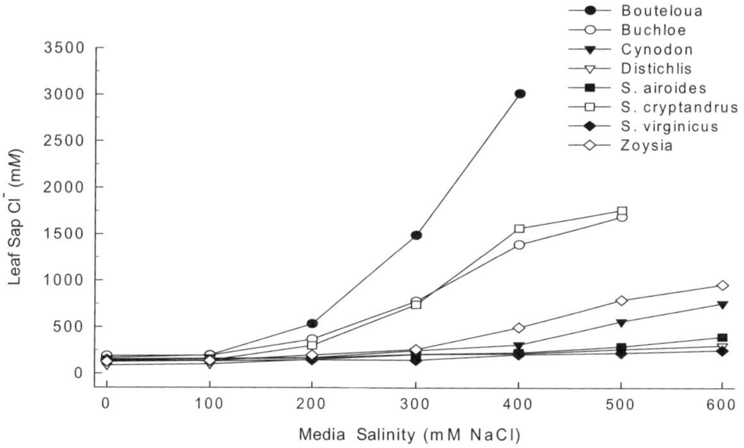

Figure 2. *Leaf sap Chloride⁻ levels of grasses exposed to increasing salinity levels in solution culture.*

Saline ion exclusion also appears to be an important factor influencing intraspecies salinity tolerance, i.e. at the cultivar or accession level. For example, salt-sensitive populations were found having, at a given test salinity, higher shoot Na^+ and Chloride than coastal (or other saline-site) salt-tolerant accessions in *Festuca rubra* L. (Hannon & Barber, 1972; Khan & Marshall, 1981), *Cynodon* (Ramakrishnan & Nagpal, 1972), and *Agrostis stolonifera* L. (Wu, 1981). Relative salinity tolerance of *Zoysia* cultivars and accessions have successfully been predicted on the basis of shoot Na^+ concentrations occurring under salt stress (Marcum et al., 1998; Marcum, 2003).

1.2.2. Osmotic Adjustment and Ion Regulation

Osmotic stress due to lack of osmotic adjustment, resulting in reduced water absorption and physiological drought, has long been considered a major cause of salinity injury in plants (Bernstein & Hayward, 1958; Levitt, 1980; Harivandi et al., 1992). Maintenance of cell turgor and plant growth requires sufficient increase in sap osmolality to compensate for external osmotic stress, the process of osmoregulation, or osmotic adjustment (Hellebust, 1976; Levitt, 1980). In a saline environment, osmotic adjustment is needed to avoid osmotic stress, yet this may result in ion toxicity (Yeo, 1983; Gorham et al., 1985).

It has been noted that monocots (relative to salt-tolerant dicots), including Poaceae, tend to restrict saline ion uptake. This has been suggested to cause cell dehydration and reduced growth under saline conditions, due to lack of osmotic adjustment (Albert & Popp, 1977; Gorham et al., 1980; Gorham, 1985). Indeed, declining shoot water content is commonly observed in grasses under salinity stress (Greenway et al., 1966; Greenway & Munns, 1980; Weimberg & Shannon, 1988; Marcum & Murdoch, 1990), though slight increase in shoot succulence under moderate salinity has been noted in some grass halophytes (Blits & Gallagher, 1991; Marcum & Murdoch, 1992; Khan et al., 1999). However, complete osmotic adjustment occurred in all eight grasses, sap osmolalities being maintained below (more negative than) media osmolality (Figure 3). In fact, salt-sensitive grasses osmotically adjusted to a much greater degree than salt-tolerant ones. Among the eight grasses, shoot sap osmolality was highly negatively correlated with salinity tolerance and root growth under salt stress ($r > -0.8$). Complete osmotic adjustment under salinity stress has been reported previously in a range of grasses (Peacock & Dudeck, 1985; Wyn Jones & Gorham, 1989; Marcum & Murdoch, 1990). In these studies, shoot sap osmolality level was negatively correlated with salinity tolerance. In other words, in salt tolerant grasses, osmotic adjustment, though complete, is nevertheless minimized, i.e. shoot sap osmolality is maintained Chlorideose to saline media levels.

Though salinity tolerance in grasses is clearly associated with saline ion exclusion, Na^+ and Chloride$^-$ have been instrumental for shoot osmotic adjustment in a number of studies, comprising the majority of osmotically active solutes (Marcum & Murdoch, 1990; Warwick & Halloran, 1991; Marcum & Murdoch, 1992; Khan et. al., 1999). Among these eight grasses, shoot Na^+ and Cl^- concentrations were highly correlated with osmotic adjustment ($r=0.9$). Therefore, though saline ion exclusion is clearly critical for salinity tolerance in grasses, saline ion regulation, rather than exclusion, may be a more apt description of the salinity tolerance mechanism operating in grasses.

Saline ion regulation in grasses may occur in several ways. Selectivity for K^+ over Na^+ may occur by selective K^+ absorption-vacuolar Na^+ compartmentation in root cortical cells or endodermis, or by selective saline ion excretion through specialized salt glands or bladders (Levitt, 1980; Kramer, 1984; Jeschke, 1984; Daines & Gould, 1985; Garbarino & Dupont, 1988). In glycophytic grasses, tissue Na^+ may be reabsorbed from the xylem via mature xylem parenchyma cells in roots or shoots, and translocated back to soil (Yeo et al., 1977; Jeschke, 1979; Taleisnik, 1989). Alternately, ion partitioning may occur, whereby saline ions are redistributed to mature, senescing leaves or other organs (Lessani & Marschner, 1978; Yeo & Flowers, 1984; Bhatti et al., 1993; Jeschke et al., 1995).

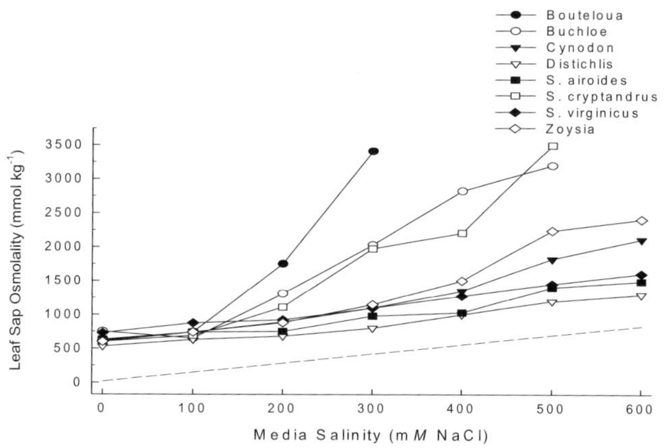

Figure 3. *Leaf sap osmolality of grasses exposed to increasing salinity levels in solution culture.*

1.2.3. Glandular Ion Excretion

Salt glands or bladders are present in a number of salt-adapted species, which eliminate excess saline ions from shoots by excretion (Waisel, 1972; Liphschitz & Waisel, 1982; Fahn, 1988). Multicellular epidermal salt glands are present in several families of dicotyledons, e.g. Frankeniaceae, Plumbaginaceae, Aviceniaceae, and Tamaricaceae (Waisel, 1972; Fahn, 1988). Within the Poaceae, bicellular epidermal salt glands have been reported to occur in over 30 species within the tribes Chlorideae, Eragrosteae, Aeluropodeae, and Pappophoreae (Liphshchitz & Waisel, 1974; Taleisnik & Anton, 1988; Amarasinghe & Watson, 1989), members of the subfamily Chloridoideae (Gould & Shaw, 1983; Chlorideayton & Renvoize, 1986).

Salt glands of the Poaceae are, in outward appearance, similar to leaf epidermal bicellular microhairs. Though microhairs resembling salt glands have been observed in all grass subfamilies except Pooideae (Liphschitz & Waisel, 1982; Amarasinghe & Watson, 1988), functioning salt glands have been found only within the subfamily Chloridoideae (Amarasinghe & Watson, 1988; Amarasinghe & Watson, 1989). This is probably due to an ultrastructural modification hypothesized to be responsible for salt excretion in Chloridoid grasses: a series of parallel, invaginated plasma membrane channels within the gland's basal cell (Liphschitz & Waisel, 1982; Oross & Thomson, 1982a; Oross & Thomson, 1982b). These membranes are actually infoldings of the plasmalemma that originate adjacent to the wall separating the cap and basal cells, forming open channels in the direction of ion flow. Ultracytochemical localization of ATPase activity within salt gland basal cells of *S. virginicus* supports the hypothesis of active ion loading at these sites (Naidoo & Naidoo, 1999). In addition, there are numerous mitochondria associated with the parallel membranes, probably involved in providing an energy supply for channel ion loading (Levering & Thomson, 1971; Naidoo & Naidoo, 1999).

Salt glands in the Poaceae, structurally distinct from the multicellular glands of dicots, consist of a basal cell, attached, or imbedded, into the leaf epidermis, and a cap cell (73, 74) (Figure 5A). The glands are characterized by cutinized cell walls, and are often surrounded by papillae. Though the basic, bicellular structure is the same in all Chloridoid species, their appearance varies (Liphschitz & Waisel, 1982) (Figure 5). In some species, glands are sunken into the epidermis, with the basal cell totally imbedded, ex. *Distichlis* (Figure 5H). In others, the basal cell is semi-imbedded, ex. *Cynodon* (Figure 5G). Finally, the basal cell may extend out from the epidermis, with the gland lying recumbent to the leaf surface, ex. *Bouteloua* (Figure 5B). Salt glands of Poaceae are quite small (usually 25-70 μm in length), though size may vary substantially, from imbedded to elongated, protruding types. Glands range in size from 15 μm in length in *Distichlis* (Marcum, 1990), 35 μm in Zoysia (Marcum & Murdoch, 1996), to 70 μm in *Buchlon* (Marcum, 1999) (Figure 5). Salt glands have been found on both abaxial and adaxial leaf surfaces of excreting species (Marcum & Murdoch, 1996; Liphshchitz & Waisel, 1974; Marcum, 1999). Glands are longitudinally arranged in parallel rows atop intercostal regions of leaves, adjacent to rows of stomates (Figure 5A).

Evidence that salt gland ion excretion is an active, metabolically driven process is varied, including effects of temperature (Pollak & Waisel, 1979), light (Pollak & Waisel, 1970), oxygen pressure (Liphschitz & Waisel, 1982), and metabolic inhibitors (84) on excretion rate, as well as selectivity of ion excretion. Excretion is typically highly selective for Na^+ and Cl^- (Wieneke et al., 1987; Arriaga, 1992; Worku & Chapman, 1998), though other ions may be excreted in minute amounts, such as K^+, Ca^{2+}, and Mg^{2+} (Liphschitz & Waisel, 1982; Marcum & Murdoch, 1990, 1994; Naidoo & Naidoo, 1998). Comparison of salt gland excretion rates among studies is difficult, due to the varying influence of environmental factors, such as light and temperature, cumulative days of exposure to salt stress, and plant factors, such as leaf age (Jeschke et al., 1995). Also, units of measurement differ, one fundamental difference being whether excretion rates are based on leaf area or leaf weight. Finally, excretion rate is not static, but is influenced by saline ion concentrations in the growing media. Increasing media salinity generally stimulates excretion up to an optimal level, above which excretion rate may decline (Liphschitz & Waisel, 1982). Maximum excretion rate was reported to occur at 150 to 200 mM Na Chloride (8-13 dSm^{-1}) in moderately tolerant Chloridoid species such as *Cynodon*, *Rhodesgrass*, goosegrass [*Eleusine indica* (L.) Gaertn.], and Kallargrass (Wieneke, 1987; Liphshchitz & Waisel, 1994; Worku & Chapman, 1998). However, excretion was maximal at 200 mM Na Chloride (17 dSm^{-1}) in *Distichlis* and *Spartina* spp. (Liphschitz & Waisel, 1982), and 300 mM Na Chloride (23 dSm^{-1}) in *S. virginicus* (Marcum & Murdoch, 1992).

Among the C_4 grasses reported here, shoot Na^+ and $Chloride^-$ concentrations were negatively correlated, while salt tolerance was positively correlated with salt gland Na^+ and Cl^- excretion rates. (Table 2) shows ion excretion rates for the eight grasses. Note that *S. virginicus* had Na^+ and Cl^- excretion rates 35 and 38 times higher, respectively, than *Buchlon*. Similar strong correlations between salt gland excretion rates, shoot Na^+ and Cl^- concentrations, and salinity tolerance were observed among three C_4 grasses in another study (Marcum & Murdoch, 1994).

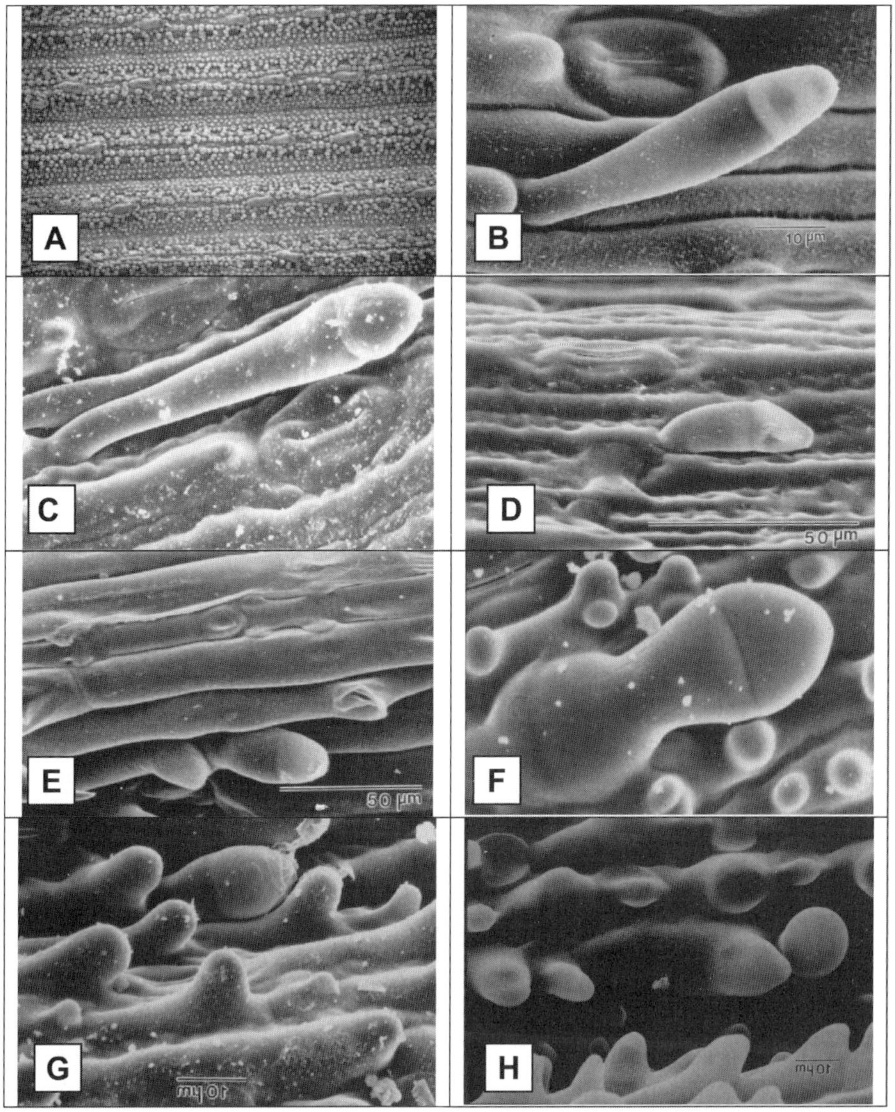

Figure 4. Scanning electron micrographs of adaxial leaf surfaces. A: Overview of Zoysia (Zoysia matrella *(L.) Merr.) leaf surface, showing location of salt gland relative to other structures. B: Salt gland of* Bouteloua curtipendula. *C: Salt gland of* Buchlon dactyloides. *D: Salt gland of* Sporobolus cryptandrus. *E: Salt gland of* Sporobolus airoides. *F: Salt gland of* Zoysia japonica. *G: Salt gland of* Cynodon dactylon. *H: Salt gland of* Distichlis spicata.
Key to photo labels: B = basal cell, C = cap cell, G = salt gland, I = Intercostal zone of leaf epidermis, P = papilla, S = stomate/stomata.

Sodium and Chloride$^-$ excretion rates were negatively correlated to shoot concentrations, but positively correlated to leaf salt gland density and salinity tolerance among 57 *Zoysia* grasses species (Marcum & Murdoch, 1990; Marcum et al., 1998). Excretions rates of various *Zoysia* spp. reported range from 130 μmol Na^+/g leaf dry wt./week in salt-sensitive *Zoysia japonica* to 730 μmol Na^+/g leaf dry wt./week in salt-tolerant *Zoysia matrella*, with gland densities ranging from 28/mm^2 leaf surface in salt-sensitive *Zoysia japonica* to 100/mm^2 in salt-tolerant *Zoysia macrostachya* Franch. & Sav.

Table 2. *Leaf salt gland Chloride⁻ and Na^+ excretion rates[a] of three* Chloridoid grasses. *Ion excretion measured in plants exposed to 200 mM Na Chloride.*

Grass	*Chloride⁻*	*Na^+*
S. virginicus	2104	2540
Distichlis	1268	1130
S. airoides	563	565
Zoysia	423	130
Cynodon	394	87
S. cryptandrus	85	65
Bouteloua	56	65
Buchlon	42	51
$LSD_{0.05}$	56	72

[a]*Excretion rates in μmol ion/g leaf dry wt./week.*

1.2.4. Ion Compartmentation and Compatible Solutes

In vitro studies have shown that enzymes of both glycophytes and halophytes have similar sensitivities to salt, being inhibited at concentrations above 100-200 mM (approximately 8-17 dSm^{-1}) (Wyn Jones et al., 1979; Greenway & Munns, 1980). Therefore, salt-tolerant plants growing under saline conditions must restrict the level of ions in the cytoplasm. As data above has illustrated, salt tolerant grasses utilize inorganic ions for a large part of their osmotic adjustment under saline growing conditions, as the ability to accumulate organic solutes on a whole cell basis is metabolically expensive, and therefore limited (Levitt, 1980; Kramer,1984). Salt tolerant plants that successfully accumulate saline ions for osmotic adjustment above concentrations of 100-200 mM do so by compartmentalizing them within the vacuole, which typically makes up 90 to 95% of mature plant cell volume (Flowers, 1985). Evidence exists for salinity inducing a K^+/Na^+ exchange across the tonoplast mediated by Na^+/H^+ antiport activity, resulting in saline ion compartmentation in vacuoles (Jeschke, 1984; Garbarino & Dupont, 1988). Under these conditions, osmotic potential of the cytoplasm is maintained by the accumulation of organic solutes that are compatible with enzyme activity, termed "compatible solutes" (Wyn Jones & Gorham, 1983; Wyn Jones, 1984). Under highly saline conditions, relatively few organic solutes, including glycinebetaine, proline, and certain polyols and cyclitols, can be accumulated in sufficient concentrations to osmotically adjust the cytoplasm without inhibiting enzymes (Gorham, 1996). Evidence exists for the cytoplasmic localization of these compounds (Leigh et al., 1981; Aspinall & Paleg, 1981; Wyn Jones, 1984). Of these, glycinebetaine and proline typically accumulate in grasses (Rhodes & Hanson, 1993).

Total leaf Na^+ + $Chloride^-$ levels exceeded 200 mM in all three Chloridoid grasses grown under moderate to high salinity (Figure 2), necessitating vacuolar ion compartmentation for survival. Glycinebetaine levels increased under salinity in all grasses, reaching highest levels (62 mM) in *Distichlis* (Table 3).

Table 3. *Leaf sap glycinebetaine and proline levels (mM) of grasses exposed to 0 and 300 mM Na Chloride.*

	Glycinebetaine		*Proline*	
Grass	*0 mM*	*300 mM*	*0 mM*	*300 mM*
Distichlis	12	62	1	2
S. virginicus	36	60	1	2
Zoysia	51	13	7	1
S. airoides	50	14	1	1
Cynodon	38	6	3	1
S. cryptandrus	36	13	3	1
Buchlon	20	8	6	2
Bouteloua	12	4	6	1
$LSD_{0.05}$	8	5	2	N.S.

Though proline concentrations also increased under salinity, maximum levels occurred in salt-sensitive BuchloN, reaching only 6 mM. Assuming that glycinebetaine and proline are located in the cytoplasm (see above), which occupies 10% of total cell volume, the contributions of glycinebetaine and proline to cytoplasmic osmotic adjustment can be calculated (Table 4). Glycinebetaine made substantial contributions to cytoplasmic osmotic adjustment in salt tolerant grasses only. In contrast, proline contributions were insignificant in all grasses.

Table 4.*Estimated[a] contribution to cytoplasmic osmotic adjustment of glycinebetaine and proline, in mosmol* kg^{-1} *(Osml), and as a percentage (%) of total osmolality, of plants grown at 300 mM Na Chloride.*

	Glycinebetaine		*Proline*	
Grass	*Osml*	*%*	*Osml*	*%*
S. virginicus	822	80	31	3
Distichlis	625	74	17	2
S. airoides	501	53	7	1
Zoysia	513	47	67	6
Cynodon	378	39	27	3
S. cryptandrus	357	18	29	1
Buchlon	209	10	59	3
Bouteloua	125	4	56	2

[a]*Estimate assumes glycinebetaine and proline are located in the cytoplasm, comprising 10 percent of total cell volume, with an osmotic coefficient of 1.0 for each compound.*

In another study, glycinebetaine made substantial contributions to cytoplasmic osmotic adjustment in 5 of 6 grasses in the study, the exception being salt-sensitive *Eremochloa ophiuroides* (Munro) Hack. The 5 salt tolerant grasses included *Cynodon*, *Zoysia*, and *Paspalum vaginatum* Swartz. As above, proline contributions were too small to contribute to cytoplasmic osmotic adjustment (Marcum & Murdoch, 1994). Other studies involving *S. virginicus* support the importance of glycinebetaine as a compatible solute, relative to proline. Quaternary ammonium compounds (predominately glycinebetaine, and possibly other related betaines) accumulated to 48 μmol g^{-1} dry weight in shoots of *S. virginicus* grown in seawater, while proline levels reached only 1.6 μmol g^{-1} dry weight (Blits & Gallagher, 1991). Similarly, proline levels were insufficient to contribute significantly to cytoplasmic adjustment of *S. virginicus* grown in 80% seawater (Naidoo & Naidoo, 1998). In lines of tall wheatgrass grown under 20 dS m^{-1} total salinity, glycinebetaine accumulated to 45 μmol g^{-1} fresh weight in shoots, compared to only 1 μmol for proline (Weimberg & Shannon, 1988).

While glycinebetaine concentrations under salinity were positively correlated ($r^2 = 0.6$) with salinity tolerance among these eight grasses, proline concentrations were negatively correlated ($r^2 = -0.72$), suggesting that glycinebetaine, but not proline, acts as a compatible solute. Though both compounds have traditionally been considered compatible solutes, recent evidence has favored the role of glycinebetaine. For example, (i) glycinebetaine is excluded from the hydration sphere of enzyme proteins and thus tends to stabilize their tertiary structure (Yancy, 1994), (ii) corn (*Zea mays* L.) mutants lacking a critical enzyme for glycinebetaine biosynthesis also lack salt tolerance (Saneoka et al., 1995), and (iii) exogenously applied glycinebetaine has enhanced the salinity tolerance of glycophytes such as rice (*Oryza sativa* L.) (Harinasut et al., 1995). In contrast, proline accumulation has recently been considered by some investigators merely a result of plant injury, due to a universally rapid appearance following any type of stress (Colmer et al., 1995; Mumtaz et al., 1995).

2. SUMMARY

The Poaceae, represented by over 7,500 species, show extreme range in salinity tolerance, from salt-sensitive to extremely salt-tolerant (halophytic). In this chapter the range of salinity tolerance, and physiological adaptations to salinity present in grasses was described, focusing on eight grass species representing the range of salt tolerance present in the Poaceae. Salinity tolerance, indicated by 50% growth reduction, ranged from 150 mM in *Bouteloua* to >600 mM (seawater is approx. 550 mM) in *Distichlis* and *S. virginicus*. Though shoot growth decline with increasing salinity is typical, shoot growth may be stimulated by moderate salinity in highly salt-tolerant or halophytic grasses. However, root growth stimulation under moderate salinity is much more common in salt-tolerant grasses, resulting in increased root/shoot ratios, and therefore increased water absorption/transpiration area, which may be an adaptive mechanism to saline osmotic stress.

It has long been accepted that the major causes of plant growth inhibition under salinity stress are osmotic stress (osmotic inhibition of plant water absorption), and specific ion effects, including toxicities and imbalances. In a number of studies salinity tolerance in the Poaceae has been related to shoot saline ion exclusion. However, studies have shown that complete osmotic adjustment does occur under salt stress, even in salt-sensitive grasses. Since the predominant osmotica utilized are typically saline ions, ion regulation, rather than ion exclusion, may be a more apt description of the mechanism of salt tolerance occurring in the Poaceae. Grasses regulate saline ion concentrations by vacuolar ion compartmentation at the root or shoot or by excretion via specialized salt glands, though ion reabsorption by xylem/phloem and redistribution to roots or senescing leaves may play a minor role.

Bicellular leaf epidermal salt glands occur in a number of C_4 grasses. Basal cells have specific ultrastructural modifications, including parallel partitioning membranes, allowing active, selective saline ion excretion. Excretion rates, which may be substantial, are dependent on media salinity level, and are typically highly selective for Na^+ and Cl^-. More recently, salinity tolerance of grasses has been related to salt gland excretion rate and leaf salt gland density.

Enzymes of higher plants, salt-sensitive and tolerant alike, are inhibited by saline ion concentrations above 100-200 mM. Under salt stress, grasses typically accumulate saline ions to well above these levels for shoot osmotic adjustment, necessitating Na^+ and Cl^- compartmentation in vacuoles, which comprise 90-95% of mature cell volume. Remaining cytoplasmic osmotic adjustment is achieved by certain organic osmotica compatible with cell enzymes, termed "compatible solutes". Glycinebetaine and proline typically accumulate in salt-stressed grasses, and have been proposed as compatible solutes. However, recent evidence has supported glycinebetaine, not proline, as a functional compatible solute.

3. REFERENCES

Ackerson, R.C. & Youngner, V.B. 1975. Responses of bermudagrass to salinity. Agronomy Journal 67: 678-681.

Albert, R. & Popp, M. 1977. Chemical composition of halopytes from the Neusiedler Lake region in Austria. Oecologia 27: 157-170.

Amarasinghe, V. & Watson, L. 1988. Comparative ultrastructure of microhairs in grasses. Botanical Journal Linnean Society 98: 303-319.

Amarasinghe, V. & Watson, L. 1989. Variation in salt secretory activity of microhairs in grasses. Australian Journal Plant Physiology 16: 219-229.

Aronson, J.A. 1989. Haloph: A Data Base of Salt Tolerant Plants of the World. Tucson, Arizona: Office of Arid Land Studies, University of Arizona. 68-70 pp.

Arriaga, M.O. 1992. Salt glands in flowering culms of Eriochloa species (Poaceae). Bothalia 22: 111-117.

Aspinall, D. & Paleg, L.G. 1981. Proline accumulation: Physiological aspects. In: L.G Paleg & D. Aspinall (Eds.), Physiology and Biochemistry of Drought Resistance in Plants. Sydney, Australia: Academic Press. 205-241 pp.

Bernstein, L. & Hayward, H.E. 1958. Physiology of salt tolerance. Annual Review Plant Physiology 9: 25-46.

Bhatti, A.S., Steinert, S., Sarwar, G., Hilpert, A. & Jeschke, W.D. 1993. Ion distribution in relation to leaf age in *Leptochloa fusca* (L.) Kunth. (*Kallar grass*). I.K, Na, Ca and Mg. New Phytologist 123: 539-545.

Blits, K.C. & Gallagher, J.L. 1991. Morphological and physiological responses to increased salinity in marsh and dune ecotypes of *Sporoblus virginicus* (L.) Kunth. Oecologia 87: 330-335.

Butler, J.D., Fults, J.L., & Sanks, G.D. 1974. Review of grasses for saline and alkali areas. International Turfgrass Research Journal 2: 551-556.

Carrow, R.N. & Duncan, R.R. 1998. Salt-Affected Turfgrass Sites – Assessment and Management. Chelsea, Michigan:Ann Arbor Press. 83-99 pp.

Choukr-Allah, R. 1996. The potential of halophytes in the development and rehabilitation of arid and semi-arid zones. In: R. Choukr-Allah, C.V. Malcolm, & A. Hamdy, (Eds.), Halophytes and Biosaline Agriculture. New York, New York:. Marcel Dekker. 3-13 pp.

Chlorideayton, W.D. & Renvoize, S.A. 1986. Genera Graminum, Grasses of the World. London, Great Britian: HMSO Books.

Colmer, T.D., Epstein, E. & Dvorak, J. 1995. Differential solute regulation in leaf blades of various ages in salt-sensitive wheat and a salt-tolerant wheat X *Lophopyrum elongatum* (Host) A. Löve amphiploid. Plant Physiology 108: 1715-1724.

Daines, R.J. & Gould, A.R. 1985. The cellular basis of salt tolerance studied with tissue cultures of the halophytic grass *Distichlis spicata*. Journal of Plant Physiology 119: 269-280.

Dudeck, A.E., Singh, S., Giordano, C.E., Nell, T.A. & McConnell, D.B. 1983. Effects of sodium chloride on *Cynodon* turfgrasses. Agronomy Journal 75: 927-930.

Dudeck, A.E. & Peacock, C.H. 1993. Salinity effects on growth and nutrient uptake of selected warm-season turf. International Turfgrass Society Research Journal 7: 680-686

Donovan, L.A. & Gallagher, J.L. 1985. Morphological responses of a marsh grass, *Sporobolus virginicus* (L.) Kunth., to saline and anaerobic stresses. Wetlands 5: 1-13.

Fahn, A. 1988. Secretory tissues in vascular plants. New Phytologist 108: 229-257.

Flowers. T.J. 1985. Physiology of halophytes. Plant Soil 89: 41-56.

Francois, L.E. 1988. Salinity effects on three turf bermudagrasses. HortScience 23: 706-708.

Garbarino, J. & Dupont, F.M. 1988. Na Chloride induces a salinity induces K^+/Na^+ antiport in tonoplast vesi Chloridees from barley roots. Plant Physiology 86: 231-236.

Ghassemi, F., Jakeman, A.J. & Nix, H.A. 1995. Salinisation of Land and and Water Resources. Wallingford Oxon, UK: CAB International. 291-335 pp.

Gorham, J., Wyn Jones, R.G. & McDonnell, E. 1985. Some mechanisms of salt tolerance in crop plants. Plant Soil 89: 15-40.

Gorham, J., Hughes, L.L., & Wyn Jones, R.G. 1980. Chemical composition of salt-marsh plants from Ynys-Mon (Anglesey): the concept of physiotypes. Plant, Cell and Environment, 3: 309-318.

Gorham, J., Randall, P.J., Delhaize, E., Richards, R.A., & Munns, R. 1993. Genetics and physiology of enhanced K/Na discrimination. Genetic Aspects of Plant Mineral Nutrition: Developments in Plant and Soil Sciences, 50: 151-158.

Gorham, J. 1996. Mechanisms of salt tolerance of halophytes. In R. Choukr-Allah, C.V. Malcolm & Hamdy, (Eds.). Halophytes and Biosaline Agriculture. pp 31-53. New York, New York:Marcel Dekker,..

Gould, F.W. & Shaw, R.B. 1983. Grass Systematics, 2nd ed. College Station, TX:Texas A&M University Press. 1-15 pp.

Greenway, H., Gunn, A., & Thomas, D.A. 1966. Plant response to saline substrates. VIII. Regulation of ion concentration in salt sensitive and halophytic species. Australian Journal Biological Science, 19: 741-756.

Greenway, H., & Munns, R. 1980. Mechanisms of salt tolerance in nonhalophytes. Annual Review Plant Physiology, 31: 149-190.

Harinasut, P., Tsutsui, K ., Takabe, T., Nomura, M., Kishitani, S., & Takabe, T. 1994.Glycinebetaine enhances rice salt tolerance. In P. Mathis (Ed.). Photosynthesis: From Light to Biosphere, Vol IV. pp 733-736. Dordrecht, The Netherlands:.Kluwer Academic Press.

Hamdy. A. 1996. Saline irrigation: Assessment and management techniques. In R Choukr-Allah, C. V. Malcom, & A. Hamdy, (Eds.) Halophytes and Biosaline Agriculture. pp 147-180. New York, New York: Marcel Dekker.

Hannon, N.J., & Barber, H.N. 1972. The mechanism of salt tolerance in naturally selected populations of grasses. Search, 3: 259-260.

Harivandi, M.A., Butler, J.D., & Wu, L. 1992. Salinity and turfgrass culture. In D. V. Waddington, R. N. Carrow.,& R. C. Shearman (Eds.). Turfgrass. pp 207-229. Madison, Wisconsin: ASA, CSSA, and SSSA..

Hellebust, J.A. 1976. Osmoregulation. Annual Review Plant Physiology, 27: 485-505.

Hitchcock, A.S. 1971. Manual of the Grasses of the United States, 2nd ed., pp 1-14. New York, New York:.Dover Pub., Inc.

Jeschke, W.D. 1979. Univalent cation selectivity and compartmentation in cereals In D.L. Laidman,. & R.G. Wyn Jones. (Eds.). Recent Advances in the Biochemistry of Cereals. . pp 37-61. New York, New York: Academic Press.

Jeschke, W.D. 1984. K^+-Na^+ exchange at cellular membranes, intracellular compartmentation of cations, and salt tolerance. In R.C. Staples & G.H. Toenniessen (Eds.). Salinity Tolerance in Plants. pp 37-66. New York, New York: John Wiley & Sons.

Jeschke, W.D., Klagges, S., Hilpert, A., Bhatti, A.S., & Sarwar, G. 1995. Partitioning and flows of ions and nutrients in salt-treated plants of *Leptochloa fusca* L. Kunth. I. Cations and chloride. New Phytologist, 130: 23-35.

Johnson, R.C. 1991. Salinity resistance, water relations, and salt content of crested and tall wheatgrass accessions. Crop Science, 31: 730-734.

Kemp, P.R., & Cunningham, G.L. 1981. Light, temperature and salinity effects on growth, leaf anatomy and photosynthesis of *Distichlis spicata* (L.) Greene. American Journal Botany, 68: 507-516.

Khan, M.A., Ungar, I.A., & Showalter, A.M. 1999. Effects of salinity on growth, ion content, and osmotic relations in *Halopyrum mucronatum* (L.) Stapf. Journal Plant Nutrition, 22, 191-204.

Khan, A.H. & Marshall, C. 1981. Salt tolerance within populations of chewing fescue (*Festuca rubra* L.). Communications Soil Science Plant Analyist, 12(12): 1271-1281.

Kramer, D. 1984. Cytological aspects of salt tolerance in higher plants. In , R.C Staples. & G.H. Toenniessen, (Eds.) Salinity Tolerance in Plants. pp 3-15. New York, New York:John Wiley & Sons.

Leigh, R.A., Ahmad, N., & Wyn Jones, R.G. 1981. Assessment of glycinebetaine and proline compartmentation by analysis of isolated beet vacuoles. Planta, 153: 34-41.

Lessani, H., & Marschner, H. 1978. Relation between salt tolerance and long-distance transport of sodium and chloride in various crop species. Australian Journal Plant Physiology, 5: 27-37.

Levering, C.A., & Thomson, W.W. 1971. The ultrastructure of the salt gland of *Spartina foliosa*. Planta, 97: 183-196.

Levitt, J. 1980. Responses of plants to environmental stresses, Vol. II. pp 35-50. New York, NewYork: Academic Press.

Liphshchitz, N., & Waisel, Y. 1974. Existence of salt glands in various genera of the Gramineae. New Phytologist , 73: 507-513.

Liphschitz, N., & Waisel, Y. 1982. Adaptation of plants to saline environments: salt excretion and glandular structure. In D.N Sen,. & K.S. Rajpurohit, (Eds.). Tasks for Vegetation Science, Vol. 2: Contributions to the Ecology of Halophytes. pp 197-214. The Hague,Netherlands:. W. Junk Publisher.

Maas, E.V. 1986. Salt tolerance of plants. Applied Agriculture Research, 1: 12-26.

Maas, E. V. & Hoffman, G .J. (1977). Crop salt tolerance-current assessment. Journal of Irrigation Drainage Div ASCE, 103: 115-132.

Marcum, K.B. 1999. Salinity tolerance mechanisms of grasses in the subfamily Chloridoideae. Crop Science, 39: 1153-1160.

Marcum, K.B. 2003. USGA Turf and Environmental Research Online.at http://www.usga.org/turf/ Marcum, K.B., & Murdoch, C.L. 1992. Salt tolerance of the coastal salt marsh grass, *Sporobolus virginicus* (L.) Kunth. New Phytologist, 120: 281-288.

Marcum, K.B., & Murdoch, C.L.1990a Growth responses, ion relations, and osmotic adaptations of eleven C_4 turfgrasses to salinity. Agronomy Journal, 82: 892-896.

Marcum, K.B., & Murdoch, C.L. 1990b. Salt glands in the Zoysieae. Annuals Botany, 66: 1-7.

Marcum, K.B. & Murdoch, C.L. 1994. Salinity tolerance mechanisms of six C_4 turfgrasses. Journal American Society Horticultural Science, 119: 779-784.

Marcum, K.B., Anderson, S.J., & Engelke, M.C. 1998. Salt gland ion secretion: A salinity tolerance mechanism among five zoysia grass species. Crop Science, 3: 806-810

Mumtaz, S., Maqvi, S.S.M., Shereen, A., & Khan, M.A. 1995. Proline accumulation in wheat seedlings subjected to various stresses. Acta Physiology Plant, 17: 17-20.

Oross, J. W., & Thomson. W.W. 1982a. The ultrastructure of *Cynodon* salt glands: the apoplast. European Journal Cell Biollogy, 28: 257-263.

Oross, J.W. & Thomson, W.W. 1982b. The ultrastructure of the salt glands of *Cynodon* and *Distichlis* (Poaceae). American Journal Botany, 69(6): 939-949.

Naidoo, G., & Naidoo, Y. 1998. Salt tolerance in *Sporobolus virginicus*: the importance of ion relations and salt secretion. Flora-Jena, 193: 337-344.

Naidoo, Y., & Naidoo, G. 1999. Cytochemical localisation of adenosine triphosphatase activity in salt glands of *Sporobolus virginicus* (L.) Kunth. South African Journal Botany, 65: 370-373.

Oldeman, L.R., van Engelen, V.W. P., & Pulles, J.H.M. 1991. The extent of human-induced soil degradation. In L.R. Oldeman, R.T.A. Hakkeling, & W.G Sombroek,. (Eds.) World Map of the Status of Human-Induced Soil Degradation: An Explanatory Note. pp 27-33. Wageningen, Netherlands: International Soil Reference and Information Centre,

O'Leary, J.W. 1971. Physiological basis for plant growth inhibition due to salinity. In W.G. McGinnies, B.J. Goldman, & P. Paylore. (Eds.). Food, Fiber and the Arid Lands. pp 331-336. Tucson, Arizona:University of Arizona Press.

Parrondo, R.T., Gosselink, J.G., & Hopkinson.C.S. 1978. Effects of salinity and drainage on the growth of three salt marsh grasses. Botanical Gazzette. 139: 102-107.

Peacock, C.H., & Dudeck, A.E. 1985. A comparative study of turfgrass physiological responses to salinity. International Turfgrass Research Journal, 5: 821-829.

Pollak, G., & Waisel, Y. 1970. Salt secretion in *Aeluropus litoralis* (Willd.) Parl. Annuals Botany. 34: 879-888.

Pollak, G., & Waisel, Y. 1979. Ecophysiological aspects of salt excretion in *Aeluropus litoralis*. Physiology Plant, 47: 177-184.

Ramakrishnan, P.S., & Nagpal, R. 1973. Adaptation to excess salts in an alkaline soil population of *Cynodon dactylon* (L.) Pers. Journal Ecology, 61: 369-381.

Reid, S.D., Koski, A.J., & Hughes, H.G. 1993. Buffalograss seedling screening invitro for Na Chloride tolerance. Horticulture Science, 28: 536.

Rhodes, D., & Hanson,A.D. 1993. Quaternary ammonium and tertiary sulfonium compounds in higher plants Annual Review Plant Physiology Plant Molecular Biology, 44: 357-384.

Saneoka, H.C., Nagasaka, C., Hahn, D.T., Yang, W.J., Premachandra, G.S., Joly, R.J., & Rhodes, D. 1995. Salt tolerance of glycinebetaine-deficient and -containing maize lines. Plant Physiology, 107: 631-638.

Szabolcs. I. 1989 Salt-Affected Soils. pp 5-30. Boca Raton, FL: CRC Press.

Taleisnik, E.L. 1989. Sodium accumulation in P*appophorum* I. Uptake, transport and recirculation. Annuals Botany, 63: 221-228.

Taleisnik, E.L. & Anton, A.M. 1988. Salt glands in *Pappophorum* (Poaceae). Annuals Botany, 62: 383-388.

Torello, W.A., & Symington, A.G. 1984. Screening of turfgrass species and cultivars for Na Chloride tolerance. Plant Soil, 82: 155-161.

U.S. Salinity Laboratory Staff. 1954. Diagnosis and improvement of saline and alkali soils. In L.A. Richards (Ed.) USDA Handbook 60. pp 100-130. Washington, DC: U.S. Gov. Printing Office.

Waisel, Y. 1972. Biology of Halophytes. pp. 141-165. New York, New York: Academic Press.

Waisel, Y. 1985. The stimulating effects of Na Chloride on root growth of Rhodes grass (*Chloris gayana*). Physiology Plant, 64: 519-522.

Warwick, N.W. M. & Halloran, G.M. 1991. Variation in salinity tolerance and ion uptake in accessions of brown beetle grass [*Diplachne fusca* (L.) Beauv.]. New Phytologist, 119: 161-168.

Weimberg, R., & Shannon, M.C. 1988. Vigor and salt tolerance in 3 lines of tall wheatgrass. Physiology Plant, 73: 232-237.

de Wet, J. M.J ,. & Harlan, J.R. 1970. Biosystematics of *Cynodon* L.C. Rich. (Gramineae). Taxon, 19: 565-569

Wieneke, J., Sarwar, G., & Roeb, M. 1987. Existence of salt glands on leaves of Kallar grass (*Leptochloa fusca* L. Kunth.). Journal Plant Nutrition, 10: 805-820.

Wood, J.N., & Gaff, D.F. 1989. Salinity studies with drought-resistant species of *Sporobolus*. Oecologia 78:559-564.

Worku, W., & Chapman, G.P. 1998. The salt secretion physiology of a Chloridoid grass, *Cynodon dactylon* (L.) Pers., and its implications. Sinet, 21: 1-16.

Wu, L. 1981. The potential for evolution of salinity tolerance in *Agrostis stolonifera* L. and *Agrostis tenuis* Sibth. New Phytologist, 89: 471-486.

Wu, L., & Lin, H. 1993. Salt concentration effects on buffalograss germplasm, seed germination, and seedling establishment. International Turfgrass Research Journal, 7: 883-828.

Wyn Jones, R.G. 1984. Phytochemical aspects of osmotic adaptation. In B.N. Timmerman. (Ed.). Recent Advances in Phytochemistry, Vol. 18, Phytochemical Adaptations to Stress. pp 55-78. New York, New York: Plenum Press.

Wyn Jones, R.G., & Gorham, J. 1983. Osmoregulation. In O.L. Lange, P.S Nobel, C.B Osmond & H. Ziegler. (Eds.). Physiological Plant Ecology III. Responses to the Chemical and Biological Environment. pp 35-58. Berlin, Germany:Springer-Verlag

Wyn Jones, R.G., & Gorham, J. 1989. Use of physiological trains in breeding for salinity tolerance. In F.W.G Barer.. (Ed.) Drought Resistance in Cereals. pp 95-106, Wallingford, UK:CAB International.

Wyn Jones, R.G., Brady, C.J., & Speirs, J. 1979. Ionic and osmotic relations in plant cells. In D.L Laidman., & R.G. Wyn Jones, (Eds.). Recent Advances in the Biochemistry of Cereals. pp 63-103. New York, New York: Academic Press..

Yancy, P.H. 1994. Compatible and counteracting solutes. In K. Strange, (Ed.) Cellular and Molecular Physiology of Cell Volume Regulation. pp. 81-109. Boca Raton, Florida: CRC Press.

Yang, Y.W., Newton, R.J., & Miller, F.R. 1990. Salinity tolerance of Sorghum:I. Whole plant response to sodium chloride in *S. bicolor* and *S. halepense*. Crop Science, 30: 775-781.

Yeo, A.R. 1983. Salinity resistance: Physiologies and prices. Physiology Plant, 58: 214-222.

Yeo, A.R., & Flowers, T.J. 1984. Mechanisms of salinity resistance in rice and their role as physiological criteria in plant breeding. In R.C. Staples & G.H. Toenniessen (Eds.). Salinity Tolerance in Plants. pp 151-171. New York, New York: John Wiley & Sons.

Yeo, D. Kramer, A.R., Lauchli, A., & Gullasch, B. 1977. Ion distribution in salt-stressed mature *Zea Mays* roots in relation to ultrastructure and retention of sodium. Journal Experimental Botany. 28: 17-30.

CHAPTER 12

LOCALIZATION OF POTENTIAL ION TRANSPORT PATHWAYS IN THE SALT GLANDS OF THE HALOPHYTE SPOROBOLUS VIRGINICUS

Y. NAIDOO[1] AND G. NAIDOO

School of Biological and Conservation Sciences, University of KwaZulu-Natal, P/Bag X54001, Durban, 4000, South Africa
[1]Corresponding author:naidoogn@ukzn.ac.za

Abstract. Salt glands in grasses are typically bicellular, being composed of a flask-shaped basal cell which is embedded in the leaf and a dome-shaped cap cell that is exposed to the leaf surface. In this study we attempted to localize potential ion transport pathways through the bicellular salt glands of *Sporobolus virginicus*, a halophytic, perennial, C_4 grass, using a variety of ultracytochemical techniques. Differential membrane staining, using zinc iodide-osmium tetra oxide, was used to determine the origin of membranes associated with secretory activity. In situ precipitation of lanthanum was used to locate barriers to apoplastic transport. Chloride was localised following silver precipitation. The fluorochromes, Cellufluor and uraniun, were used to determine potential apoplastic and symplastic pathways respectively. The data suggested that the ion transport pathway from the leaf veins to the basal cell of the gland may be apoplastic or symplastic. Apoplastic transport of ions was supported by the absence of suberin or cutin and by the localisation of chloride and in situ precipitation of lanthanum in the walls of the basal cell of the salt gland. Symplastic transport was suggested by the presence of numerous plasmodesmata between the basal cell of the gland and adjacent cells, the close association of chloride with plasmodesmata and the localisation of uraniun within the symplast. Transport of ions from the basal to the cap cell of the gland appeared to be symplastic, via thc abundant plasmodesmata interconnecting these cells. In the cap cell, ions appeared to be transported within small vacuoles and released into the cuticular chamber, which eventually ruptures to release the salt outside the leaf.

1. INTRODUCTION

Salt secretion is a common phenomenon in many halophytes belonging to diverse plant families (Breckle, 1995). Species with salt glands usually occur in saline areas and are highly salt tolerant. In the Poaceae, the salt glands are typically bicellular and their structure has been extensively investigated (Levering & Thomson, 1971; Liphschitz & Waisel, 1974; Taleisnik & Anton, 1988; Amarsinghe & Watson, 1989; Naidoo & Naidoo, 1998b; Somaru et al., 2002).

M.A. Khan and D.J. Weber (eds.), Ecophysiology of High Salinity Tolerant Plants, 173-185.

The bicellular salt gland of grasses is generally composed of a voluminous, flask-shaped basal cell embedded in the leaf and a smaller dome-shaped cap cell that protrudes beyond the epidermis. The cuticle that overlies the gland is often porose. The cytoplasm of the basal cell is generally dense and partitioned into channels by a system of paired membranes which originate from invaginations of the plasma membrane, adjacent to the common wall between basal and cap cells. Numerous mitochondria and microtubules are closely associated with the partitioning membranes. The basal cell is intimately connected to the adjacent cap, epidermal and mesophyll cells by numerous plasmodesmata. The cytoplasm of the cap cell lacks partitioning membranes, but contains numerous small vacuoles and a large number of organelles in close proximity to the outer surface of the gland. A cuticular chamber is often present between the cap cell and the overlying cuticle (Levering & Thomson, 1971; Naidoo & Naidoo, 1998b; Somaru et al., 2002).

There is general agreement that salt glands remove by secretion, excess ions delivered to the leaves in the transpiration stream (Fahn, 1979; Naidoo & Naidoo, 1998a). It is accepted that ions are transported into the mesophyll via the vascular system (Ish-Shalom-Gordon & Dubinsky, 1990). The transport pathway from the mesophyll via the glands to the exterior, however, remains unresolved. The aim of this study was to determine the potential pathways of ion movement through salt glands using a variety of ultracytochemical techniques. For our experimental material we selected *Sporobolus virginicus* L. (Kunth), a halophytic, perennial, C_4 grass that possesses typical bicellular salt glands predominantly on the adaxial leaf surface (Naidoo & Naidoo, 1998b).

Differential membrane staining, using zinc iodide-osmium tetroxide (ZI0), which enhances the contrast of endomembranes (Harris, 1978), was used to determine the origin of those membranes associated with secretory activity. In situ precipitation of lanthanum (La^{3+}) was used to locate barriers to apoplastic transport (Campbell et al., 1974). Chloride (Cl^-) was localized following silver precipitation (Harvey et al., 1976) while the flurochromes, Cellufluor and uranium were used to map out transport pathways.

2. MATERIALS AND METHODS

Flag leaves (i.e. the last fully expanded leaves) of terminal rhizomes of *S. virginicus* were collected from a saline site at the Beachwood Mangroves Nature Reserve, Durban, South Africa. For light, fluorescence and electron microscopy samples were taken from the mid-region of flag leaves.

2.1. Zinc iodide-osmium tetraoxide (ZI0) impregnation

Differential membrane staining by Z10 (as a post-fixative) was used to determine the origin of those membranes, which were associated with glandular secretory activity. Pieces of leaf tissue (1 mm^2) were fixed in 3% gluteraldehyde buffered with 0.1 M sodium cacodylate under vacuum (pH 7.2) for 3-4h. The material was rinsed in double-distilled water for 1h. Tissue was then post-fixed in 2% aqueous ZI0 (Harris, 1978) for 24h. After ZI0 treatment, the tissue was washed with double-distilled water for 1.5h and then dehydrated through a graded ethanol series and embedded in a low viscosity resin (Spurr, 1969). Sections with dark gold to purple interference colors were collected on copper grids and viewed without post-staining with a Philips 301 Transmission Electron Microscope (TEM) at 80 kV.

2.2. *Chloride localization*

Chloride was cytochemically localised by precipitation with silver acetate (Harvey et al., 1976). Small pieces of leaf (1 mm^2) were fixed overnight at 4°C in 1% osmium tetroxide (O_sO_4) with or without 30mM silver acetate buffered at pH 7.2 with 0.1M sodium cacodylate in vials wrapped in aluminum foil. Dehydration and embedding procedures were undertaken under red light to prevent reduction of Ag^+ (Smith et al., 1982). The tissue was rinsed under red light at 4°C with 0.05M sodium cacodylate buffer for 3 periods of 20s before rapid penetration of resin. The material was then dehydrated through a graded acetone series. One drop of a solution of 1mM HNO_3 was included in the 60% acetone to remove non-specific deposits of silver phosphate and carbonate. Material was infiltrated with Pallaghy's (1973) low Cl^- modification of Spurr's (1969) low viscosity resin and polymerised at 60° for 24-72h. Ultrathin sections were viewed unstained.

2.3. *Lanthanum precipitation*

Barriers to apoplastic transport were determined by the in situ precipitation of La^{3+} (Campbell et al., 1974). Flag leaves were rinsed to remove external salt and the basal cut ends placed in a medium prepared by using 1.2% La^{3+} made by raising the pH of a La $(NO_3)_3$ solution to 7.2 by slowly adding 0.2 N NaOH (Revel & Karnovsky, 1967). For controls, cut ends of leaves were placed in distilled water. The leaves were exposed to the La $(NO_3)_3$ solution for a 15h period in the light. After La^{3+} application the leaves were prepared for electron microscopy. Small pieces of leaf tissue (1 mm^2), excised about 10 mm above the solution, were fixed in 2.5% phosphate-buffered glutaraldehyde at pH 7.0 for 2h and post-fixed in 1% phosphate buffered $0s0_4$ (pH 7.0). After dehydration with acetone, the tissue was embedded in Spurr's (1969) epoxy resin. Ultrathin sections were viewed without post-staining.

2.4. *Fluorescence microscopy*

For fluorescence microscopy studies, thin, hand-cut sections of leaf blade material were exposed to one of two fluorescence probes. Cellufluor was used to determine possible apoplastic pathways while uranium probed symplastic transport.

2.5. Cellufluor treatment

The basal cut ends of flag leaves (3 mm), removed from salt-treated plants were placed in 0.01% solution of Cellufluor for 5h. Cellufluor at this concentration is non-toxic to living plant cells (Peterson & Perumalla, 1984; Moon et al., 1986). After Cellufluor application, leaves were washed in running tap water for approximately 30 min. to remove any unbound tracer. Free hand cross-sections, taken from near the base of the leaf blades were examined with a Zeiss Axiophot epifluorescence microscope fitted with a violet filter assembly (excitation wavelengths 400-440 nm, beam sputter 460 nm and barrier filter 470 nm). Sections of living untreated material were also viewed under violet light and used as controls because of auto fluorescence of walls of some leaf tissues.

2.6. Uranium treatment

Excised flag leaves were treated with uranium, following a method modified from Peterson & Perumalla (1984). The cut ends of the leaves were placed in a 0.01% solution of uranium in 0.5mM KH_2PO_4 (pH 5.3) for 6h in a humid chamber. Uranium at this concentration is non-toxic to living plant cells (Peterson & Perumalla, 1984). After treatment, the leaves were rinsed briefly (5-15s) in flowing tap water and freehand cross-sections made. Excess uranium, released by cells injured during sectioning, was removed by floating the sections in a solution of 0.5 mM KH_2PO_4. The sections were viewed with UV light, using the epifluorescence microscope described for Cellufluor application. The same controls that were used for the Cellufluor treatment were utilized for evaluating results obtained after treatment with uranium.

2.7. Light microscopy

Historesin-embedded leaf sections were treated with Sudan black B (O'Brien & McCully, 1981) to determine the presence of suberin and /or cutin in gland cells.

2.8. Energy dispersive X-ray microanalysis (EDX)

Ultrathin sections, prepared from leaf tissues processed for Cl^- localization, and La^{3-} precipitation, were collected on copper grids and viewed with a Philips CM 120 Biotwin TEM. In each case the electron-dense reaction product associated with the walls of the basal cell of salt glands was qualitatively analyzed by the EDX microanalysis using the EDAX system interfaced with the Philips CM 120 Biotwin operating at 80kV. Spot (200 nm) analyses were performed on the electron–dense deposits associated with walls of glandular cells.

3. RESULTS

The salt glands of *S. virginicus* are typically bicellular. Each gland is composed of a large, flask-shaped basal cell, which is embedded in the leaf tissue and a smaller cap cell that extends beyond the epidermis. The cytoplasm of the basal cell is partitioned into channels by a system of paired membranes, which appear to originate from invaginations of the plasma membrane. Closely associated with the partitioning membranes are numerous mitochondria, microtubules and cisternae of endoplasmic reticulum. The basal cell is intimately connected to adjacent cells by numerous small plasmodesmata. The dense cytoplasm of the cap cell contains numerous small vacuoles and a concentration of organelles in close proximity to the outer surface of the gland (Figures 1 to 4).

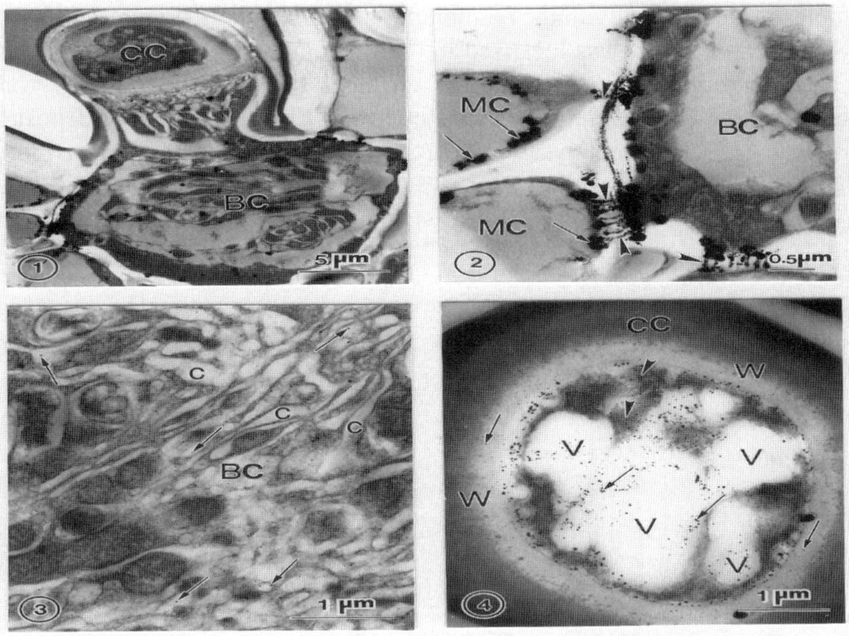

Figure 1. *Longitudinal section (L.S.) of bicellular salt gland of* S. virginicus *with flask-shaped basal cell (BC) small dome-shaped cap cell (CC). and constricted bottleneck (N) region.*
Figure 2. *Chloride cytochemistry produces a finely-localized dense reaction product presumably silver chloride in peripheral cytoplasm of mesophyll cells (MC, arrows) and in plasmodesmata (arrowheads) interconnecting basal (BC) and mesophyll cells.*
Figure 3.*Section of basal cell cytoplasm showing tubular network of membrane-bound channels (c) containing particulate electron-dense deposits (arrows).*
Figure 4. *Cross-section of cap cell (CC) of salt gland showing electron-dense deposits in wall (W, arrows) vacuoles (V, arrows) and cytoplasm (arrowheads).*

In all staining procedures at least four micrographs were analyzed from each of three replicate plants to determine the location of staining patterns or electron-dense deposits. The micrographs presented were selected for being the most representative.

3.1. Cl^- localization

Chloride was localized by precipitation with a solution of silver acetate. The reaction product, $AgCl_2$, appears as an electron-dense deposit. These deposits suggest potential pathways of Cl^- transport. Electron-dense reaction products indicating sites of Cl^- localisation were prominent in the basal cell of the salt gland. These electron-dense deposits were mainly located in the vicinity of plasmodesmata, (Figure 2) interconnecting adjacent mesophyll and epidermal cells. Within the basal cell, electron-dense deposits occurred scattered in the cytoplasm and especially in the membrane- bound channels (Figure 3).

In the cap cell, Cl^- was localised in the walls and vacuoles (Figure 4) and was especially prominent in the cuticular chamber between the cap cell wall and overlying cuticle. The elemental constituents of the electron-dense deposits were verified by EDX spot analyses to be Ag^+ and Cl^- (Figure 5). Furthermore, leaf samples fixed only in osmium and used as controls exhibited no electron-dense deposits.

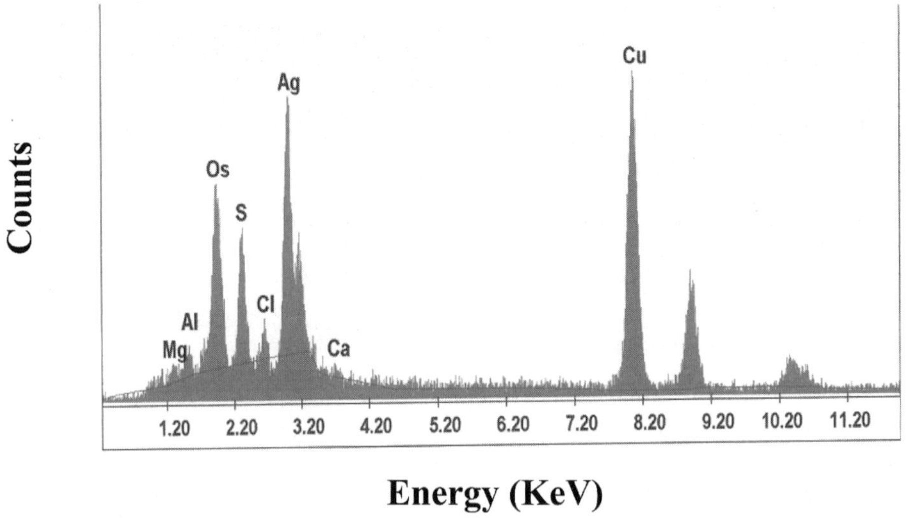

Figure 5. *X-ray emission spectrum of electron- dense deposits (produced by chloride cytochemistry) associated with basal cell wall of salt gland of S. virginicus (Ag = silver, Al = aluminium, Ca = calcium, Cl = chloride, Cu = copper, Mg = magnesium, Os = osmium, S = sulphur).*

3.2. ZI0 impregnation

The ZI0 staining technique, which enhances the contrast of endomembranes (Dauwalder & Whaley, 1973), suggested that ER is a prominent component of the cytoplasm of the gland cells. The cap cell exhibited large amounts of lamellar or cisternal ER, especially in close proximity to the plasmalemma (Figure 7). The basal cell on the other hand lacked intense ZI0 reaction, although some ER cisternae and mitochondria showed marked staining (Figure 6).

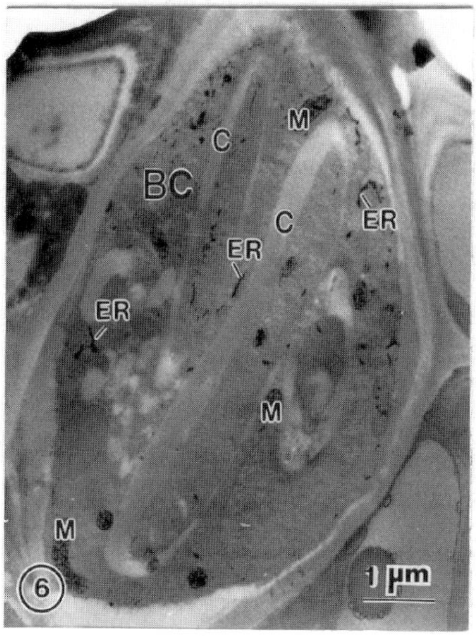

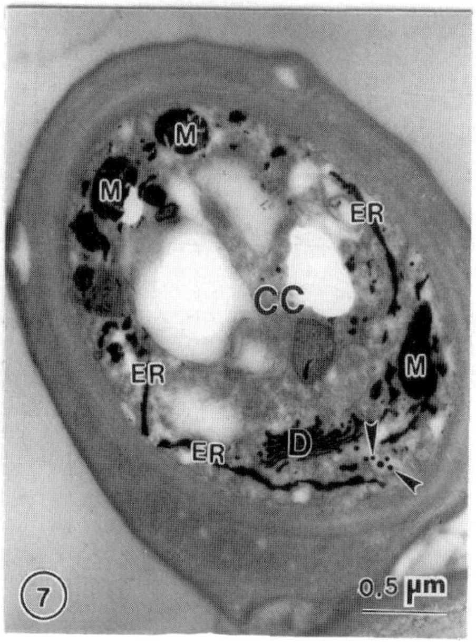

Figure 6. *Section of basal cell (BC) of salt gland (Z10 post-fixation) with stained cisternal ER interspersed between membrane-bound channels (C). Mitochondria (M) show distinct staining.*
Figure 7. *Section of cap cell of salt gland (Z10 post-fixation) showing distinct staining of ER, mitochondria (M), dictyosome (D) and related vesicles (arrowheads).*

3.3. La^{3+} precipitation

There was marked accumulation of La^{3+} in the apoplast of salt glands. Deposits of La^{3+} were closely associated with the plasmalemma, plasmodesmata partitioning membranes and membrane bound channels of the basal cell of the gland (Figure 8). The apoplast of the cap cell lacked La^{3+} deposits (Figure 9). The electron-dense deposits were confirmed to be La^{3+} by EDX spot microanalysis (Figure 10). Control specimens lacked electron-dense deposits.

3.4. Fluorescence microscopy

The bright blue fluorescence of the apoplastic tracer, Cellufluor, was evident in the apoplast of the vascular bundles, walls of the bundle sheath, mesophyll, epidermis and sclerenchyma cells (Figure 11). The tracer also penetrated the apoplast of the salt gland cells. The yellow fluorescence of the symplastic tracer, uranium was observed in the symplasm including vascular bundles, bundle sheath cells, mesophyll, epidermis and salt gland cells (Figure 12). Sections not stained with fluorochrome exhibited strong orange-yellow autofluorescence, which was quite distinct from the blue and yellow fluorescence of Cellufluor and uranium respectively.

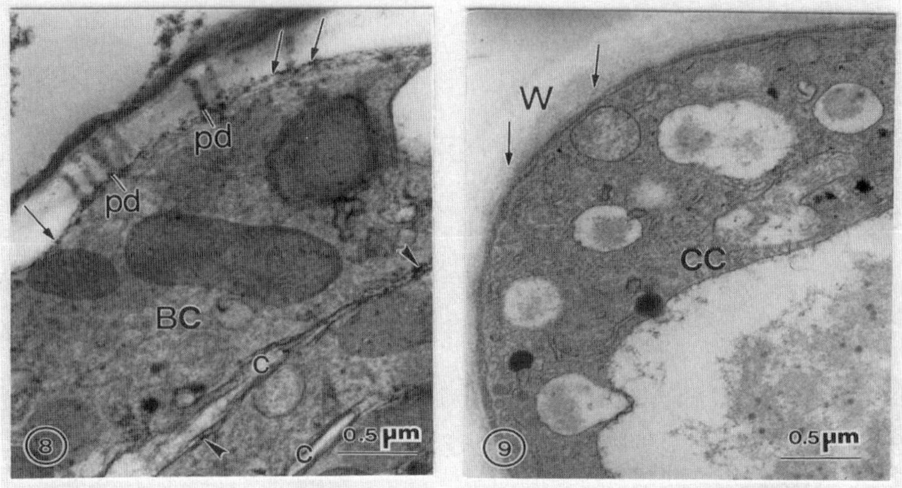

Figure 8. *Part of basal cell (BC) of salt gland with La^{3+} deposits closely associated with the plasmalemma (arrows), plasmodesmodata (Pd) and the partitioning membranes (arrowheads) of the channels (c).*
Figure 9. *Section of cap cell (CC) of salt gland showing absence of La^{3+} deposits in wall (W, arrows).*

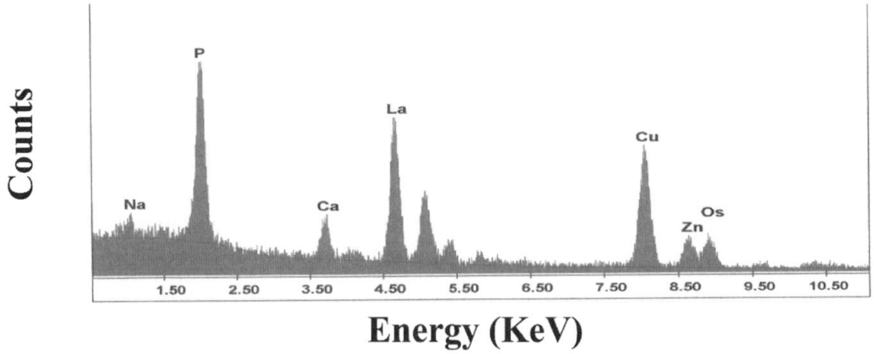

Figure 10. *X-ray emission spectrum of electron - dense deposits (after La^{3+} precipitation) associated with basal cell walls of salt gland of* S. virginicus *(Na = sodium, P = phosphorus, Ca = calcium, La = lanthanum, Cu = copper, Zn = zinc, Os = osmium).*

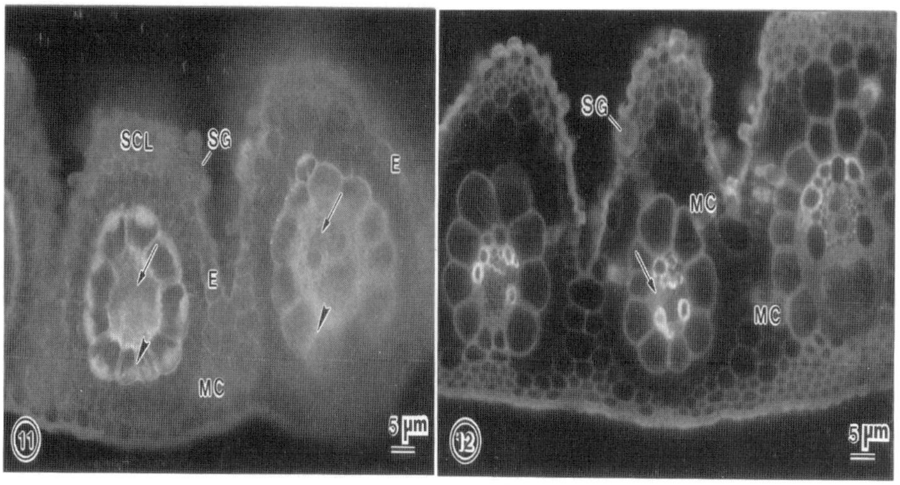

Figure 11. *Fluorescence micrograph of cross-section of living leaf of* S. virginicus *viewed under violet light. After treatment with cellufluor (indicated by bright blue fluorescence), the fluorochrome penetrated into the apoplast of veins (arrows), walls of bundle sheath (arrowheads), mesophyll (MC), epidermis (E) and sclerenchyma (SCL) tissue. Note penetration of tracer into salt gland cells (SG).*

Figure 12. *Fluorescence micrograph of cross-section of living leaf of* S.virginicus *viewed under ultraviolet light. After exposure to uranium, the fluorochrome (indicated by bright yellow fluorescence) penetrated the vascular bundles of veins (arrows), mesophyll (MC) and salt gland cells (SG).*

4. DISCUSSION

Ions transported to the leaves in the transpiration stream are separated from the epidermal salt glands by a distance of about 6 to 8 cells (Figures 11 & 12). From the leaf veins, ions may be conveyed to the glands apoplastically, symplastically or by a combination of both pathways.

Staining of historesin sections with Sudan black B suggested no suberization or lignification of the cell walls of the basal cell. Similar results were reported for salt glands of *Cynodon* and *Distichlis* (Oross & Thompson, 1982a, 1982b). On a structural basis therefore, there appears to be apoplastic continuity in *S. virginicus*, *Cynodon* and *Distichlis* from the leaf veins to the basal cell of salt glands. These results however, contradict those of Ruhland (1915), Arisz et al. (1955) and Helder (1956). Convincing evidence for apoplastic transport in the basal cell is indicated by the localisation of La^{3+} in the cell walls and membrane-bound channels of the basal cell of *S. virginicus* (Figure 8). In other studies on salt secretion, La^{3+} deposits were located throughout the apoplast in the leaves of *Atriplex*, *Tamarix*, *Limonium* (Campbell et al., 1974) and *Cynodon* (Oross & Thompson, 1982b). The bright blue fluorescence of the apoplastic tracer, Cellufluor, within the apoplast of the gland is additional evidence supporting the potential for apoplastic transport from the leaf veins to the salt glands (Figure 11).

The localization of electron-dense deposits, presumably $AgCl_2$ in the mesophyll cell walls, in the walls of the basal cell in the bottleneck region (transfusion zone) as well as in the membrane-bound extracellular channels (Figures 2 & 3) supports an apoplastic transport pathway. Strong evidence for apoplastic transport was reported in the salt glands of *Tamarix* and *Frankenia* using chloride localization (Campbell & Thomson, 1975, 1976) and in *Sporobolus arenarius* using electron microprobe (Ramati et al., 1976). The results reported here, as well as those of others (Campbell & Thomson, 1975; Ramati et al., 1976), suggest that the extracellular channels are connected apoplastically with the mesophyll cells and may be involved in the elimination of ions. According to this view, ions are actively pumped into the channels near their closed ends. As a consequence of localised ion loading, a standing osmotic gradient would be established within the channels that drives a vectorial flow of solutes toward their open ends near the wall separating the basal and cap cells (Levering & Thomson, 1971, 1972).

The absence of La^{3+} in the apoplast of the cap cell wall in *S. virginicus* (Figure 9) and in the apoplast of the outer secretory cells of multicellular glands (Thomson et al., 1988) suggests apoplastic discontinuity between collecting and secretory cells. The wall of the cap cell in this region is very thick and suberised (Naidoo & Naidoo, 1998b).

Several lines of evidence suggest that the ion transport pathway from the mesophyll to the basal cell may also be symplastic. Plasmodesmatal connections, which serve as symplastic transport pathways, were numerous between the basal and adjoining epidermal and mesophyll cells (Figures 2 & 8). The association of Cl^- with the plasmodesmata at the transfusion zone and membrane bound channels of the basal cell (Figures 2 & 3) suggests symplastic transport. Additional support for symplastic transport was indicated by the location of uranium in the vascular bundles, bundle sheath, mesophyll, epidermis and salt gland cells (Figure 12). In several other studies, Cl^- was also associated with plasmodesmata in salt glands of *Limonium* (Ziegler & Lüttge, 1967), Aegiceras (van Steveninck et al., 1978) and *Tamarix* (Campbell & Thomson, 1975). In *Tamarix,* however, Cl^- was also detected in the cell walls suggesting that ion transport in this species was symplastic as well as apoplastic (Campbell & Thomson, 1975).

The abundance of electron-dense AgCl deposits in the walls of the cap cell as well as in the cuticular chamber suggests that ions are secreted directly into these compartments prior to elimination to the exterior. Similar accumulation of salt in collecting chambers in the glands of *Limonium* (Ziegler & Lüttge, 1967), *Tamarix* (Campbell & Thomson, 1975), *Frankenia* (Campbell & Thomson, 1976) and *Cynodon* (Oross & Thomson, 1982b) has also been reported. The high dissociation constant of NaCl would ensure that Na^+ and Cl^- are transported separately into these compartments prior to elimination.

The highly organised ultrastructure of the basal cell, characterised by a high proportion of cisternal ER and lipid bodies suggests that ion loading of the channels occurs via the partitioning membranes. The lipid bodies are probably involved in the synthesis of the plasmalemma-derived partitioning membranes.

Ultrastructural and cytochemical evidence suggests that salt secreted by the glands of *S. virginicus* is derived from the apoplast as well as the symplast. Ions enter the basal cell protoplast across the apoplast-symplast interface formed by the partitioning membranes. Loading of the channels is probably active, energy for ion pumps being provided by the numerous mitochondria in close proximity to the partitioning membranes (Lüttge, 1971; Naidoo & Naidoo, 1998b). Intense ATPase activity in the vicinity of the plasmalemma and partitioning membranes of the channels in *S. virginicus* suggested that these are the sites of active ion influx into the basal cell (Naidoo & Naidoo, 1999).

Ion transport from the basal to the cap cell appears to be symplastic via the abundant plasmodesmata interconnecting these cells. The absence of La^{3+} deposits in the cap cell supports this conclusion. The localisation of Cl^- in the small vacuoles of the cap cell suggests that ions may be transported in this compartment prior to release into the cuticular chamber. The occurrence of abundant cisternal ER in close association with the plasmalemma of the cap cell suggests that unloading of the gland occurs via these two membranes. Unloading is probably an active process as indicated by intense ATPase activity in the vicinity of the plasmalemma of the cap cell (Naidoo & Naidoo, 1999).

The data obtained in this study, using a variety of ion localization techniques and apoplastic and symplastic tracers, suggest that the ion transport pathway from the vascular bundles to the salt glands involves a combination of apoplastic and symplastic pathways. The salt glands are ultrastructurally adapted to receive, accumulate and efficiently eliminate ions from the leaves.

5. SUMMARY

In this study, a variety of ultracytochemical techniques were used to determine potential ion transport pathways through the bicellular salt glands of the halophytic grass, *Sporobolus virginicus*. Zinc iodide-osmium tetroxide, a differential membrane stain, was used to stain membranes involved in secretory activity. Lanthanum precipitation was used to indicate possible apoplastic barriers to ion transport. Chloride was localized by silver precipitation. The fluorochromes, Cellufluor and uranium were used to indicate potential apoplastic and symplastic ion transport pathways respectively.

The results indicated that ions may be transported to the salt glands apoplastically or symplastically. Support for apoplastic transport included the localization of Cl^- and La^{3+} and the absence of suberin and cutin in the walls of the basal cell of the gland. Evidence for symplastic transport included the presence of numerous plasmodesmatal connections between the basal cell of the gland and adjacent cells, the presence of Cl^- within plasmodesmata and the localization of uranium.

Within the salt gland, ion transport from the basal to the cap cell is symplastic, probably via the abundant plasmodesmata between these cells. Ion transport in the cap cell appears to occur within small vacuoles. The contents of these vacuoles appear to be released into the cuticular chamber. Accumulation of ions in this compartment results in its eventual rupture and release of salt to the leaf exterior.

6. ACKNOWLEDGEMENTS

This research was funded by the University of Durban-Westville and the National Research Foundation. The authors thank V. Bandu, EM Unit, and University of Natal for EDX microanalysis, A.Rajh for photographic assistance and B. Bhiman and S. Chetram for preparing the manuscript.

7. REFERENCES

Amarasinghe, V. & Watson, L. 1989. Variation in salt secretory activity of microhairs in grasses. Australian Journal of Plant Physiology 16: 219-229.

Arisz, W.H., Camphuis, I.J., Heikens, H. & van Tooren, A.J. 1955. The secretion of the salt glands of Limonium latifolium Ktze. Acta Botanica Neerlandica 4: 322-338.

Breckle, S.W. 1995. How do halophytes overcome salinity? In: M.A. Khan and I.A. Ungar (Eds.), Biology of Salt Tolerant Plants, Michigan: Book Crafters. 199-213 pp.

Campbell, N. & Thomson, W.W. 1975. Chloride localization in the leaf of *Tamarix*. Protoplasma 83: 1-14.

Campbell, N. & Thomson, W.W. 1976. The ultrastructural basis of chloride tolerance in the leaf of *Frankenia.* Annals of Botany 40: 687-693.

Campbell, N., Thomson, W.W. & Platt, K. 1974. The apoplastic pathway of transport to salt glands. Journal of Experimental Botany 25: 61-69.

Dauwalder, M. & Whaley, W.G. 1973. Staining of cells of Zea mays root apices with the osmium-zinc iodide and osmium impregnation techniques. Journal of Ultrastructral Research 45: 279-296.

Fahn, A. 1979. Secretory tissues in plants. London, UK: Academic Press. 302 pp.

Harris, N. 1978. Nuclear pore distribution and relation to adjacent cytoplasmic organelles in cotyledon cells of developing *Vicia faba*. Planta 141: 121-128.

Harvey, D.M.R., Flowers, T.J. & Hall, J.L. 1976. Localization of chloride in leaf cells of the halophyte *Suaeda maritima* by silver precipitation. New Phytologist 77: 319-323.

Helder, R.J. 1956. The loss of substances by cells and tissues (salt glands). In: Ruhland, H. (Ed.), Encyclopedia of Plant Physiology, Vol. 11, Part B. Berlin, Germany: Springer Verlag, 225-239 pp.

Ish-Shalom-Gordon, N. & Dubinsky, Z. 1990. Possible modes of salt secretion in *Avicennia marina* in the Sinai. Plant Cell Physiology 31: 27-32.

Levering, C.A. & Thomson, W.W. 1971. The ultrastructure of the salt glands of *Spartina foliosa*. Planta 97: 183-196.

Levering, C.A. & Thomson, W.W. 1972. Studies on the ultrastructure and mechanism of secretion of the salt gland of the grass *Spartina*. Proceedings of the 30th Electron Microscopy Society Congress. 222-223 pp.

Liphschitz, N. & Waisel, Y. 1974. Existence of salt glands in various genera of the Gramineae. New Phytologist 73: 507-513.

Lüttge, U. 1971. Structure and function of plant glands Annual Review of Plant Physiology 22: 23-44.

Moon, G.J., Clough, B.F., Peterson, C.A. & Allaway, W.G. 1986. Apoplastic and symplastic pathways in *Avicennia marina* roots revealed by fluorescent tracer dyes. Australian Journal of Plant Physiology 13: 637-648.

Naidoo, G. & Naidoo, Y. 1998a. Salt tolerance of *Sporobolus virginicus*: the importance of ion relations and salt secretion. Flora 193: 337-344.

Naidoo, Y. & Naidoo, G. 1998b. *Sporobolus virginicus* leaf salt glands: morphology and ultrastructure. South African Journal of Botany 64: 198-204.

Naidoo, Y. & Naidoo, G. 1999. Cytochemical localization of adenosine triphosphatase activity in salt glands of *Sporobolus virginicus*. South African Journal of Botany 65: 370-373.

O'Brein, T.P. & McCully, M.E. 1981. The study of plant structure: structure and selected methods. Termarcasphi Pty. Ltd. Melbourne.

Oross, J.W. & Thomson, W.W. 1982a. The ultrastructure of the salt glands of Cynodon and Distichlis (Poaceae). American Journal of Botany 69: 939-949.

Oross, J.W. & Thomson, W.W.1982b.The ultrastructure of *Cynodon* salt glands secreting and non secreting. Journal of Cellular Biology 34: 287-291.

Peterson, C.A. & Perumalla 1984. Development of the hypodermal casparian band in corn and onion roots. Journal of Experimental Botany 35: 51-57

Ramati, A., Liphschitz, N. & Waisel, Y. 1976. Ion localisation and salt secretion in *Sporobolus arenarius* (Gov.) Duv-Jouv. New Phytologist 76: 289-294.

Revel, J.P. & Karnovsky, M.J. 1967. Hexagonal array of subunits in intercellular junctions of the mouse heart and liver. Journal of Cellular Biology 33: C7-C12.

Rühland, N. 1915. Untersuchungen über die Hautdrüsen der Plumbaginaceen. Jahrb. Wiss. Botany 55: 409-498.

Smith, M.M., Hodson, M.J., OP, K.H. & Wainwright, S.J. 1982. Salt-induced ultrastructural damage to mitochondria in root tip of a salt-sensitive ecotype of *Agrostis stolonifera*. Journal of Experimental Botany 33: 886-895.

Somaru, R., Naidoo, Y. & Naidoo, G. 2002. Morphology and ultrastructure of the leaf salt glands of *Odyssea paucinervis* (Stapf) (Poaceae). Flora 197: 67-75.

Spurr, A.R. 1969. A low-viscosity epoxy resin embedding medium for electron microscopy. Journal of Ultrastructural Research 26: 31-43.

Taleisnik, E.L. & Anton, A.M. 1988. Salt glands in Pappophorum (Poaceae). Annals of Botany 62:383-88.

Thomson, W.W., Faraday, C.D. & Oross, J.W. 1988. Salt glands. In: D.A. Baker & J.L. Hall. (Eds.), Solute Transport in Plant Cells and Tissues, Essex, UK: Longman Scientific and Technical, 498-537 pp.

Van Steveninck, R.F.M., Van Steveninck, M.E., Peters, P.D. & Hall, T.A. 1976. Ultrastructural localization of ions. Journal of Experimental Botany 27: 1291-1312.

Ziegler, H. & Lüttge, U. 1967. Dei Salzdrusen von Limonium vulgare. 11. Die lokalisierung des chloride. Planta 74: 1-17.

CHAPTER 13

CELLULAR RESPONSES TO SALINITY OF TWO COASTAL HALOPHYTES WITH DIFFERENT WHOLE PLANT TOLERANCE: KOSTELETZKYA VIRGINICA (L.) PRESL. AND SPOROBOLUS VIRGINICUS (L.) KUNTH

XIANGGAN LI[1], DENISE M. SELISKAR[2] AND JOHN L. GALLAGHER[2]

[1]*Syngenta, P.O. Box 12257, 3054 Cornwallis Road, Durham, NC 27709-2257, USA*
[2]*Halophyte Biotechnology Center, College of Marine Studies, 700 Pilottown Road, University of Delaware, Lewes, Delaware, 19958, USA*

Abstract. At the whole plant level, *Sporobolus virginicus* is more salt-tolerant than *Kosteletzkya virginica*. Cellular level (callus and protoplast) responses to salinity are reported here. The callus of *Kosteletzkya* had higher relative growth rates than *Sporobolus*, particularly at the highest salinity, however survival rate at 170 and 340 mM NaCl was similar. Upon salinization, *Kosteletzkya* had higher cell wall digestibility and membrane integrity than *Sporobolus*, which may reflect differences in composition. *Kosteletzkya* protoplast viability was higher than *Sporobolus* at salinities up to 340 mM NaCl. *Kosteletzkya* demonstrated a consistent pattern of salt tolerance between cellular and whole plant levels, however *Sporobolus* had a relatively higher salt tolerance at the whole plant level, perhaps attributable to salt glands in its leaves. *Kosteletzkya* has no such glands.

1. INTRODUCTION

As the acreage of salinized land increases worldwide, an understanding of the mechanisms of salt tolerance operating on cellular, organ, and whole plant levels becomes increasingly important. Salt tolerance in halophytes is achieved at the whole plant level by integrating several mechanisms, such as active excretion of ions by salt glands (Naidoo et al., 1992), exclusion of ions due to membrane impermeability (Ball, 1988; Wu et al., 1998), ion dilution by rapid growth associated with an increase in water content (Blits & Gallagher, 1993), compartmentalization of ions (Breckle, 1995), and production of organic osmotica (Rhodes & Hanson, 1993). The mechanisms of salt tolerance operating on the

M.A. Khan and D.J. Weber (eds.), Ecophysiology of High Salinity Tolerant Plants, 187-200.

whole plant level may or may not be related to the tolerance at the cellular level in vitro. In some cases, no correlation regarding salt tolerance is found between the whole plant and callus originated from the species. In such plants salt tolerance depends not on cellular aspects, but on the physiological and anatomical integrity of the whole plant (Tal, 1984). However, a positive correlation between the response of the whole plants and the callus was found in some halophytes, such as *Spartina pectinata* (Warren et al., 1985) and *Spartina patens* (Li et al., 1995). These disparities are attributed to differences in the relative importance of cellular tolerance to the overall tolerance of morphological and physiological mechanisms operating at the whole plant level of organization (Smith & McComb, 1981; von Hedenstrom & Breckle, 1974).

Halophytes have evolved adaptive mechanisms to cope with salt and have the salt-tolerance genetic materials regarding those mechanisms. This research examines the effect of a range of NaC1 concentrations on callus growth, cell viability, and protoplast tolerance of cell cultures of two halophytes, one dicot *Kosteletzkya virginica* (L.) Presl. and one monocot, *Sporobulus virginicus* (L.) Kunth. *Kosteletzkya* virginica requires low salinity water for germination, but once established it could tolerate coastal seawater (30 g kg^{-1}, in contrast to open ocean seawater, 35 g kg^{-1}). This dicot species has good grain yield, up to 1.46 tons per hectare (Gallagher, 1985). As a member of the Malvacae, it is a relative of both cotton and kenaf. *Sporobolus virginicus* is a monecious rhizomatous perennial monocot of the Poaceae family. Forage yields of 8.75 tons per hectare were obtained in saline conditions under drip irrigation in the Egyptian desert (Gallahger, 1985). This species can tolerate irrigation with water up to 80 g kg^{-1} salinity (Gallagher, 1979) and is a cosmopolitan species on sub-tropical to low latitude temperate beaches and salt marshes.

Kosteletzkya virginica seedlings and adult plants are less salt-tolerant than *S. virginicus* (Blits & Gallagher, 1990ab) and their response to salt is different. In contrast to *K. virginica, S. virginicus* has specialized salt glands to excrete salts onto the leaf surface where it crystallizes and either washes off in rain or tidal water or falls off when large crystals build up. Since *S. virginicus* has foliar salt glands to remove salt that enters the plant, it might be expected that the whole plant would be more salt-tolerant than the cells. Knowing the degree of cellular salt tolerance in these two halophytic species is of further interest in order to identify the best technology to use in efforts to improve these halphytic crops (Gallagher & Seliskar, 1993). For example, selection for highly salt-tolerant cell lines from tissue cultures will be more productive if cellular mechanisms are an important component in whole plant tolerance. A better understanding of cellular salt tolerance of halophytes may furnish insights into what characteristics glycophytes, such as rice and cotton, should acquire in order to be highly salt-tolerant (Wei et al., 2001; Dubouzet et al., 2003; Kumar et al., 2004).

2. MATERIALS AND METHODS

Cell suspension cultures of *Kosteletzkya virginica*, seashore mallow, were initiated from callus derived from embryo explants (Cook et al., 1989). Suspension cultures of *K. virginica* were maintained in 250-mL Erlenmeyer flasks containing 50 mL of nutrient medium composed of 30.0 g L^{-1} sucrose, 2 mg L^{-1} indoleacetic acid (IAA), and 1 mg L^{-1} 2,4-dichlorophenoxyacetic acid (2,4-D) plus Murashige and Skoog's salt (Murashige & Skoog, 1962), with addition of 100 mg L^{-1} inositol and 0.1 mg L^{-1} thiamine-HCL. The pH of the medium was adjusted to 5.7 before autoclaving. The cultures were grown on a rotary shaker at 21°C under continuous low intensity fluorescent light (30 μE m^{-2} s^{-1}) and were subcultured at two-week intervals by transferring 10 mL of cell suspension to fresh medium. Cultures in this study were harvested during log-phase growth, seven days after subculture (Blits & Gallagher, 1990c).

Unemerged inflorescences were collected from individual greenhouse-grown plants of *Sporobolus virginicus* and dissected out of the surrounding sheath under sterile conditions. Explants were cultured on medium, which contained Murashige, and Skoog revised MS salts plus 30 g L^{-1} sucross, 0.5 mg L^{-1} 6-benzylaminopurine (BAP), 1.0 mg L^{-1} naphthaleneacetic acid (NAA), 0.5 mg L^{-1} 2, 4-D and 50 mL L^{-1} coconut water (Straub et al., 1992). The pH was set to 5.7 and medium was dispensed in 25 mL aliquots to 25 X 150 mm tubes that were capped and autoclaved at 18 psi for 20 minutes. Developing callus was excised from the explants after one month in culture, and calli of different morphologies were cultured separately.

Protoplasts were prepared from cultures of *S. virginicus* and *K. virginica* by digesting them in an enzyme mixture (Table 1), followed by filtering through a 50 μm mesh screen. The protoplasts were pelleted by centrifugation at 100 g for 5 min, and cell debris removed by centrifuging the resuspended protoplasts in 21% sucrose or 12% Ficoll (400,000 FW) plus 2 g $CaC1_2$ and 25 g KC1 L^{-1}.

Table 1. *The recipe of the enzyme solutions used for the preparation of protoplasts from the halophytes,* Kosteletzkya virginica *and* Sporobolus virginicus.

Compound	*High Conc. (HC) (100 mL $^{-1}$)*	*Low Conc. (LC) (100 mL $^{-1}$)*
Cellulase-RS	2.0 g	1.0 g
Macerozyme-R10	0.5 g	0.5 g
Pectolyase-Y23	0.1 g	0.05 g
Polyvinl-pyrrolidone	0.1 g	0.1 g
Ascorbic acid	35 mg	35 mg
Trypsin inhibitor	100 mg	100 mg
Catalase	250 mg	250 mg
CPW salts*	1 x	1 x
Sorbitol	0.04 mole	0.04 mole
MES	500 μmole	500 μmole
pH	5.7	5.7

**For CPW salts see Frearson, 1973*

In all experiments, the viability of cells and protoplasts was assessed by staining with FDA (Fluorescein Diacetate, Widholm, 1972). FDA, stored in an acetone stock solution (5 mg ml^{-1}) at 4°C, was added to the protoplasts or suspension cells to give a final concentration of 0.01%. After 5 min at 21°C the cells and protoplasts were examined for fluorescence. The diameter of protoplasts and intactness of protoplasmic membrane was assessed under an Olympus C-2 Image Analyzer. The salt tolerance of calli, cells, and protoplasts was determined either in the medium without NaCl (unsalinized control) or in that same medium supplemented with NaCl (salinized treatment). Relative Growth Rate (RGR) was calculated from dry weight values according to the equation: RGR = $(W_2-W_1)/W_1/(T_2-T_1)$, where W refers to the dry weight, T the time, and the subscripts to the specific harvests. The experimental design was a Completely Randomized Block with three replications. Each replicate consisted of at least 1,000 observations to determine cell viability, protoplast diameter, protoplast density, and protoplast viability.

To determine the rate of salt excretion of *S. virginicus*, a comparative growth chamber study was used, where temperature and light/dark periods were varied, but the salinity of the nutrient solution was held at 30 g L^{-1}. The excreted salt was collected by rinsing the leaves with distilled water. The collected solution was dried at 60°C and the quantity of salt measured.

3. RESULTS

The relative growth rate (RGR) of *Kosteletzkya virginica* callus was not inhibited with the 85 mM NaCl treatment than in the unsalinized control (Figure 1).

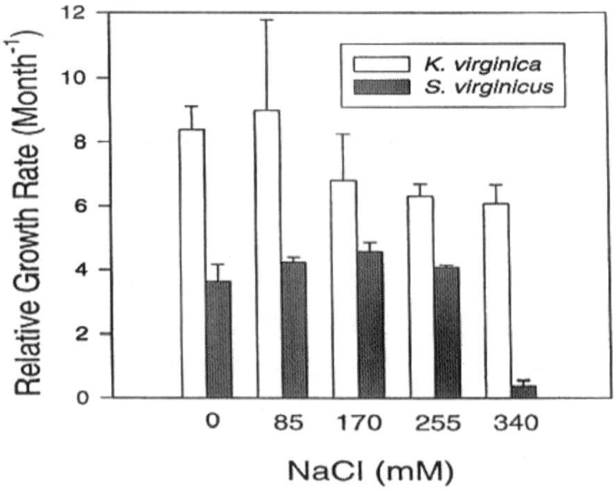

Figure 1. *Relative growth rate (month-1) of callus for* Kosteletzkya virginica *and* Sporobolus virginicus *treated with various salinities. Cultures were grown on nutrient media containing 0, 85, 170, 255, and 340 mM NaCl respectively. Values are means of dry weight ± SD of three replicate cultures.*

When salinity was further increased, the RGR of *K. virginica* callus decreased to 81%, 75%, and 71% of the control values respectively at 170, 255, and 340 mM NaCl. *Sporobolus virginicus* callus growth was stimulated by the moderate salt treatment. By the end of the experiment, mean dry weights of salt grown callus of S. virginicus were about 120%, 127%, and 113% of the control values respectively at 85 mM, 170 mM, and 255 mM NaCl treatments. However, *S. virginicus* callus growth was severely inhibited at the 340 mM NaCl treatment. At this salinity, the callus growth was only 11% of the control value. The RGR of *K. virginica* under the five treatments were significantly higher than those of *S. virginicus*, based on the paired t-test (Figure 1). The difference was particularly large at the 340 mM NaCl where the RGR of *K. virginica* was 15 times higher than that of *S. virginicus*.

Further comparison of cell survival rate after salt treatment was made for both species by comparing cell viability at the time of sampling with the viability at the beginning of the experiment. The cell viability of both *K. virginica* and *S. virginicus* declined to approximately 50% and 40% of their initial values, respectively at 170 mM salinity and to 25% and 20% of their initial values, respectively, at 340 mM NaCl (Figure 2). The lower viability after eight days indicated an initial cell death upon exposure to salt, with the most severe occurring at 340 mM for both species.

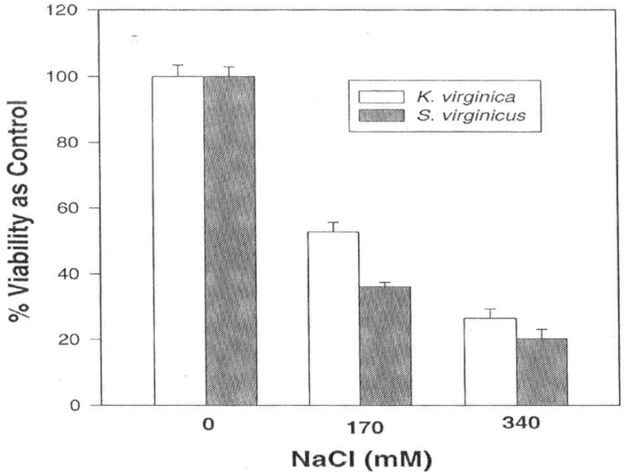

Figure 2. *The percentage of cell viability relative to the control value for* Kosteletzkya virginicia *and* Sporobolus virginicus *under various NaCl treatments. Cells were grown on nutrient solutions containing 0, 170, or 340 mM NaCl for eight days. Values are means of the viability ± SE of three replicates. Each replicate consists of at least 1,000 observations.*

In order to compare the response of protoplasts of two species to salinization, experiments were designed to find conditions at which both species yielded a similar number of viable protoplasts from unsalinized cultures. Obviously, if digestion conditions were favorable for one and not for another, the viability data from saline experiments would not be comparable for two species, one a monocot and the other a dicot. When the cells were treated with the enzyme digestion solution, the cells became plasmolyzed due to water loss and the cell wall was separated from FDA-staining living protoplasts (Figure 3a). After complete digestion, the protoplasts were purified for further saline experiments. The protoplast population of *K. virginica* is shown in (Figure 3b), indicating pure protoplasts without cell debris. In contrast, unpurified protoplasts were mixed with undigested cell (Figure 3c). Viable protoplasts displayed green color after FDA staining (Figure 3d). Dead protoplasts did not produce the green fluorescence and were distinctive from living protoplasts when viewed under a mixed violet and tungsten light (Figure 3e). Noticeably, dead protoplasts remained visually intact after death (Figure 3e).

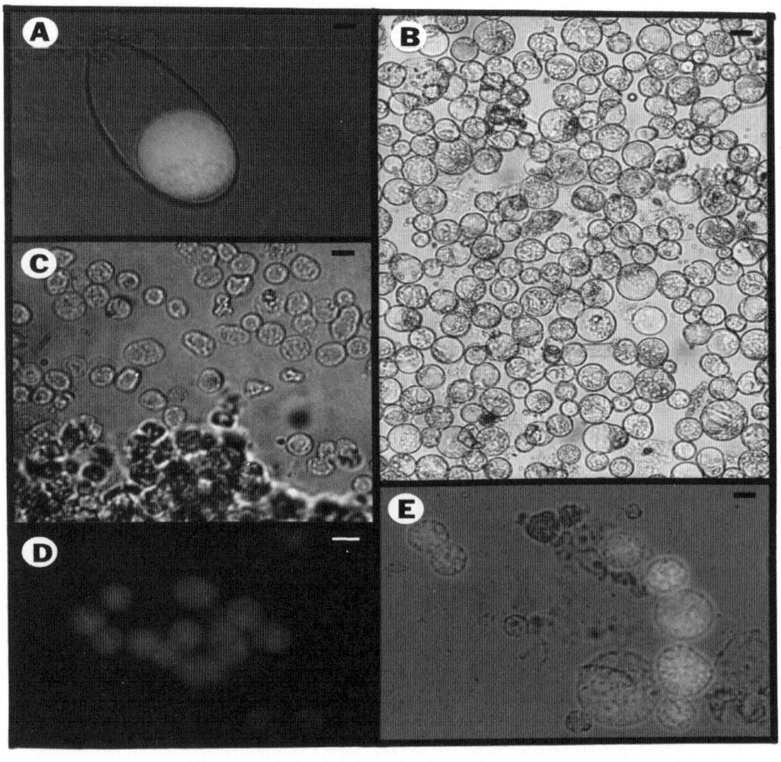

Figure 3. *Preparations and release of protoplasts. The cell plasmolyzed in the enzyme solution for release of* Kosteletzkya virginica *protoplasts (3a, bar = 9.5 μm) and protoplast population after purification (3b, bar = 28 μm). Photo 3a was taken after staining with fluorescein diacetate for 10 minutes. Photo 3b was taken before the protoplasts were used in saline experiments. Photo 3c indicates the release of protoplasts from cell materials of* Sporobolus virginicus *before purification (Bar = 12 μm). Photo 3d represents FDA-stained viable protoplasts of* Sporobolus virginicus *(Bar = 12 μm). Photo 3e shows viable protoplasts (green-FDA fluorescence) and dead protoplasts (non-FDA fluorescence) under mixed violet and tungstein lights (Bar = 14 μm).*

Both ambient temperature and concentration of the enzyme in digestion solution were compared for both species. Both the high concentration (HC) and low concentration (LC) of the enzyme mixture (Table 1) yielded a substantial number of protoplasts for both species (Figure 4). Within the initial 1.5 hours, *S. virginicus* cells produced fewer protoplasts (Figure 4a) than *K. virginica* (Figure 4b). During six to 12 hours, the yield of protoplasts from both cell cultures reached the maximum (Figure 4a & 4b). In both species the high concentration (HC) of the enzyme mixture was superior to the low concentration (LC) for isolation at 21 °C, but not at 28 °C.

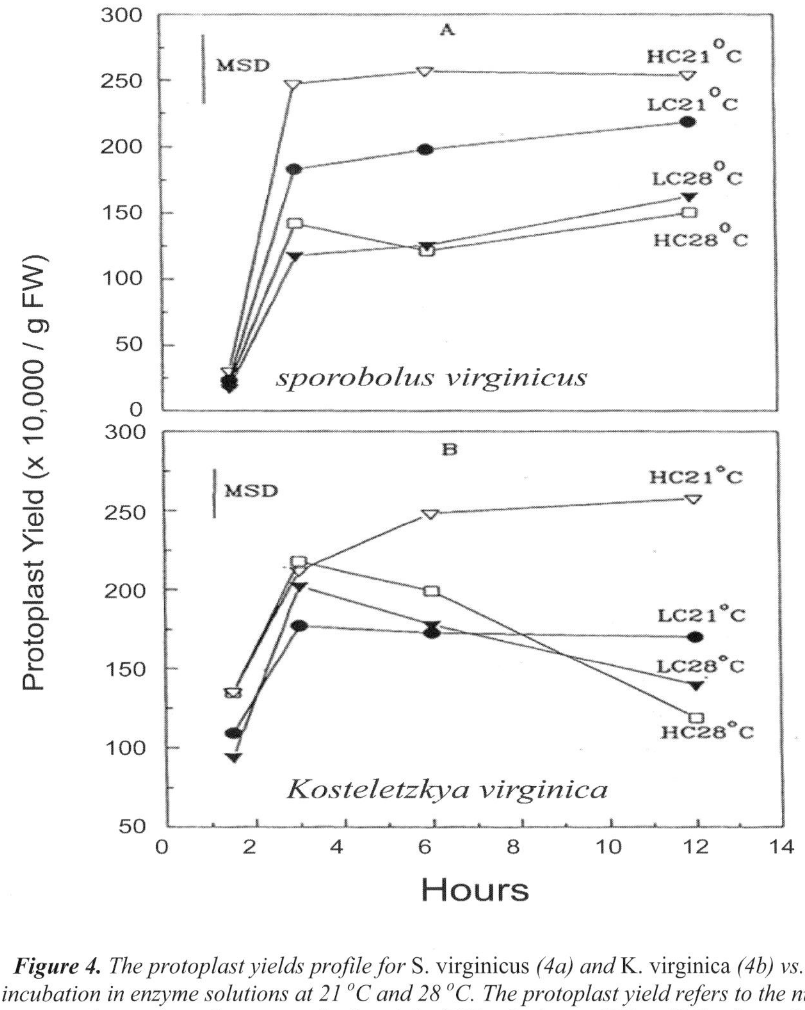

Figure 4. *The protoplast yields profile for* S. virginicus *(4a) and* K. virginica *(4b) vs. time of incubation in enzyme solutions at 21 °C and 28 °C. The protoplast yield refers to the number of protoplasts released per gram fresh weight (FW) of culture. HC and LC refer to the high concentration and low concentration of enzyme solution (see TABLE 1). MSD = Maximum Standard Deviation from three replicate preparations.*

We further investigated the impact of salt adaption on protoplast release since the biochemical changes in cell walls have been reported for plants exposed to salinity (Seliskar, 1985b). In general, the efficiency of protoplast release from cultures (digestibility) was sharply reduced over an eight-day NaCl treatment, although the digestion conditions were optimum for both species. The protoplast release from the salinized culture relative to the unsalinized control for *K. virginica* declined to 76%, 69%, 50%, and 30% after exposed to 170 mM NaC1 respectively for two, four, six, and eight days (Figure 5). The different degree of reduction in the digestibility of cell wall after salinization between *K. virginica* and *S. virginicus* might be attributed to changes in cell wall composition (Iraki et al., 1989ab), which might be relative to their taxonomic difference, one a dicot and the other a monocot grass.

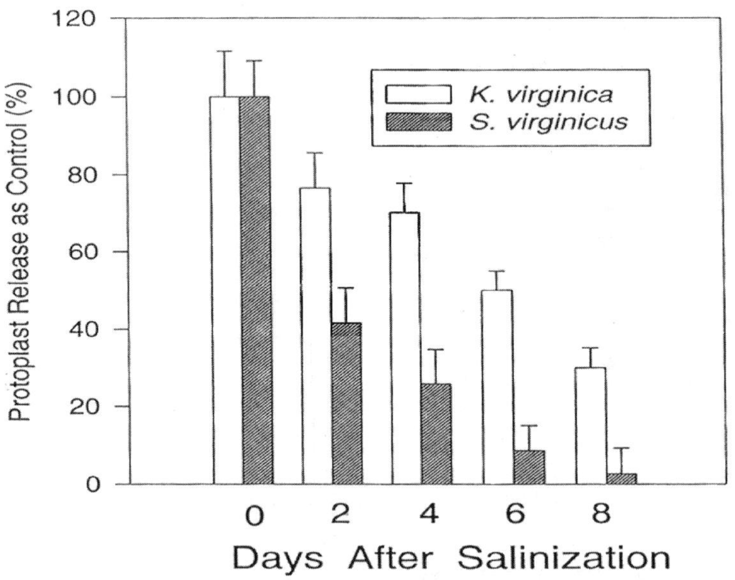

Figure 5. *Protoplast releases in relation to the duration (days) the cell cultures of* Kosteletzkya virginica *and* Sporobolus virginicus *were previously exposed to NaC1. The Y-axis represents the % of protoplasts released from salinized cultures relative to the unsalinized control. The cell cultures used for protoplast isolation were treated with 170 mM NaC1 for 0, 2, 4, 6, and 8 days.*

The above reliable procedure to produce intact and viable protoplasts made it possible to further investigate the response of halophyte protoplasts to salinization for both species. Protoplasts isolated with the high concentration of enzyme solution at 21°C were immediately used for saline tests. Both *K. virginica* and *S. virginicus* protoplasts decreased their diameter after exposure to NaC1 for four hours (Table 2).

Table 2. *The volume change of* Kosteletzkya virginica *and* Sporobolus virginicus *protoplasts in response to salinities. All treatments contain Kao salts (Kao & Michayluk, 1975), 0.4 M mannitol, 3% sucrose, and 2.0 mg L* $^{-1}$ *NAA. Values are means of three replicates after protoplasts exposed to salinity for four hours. Each replicate consists of at least 100 observations.*

NaCl (mM)	*Diameter (D) (μm)*	*% of Control D(%)*	*Volume (V) (μm³)*	*% of Control V(%)*
		Kosteletzkya virginica		
0	40.0	100	33493	100
43	38.4	96	29633	89
85	36.4	91	25240	75
128	35.3	88	23020	69
170	33.5	84	19675	59
		Sporobolus virginicus		
0	17.3	100	2688	100
43	16.7	96	2417	89
85	15.9	92	2093	78
128	15.2	88	1838	68
170	14.4	80	1561	58

When 170 mM NaC1 was incorporated into the medium, *K. virginica* protoplast diameter and volume relative to the control were reduced to 84% and 59%, respectively (Table 2). *Sporobolus virginicus* had a similar response. It is interesting to note that *K. virginica* protoplasts have a larger diameter (40μm) under unsalinized conditions than do *S. virginicus* protoplasts (17.3 μm). A similar change in the percentage of the volume for both species indicated similar dehydration for both species when exposed to salinity.

Similar profiles of protoplast density per ml of culture were observed between the unsalinized control and salinized treatment for *K. virginica* (Figure 6). But a similar comparison with *S. virginicus* indicated a sharp reduction in protoplast density with time in the 340 mM NaC1 treatment (Figure 6). Although the non-salinized decline in protoplast density in both species was similar, the response to salinity was greatly different.

We further tested the protoplast viability profiles after saline treatments for both species by running a series of salinity trials. The protoplast viability for the unsalinized control declined for both species during the time-course experiment (Table 3). Trends of decrease in protoplast viability throughout experiments for both species when unsalinized were similar to that of Cupples et al., 1991. He reported that guard cell protoplasts of *Nicotiana glauca* either grew and deposited new cell walls or they died, where the percentage of surviving tobacco cells declined to 5-10% of the initial number of cells after 72 hours and to 1% of the initial number after six days in culture. In our case, the viability of both *S. virginicus* and *K. virginica* in the unsalinized control declined to 43.2% and 13.6%, respectively, of initial viability after eight days in culture.

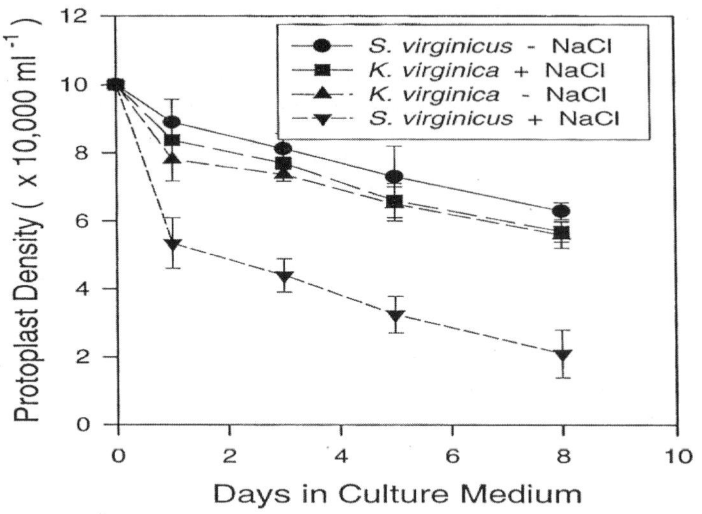

Figure 6. *The protoplast density profile (number of visible intact protoplasts per mL) during the time-course in the culture medium.* Kosteletzkya virginica *protoplasts supplemented with 340 mM NaC1 and without NaC1 is in comparison to* Sporoblus virginicus *protoplasts supplemented with 340 mM NaC1 and without NaC1. Values are mean ± SE of three replicates; each replicate consists of at least 1,000 observations.*

Table 3. *The protoplast viability (%) of* Kosteletzkya virginica *and* Sporobolus virginicus *in response to salinization during eight-days in culture. All treatments contain Kao salts (Kao & Michayluk, 1975), 3% sucrose, 0.4 M mannitol, and 2.0 mg L^{-1} NAA in addition to various amount of NaC1. Values are means ± SE of three replicates. Each replicate consists of at least 1,000 observations.*

NaC1 (mM)	*Days* 0	2	5	8
		Kosteletzkya virginica		
0	51.5 ± 2.8	35.7 ± 4.0	16.6 ± 0.8	7.0 ± 0.8
43	51.5 ± 2.8	26.5 ± 3.3	16.0 ± 3.8	9.7 ± 0.9
85	51.5 ± 2.8	30.7 ± 2.8	17.2 ± 6.8	8.0 ± 3.0
170	51.5 ± 2.8	29.1 ± 0.3	17.3 ± 1.4	8.6 ± 3.1
340	51.5 ± 2.8	25.1 ± 8.8	*13.5 ± 1.1*	6.9 ± 2.0
		Sporoblus virginicus		
0	61.1 ± 3.1	47.8 ± 1.0	39.8 ± 2.5	26.4 ± 0.6
43	61.1 ± 3.1	44.7 ± 2.1	28.2 ± 2.1	7.8 ± 0.7
85	61.1 ± 3.1	42.4 ± 2.1	19.1 ± 1.2	0.4 ± 0.1
170	61.1 ± 3.1	40.2 ± 2.2	14.6 ± 0.8	0.1 ± 0.1
340	61.1 ± 3.1	37.6 ± 2.4	10.2 ± 1.9	0.9 ± 0.2

Kosteletzkya virginica protoplasts neither showed much difference in viability between the unsalinized control and salinity treatments nor demonstrated a difference in viability among the salinity treatments (Table 3), with percent viability of the eight-day culture virtually the same at 0 and 340 mM NaCl (13% of initial values). *Sporobolus virginicus* protoplasts did show a difference in viability between control and salinity treatments. After eight days in culture, the viability of salinized protoplasts decreased to 13% of the initial values at 43 mM NaC1 and few viable protoplasts of *S. virginicus* remained at the salinities greater than 43 mM NaC1 (Table 3). The viability of the two species was similar at 43 mM salinity after the eight-day treatment (Table 3). Compared to the initial value, *K. virginica* protoplasts maintained their viability under salinization, further indicating that this halophyte species was able to cope with salinity at the protoplast level.

When whole plants are considered, cellular mechanisms are integrated with the whole plant physiological process. In order to get a measure of the whole plant process, salt excretion from the leaves of *S. virginicus* was measured. *Kosteletzkya virginica* does not have salt glands. *Sporobolus virginicus* excreted as much as 0.61 $\mu g\ mm^{-2}\ hr^{-1}$ salts to the surface of its leaves (Table 4).

Table 4. *Salt excretion ($\mu g\ mm^{-2}\ hr^{-1} \pm 1$ SEM) from the leaves of* Sporobolus virginicus *in relation to the temperature and light and dark cycles (L/D).*

Temperature (°C)	*L/D (hr)*	*Salt excreted ($\mu g\ mm^{-2}\ hr^{-1}$)*
38/28	16/8	0.45 ± 0.07
28/18	16/8	0.61 ± 0.02
28/18	10/14	0.43 ± 0.08

4. DISCUSSION

Sporobolus virginicus seedlings are extremely resistant to high salinity, and 80% of the plants survived 2.5 times the salinity of seawater for the duration of an eight-month experiment (Gallagher, 1979). Even though little growth occurred under high salinity conditions, mere survival until rainfall or spring tides dilute the soil solution is an important adaptive feature for *S. virginicus* plants living in saltpan fringe areas (Gallagher, 1979). Root growth increased linearly with increasing salinity up to 450 mM NaC1 in whole plant experiments reported by (Marcum & Murdoch, 1992). A stimulation of *S. virginicus* root growth by seawater was observed by Blits & Gallagher (1991).

In contrast, *Kosteletzkya virginica* seedlings did not survive in 66% strength seawater although seeds could be germinated in this medium (Poljakoff-Mayer et al., 1992, 1994). Root and shoot growth of *K. virginica* were inhibited at external NaC1 concentrations greater than 85 mM with root and shoot dry weights reduced to about 40% (170 mM) and 20% (255 mM), respectively, of the control values (Blits & Gallagher, 1990ab). Under salt-water irrigation (25%) in field conditions mature plants do flower and set seed (Gallagher, 1985).

Compared to protoplasts, the osmotic adjustment in whole *S. virginicus* plants was largely active, tissue dehydration playing only a minor role, as indicated by shoot fresh weight vs. dry weight ratios, which were relatively constant across salinity levels (Marcum & Murdoch, 1992). However, the water content of salinized *K. virginica* leaves, even under conditions producing maximum growth, was somewhat lower than in the controls. Water loss in this case may serve to concentrate solutes at levels required to sustain growth under saline conditions (Blits & Gallagher, 1991; Romero-Aranda et al., 2001).

If the whole plants are considered, the greater salt tolerance of *S. virginicus* may be in part attributed to the salt glands in the leaves, which excrete salts. In contrast, *K. virginica* has no such salt gland to excrete salts. Hence the integrated whole plant characteristics play a major role in the salinity tolerance of *S. virginicus*.

Our observations indicate: (1) higher protoplast viability of *K. virginica* than that of *S. virginicus* after salinization, and (2) both species experienced a similar drop in cell viability upon salinization. It appears, although *S. virginicus* whole plants demonstrate higher salt tolerance than *K. virginica*, the callus, individual cells, and protoplasts of *K. virginica* expressed higher or similar tolerance to that of *S. virginicus*.

These observations on protoplasts, cell suspension, callus cultures, and whole plants also support the concept of the integrated nature of salt tolerance based on various levels of organizations, including morphometric plasticity (Seliskar, 1985a, 1985b), and biochemical compositions of cell walls and protoplasmic membranes (Wu et al., 1998, 2005). The different changes in protoplast density upon salinity treatment (Figure 6) might indicate a different biochemical adaptation in addition to a divergent plasma membrane composition in the two species. Alteration of the chemical structure of the cell wall was reported by Iraki et al., (1989a, 1989b). *Sporobolus virginicus* protoplasts could die via collapse of the protoplasmic membrane, while *K. virginica* could die via loss of permeability of the protoplasmic membrane. Noticeably, both species had certain numbers of protoplasts that remained visually intact after death (Figure 3e).

Therefore, the salt tolerance of *K. virginica* reflects closely its cellular tolerance pattern, while the salt tolerance of *S. virginicus* is the result of combining cellular salt-tolerance at moderate levels of salinity with whole plant mechanisms (such as salt glands), which maintain internal salinities at levels within the cellular tolerance of the plant.

5. SUMMARY

Kosteletzkya virginica is a potential grain producer and yields up to 1.46 tons of edible seeds per ha under irrigation with 70% strength seawater. *Sporobolus virginicus* can survive salinities as high as 2.5 times seawater and may yield 8.75 tons of forage per ha with coastal water irrigation. At the whole plant level, *S. virginicus* is more salt-tolerant than *K. virginica*. Their comparative in vitro cellular responses to salinity are reported here. The callus of *K. virginica* had higher relative growth rates than that of *S. virginicus* within the range of 0 to 340 mM NaC1 salinities, particularly at the highest salinity tested. Both species had a similar number of cells survive at the salinities of 170 and 340 mM NaC1. Upon salinization, *K. virginica* had higher cell wall digestibility and protoplasmic membrane intactness of than did *S. virginicus*, which might reflect the biochemical differences in the composition of the cell wall and protoplasmic membrane. *K. virginica* protoplast viability was higher than that of *S. virginicus* at higher salinities up to 340 mM NaC1. *Kosteletzkya virginica* demonstrated a consistent pattern of salt tolerance between cellular and whole plant levels but *S. virginicus* did not. The relatively higher salt tolerance of *S. virginicus*, at the whole plant level, maybe in part attributed to the presence of salt glands in its leaves. *Kosteletzkya virginica* has no such glands.

6. ACKNOWLEDGEMENTS

The authors thank Michael League for his help with the manuscript. Support for this research came from the University of Delaware Sea Grant College Program under Grant NA16RG0162, Project R/B-22 from the Office of Sea Grant, National Oceanic and Atmospheric Administration (NOAA), US Department of Commerce, and from the Coastal Ocean Program Office of NOAA through Grant NA90AA-D-SG457 to the University of Delaware Sea Grant College Program.

7. REFERENCES

Ball, M.C. 1988. Ecophysiology of Mangroves. Trees-Structure and Function 2: 129-142.

Blits, K.C. & Gallagher, J.L. 1990a. A salinity tolerance of *Kosteletzkya virginica* I. shoot growth, ion, and water relations. Plant Cell and Environment 13: 409-418.

Blits, K.C. & Gallagher, J.L. 1990b. Salinity tolerance of *Kosteletzky virginica* II. Root growth, lipid content, ion and water relations. Plant Cell and Environment 13: 419-425.

Blits, K.C. & Gallagher, J.L. 1990c. Effect of NaC1 on lipid content of plasma membranes isolated from roots and cell suspension cultures of the dicot halophyte Kosteletzkya virginica (L.) Presl. Plant Cell Reports 9: 156-159.

Blits, K.C. & Gallagher, J.L. 1991. Morphological and physiological responses to increased salinity in marsh and dune ecotypes of *Sporobolus virginicus*. Oecologia 87: 330-335.

Blits, K.C., Cook, D.A. & Gallagher, J.L. 1993. NaC1 tolerance in cell suspension cultures of the halophyte *Kosteletzkya virginica*. Journal of Experimental Botany 44: 681-686.

Breckle, S.W. 1995. How do halophytes over some salinity? In: Khan, A.A. & Ungar, I.A. (Eds.), Biology of Salt Tolerant Plants, Chelsea, Michigan: Book Crafters. 199-214 pp.

Cook, D.A., Decker, D.M. & Gallagher, J.L. 1989. Regeneration of *Kosteletzkya virginica* from callus cultures. Plant Cell Tissue and Organ Culture 17: 111-119.

Cupples, W., Lee, J. & Tallman, G. 1991. Division of guard protoplasts of *Nicoltiana glauca* (Graham) in liquid cultures. Plant Cell and Environment 14: 691-697.

Dubouzet, J.G., Sakuma, Y., Ito, Y., Kasuga, M., Dubouzet, E.G., Miura, S., Seki, M., Shinozaki, K. & Yamaguchi-Shinozaki, K. 2003. OsDREB genes in rice, Oryza sativa L., encode transcription activators that function in drought-, high-salt- and cold-responsive gene expression. Plant Journal 33: 751-63.

Frearson, E.M., Power, J.B. & Cocking, E.C. 1973. The isolation, culture and regeneration of Petunia leaf protoplasts. Developmental Biology 3: 130-137.

Gallagher, J.L. 1979. Growth and element compositional responses of *Sporobolus virginicus* (L.) Kunth to substrate salinity and nitrogen. American Midland Naturalist 102: 68-75.

Gallagher, J.L. 1985. Halophytic crops for cultivation at seawater salinity. Plant and Soil 89: 323-336.

Gallagher, J.L. & Seliskar, D.M. 1993. Selecting halophytes for agronomic value: Lessons from whole plants and tissue cultures. In: L. Moncharoen (Ed.), Utilizing Salt-affect lands, Bangkok, Thailand: Funny Publishing Limited Partnership. 414-425 pp.

Iraki, N.M., Bressan, R.A., Hasegawa, P.M. & Carpita, N.C. 1989a. Alteration of the physical and chemical structure of the primary cell wall of growth-limited plant cells adapted to osmotic stress. Plant Physiology 91: 39-47.

Iraki, N.M., Bressan, R.A., Hasegawa, P.M. & Carpita, N.C. 1989b. Extracellular polysaccharides and proteins of tobacco cell cultures and changes in composition associated with growth-limiting adaptation of water and saline stress. Plant Physiology 91: 54-61.

Kao, K.N. & Michayluk, M.R. 1975. Nutritional requirements for growth of Vicia hajastana cells and protoplasts at a very low population density in liquid media. Planta (Berl.) 126: 105-110.

Kumar, S., Dhingra, A. & Daniell, H. 2004. Plastid-expressed betaine aldehyde dehydrogenase gene in carrot cultured cells, roots, and leaves confers enhanced salt tolerance. Plant Physiology 136: 2843-54.

Li, X., Seliskar, D.M., Moga, J.A. & Gallagher, J.L. 1995. Plant regeneration from callus cultures of salt marsh hay, Spartina patens, and its cellular-based salt tolerance. Aquatic Botany 51: 103-113.

Marcum, K.B. & Murdoch, C.L. 1992. Salt tolerance of the coastal salt marsh grass, *Sporobolus virginicus* (L.). New Phytologist 120: 281-288.

Murashige, T. & Skoog, F. 1962. A revised medium for rapid growth and bioassay with tobacco tissue cultures. Physiologia Plantarum 15: 473-497.

Naidoo, U., Barnabas, A.D., Lawton, J.R. & Naidoo, G. 1992. Salt glands on the leaves of *Sporobolus virginicus* (L.) Kunth. Electron Microscopy 3: 457-458.

Polijakoff-Mayer, A., Somers, G.F., Werker, E. & Gallagher, J.L. 1992. Seeds of *Kosteletzkya virginica* (Malvaceae): their structure, germination, and salt tolerance. I. Seed Structure and dormancy. American Journal of Botany 79: 249-256.

Polijakoff-Mayer, A., Somers, G.F., Werker, E. & Gallagher, J.L. 1994. Seeds of *Kosteletzkya virginica* (Malvaceae): their structure, germination, and salt tolerance. II. Germination and salt tolerance. American Journal of Botany 81: 54-59.

Rhodes, D. & Hanson, A.D. 1993. Quaternary ammonium and tertiary sulfonium compounds in higher plants. Annual Review of Plant Physiology and Plant Molecular Biology 44: 357-384.

Romero-Aranda, R., Soria, T. & Cuartero, J. 2001. Tomato plant-water uptake and plant-water relationships under saline growth conditions. Plant Science 160: 265-272.

Seliskar, D.M. 1985a. Morphometric variations of five marsh halophytes along environmental gradients. American Journal of Botany 72:1340-1352.

Seliskar, D.M. 1985b. Effect of reciprocal transplanting between extremes of plant zones on morphometric plasticity of five plant species in an Oregon salt marsh. Canadian Journal of Botany 63: 2254-2262.

Smith, M.K. & McComb, J.A. 1981. Effect of NaC1 on the growth of whole plants and their corresponding callus cultures. Australian Journal of Plant Physiology 8: 267-275.

Straub, P.F., Decker, D.M. & Gallagher, J.L. 1992. Characterization of tissue culture initiation and plant regeneration in *Sporobolus virginicus*. American Journal of Botany 79: 1119-1125.

Tal, M. 1984. Salt tolerance of tissue cultures. In: R. Staples (Ed.), Salinity tolerance in plants, New York, NY: Academic Press. 301-320 pp.

von Hedenstrom, J. & Breckle, S. 1974. Obligate halophytes? A test with tissue culture methods. Zeitzchrift für Pflanzenphysiologie 107: 347-356.

Warren, R.S., Baird, L.M. & Thompson, A.K. 1985. Salt tolerance in cultured cells of *Spartina pectinata* Plant Cell Reports 4: 84-87.

Wei, Y., Guangmin, X., Daying, Z. & Huimin, C. 2001. Transfer of salt tolerance from *Aeleuropus littorulis* sinensis to wheat (*Triticum aestivum* L.) via asymmetric somatic hydridization. Plant Science 161: 259-266.

Widholm, J.M. 1972. The use of fluorescein diacetate and phenosafranine for determining viability of cultured plant cells. Stain Technology 47: 189-194.

Wu, J., Seliskar, D.M. & Gallagher, J.L. 1998. Stress tolerance in the marsh plant *Spartina patens*: Impact of NaCl on growth and root plasma membrane lipid composition. Physiologia Plantarum 102: 307-317.

Wu, J., Seliskar, D.M. & Gallagher, J.L. 2005. The response of plasma membrane lipid composition in callus of the halophyte *Spartina patens* (Poaceae) to salinity stress. American Journal of Botany 92: 852-858.

CHAPTER 14

ECO-PHYSIOLOGICAL STUDIES ON INDIAN DESERT PLANTS: EFFECT OF SALT ON ANTIOXIDANT DEFENSE SYSTEMS IN ZIZIPHUS SPP.

N. SANKHLA[1], H. S. GEHLOT[2], R. CHOUDHARY[2], S. JOSHI[2] AND R. DINESH[2]

[1]*Texas A&M University, Agriculture Research Center, Dallas (TX), USA 75252*
[2]*Botany Department, J.N. Vyas University, Jodhpur, India 342001*

Abstract. *Ziziphus* spp. are extensively used for fruit production in hot, arid and saline tracts of Indian desert. In order to gain an insight into the role of salt on antioxidant activity, levels of proline, H_2O_2, MDA, ASC, GSH, and the activities of enzymes involved in detoxification of reactive oxygen species (CAT, PER, SOD, APX, MDHAR, DHAR, GR) were monitored in *Z. nummularia* and *Z. rotundifolia* seedlings grown under salt stress (100 mM, 200 mM). High salinity caused a promotion in MDA, H_2O_2 and proline content and inhibited the activities of CAT, PER and SOD in both the species, but increased the activities of APX, MDHAR, DHAR and GR in *Z. rotundifolia*, and MDHAR and GR in *Z. nummularia*. In *Z. nummularia*, salinity also inhibited the activity of APX. At low saliniy, both the species indicated elevated levels of ACS and GSH. In *Z. rotundifolia*, high antioxidant level was maintained even under high salinity. These results indicate that in *Ziziphus* SOD is not crucial in detoxification of ROS, and ASC and GSH play an important role in ROS scavenging. An evaluation of current results, as well as earlier reports relating to growth and physiological characteristics, suggests that although both the species of *Ziziphus* are moderately salt tolerant, at high salinity *Z. rotundifolia* may have an edge over *Z. nummularia* with respect to its performance.

1. INTRODUCTION

The North-West Indian Thar desert (20-30°N and 68-70°E) accounts for nearly 63% of the arid zone of the country of India. The region is characterized by an acute shortage of water due to highly erratic and uncertain precipitation, extremes of temperatures, high photon irradiance, high thermal load, and at specific sites, inimical levels of salt in the soil (Sankhla et al., 1975). There is a notable lack of specific knowledge about the role of environmental constraints in growth, development and yield of Indian desert plants, and equally apparent is the paucity of information on the adaptive responses of plants to sub-optimal conditions.

Zizipus spp. (locally knows as ber) constitute an important source of fresh edible fruits in arid regions of India (Pareek, 2001). Although a large number of

M.A. Khan and D.J. Weber (eds.), Ecophysiology of High Salinity Tolerant Plants, 201-213.

wild and domestic species of ber are available in the Indian deserts, but in the North Western Indian desert traditionally three species (*Z.nummularia, Z. rotundifolia, Z. mauritiana*) are used for commercial fruit production. Besides fruits, these plants also yields timber of marginal value, brushwood, fuelwood, and leaf fodder (Pareek, 2001). Thus, ber is an important multi-purpose species suitable for integration into agroforestry systems.

Ziziphus species exhibit a combination of drought avoidance and drought tolerance mechanisms, including effective photoprotective mechanisms and stimulation of antioxidative metabolic pathways, osmotic adjustment and sensitive stomatal closure under stress, to effectively postpone dehydration and minimize damage to survive in hostile environments with unpredictable precipitation (Clifford et al., 1998; Sankhla, 1998; Arndt et al., 2001; Pareek, 2001). Leaves of *Z. rotundifolia* have high mucilage concentration (Clifford et al., 2002) which may serve as a source for the remobilization of solutes for osmotic adjustment, thus enabling more effective water uptake and assimilate redistribution into roots and stems prior to defoliation as the drought stress intensifies. In particular, *Z. rotundifolia* has a high degree of plasticity in response to water deficits. It appears likely that the extensive root systems and readiness to shed leaves under severe drought constitute the main mechanism of success of *Ziziphus* species in extremely hot and arid environments (Sankhla, 1998; Jones, 1999).

Ziziphus is also considered a moderately salt tolerant species (Pareek, 2001). However, the different species vary with respect to their salt tolerance. For instance, *Z. nummularia* seedlings did not survive at 15 dSm^{-1} (Meena et al., 2003), while the established plants are reported to grow even at 21 dSm^{-1} salinity (Dhankhar et al., 1980). Hooda et al. (1990) reported that the growth of *Z.mauritiana* trees was reduced at 5 dSm^{-1}, fruit set was affected at 15 dSm^{-1} resulting in 70% yield reduction, but at 20 dsm^{-1} salinity level, not a single plant survived on sandy loam soil. In recent years, by budding elite *Z. mauritiana* clones on root stocks of wild drought hardy *Z. nummularia* and *Z. rotundifolia*, have been introduced in the arid regions of Indian desert (Pareek, 2001). However, there is a notable lack of specific information on basic aspects of eco-physiology, adaptive mechanisms and relative salt tolerance of these species.

Abiotic stresses, especially salinity and drought, are among the primary causes of loss in crop productivity and quality. As with other abiotic stresses, plant responses to excessive salt represent a complex whole plant syndrome with physiological and biochemical functions controlled by multiple genes, and environmental and soil factors having influence on its expression (Bohnert & Jensen, 1996; Hasegawa et al., 2000). Although the specific mechanisms imparting salt tolerance are not completely understood, spectacular progress has been made in our understanding and improvement of salt tolerance in plants (Chinnusamy et al., 2005; Vinocur & Altman, 2005; Tester & Bacic, 2005). Salt tolerance is thought to be mediated by stress adaptation effectors that influence ion homeostasis, osmolyte biosynthesis, activated oxygen radical scavenging, signaling pathways, water transport and long distance co-ordination (Hasegawa et al., 2000; Mittler, 2002). Salinity induces both hyperosmotic and hyperionic effects in plants which leads to membrane disorganization, increase in activated oxygen species production and metabolic toxicity through oxidative damage to pigments, lipids, proteins and nucleic acids. Although plants have evolved both enzymatic and non-enzymatic mechanisms for scavenging of reactive oxygen species, imbalance between production and quenching of ROS leads to plant toxicity. A correlation between antioxidant capacity and salt tolerance has been demonstrated in a wide array of plant species (Gossett et al., 1994; Rout & Shaw, 2001; Shalata et al., 2001; Bor et al., 2003; Vaidyanathan et al., 2003; Amor et al., 2005).

The specific objective of the current investigation was to study the effect of salinity on the antioxidant system in *Ziziphus* spp. As such, the activities of some enzymes related to ROS scavenging, ascorbate-glutathione cycle, as well as the changes in the level of water soluble antioxidants (ascorbate, glutathione), H_2O_2, lipid peroxidation and proline were monitored in *Ziziphus* seedlings grown under NaCl salinity.

2. MATERIALS AND METHODS

2.1. Plant material and treatments

Seeds of *Ziziphus nummularia* and *Z. rotundifolia* were obtained from mature fruits by breaking the stony endocarp. Seedlings were initiated in plastic pots from seeds in a growth chamber equipped with a light bank consisting of fluorescent and incandescent lamps (600 $\mu mol.m^{-2}.s^{-1}$). During growth the seedlings were maintained at 30/25°C (day/night) under a 16 hour photo period. Forty days old seedlings were stressed with varying concentrations of sodium chloride (100 and 200 mM). Physiological parameters were evaluated after 12 days of salt stress.

2.2. Lipid peroxidation

Lipid peroxidation in the leaf tissue was measured in terms of malondialdehyde (MDA), a decomposition product of the oxidation of polyunsaturated fatty acids, as thiobarbituric acid reactive material as outlined in an earlier publication (Upadhyaya et al., 1985).

2.3. Estimation of proline

Proline was extracted and measured as per the procedure of Bates et al. (1973).

2.4. Determination of hydrogen peroxide

Leaves were homogenized in 50 ml potassium phosphate, pH 6.5, and centrifuged at 10000 g for 25 min. An aliquot was mixed with 1% titanium chloride prepared in conc. HCl. The absorbance of supernatant was measured at 410 nm and H_2O_2 content calculated according to Jana and Choudhri (1981).

2.5. Reduced (ASC) and oxidized (DHA) ascorbate

Ascorbate (ASC) and dehydroascorbate (DHA) were determined following the method used by Wang et al. (1991). This essay is based on the reduction of ferric ion to ferrous ion with ascorbic acid in acid solution followed by formation of the red chelate between 4,7-diphenyl-1,10-phenenthrolin (bathrophenanthrolin) and ferrous ion that absorbs at 534 nm. Total ascorbate was determined through a reduction of DHA to ASC by dithiothreitol.

2.6. Reduced (GSH) and oxidized (GSSG) glutathione

Glutathione was determined after Smith (1985). Plant material was homogenized in 5% (w/v) sulphosalicylic acid. GSSG and total glutathione (GSSG+GSH) content was determined by the 5, 5'-dithio-bis (2-nitrobenzoic acid)-GSSG reductase recycling. GSH content was then estimated from the difference between total glutathione and GSSG.

2.7. ROS scavenging enzymes

Activity of catalase(CAT) was determined by monitoring the decrease in absorbance due to H_2O_2 reduction at 240 nm, whereas peroxidase(PER) activity was measured following the increase in absorbance at 420 nm due to formation of tetraguaiacol (Vaidyanathan et al., 2003). Superoxide dimutase (SOD) activity was measured by recording the inhibition of photochemical reduction of nitroblue tetrazolium(NBT) at 560 nm (Giannopolitis & Ries, 1977).

Ascorbate peroxidase (APX) was extracted in 50 mM potassium phosphate buffer (pH 7) containing 1mM ASC, and its activity was measured by monitoring the oxidation of ASC at 290 nm (Amako et al., 1994), while dehyroascorbate reductase (DHAR) activity was assayed by monitoring the increase in absorbance at 265 nm due to ASC formation. The activities of monohydroascorbate reductase(MDHAR) and glutathione reductase(GR) were measured following the oxidation of NADH at 340 nm (Vaidyanathan et al., 2003).

3. RESULTS

3.1. Lipid peroxidation

Higher MDA content generally reflects increased lipid peroxidation which is symptomatic of cellular damage. Membrane lipid peroxidation in the leaves of *Z. nummularia* and *Z. rotundifolia* were assessed in terms of MDA content (Table 1). MDA content was significantly higher in *Z. nummularia* than in *Z. rotundifolia* even under controlled conditions. The rate of increase in MDA content in the presence of salt was 20-40 % in *Z.nummularia* and 12-20% in *Z. rotundifolia.* Thus, salt induced enhancement in lipid peroxidation was relatively much higher in *Z.nummularia* that in *Z. rotundifolia.*

Table 1. *Effect of salt on MDA, H_2O_2 and proline content in* Ziziphus *spp.*

Salt	*MDA*	*H_2O_2*	*Proline*
Z. nummularia			
Control	35	100	238
Salt 100 mM	42	108	320
Salt 200 mM	49	133	410
Z. rotundifolia			
Control	25	100	134
Salt 100 mM	28	105	284
Salt 200 mM	30	120	346

**Values of MDA are in $nmol^{-1}$ F.Wt, H_2O_2 in % and proline in $\mu g.g^{-1}$ F.Wt*

3.2. Proline accumulation

Both the species of *Ziziphus* accumulate proline in response to salinity stress (Table 1). Although *Z. nummularia* accumulated relatively higher levels of proline in the presence of salt in comparison to *Z.rotundifolia*, the rate of proline accumulation in response to salinity was higher in *Z. rotundifolia.*

3.3. H_2O_2 content

The production of H_2O_2 under stress is known to increase both by enzymatic as well as non-enzymatic processes. In our study, a significant increase in H_2O_2 content in the leaves of *Ziziphus* was observed only at higher salt concentration (Table 1).

3.4. Antioxidant Pool

The tripeptide glutathione and ascorbate are two important antioxidants found in plants that efficiently scavenge oxy-radicals (Alscher & Hess, 1993). We investigated the changes in the level of these antioxidants in response to salinity stress in leaves of *Ziziphus* (Tables 2 & 3).

Table 2. *Effect of salt on ascorbate content in* Ziziphus *spp.*

Treatment	*ASC(T)*	*ASC*	*DHA*	*ASC:DHA*
Z. nummularia				
Control	1576	940	636	1.47
Salt 100 mM	1700	1180	520	2.26
Salt 200 mM	1490	900	590	1.52
Z. rotundifolia				
Control	2730	1780	950	1.87
Salt 100 mM	2980	2100	880	2.38
Salt 200 mM	3120	2300	820	2.80

*Values are in $\mu g.g^{-1}$ F.Wt.

3.4.1. Reduced (ASC) and oxidized (DHA) ascorbate.

During growth at low salt level both *Z. nummularia* and *Z. rotundifolia* exhibited increased ASC and decreased DHA content (Table 2). At high salinity level, however, only *Z. rotundifolia* maintained enhanced ASC level.

3.4.2. Reduced (GSH) and Oxidized (GSSG) glutathione

In presence of low level of salt, both the species of *Ziziphus* showed increased level of GSH, However, at high salt concentration, *Z. rotundifolia* exhibited much higher GSH content than *Z. nummularia* and exhibited a high GSH: GSSG ratio (Table 3).

Table 3. *Effect of salt on glutathione content in* Ziziphus *spp.*

Treatment	*GLUT(T)*	*GSH*	*GSSG*	*GSH:GSSG*
Z. nummularia				
Control	168	116	52	2.23
Salt 100 mM	262	187	75	2.49
Salt 200 mM	300	210	90	2.33
Z. rotundifolia				
Control	255	197	58	3.40
Salt 100 mM	368	298	70	4.25
Salt 200 mM	405	325	80	4.06

* *Values in $\mu g.g^{-1}$ F.Wt.*

3.5. Activities of ROS scavenging enzymes

Under salt stress, the activities of enzymes related to active oxygen detoxification exhibited different responses in the two species of *Ziziphus*. In *Z. rotundifolia*, the activity of CAT increased under mild salt stress, but this increase was not maintained in the presence of high salt (Figures 1 & 2).

In contrast, in *Z. nummularia*, salt brought about a drastic inhibition in the activity of CAT (Figure 1). The activity of PER was also found to be lower than control in the presence of salt in *Z. nummularia*. In *Z. rotundifolia*, decreased activity of PER was recorded only in the seedlings grown in the presence of high salt concentration.

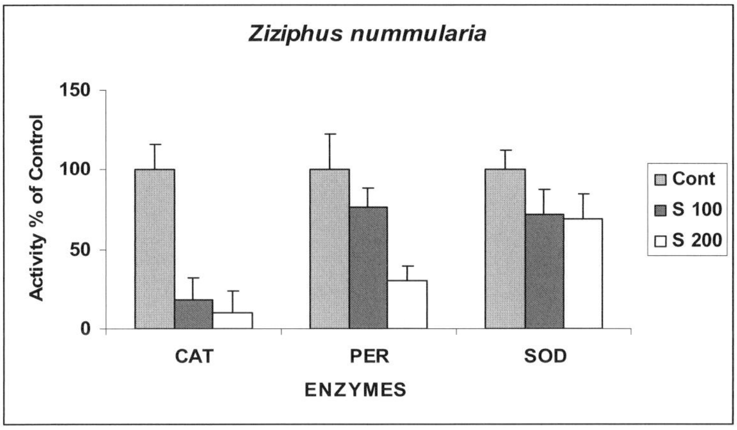

Figure 1. *Effect of salt on the activities of CAT, PER and SOD in* Z. nummularia.
Activity in control: CAT=23.7; PER= 0.35(μM.min^{-1}.g^{-1}F.Wt.); SOD= 35.8(units.g^{-1}.F.Wt.).

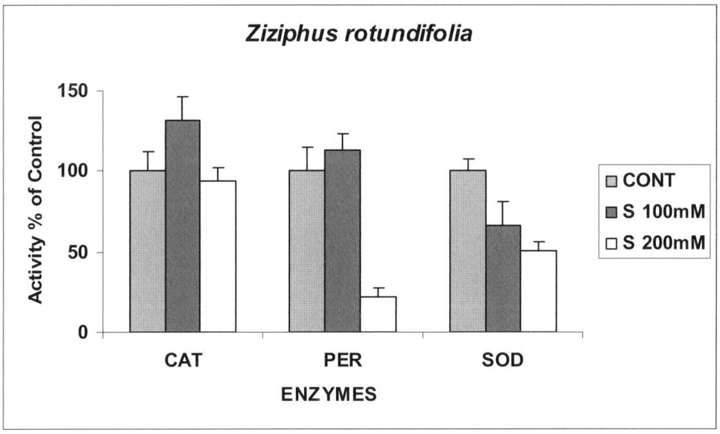

Figure 2. *Effect of salt on the activities of CAT, PER and SOD in* Z. rotundifolia.
Activity in control: CAT=17.7; PER=0.53(μM.min^{-1}.g^{-1}F.Wt.); SOD=19.7(units.g^{-1}.F.Wt.).

Although in many plants SOD directly modulates the level of reactive oxygen species, in *Ziziphus* following growth in salt an inhibition in the activity of SOD was clearly evident in both the species (Figures 1 & 2).

3.6. Activities of enzymes of ascorbate-glutathione cycle

Results relating to salt-induced alteration in the activities of various enzymes involved in ASC-GSH cycle in *Ziziphus* are presented in (Figures 3 & 4).

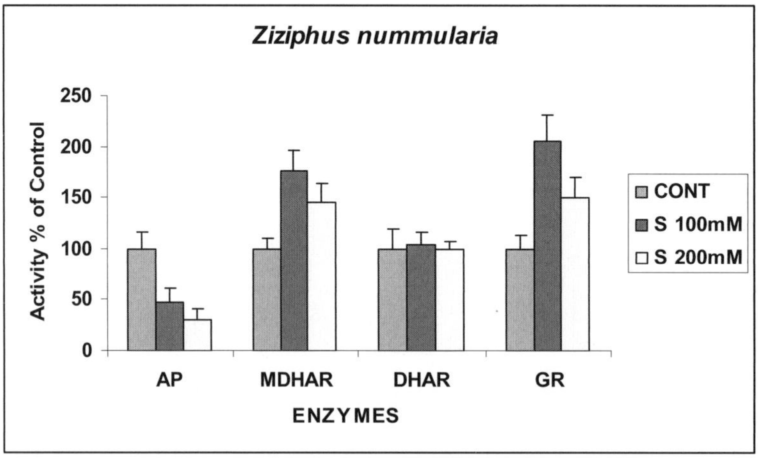

Figure 3. *Effect of salt on the activities of APX, MDHAR, DHAR and GR in* Z. nummularia. *Activity in control: APX=2.5; MDHAR=5.6; DHAR=1.5; GR=0.38 ($\mu M.min^{-1}.g^{-1}F.Wt.$).*

In *Z. rotundifolia,* activities of APX, MDHAR, DHAR, and GR increased significantly at low salt concentration, and continued to show elevated activities than the control at high salt concentration (Figure 4).

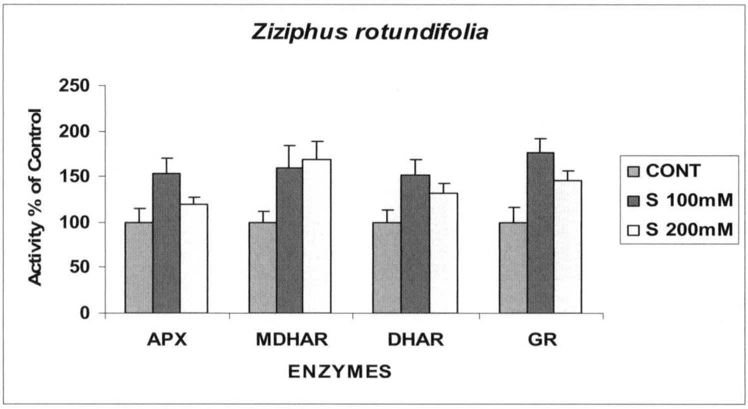

Figure 4. *Effect of salt on the activities of APX, MDHAR, DHAR and GR in* Z. rotundifolia. *Activity in control: APX=3.2; MDHAR=6.2; DHAR=1.8; GR=0.49($\mu M.min^{-1}.g^{-1}F.Wt.$).*

In contrast, in Z. *nummularia*, the activities of APX and DHAR did not show any appreciable increase in the presence of salt. However, a promotion was observed in the activities of MDHAR as well as GR (Figure 3).

4. DISCUSSION

Toxic reactive oxygen species such as superoxide radicals, H_2O_2 and singlet oxygen are formed regularly in many cellular reactions when electrons are misdirected and donated to oxygen (Bowlar et al., 1994). An increase in MDA and H_2O_2 content is indicative of salinity induced oxidative stress in *Ziziphus* seedlings. Plants have evolved both enzymatic and non-enzymatic mechanisms for ROS scavenging. Enzymes such as superoxide dismutases (SOD) react with superoxide radicals at almost diffusion limited rates to produce H_2O_2 which may be disposed by catalase and peroxidase. H_2O_2 may also be scavenged through ASC-GSH cycle also referred to as Hallowell-Asada active oxygen detoxification pathway (Halliwell, 1997; Asada, 1999). In this pathway, the initial reaction in the scavenging of H_2O_2 is catalysed by ascorbate peroxidase. Subsequent generation of ascorbate proceeds via activities of monodehydraascorbate reductase (MDHAR),dehydraascorbate reductase (DHAR) and glutathione reductase(GR). Additionally, plant cells contain high levels of antioxidants such as ascorbate (ASC) and gluthione(GSH) which are efficient oxyradical scavengers. In order to better evaluate antioxidant potential of *Ziziphus* species we have monitored activities of all the enzymes involved in ROS scavenging, as well as the levels of ascorbate, glutathione, MDA, proline and H_2O_2 in seedlings subjected to salinity stress.

Under salt stress, the two species of *Ziziphus* exhibited different responses with respect to the activities of CAT, PER and SOD. In *Z. rotundifloia*, the activity of CAT increased under mild salt stress, but this increase was not maintained at high salinity. In *Z. nummularia*, on the other hand, salt drastically reduced the activity of CAT. The activity of PER under salt stress also remained lower than the control in both the species. A clear cut reduction was also observed in the activity of SOD. Both increase as well as decrease in the activity of SOD has been reported in plants in response to salinity stress. (Dionisio-Sese & Tobita, 1998; Shalata et al., 2001; Rout & Shaw, 2001). It appears that the activity of SOD under salinity varies depending upon plant species, its age, organ analyzed as well as the level of salinity (Chaparzadeh et al., 2004). The decreased SOD activity following growth in salt indicates that in *Ziziphus* this enzyme is not crucial for detoxification of activated O_2, and non-enzymatic routes for conversion of O_2^- to H_2O_2 via antioxidants such as GSH and ASC may compensate for reduced activity of SOD. This view is supported from our results on GSH and ASC level under salt stress. At low salt level, both the species of *Ziziphus* exhibited enhanced level of antioxidants, and even at high salinity *Z. rotundifolia* continued to have significantly higher level of antioxidants.

Ascorbate plays a pivotal role in protection of plants against oxidative stress (Smirnoff & Wheeler, 2000). It is an excellent electron donor with a wide range of physiological functions. In *Ziziphus,* following salt treatment, both the species of *Ziziphus* exhibited enhanced levels of ASC. However, at high salinity level only *Z. rotundifolia* maintained increased ASC level and ASC: DHA ratio. The higher ASC content observed at high salinity could suggest that ASC synthesis was stimulated or ASC catabolism was inhibited. High ASC content has also been reported previously in tomato plants acclimated to salinity (Shalata et al., 2001).

Ascorbate peroxidase has been characterized as one of the most potent H_2O_2 scavengers in plants (Shigeoka et al., 2002). It ulilizes ascorbate in reduced form as a reductant, and therefore, its redox status directly affects the functioning of ascorbate-glutathione cycle.

APX utilizes ascorbate and forms ascorbyl radical (MDHA), which in turn, may be oxidized to DHA depending on the cellular redox status (Potters et al., 2002). Thus, the changes in the ratio of ASC to DHA are crucial for efficient cellular functioning. A higher MDHAR activity, results in inhibition of MDHA disproportionation as well as an efficient conversion of DHA to reduced ascorbate (Gara et al., 2000). In fact, the conversion of ascorbate from DHA is coupled to extensive utilization of GSH.

Glutathione, an important component of thiol pool, regulates protein thiol-disulphide redox status and protects against oxidative stress by its involvement in the ascorbate-glutathione cycle (Alscher & Hess, 1993). Both the species of *Ziziphus* showed increased level of GSH at low salt concentration, while in *Z. rotundifolia* a high GSH: GSSG was maintained even under high salinity. This was accompanied by an increase in the activity of GR.

DHAR and MDHAR are involved in enzymatic regeneration of ASC from DHA and MDHA. In *Z. rotundifolia*, activities of APX, MDHAR, DHAR and GR increased significantly at low salt concentration, and continued to exhibit elevated activities than the control even at high salt concentration. This suggests that enzymatic disproportionation of MDHA to ASC might have taken place. In *Z. nummularia* also a promotion was observed in the activities of MDHAR as well as GR under salinity stress.

Proline accumulation in higher plants entails precursor formation, proline biosynthesis and catabolism, and amino acid inter-conversions (Rus-Alvarez & Guerrier, 1994). In many plants, one of the most striking responses to a variety of stresses including salinity stress is manifested in a dramatic accumulation of proline (Mansour, 2000). Proline accumulation under stress has been assigned several roles. These include its role as a compatible cytosolute, an osmoprotectant, a protectant of organelles and cytosolic enzymes, a reserve source of carbon and nitrogen, a sink for excess reductants providing the NAD^+ and $NADP^+$ required for respiratory and photosynthetic processes, a hydroxyl radical scavenger and redox buffer/shuttle (Delauney & Verma, 1993; Rus-Alvarez & Guerrier, 1994; Hare & Cress, 1997). However, it has been reported that in rice proline accumulation during salt stress merely reflects injury symptoms (Lutts et al.,

1999). Although both the species of *Ziziphus* accumulate proline in response to salinity, it is not clear whether this osmolyte significantly influences salinity-induced oxidative stress response in *Ziziphus*. However, earlier it has been reported that in severely water-stressed *Ziziphus* leaves, high leaf nitrate reductase activities were paralleled by increases in proline concentration, suggesting an osmoprotective role for proline (Arndt et al., 2001).

Overall, the results of the current investigation point out that under moderate salt stress the antioxidant system in *Z. rotundifolia* is able to dispose off ROS much more efficiently than that of *Z. nummularia*, although in both the species a robust non-enzymatic system involving ascorbate and glutathione is operative for the disposition of toxic oxygen species.

Our earlier results (Choudhary, 1992; Choudhary et al., 2003) indicate that although treatment with salt resulted in an immediate inhibition of net photosynthetic rate, but at low concentrations of salt the plants exhibited considerable recovery with time. However, at very high salt concentration, in addition to photosynthesis, as a result of lowered efficiency of photochemistry and the highly reduced state of Q/A, quantum yield of electron transport also decreased. Eventually, photoinhibition of assimilation rate coupled to the toxic effects of salt on cellular metabolism resulted in death of plants. These results are in accordance with the findings of Gupta et al. (2002) who reported that although both *Z. nummularia* and *Z. rotundifolia* are quite tolerant to salinity, at high salinity *Z. rotundifolia* performed better owing to a high rate of photosynthesis, osmotic adjustment, chlorophyll retention and restricted translocation of sodium from root to shoot, and large accumulation of potassium in the leaves (Meena et al., 2003).

5. REFERENCES

Alscher, R.G. & Hess, J.L. 1993. Antioxidants in Higher Plants. Boca Raton, Florida: CRC Press. 272 pp.

Amako, K., Chen, G.X. & Asada, K. 1994. Separate assays specific for ascorbate peroxidase and guaicol peroxidase and for the chloroplastic and cytosolic isozymes of ascorbate peroxidases in plants. Plant Physiology 35: 497-504.

Amor, N.B., Hamed, K.B., Debez, A., Grignon, C. & Abdelly, C. 2005. Physiological and antioxidant responses of the perennial halophyte *Crithmum maritimum* to salinity. Plant Science 168: 889-899.

Arndt, S.K., Clifford, S.C., Wanek, W., Jones, H.G. & Popp, M. 2001. Physiological and morphological adaptations of the fruit tree *Ziziphus rotundifolia* in response to progressive drought stress. Tree Physiol 21: 705-715.

Asada, K. 1999. The water-water cycle in chloroplasts: scavenging of active oxygens and dissipation of excess photons. Annual Review Plant Physiology Plant Molecular Biology 50: 601-639.

Bates, L.S., Waldren, R.P. & Teare, I.D. 1973. Rapid determination of free praline for water stress studies. Plant Soil 39: 205-207.

Bohnert, H.J. & Jensen, R.G. 1996. Strategies for engineering water stress tolerance in plants.Trends Biotechnology 14: 89-97.

Bor, M., Ozdemir, F. & Turkan, I. 2003. The effect of salt stress on lipid peroxidation and antioxidants in leaves of sugar beet *Beta vulgaris* L. and wild beet *Beta maritime* L. Plant Science 164: 77-84.

Bowler, C., Van Camp, W., Van Montagu, M. & Inze, D. 1994. Superoxide dismutase in plants. Critical Review Plant Sciences 13: 199-218.

Chaparzadeh, N., D'Amico, M.L., Khavari-Nejad, R.A., Izzo, R. & Navari-Izzo, F. 2004. Antioxidative responses of *Calendula officinalis* under salinity conditions. Plant Physiology Biochemisty 42: 695-701.

Chinnusamy, V., Jagendorf, A. & Zhu, J.K. 2005. Understanding and improving salt tolerance in plants. Crop Science 45: 437-448.

Choudhary, R. 1992. Studies on the role of environmental stress on growth, metabolism and photosynthesis of Indian desert plants. Ph.D. Thesis, Jodhpur, India: J. N. Vyas University.

Choudhary, R., Joshi, S., Gehlot, H.S. & Sankhla, N. 2003. Chlorophyll fluorescence, photosynthesis and enzyme activities in ber (*Ziziphus rotundifolia*) under salinity stress. Proceedings Plant Growth Regulators Society America 30: 119-121.

Clifford, S.C., Arndt, S.K., Corlett, J.E., Joshi, S., Sankhla, N., Popp, M. & Jones, H.G. 1998. The role of solute accumulation, osmotic adjustment and changes in cell wall elasticity in drought tolerance in *Ziziphus mauritiana* (Lamk.). Journal Experimental Botany 49: 967-977.

Clifford, S.C., Arndt, S.K., Popp, M. & Jones, H.G. 2002. Mucilages and polysaccharides in *Ziziphus* species (Rhamnaceae): location, composition and physiological roles during drought-stress. Journal Experimental Botany 53: 131-138.

Delauney, A.J. & Verma, D.P.S. 1993. Proline biosynthesis and osmoregulation in plants. The Plant Journal 4: 215-223.

Dhankhar, O.P., Makhija, M. & Singhrot, R.S. 1980 Effect of salinity levels on germination of seeds and growth of transplanted seedlings of ber (*Ziziphus rotundifolia*). In: H. S. Mann. (Ed.), Proceedings International Symposium of the Arid Zone Research Development, Jodhpur, India:Syposium Proceedings. 351-356 pp.

Dionisio-Sese, M.L. & Tobita, S. 1998. Antioxidant responses of rice seedlings to salinity stress. Plant Science 135: 1-9.

Gara, L.D., Paciolla, C., Tullio, M.C.D., Motto, M. & Arrigioni, O. 2000. Ascorbate-dependent hydrogen peroxide detoxification and ascorbate regeneration during germination of a highly productive maize hybrid: evidence of an improved detoxification mechanism against reactive oxygen species. Physiology Plant 109: 7-13.

Giannopolitis, C.N. & Ries, S.K. 1977. Superoxide dismutase. I. Occurrence in higher plants. Plant Physiology 59: 309-314.

Gossett, D.R., Millhollon, E.P. & Lucas, C. 1994. Antioxidant response to NaCl stress in salt-tolerant and salt-sensitive cultivars of cotton. Crop Science 34: 706-714.

Gupta, N.K., Meena, S.K., Gupta, S. & Khandelwal, S.K. 2002. Gas exchange, membrane permeability, and ion uptake in two species of Indian jujube differing in salt tolerance. Photosynthetica 40: 535-539.

Halliwell, B. 1987. Oxidative damage, lipid peroxidation, and antioxidant protection in chloroplasts. Chemistry Physics Lipids 44: 327-340.

Hare, P.D. & Cress, W.A. 1997. Metabolic implications of stress induced praline accumulation in plants. Plant Growth Regulators 21: 79-102.

Hasegawa, P.M., Bressan, R.A., Zhu, J.K. & Bohnert, H.J. 2000. Plant Cellular and molecular responses to high salinity. Annual Review Plant Physiology Plant Molecular Biology 51: 463-499.

Hooda, P.S., Sindhu, S.S., Mehta, P.K. & Ahlawat, V.P. 1990. Growth, yield and quality of ber (*Zizyphus mauritiana* Lamk.) as affected by soil salinity. Journal Horticulture Science 65: 589-593.

Jana, S. & Choudhuri, M.A. 1981. Glycolate metabolism of three submerged aquatic angiosperms during aging. Aquatic Botany 12: 345-354.

Jones, H.G. 1999. Selection of drought-tolerant fruit trees for summer rainfall regions of Southern Africa and India. European Commission STD-3, Brussels, Belgium: CTA.

Lutts, S., Majerus, V. & Kinet, J.M. 1999. NaCl effects on proline metabolism in rice (*Oryza sativa*) seedlings. Physiology Plant 105: 450-458.

Mansour, M.M.F. 2000. Nitrogen containing compounds and adaptation of plants to salinity stress. Biology Plant, 43 491-500.

Meena, S.K., Gupta, N.K., Gupta, S., Khandelwal, S.K. & Sastry, E.V.D. 2003. Effect of sodium chloride on the growth and gas exchange of young *Ziziphus* seedling rootstocks. Journal Horticulture Science & Biotechnology 78: 454-457.

Mittler, R. 2002. Oxidative stress, antioxidants and stress tolerance. Trends Plant Science 7: 405-410.

Pareek, O.P. 2001. Ber. International Centre for Underutilised Crops, Southhampton, UK: University of Southhampton.

Potters, G., Gara, L.D., Asard, H. & Horemans, N. 2002. Ascorbate and glutathione:guardians of cell cycle, partners in crime ? Plant Physiology Biochemistry 40: 537-548.

Rout, N.P. & Shaw, B.P. 2001. Salt tolerance in aquatic macrophytes: possible involvement of antioxidant enzymes. Plant Science 160: 415-423.

Rus-Alvarez, A. & Guerrier, G. 1994. Proline metabolic pathways in calli from *Lycopersicon esculentum* and *L. pennellii* under salt stress. Biology Plant 36: 277-272.

Sankhla, N. 1998. Work done in India. In; Selection of drought tolerant fruit trees for summer rainfall regions of Southern Africa and India (Co-ordinator: H. G. Jones). Brussels, Belgium: EU-STD.

Sankhla, N., Ziegler, H., Vyas, O.P., Stichler, W. & Trimborn, P. 1975. Eco-physiological studies on Indian arid zone plants II. A screening of some species for the C_4-pathway of photosynthetic CO_2-fixation. Oecologia 21: 123-129.

Shalata, A., Mittova, V., Volokita, M., Guy, M. & Tal, M. 2001. Response of the cultivated tomato and its wild salt-tolerant relative *Lycopersicon pennellii* to salt-dependent oxidative stress: the root antioxidative system. Physiology Plant 112: 487-494.

Shigeoka, S., Ishikawa, T., Tamoi, M., Miyagawa, Y., Takeda, T. & Yabuta, Y. 2002. Regulation and function of ascorbate peroxidase isozymes. Journal Experimental Botany 53: 1305-1319.

Smirnoff, N. & Wheeler, G.L. 2000. Ascorbic acid in plants: biosynthesis and function. Critical Review Plant Science 19: 267-290.

Smith, I.K. 1985. Stimulation of glutathione synthesis in photorespiring plants by catalase inhibitors. Plant Physiology 19: 1044-1047.

Tester, M. & Bacic, A. 2005. Abiotic stress tolerance in grasses. From model plants to crop plants. Plant Physiology 137: 791-793.

Upadhyaya, A., Sankhla, D., Davis, T.D., Sankhla, N. & Smith, B.N. 1985. Effect of paclobutrazol on the acticities of some enzymes of activated oxygen metabolism and lipid peroxidation in senescing soybean leaves. Journal Plant Physiology 121: 453-461.

Vaidyanathan, H., Sivakumar, P., Chakrabarty, R. & Thomas, G. 2003. Scavenging of reactive oxygen species in NaCl-stressed rice (*Oryza sativa* L.)—differential response in salt-tolerant and sensitive varieties. Plant Science 165: 1411-1418.

Vinocur, B. & Altman, A. 2005 Recent advances in engineering plant tolerance to abiotic stress: achievements and limitations. Current Opinions Plant Science 16: 1-10.

Wang, S.Y., Jiao, H.J. & Faust, M. 1991. Changes in ascorbate, glutathione, and related enzyme activities during thidiazuron-induced budbreak of apple. Physiology Plant 82: 231-236.

CHAPTER 15

SABKHA EDGE VEGETATION OF COASTAL AND INLAND SABKHAT IN SAUDI ARABIA

HANS-JÖRG BARTH

Department of Physical Geography, University of Regensburg, Germany, Hjbarth@Gmx.Net

Abstract. The sabkhat are a very typical geomorphologic feature in the Eastern Province of Saudi Arabia as well as in Kuwait, Bahrain, Qatar and the United Arab Emirates. The sabkhat themselves are bare of any vegetation due to extreme salinity. But closer to the sabkha edges there is a very distinct series of halophytic vegetation controlled by salinity and soil moisture. The characteristic vegetation types located at coastal sabkha edges are: *Halocnemum-Ghuspan*, *Halopeplis-Zygophyllum*, *Zygophyllum,* and *Zygophyllum-Cyperus*. At inland sabkha edges, slightly different hydrological and climatic aspects lead to different vegetation zones which are as follows: *Aeluropus lagopoides*, 1-3 different *Zygophyllum qatarense* types, *Panicum turgidum* and typical inland vegetation. A very distinct vegetation type which occurs at both inland and coastal sabkhat is the *Phoenix dactylifera* / *Tamarix* sp type. These sites are highly valued by recreation seeking people because they provide shade and green in the otherwise unspectacular landscape.Generally there are much more locations along the sabkha edges where the environmental conditions would support *Phoenix/Tamarix* communities. It is strongly suggested that these communities be cultivated in order to provide more recreational possibilities especially near urban settlements.

1. INTRODUCTION

The coastal lowlands of Saudi Arabia, Qatar and the UAE are presently characterized by strong desert conditions providing habitat to Sahara Arabian open shrub lands. During the last few centuries traditional nomadic husbandry had some influence on the desert ecosystems, especially on species composition and reduction of plant cover. In the 20th century land degradation was intensified by the rapid social and economic changes that occurred due to extensive oil exploitation in the whole region. Most of these changes though concentrated on the non-halophytic inland vegetation, which is primarily grazed by livestock. The halophytic plant communities of the coastal flats and around inland sabkhat are least affected. Due to the growing attention which is given to salt tolerant species and their potential regarding seawater irrigation, rehabilitation of saline areas and industrial application (nicely reviewed by Dagar, 2003) this paper presents an

M.A. Khan and D.J. Weber (eds.), Ecophysiology of High Salinity Tolerant Plants, 215-224.

overview of the sabkha edge vegetation occurring at Saudi Arabian coastal and inland sabkhat.

1.1. General geographic description of the surveyed area

The study area is located along the Arabian Gulf coast, north of Jubail Industrial City. It is part of the central coastal lowlands of the Eastern Province of Saudi Arabia. The relief is weak, although some limestone exposures in the form of minor escarpments and small domes are quite common. General features in the topography of the area are flat and extensive sand sheets, and widely rolling inactive dune systems, which are stabilized by vegetation and therefore referred to as "fossil" dunes. The longitudinal sand ridges, reaching heights of 5 to 10 m and lengths of more than 100 m are oriented in a general north-south direction. The flat low lying sand sheets are undisturbed layers of wind blown sand and still in most cases covered by protective vegetation. Although increasing grazing pressure reduces the vegetation cover in several areas below critical levels, thus reactivating active sand dynamics. This is generally observed where the vegetation cover dropped below 5 %. Besides sand sheets and dunes, sabkhat are a common feature in the Eastern lowlands. Coastal sabkhat are characteristically flooded by seawater periodically. Their groundwater is of marine origin and so are most of the sediments. Inland sabkhat are only flooded episodically by precipitation and they are mostly aeolian in nature (Barth & Böer, 2002).

The Eastern Province belongs to the arid part of the subtropical desert belt. Precipitation is confined exclusively to the winter season and varies greatly in volume and spatial occurrence. The long-term average annual rainfall in the area is approximately 80 mm/a. The mean annual temperature is 26°C, with extreme values of 48°C in summer and 3°C in winter.

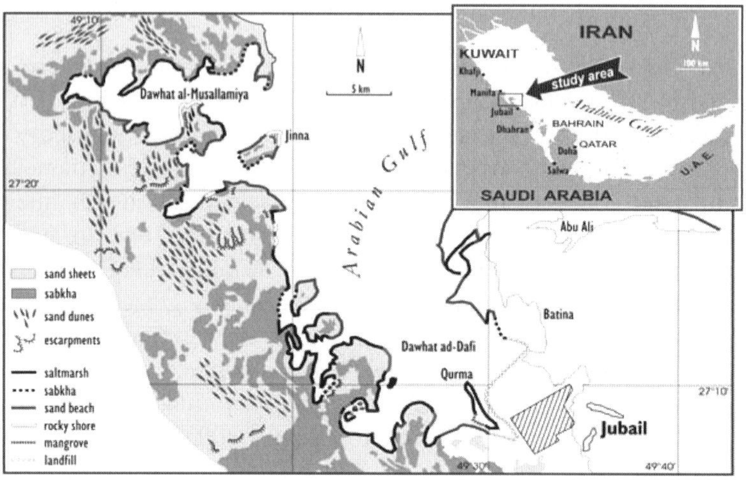

Figure 1. *Location of the Area of interest map.*

2. RESULTS AND DISCCUSION

2.1. The vegetation associated with coastal sabkhat

The sabkhat itself is bare of any vegetation due to severe environmental conditions (wind exposure, high temperatures and above all: extreme salinity). Near the sabkha edge the transition towards the adjacent sand sheets is usually characterized by a series of well-defined zones, each occupied by a different plant community. The typical vegetation zones are controlled by groundwater salinity and the thickness of the overlying sand and its composition (grain size, carbonate and sulphate content). The schematic transect in figure 2 displays the different vegetation zones of coastal sabkhat edges.

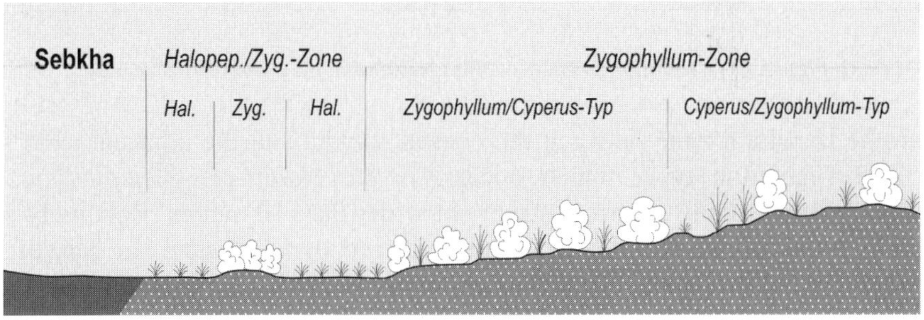

Figure 2. *Schematic transect displaying the different vegetation zones occurring at coastal sabkhat edges.*

A phenomenon almost exclusively occurring at coastal sabkhat in the Eastern Province are areas of Halocnemum-Ghuspan (although they were observed associated to inland sabkhat in the UAE). Ghuspan is a local term for "little dune". In the literature these dunes are often referred to as "nebkhat". These little dunes are associated with *Halocnemum strobilaceum* (Pall.) M.B. and they are oriented according to the main wind direction with their axis in NNW-SSE direction (Figure 3). These only 10 to 40 cm high plants collect moving sand and built up to 60 cm high-elongated sand hills around them seldom longer than 2 metres. The Halocnemum-Ghuspan always occurs in colonies of ten to more than hundred nebkhat and reach up to 200 m into the coastal sabkhat. This is the threshold where the amount of saltating sand drops below the level where the little dunes can be accumulated (Fryberger et al., 1984). Without the elevation (due to the accumulation of aeolian sand) the plants are not able to tolerate the extreme salinity of the sabkhat surface material as well as the groundwater below (Barth, 1998). The general mineralisation (the water soluble fraction) of the soil within Halocnemum-Ghuspan (10 cm below surface) is clearly the highest compared to all other vegetation types associated with sabkhat. The Cl^- concentration is generally above 1000 mg/l and Na^+ more than 600 mg/l (Table 1).

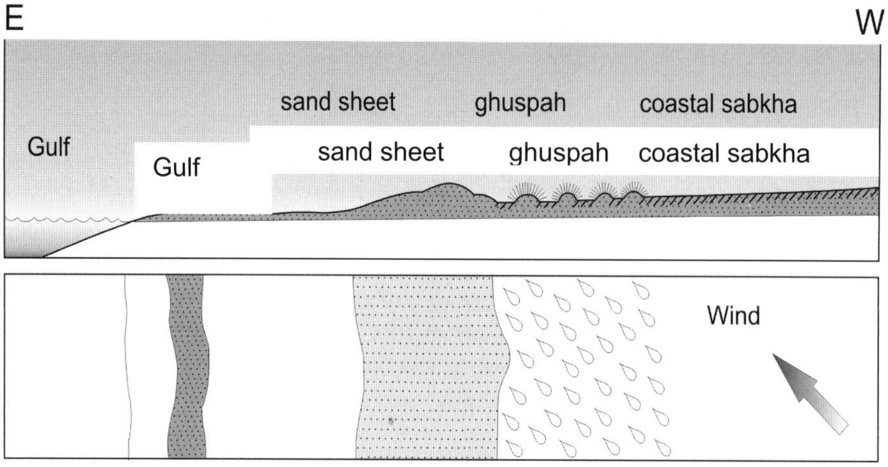

Figure 3. *Orientation of* Halocnemum *Nebkhat in the main wind direction.*

In the transition zone between the coastal sabkha and the adjacent sand sheet the first vegetation type, which occurs, is the *Halopeplis-Zygophyllum* type (Figure 2). It is usually a narrow stripe not wider than 10 metres. Relatively small *Halopeplis perfoliata* individuals occur scattered or in clusters. In between the *Halopeplis* clusters small (not higher than 15 cm) *Zygophyllum qatarense* individuals are located on relatively dryer sites (compared to the *Halopeplis* stands close by) mostly composed of shell fragments and wind blown sand. The thickness of the dry sand correlates with the size of the *Zygophyllum* individuals. Associated plants are *Suaeda vermiculata* and *Limonium axillare* (but they are of little importance when compared to the abundance of *Zygophyllum* and *Halopeplis* respectively). Apart from the soil water content the most obvious difference between the *Zygophyllum* and *Halopeplis* stands is the significantly higher concentration of Mg^{2+} at the *Halopeplis* sites (450 mg/l compared to 10 mg/l, see Table 1). All other components show relatively similar values.

With increasing thickness of dry sand above the capillary fringe *Halopeplis perfoliata* disappears and *Zygophyllum qatarense* becomes the dominant species. Therefore this zone is referred to as *Zygophyllum* type (Figure 2). According to co-dominant species the *Zygophyllum* zone can be differentiated into *Zygophyllum*, *Zygophyllum-Cyperus* (Table 2) or even *Cyperus-Zygophyllum* type (Table 3). It is very obvious, that the *Zygophyllum* individuals at the edge of coastal sabkhat are generally much smaller than their counterpart at inland sabkha edges where they often reach a size of more than 3 m^2. *Cyperus conglomerates* is frequently co-dominant or it may even become dominant (Table 3). Other associated plants are *Sporobolus ioclados*, *Aeluropus lagopoides*, *Fagonia olivieri*, *Cornulaca monacantha* and *Anabasis setifera*. In spring several annual species may be present (Table 4).

Table 1. *Ionisation (water soluble fraction) of soil at different vegetation types (10 g dry soil in 100 ml aqua dist. Mounted on a shaker for 24h). Each value is the mean value of 3 samples from 5 different locations. Sample depth 10 cm if not mentioned otherwise.*

Vegetation type	Cl^-	Na^+	Mg^{2+}	Ca^{2+}	K^+	SO_4^{2-}	*pH*
Phoenix dact. 20 cm soil depth	115	93	390	264	6.9	1280	8.4
Phoenix dact. 40 cm soil depth	50	38	380	262	7.3	960	8.3
Halocnemum Ghuspan	1350	720	570	330	34.5	1700	9
Halopeplis	530	375	450	421	10	570	8.95
Zygophyllum assoc. Halopep.	650	265	10	513	16	420	9.14
Zygophyllum dry sand 20 cm above capillary fringe	33	18.8	350	355	1.2	630	8.56
Zygophyllum/Cyperus	19	17.6	512	293	1	660	8.4
Calligonum	6	13.5	375	319	3.14	600	8.5
Stipagrostis	1	13.5	300	339	0.7	630	8.4
Panicum	3	5	10	242	0.36	180	8.85
Penisetum	14	6.5	225	290	0.68	100	8.85
Rhanterium	4	2.9	8	240	0.29	100	9.1

Table 2. *Vegetation test square at representative coastal sabkha edge (10-20 cm dry sand above capillary fringe). Location: 27°09'40''N/49°19'37''E, Oct. 7th 1994.*

10x10 m square *species*	*number of individuals*	*total cover in %*	*mean cover of individual*	*min./max. cover of Individual*	*min./max. height of individual (cm)*
Zygophyllum qatarense	15	1.35	0.09	0.01/0.25	15/35
Cyperus conglomeratus	80	0.64	0.008	0.002/0.01	5/20
Fagonia olivieri	2	0.04	0.02	0.01/0.035	5/9

Table 3. *Vegetation test square at representative coastal sabkha edge (30-40 cm dry sand above capillary fringe). Location: 27°11'32''N/49°17'16''E, Oct. 7th 1994.*

10x10 m square *species*	*number of individuals*	*total cover in %*	*mean cover of individual*	*min./max. cover of Individual*	*min./max. height of individual (cm)*
Cyperus conglomeratus	60	0.48	0.008	0.001/0.01	5/15
Zygophyllum qatarense	3	0.21	0.07	0.07	20
Fagonia olivieri	1	0.024	0.024	0.024	5
Cornulaca monacantha	3	0.015	0.005	0.005	8/12

Table 4. *Vegetation test square at representative coastal sabkha edge (30 cm dry sand above capillary fringe). Location: 27°10'19''N/49°19'38''E, March 28th 1995.*

10x10 m square species	number of individuals	total cover in %	mean cover of individual	min./max. cover of individual	min./max. height of individual (cm)
Zygophyllum qatarense	17	0.586	0.035	0.008/0.07	10/30
Plantago boissieri	145	0.3	0.002	0.0005/0.004	5/14
Launaea angustifolia	11	0.08	0.007	0.004/0.09	20/40
Helianthemum lippii	2	0.005	0.0025	0.0025	12
Oligomeris linifolia	12	0.003	0.00025	0.00025	4/15
not identified annual	10	0.001	0.0001	0.0001	4/10
Erodium deserti	17	0.002	0.00012	0.00012	1/5
Monsonia nivea	4	0.0004	0.0001	0.0001	1/2

Further inland (adjacent to the *Zygophyllum type*) the vegetation shows no difference compared to the inland sabkha sites.

2.2. The vegetation associated with inland sabkhat

In the inland sabkhat the first vegetation to occur (towards the sabkha edge) are clusters of *Aeluropus lagopoides*. The groundwater salinity threshold (where *Aeluropus* survives) is about 45 mS/cm. This little grass is very common along the direct transition from sabkha to dry sand of the sabkha edge. In most cases this zone is not wider than 2-10 metres (Figure 5). The cover of the grass can be up to 30%. As soon as the dry sand becomes thicker than 10 cm the grass occurs only occasionally.

Adjacent to the *Aeluropus* belt there are two possible vegetation types. Most common the *Zygophyllum qatarense* type and occasionally the *Cyperus conglomeratus* type. The size of *Zygophyllum* individuals which may cover 4 m^2 displaying a height of up to 100 cm turns the sabkha edges of inland sabkhat into conspicuous green belts (Figure 4, Table 5). Within this zone the total vegetation cover may reach 20%. More typical though are values of 5%. The groundwater salinity within the *Zygophyllum* zone is generally higher on the sabkha side (40 mS/cm) and decreases towards the sand sheets (to less than 20 mS/cm) (Figure 5). On the sabkha side of the *Zygophyllum* belt associated plants are *Aeluropus lagopoides* and *Sporobolus ioclados* (Table 6). Annuals are not very common within this zone. On the opposite side *Cyperus conglomeratus*, *Panicum turgidum*, *Aeluropus lagopoides* as well as several annuals are common associates (Table 7). With increasing thickness of dry sand (above the capillary fringe) and decreasing salinity (Table 1) the typical sand sheet vegetation i.e. *Calligonum comosum*, *Haloxylon salicornicum* and several perennial grasses like *Panicum turgidum*, *Stipagrostis plumosa* and *Penisetum divisum* dominate.

Figure 4. Aerial Photograph of inland sabkhat displaying a green belt of Zygophyllum qatarense.

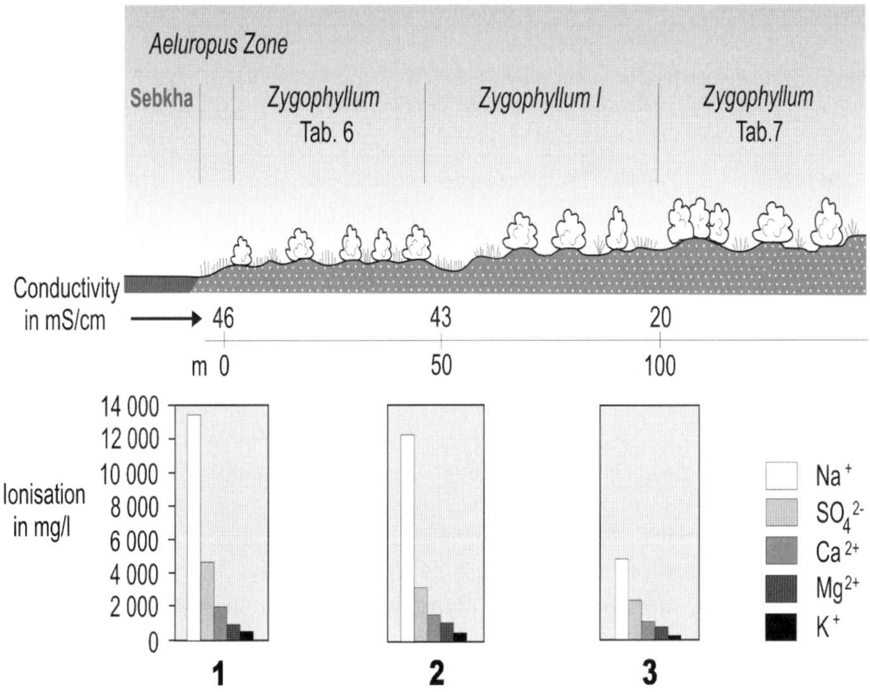

Figure 5. Transect through typical inland sabkha edge.

Table 5. *Zygophyllum type at inland sabka edge 30-40 cm above capillary fringe. Location: 27°11'08''N/49°16'40''E. Oct. 9th 1994.*

10x10 m square species	number of individuals	total cover in %	*mean cover of individual*	*min./max. cover of Individual*	*min./max. height of individual (cm)*
Zygophyllum qatarense	33	14.51	0.44	0.005/1.44	10/60
Cyperus conglomeratus	38	1.33	0.035	0.005/0.16	8/25
Panicum turgidum	9	0.85	0.094	0.01/0.16	5/30
Aeluropus lagopoides	13	0.0013	0.0001	0.0001	3/6

Table 6. *Zygophyllum type at inland sabka edge 10-20 cm above capillary fringe. Location: 27°08'51''N/49°20'03''E. Oct. 7th 1994.*

10x10 m square *species*	*number of individuals*	*total cover in %*	*mean cover of individual*	*min./max. Cover of individual*	*min./max. height of individual (cm)*
Zygophyllum qatarense	13	3.06	0.24	0.04/1	15/80
Sporobolus ioclados	18	0.424	0.024	0.01/0.04	8/12
Aeluropus lagopoides	35	0.03	0.0008	0.0008	3/7

Table 7. *Zygophyllum type at inland sabka edge 100 cm above capillary fringe. Location: 27°08'51''N/49°20'02''E. Oct. 8th 1994.*

10x10 m square *Species*	*number of individuals*	*total cover in %*	*mean cover of individual*	*min./max. Cover of individual*	*min./max. height of individual (cm)*
Calligonum comosum	6	8.83	1.47	0.25/5	80/130
Zygophyllum qatarense	33	4.92	0.15	0.04/1	20/70
Sporobolus ioclados	11	0.32	0.03	0.001/0.25	5/20
Cyperus conglomerates	6	0.012	0.002	0.002	10/15
Panicum turgidum	5	0.01	0.002	0.002	10/20
Stipagrostis plumose	3	0.003	0.001	0.001	8/20

The species composition within the *Zygophyllum* zone is generally dependant on the salinity and therefore the thickness of the sand above the capillary fringe (sabkha niveau). The following table summarizes the most abundant species according to the sand cover:

Table 8. *Species according to increasing sand thickness above capillary fringe (sabkha neveau).*

most	*sand thickness above capillary fringe (in cm)*				
dominant	*10-20*	*20-30*	*30-50*	*50-100*	*>100*
1	*Zygophyllum q.*	*Zygophyllum q.*	*Zygophyllum q.*	*Zygophyllum q.*	*Calligonum c.*
2	*Aeluropus l.*	*Sporobolus i.*	*Cyperus c.*	*Panicum t.*	*Zygophyllum q.*
3	*Sporobolus i.*	*Aeluropus l.*	*Panicum t.*	*Cyperus c.*	*Cyperus c.*
4			*Aeluropus l.*	*Sporobolus i.*	*Panicum t.*
5				*Aeluropus l.*	*Stipagrostis p.*
6					*Sporobolus i.*
	Zygophyllum vegetation types				different inland vegetation types

An other very distinctive but not very common vegetation type occurring at sabkha edges (inland as well as coastal sabkat) is the *Phoenix dactilyfera* type. The distribution of this vegetation type is mostly confined to the influx of brackish water or sweet water. At several locations along the Saudi Arabian Gulf coast artesian quells provide brackish water, which flows into the sabkhat. At the transition between the *Zygophyllum* zone and the sand sheets the facultative halophytes *Phoenix dactylifera* and *Tamarix arabica* dominate the otherwise unconspicious vegetation (Figure 6, Table 9). These are sites, which in most cases show signs of severe degradation due to intensive grazing pressure and offroad vehicle use. The shade of the trees attracts people seeking recreation as well sheep.

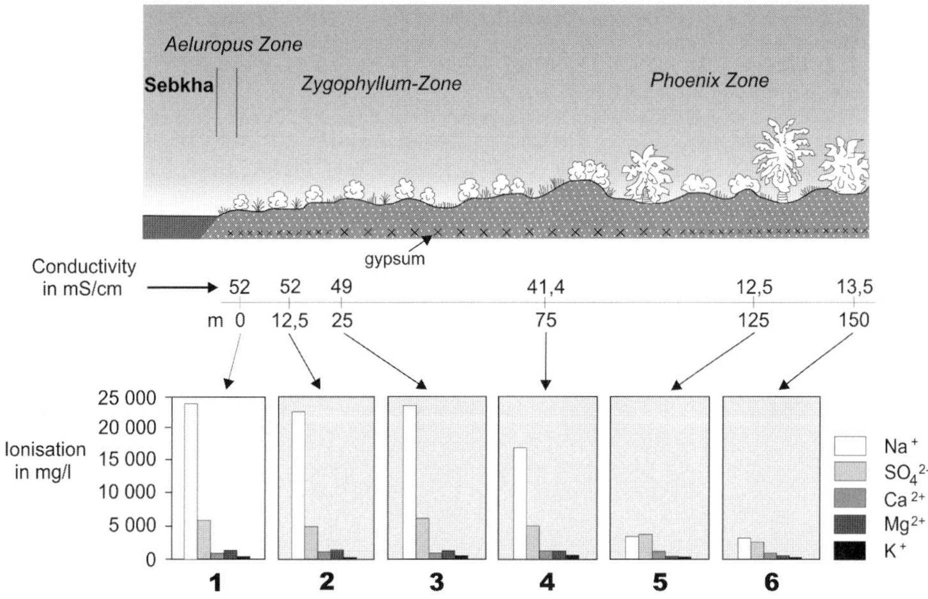

Figure 6. *Sebkha edge with Phoenix dactylifera type and groundwater mineralisation.*

Table 9. Phoenix dactylifera *type at coastal sabka edge. Location: 27°08'52''N/49°20'05''E. Oct. 8th 1994.*

35x35 m square species	*number of individuals*	*total cover in %*	*mean cover of individual (m^2)*	*min./max. cover of individual*	*min./max. height of individual (cm)*
Phoenix dactylifera	4	21.63	66	40/100	500/700
Zygophyllum qatarense	37	1.18	0.39	0.04/1	20/80
Calligonum comosum	3	0.833	3.4	1.5/7.1	90/120
Sporobolus ioclados	6	0.009	0.018	0.01/0.09	8/15
Panicum turgidum	4	0.0003	0.0008	0.0008	14/25
Stipagrostis plumosa	3	0.0002	0.0008	0.0008	20/40
Aeluropus lagopoides	4	0.0001	0.0004	0.0004	2/4

Measurements of groundwater salinity within these vegetation types showed that values between 15 and 30 mS/cm are nothing unusual. This implies that there are much more locations along the sabkha edges where the environmental conditions would support such *Phoenix/Tamarix* communities. This could be a first possibility to improve the potential value of the otherwise unattractive sabkha edges, especially close to urban areas like Jubail Industrial City. Further steps would be experiments with irrigation of seawater close to the sabkha edges in order to cultivate sabkha halophytes.

3. REFERENCES

Barth, H.-J. 1998. Sebkhas als Ausdruck von Landschaftsdegradation im Zentralen Küstentiefland der Ostprovinz Saudi Arabiens. Regensburger Geographische Schriften, Regensburg 29: 1- 279

Barth, H.-J. & BÖer, B. 2002 (Eds.), Sebkha Ecosystems, The Arabian Peninsula and Adjacent Countries. Dordrecht, Netherlands: Kluwer.

Dagar, J.C. 2003. Biodiversity of Indian Saline Habitats and Management & Utilisation of High Salinity Tolerant Plants with Industrial Application for Rehabilitation of Saline Areas. In: A.S. Alsharhan, W.W. Wood, A.S. Goudie, A. Fowler, & E.M. Abdellatif (Eds.), Desertification in the Third Millenium. Leiden, Netherlands:Swets & Zeitlinger Publishers 151-172 pp.

Fryberger, S.G., Al-sari, A.M., Clisham, T.J., Rizvi, S.A.R. & Al-Hinai, K.G. 1984. Wind sedimentation in the Jafurah sand sea, Saudi Arabia 31: 413-431

CHAPTER 16

ANALYSIS OF THE SOIL SUSTAINING SALT GRASS (DISTICHLIS SPICATA (L.) GREENE) WILD POPULATIONS IN A SEMIARID COASTAL ZONE OF MEXICO

A. ESCOBAR-HERNÁNDEZ[1], E. TROYO-DIÉGUEZ[2,4], J. L. GARCÍA-HERNÁNDEZ[2], H. HERNÁNDEZ-CONTRERAS[3], B. MURILLO-AMADOR[2], L. FENECH-LARIOS[3], R. LÓPEZ-AGUILAR [2], A. BELTRÁN-MORALES[3] AND R. VALDEZ-CEPEDA[3]

[1]CIBNOR. La Paz, B.C.S. México
[2]Centro de Investigaciones Biológicas del Noroeste (CIBNOR), S.C. Mar Bermejo # 195, Col. Playa Palo Santa Rita, La Paz, B. C. S., México CP: 23090
[3]Universidad Autónoma de Baja California Sur, La Paz, B.C.S. México
[4]Corresponding author: etroyo04@cibnor.mx.

Abstract. Wild populations of salt grass (*Distichlis spicata* (L.) Greene) are found in salt and brackish coastal marshes of Canada, USA and México. Salt grass is important for the northwestern states of Mexico, because it grows in dry and saline habitats, and because its foliage is mainly produced during periods when other grasses are not available for livestock and wild life. This study was conducted to determine the soil condition of the areas where this plant is dominant. Twenty-one soil variables were analyzed by means of multivariate analysis. Six soil samples from estuary and three soil samples from sandy beach were collected from five coastal sites of Baja California Sur, Mexico. We used the Principal Components Analysis (PCA) procedure to determine: 1) The best fitting lineal functions, 2) the variables with the highest loading, and 3) an approach to soil characterization. Two linear functions, explained 87% of the total variance, which indicated that native salt grass populations are well adapted to hyper-saline soil condition, and also to saline beach dry soil. Electrical conductivity (EC), permanent wilting point (PWP), and soil saturation (SAT), were the variables with the highest loading values as they gave the strongest numerical consistency to the linear functions. The first function was defined as a hyper-saline condition associated to the saturation point of the soil. The second was defined as a dry soil function. After applying the obtained functions to the collected data represented on axis, the existence of three soil groups with different saline condition were detected: Hyper-saline soils from estuary with a humidity near to the permanent wilting point value, hyper-saline soils from estuary with a humidity close to the soil saturation point, and saline beach dry soil.

1. INTRODUCTION

Wild populations of the medium-tall salt grass (*Distichlis spicata* (L.) Greene) are found in salt and brackish coastal marshes along the coasts of Canada and USA, as well as along the seashore of the Mexican states of Baja California and Baja

M.A, Khan and D.J Weber (eds.), Ecophysiology of High Salinity Tolerant Plants, 225-237.

California Sur. Salt grass populations are found also in interior salt lakes and can be grown in large monocultures. It is an important food source for geese and other birds. This species is important for Northwest Mexico and particularly for the state of Baja California Sur, because it is dominant where soils are saline and fine textured and where the surface soils are wet at least part of the growing season. These special conditions are present in places where no other grasses can easily grow. Thus, salt grass becomes an important food source for livestock and wild life of this region (Shreve & Wiggins, 1966; Gould & Moran, 1981). For different reasons, salt grass is regarded as potential livestock forage. Although the cattle initially avoid its rustic foliage, domestic and wild animals consume this plant during the periods when leguminous crops, like alfalfa (*Medicago sativa* L.) and perennial grasses, are not available in this region (Horvath, 2000). The plant has potential social impact in the northwestern states of Mexico, which are populated by numerous poor and small-scale farmers. Literature reports this species as high-fibrous fodder with low protein content, and high mineral content (Arredondo et al., 1991). In this context, some studies indicate that many chemical values improve significantly after fire, increasing crude protein and keeping similar levels of fiber content (Uchytil, 1990; Arredondo et al., 1991; Horvath, 2000). According to Leake et al., 2002, *Distichlis* spp. grow and spread mainly through roots, this facilitates percolation and improves soil structure and soil organic content. The roots can follow the saline water table down more than a meter because they have an inner and outer 'tube' providing gas drainage to the plant enabling growth in waterlogged conditions. The plants exude salt through their leaves so that salt appears not to accumulate in the root zone. In a general sense, it is known that the development of salt grass is related to saline soils, however, it is necessary to take into account that salinity is a condition derived from the combination of several variables of the soil, such as texture, humidity, depth, minerals content, pH, and others (Arredondo et al., 1991). Therefore, the appraisal of saline soil characters is fundamental for a well understanding of the conditions required for a specific species like salt grass. In this context, multivariate analysis can be extraordinarily useful because the complexity of the variability of soil characters (Muñoz, 1995). Principal Components Analysis (PCA) is a technique of multivariate analysis that has been widely utilized in soil sciences (Johnson, 1998), given that soil heterogeneity is determinant to the spatial, and biomass variability of a vegetal species or variety (Bresler et al., 1981). Considering the soil properties as a multivariate (Arkley, 1976) continuum system (Russo & Bresler, 1981), they can be represented as spatial-continuum functions (Arkley, 1976). A principal component (PC) is a representation of the structure of original data under observation, and accordingly, PCs result from a matrix of analysis of p variables and n observations. Each PC is a new variable derived from the original variables with reduced dimensionality and with the minimum loss of information. It can be obtained as much PCs as variables are being analyzed. The most significant PC is the one with the greatest variance (eigenvalue), which holds the best representation to explain the aims of the study (Lewis & Lisle, 1998; Shaver et al., 2002). In each resultant PC, eigenvalues illustrate the participation of each variable in study for the construction or definition of lineal function (eigenvector) of study. In addition, the PCA provides values of coordinates, which construct graphic classifications that can draw the characterization of the observations that have been taken into account, and particularly those that have the highest influence in the variability of the matter under study (Lewis & Lisle, 1998; Velásquez & Colmenares, 1999; Fielding, 2002; Shaver et al., 2002). Through soil sampling, lab analysis and the application of multivariate analysis, we carried out this work with the main objective to determine the soil condition on which wild populations of salt grass grow in five coastal sites of Baja California Sur, in northwest Mexico.

2. MATERIALS AND METHODS

2.1. Sampling sites

Sites of soil sample were defined on the bases of differences in soil and vegetative characteristics. The study was limited to three sites: The estuary of Chametla, in La Paz, B.C.S, located at 24° 05′N and 110° 20′W; the estuary of San Carlos Port, in María Magdalena Bay, by the Pacific Ocean side, located at 25° 45′N and 112° 07′W; and in the Mar Azul Beach, on Elías Calles, also on the Pacific Ocean side, located at 23° 07′N and 110° 06′W. In the estuary of Chametla, samples were taken from three plots: Edge of estuary, central part of estuary, and beach of estuary. In the estuary of San Carlos, samples were taken from a plot. At the Mar Azul site, samples were also taken from a plot.

2.2. Soil sampling field procedure

When sampling from edge of estuary, three samples were taken: Horizon A, at 0 to 4.5 cm depth; horizon B, at 4.5 to 33 cm depth; and horizon C at 33 to 71 cm depth. In central estuary plot, also three samples were taken: Horizon A at 0 to 9.5 cm depth, horizon B at 9.5 to 30 cm depth, and horizon C at 30 to 42 cm depth. On beach of estuary, only one plot was sampled, at 0 to 38 cm depth. At the San Carlos site, a single plot was sampled at 0 to 26 cm depth. At the Mar Azul site, also a single plot was sampled at 0 to 28 cm depth. At the plots where a single sample of soil was taken, this corresponded to the depth by which the rhizome-mass was found.

2.3. Soil analysis

Samples were analyzed in the Soil and Water Laboratory of the Universidad Autónoma de Baja California Sur. Measured variables were: Electrical conductivity (EC), potential hydrogen (pH), organic matter (OM), Sodium (Na^{+}), carbonate (CO_3^{-2}), bicarbonate ($HCO3^{-1}$), chlorine (Cl_2^{-2}), Calcium (Ca^{+2}), Magnesium (Mg^{+2}), sulfate ($SO4^{-2}$), Phosphorus (P), total Nitrogen (TN), potassium (K), clay (CLA), silt (SIL), sandy (SAN), permanent wilting point (PWP), available humidity (AH) and saturation (SAT).

2.4. Statistical analysis

A PCA was carried out by running the SAS statistical program (SAS ver. 6.12: 1989 - 1996). For the calculation of components, we used the procedure indicated by Muñoz (1995). The original data of each variable to be analyzed were assigned in a matrix of analysis (S) as following:

$$S = n \times p$$

Where: n = 18 (soil variables) rows; p = 9 (soil sample) columns

Self-standardization of the data (where X = 0 and S = 1), in a matrix of correlation gives the difference in the scale of representation original of the data. PCs were ordered from the greatest to the smallest one, using the obtained eigenvalues, which is the variance of the new variable, and the eigenvectors ($S_{ij}A_{ij}$), which are the rows transposed of the matrix S. The resultant eigenvectors were considered as new variables (lineal combinations of the original), which concentrate the majority of the information (variance). The importance of the PCs (eigenvector) was determined by calculating the highest eigenvalues. The Principal Components Analysis results were taken as a group (the sum of the values), in order to numerically consider most of the original information.

2.5. Analysis of variables

The analysis of variables was done in order to build lineal functions, to define the variables with the highest influence, and to characterize the condition of the soils sampled in this study (Figueras, 2000; Fielding, 2002). With the eigenvector coefficient assigned to each variable at the most significant PCs, the corresponding lineal function was built. After this numerical procedure, we proceeded to identify each PC in terms of its corresponding soil condition, with the associated eigenvector and the highest determination coefficient ($R^2 = 0.90$). Finally, the condition of the soil was characterized using the distribution of tendencies on a coordinate axis, and also considering the influence (R^2) of the variables that composed each function.

2.6. Analysis of the numerical reduction

One of the properties of the PCA is the reduction of the space of study, without significant loss of information (Figueras, 2000; Fielding, 2002). On this base, a process of reduction was carried out: From the most significant PCs, the most influencing variables were chosen. These variables were reduced with $r^2 = 90$ from the matrix for standardization values, based on the criterion that a variable correlated with another, explains in that numerical proportion, the variance of the other. Subsequently, with the original data of the variables reduced, an analysis was made using the ′Proc Prin Comp′ procedure in the computer program SAS ver. 6.12 (SAS, 1998).

3. RESULTS AND DISCUSSION

3.1. Principal components

From the original data of the variables of the soils sampled for analyses, we obtained 21 PCs, equivalent to the original number of variables. The first three PCs were the most important variables indicated by the reached significance (eigenvalue over 20%), by the proportion of the variance contained, and by their participation in the accumulated variance (Fielding, 2002). (Table 1) show that the three principal components explain 92% of the total variance contained in the matrix of data. Also, it is observed that the first PC is the most important one because it explains 46% of the total variance, while the second PC explains 36%, and the third PC explains 9.8%. In accordance to previous observations (Fielding, 2002; Shaver et al., 2002), these results suggest that the 21 principal components obtained in this study, in this sense, there are 21 different projections explaining the variance of the soil condition sustaining five wild populations of *Distichlis spicata*. We found that the three first PCs were the best projections to explain the soil condition; this includes the total integration of the original variables, always considering that the first PC has the highest significance and importance, explaining 46% of the variance. Nevertheless, it is advisable to consider the two next PCs for a greater approximation in the inference about soil condition.

Table 1. *Principal Component's results of 21 variables of soils sustaining* Distichlis spicata *populations in five coastal sites of* Baja California Sur, México.

Y	*EIGENVALUE*	*DIFFERENCE*	*PROPORTION*	*ACCUMULATED*
Y1	9.78430	2.21282	0.465919	0.46592
Y2	7.57149	5.50917	0.360547	0.82647
Y3	2.06232	1.25825	0.098206	0.92467
Y4	0.80407	0.33716	0.038289	0.96296
Y5	0.46691	0.29394	0.022234	0.98519
Y6	0.17297	0.09068	0.008237	0.99343
Y7	0.08229	0.02663	0.003918	0.99735
Y8	0.05566	0.05566	0.002650	1.00000

Y = componentes principales

It is important to mention that the number of PCs to be elected, depend on the nature of the study matter. In a similar study, Lewis and Lisle (1998), who studied lines of Canola, obtained four PCs as the most important, because after their assembly they accumulated 75% of the total variance. In another application of the method of analysis, (Velásquez & Colmenares, 1999) obtained three PCs with 80% of the accumulated variance related to the differentiation of species of *Trichogramma*

3.2. Soils variables loading.

In order to explain the variance of soil condition that contains each significant PC, the load coefficient or eigenvector was used for this purpose, in the way it was assigned to each variable in the component (Shaver et al., 2002). In (Table 2) is noted that the variance represented in the first PC, by having the highest eigenvector, is mostly explained by the variable electrical conductivity (EC = 0.291095), organic matter (OM = 0.267130), chlorine (Cl_2^{-2} = 0.271504), clay (CLA = 0.268073), bulk density (BD = 0.272335), field capacity (FC = 0.280502), permanent wilting point (PWP = 0.280521), and available humidity (AH = 0.280454). In the second PC, variance was mostly explained by the variables potential hydrogen (pH = -0.344502), Calcium (Ca^{+2} = 0.342961), Sodium (Na^{+1} = 0.321954), and Magnesium (Mg^{+2} = 0.315178). For the third PC, the variables which explained the highest proportion of variance were silt proportion (SIL = -0.449692), solid (real) density (RD = 0.392072) and the saturation point (SAT = -0.358596). In general, these results indicate that the best variables to explain the variance of the soil condition, sampled at coastal sites of Baja California Sur, are: EC, PWP, FC, AH, BD, Cl_2, CLA, OM, pH, Ca, Na, Mg, SIL, RD, and SAT.

3.3. Soils conditions functions

Once known the eigenvectores for every variable in the PCs, they were used as lineal coefficients of combination (Fielding, 2002; Statsoft, 2002a); the load coefficients shown on (Table 2) were used to express the most important components as a result of the following lineal functions:

$$Y_1 = 0.291095(EC) + 0.267130(OM) + 0.271504(Cl2) + 0.268073(CLA) + 0.272335(BD) + 0.280502(FC) + 0.280521(PWP) + 0.280454(AH) \quad (1a)$$

$$Y_2 = -0.344056(pH) + 0.342961(Ca) + 0.321954(Na) + 0.315178(Mg) \quad (1b)$$

$$Y_3 = -0.449692(SIL) + 0.392072(RD) - 0.358596(SAT) \quad (1c)$$

It is important to indicate that the load coefficients offered to each variable in the function are fictitious, for such a reason it is convenient to identify each function as a soil property or soil status (Muñoz, 1995). For that reasons, the determination coefficient (R^2) that is taken, results when the coefficients of load and the original data of each variable are related, in the corresponding function

(Fielding, 2002; Shaver et al., 2002). In this way, Y_1 was defined is a function of salinity associated to the humidity. Y_2 was defined as a function of ionic exchange, while Y_3 was a function of physical soil properties.

3.4. Reduction of variables

As it was observed, lineal functions 1a, 1b and 1c, practically included all the original variables used for the analysis. This implies that any study that may be carried out in the near future should consider all these variables. Nevertheless, it is important to be aware that one of the fundamental objectives of the PCA is to reduce the space of an object of study, with a minimum loss of variance.

The functions 1a, 1b and 1c, were subjected to a process of reduction (Muñoz, 1995; Figueras, 2000; Fielding, 2002), using correlation coefficients (r^2 = 90%) from the standardization of data matrix.

Table 2. *Loading coefficients (eigenvalor) (Cc) and determination coefficients (R^2) of each soil variable in three principal components important.*

	Y1		*Y2*		*Y3*	
	Cc	*R^2*	*Cc*	*R^2*	*Cc*	*R^2*
Na ppm	0.112560	0.35209	0.321954	0.88590	0.100315	0.14406
EC dS/m	0.291095	0.91054	0.118805	0.32691	0.161304	0.23165
pH	0.038905	0.12169	-0.344056	-0.94672	0.154883	0.2242
OM %	0.267130	0.83558	0.181691	0.49995	0.116022	0.16662
CO_3 ppm	0.189176	0.59174	-0.240771	-0.66251	0.173951	0.24981
HCO_3 ppm	0.124740	0.39019	0.258202	0.71048	-0.059051	-0.08480
Cl_2 ppm	0.271504	0.84926	0.148444	0.40846	0.223590	0.32109
Ca ppm	0.065545	0.20502	0.342961	0.94370	-0.008350	-0.01199
Mg ppm	0.078406	0.24525	0.315178	0.86725	0.092300	0.13255
SO_4 ppm	0.228559	0.71493	-0.188971	-0.51998	0.286316	0.41117
NT %	0.251720	0.78738	0.186685	0.51369	0.170939	0.24548
K ppm	0.251720	0.78738	0.186685	0.51369	0.170939	0.24548
SAN %	-0.254532	-0.79617	0.149824	0.41226	0.291202	0.41819
SIL %	0.201337	0.62978	-0.117506	-0.32333	-0.449692	-0.64579
CLA %	0.268073	0.83853	-0.159986	-0.44022	0.157535	0.22623
BD g/cm	-0.272335	-0.85186	0.095138	0.26179	0.271638	0.39009
RD g/cm	-0.014376	-0.04497	-0.247557	-0.68119	0.392072	0.56305
FC %	0.280502	0.87741	-0.166188	-0.45729	-0.092566	-0.13293
PWP %	0.280521	0.87747	-0.166180	-0.45727	-0.092413	-0.13271
AH %	0.280454	0.87726	-0.166220	-0.45738	-0.093151	-0.13377
SAT %	0.166325	0.52026	0.216213	0.59494	-0.358596	-0.51497

Y = principal components, Cc = loading coefficient (eigenvector) R^2 =determination coefficient. Na = Sodium, EC = electric conductivity, pH = potential hydrogen, OM =organic matter, CO_3 = Carbonate, HCO_3 = bicarbonate, CL_2 = Chlorine, Ca = Calcium, Mg = Magnesium, SO_4 Sulfate, TN = Total Nitrogen, K = Potassium, SAN = sandy, SIL = silt, CLA = lay, BD = Bulk density, RD = Real density, FC = capacity of field, PWP = Permanent wilting point, AH = Available humidity, SAT = Saturation.

The function 1a was reduced to EC, since it was highly correlated to OM = 97% and Cl_2 = 99%; also it was reduced to PWP because it was found highly correlated to FC = 100%, AH = 100%, CLA = 91% and with BD = -91%. The function 1b was reduced to Na because it was correlated to Ca = 94%, and with Mg = 97%, at the same time Ca relates to pH = - 90%. The function 1c was reduced to RD and SAT, because SIL was related to SAN = 94%, and this with PWP = -94%. By means of this procedure, the functions 1a, 1b and 1c, were reduced to:

$$Y_1 = 0.291095(EC) + 0.280521(PWP) \quad (2a)$$
$$Y_2 = 0.321954(Na) \quad (2b)$$
$$Y_3 = 0.392072(RD) - 0.358596(SAT) \quad (2c)$$

In order to verify the existence of these functions, the original data of each variable of the 2a, 2b and 2c functions, were analyzed with PCA. The results listed on (Table 3) indicate the existence of two significant PCs that accumulate 83% of the variance of the original data matrix. The first component explains 53% of the total original variance, and the second PC explains 30% of it. With the load coefficients of each variable in the corresponding component (Table 3), it is observed that in the first PC, electrical conductivity (EC = 0.483812), saturation (SAT = 0.548651) and sodium (Na^{+1} = 0.505560), were the most influencing variables; while in the second PC, permanent wilting point (PWP = 0.730610) was the most influencing one. On the basis of these results, the two PCs can be expressed through the following lineal functions:

$$Y_1 = 0.483812 (EC) + 0.505560 (Na) + 0.523232 (SAT) \quad (3a)$$
$$Y_2 = 0.794468 (PWP) \quad (3b)$$

Once we analyzed the determination coefficients (R^2) shown on (Table 3), each variable member of the functions previously enunciated was defined according to the corresponding soil properties. Y_1 was defined as a function of sodium salinity in saturated soils, while Y_2 was defined as a dry soil function. As it is observed in the functions 3a and 3b, the number of variables was reduced to four, which suggests the need to revise its existence and the degree of participation of its variable components. After proceeding to another PCA, results listed on (Table 4) show two significant PCs accumulating 87% of the variance from the original data matrix. While the first PC explains 59% of the variance, the second PC explains 28%. With the load coefficients of each variable in the corresponding component (Table 4), it can be observed that in the first PC, the variables EC (0.607102) and SAT (0.523223) are confirmed as the most influencing variables, while in the second PC, the variable PWP is accentuated as the influencing variable. After explaining the accumulated variance with the load coefficients assigned to each variable (Table 3), now the two PCs can be expressed through the following functions:

$$Y_1 = 0.607102(EC) + 0.523223(SAT) \quad (4a)$$
$$Y_2 = 0.794468(PWP) \quad (4b)$$

After the above procedure, we identified the 4a and 4b functions using the determination coefficients (R^2) of the (Table 4); Y_1 was confirmed as a function of a saline condition related to saturated soils, and the function Y_2 was confirmed as a function of a dry soil.

Table 3. *Values obtained through a Principal Components Analysis, for first reduction of the original lineal functions.*

	Eigenvalues			
	Eigenvalue	*Difference*	*Proportion*	*Accumulate*
Y1	2.66125	1.16884	0.532250	0.53225
Y2	1.49241	0.82815	0.298482	0.83073
Y3	0.66427	0.51541	0.132853	0.96359
Y4	0.14886	0.11564	0.029772	0.99336
Y5	0.03322	.	0.006643	1.00000
	Eigenvectors			
	Y1		*Y2*	
	Cc	R^2	Cc	R^2
CE	0.483812	0.78926	0.436453	0.53319
PWP	0.182372	0.29751	0.730610	0.89255
NA	0.505560	0.82474	-0.190363	-0.23256
DR	-0.419592	-0.68449	0.482179	0.58905
SAT	0.548651	0.89503	-0.083562	-0.10208

Y = principal components, Cc = loading coefficient (eigenvector),R^2 = determination coefficient, CE = conductivity electric, PWP = permanent wilting point, Na = sodium, Dr = relative density, SAT = saturation point.

Table 4. *Values obtained through a Principal Components Analysis for a second reduction of lineal functions.*

	Eigenvalues			
	Eigenvalue	*Difference*	*Proportion*	*Accumulate*
Y1	1.95975	1.20417	0.653251	0.65325
Y2	0.75558	0.47092	0.251860	0.90511
Y3	0.28467		0.094888	1.00000
	Eigenvectors			
	Y1		*Y2*	
	Cc	R^2	*Cc*	R^2
CE	0.653101	0.91428	-0.042238	-0.03671
PWP	0.553763	0.77522	-0.654396	-0.56883
SAT	0.516531	0.72310	0.754971	0.65625

Y = principal components, Cc = Loading coefficient, R^2 = determination coefficient, CE = Electric conductivity, PWP = permanent wilting point, SAT = saturation point.

As a global result, we found that the most influencing variables to determine the condition of the soils under study were the electrical conductivity (EC), saturation (SAT), and permanent wilting point (PWP). Nevertheless, it is important to notice that EC had the highest significance, because in all the cases its coefficient of decision was estimated to be $R^2 \geq 90\%$, particularly in the functions 1a, 3a, and 4a.

3.5. Soil types characterization

PCA also provides information to know if the collected soils from the five places have the same condition or if they are different (Shaver et al., 2002). PCs can confirm if each lineal function obtained is a general measure, in this case for the five places. We brought PCs to the space of study, represented on an axis of coordinates, to reflect the variability among the observations of the study object (Fielding, 2002). To carry out this final part of the analysis, the participation of a pair of functions is required. (Figure 1) shows the clustering of soils in five coastal places of Baja California Sur, as a result of the participation of the functions 4a and 4b.

$Y1 = 0.607102(\mathbf{CE}) + 0.523232(\mathbf{SAT})$ (4a) $\quad Y2 = 0.794468(\mathbf{PMP})$ (4b)

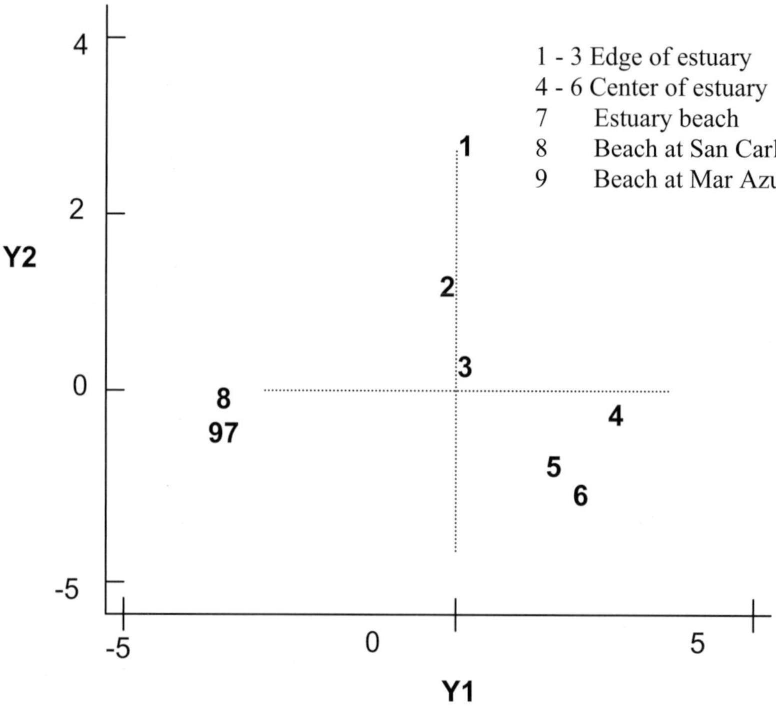

Figure 1. *Groups of soils sustaining wild populations of* Distichlis spicata *collected from five sites in Baja California Sur, Mexico, expressed by their components Y2/Y1.*

In a global sense, the tendency of clustering indicates that the condition of the study soils have an evident variability expressed through specific differences. The existence of two groups of soils with a different condition is observed, and on each one of them the existence of different types is clearly distinguished.

The first group corresponds to the soils collected in edge and central of estuary (1, 2, 3, 4, 5, 6,), that upon being far away from Y_2, a condition that suggests an influence of EC and SAT in Y_1, which can be a criterion to classify this group as a type of salinity in saturation soils. They were characterised as following:
Soils at the edge of estuary (1, 2, 3):

1: Horizon A = Silty-loam, hyper saline, with a high content of Na^{+1} and Cl^{-1}.
2: Horizon B = Loamy texture, hyper saline, with a high content of Na^{+1}, Cl^{-1}, and Ca^{+2}. 3: Horizon C = Silty-loamy texture, hyper saline, with a high content of Na^{+1}, Cl^{-1}, and Ca^{+2}.

Soils at the center of estuary (4, 5, 6):
4: Horizon A = Sandy-loamy texture, hyper saline, with a high content of Na^{+1}, Cl^{-1}, and Mg^{+2}.
5: Horizon B = Sandy-loamy texture, hyper saline, with a high content of Na^{+1}, Cl^{-1}, Ca^{+2}, and Mg^{+2}.
6: Horizon C = Sandy-loamy texture, hyper saline, with high content of Na^{+1}, Cl^{-1}, Ca^{+2}, and Mg^{+2}.

The second group that corresponds to the collected soils on beach (7, 8, 9), being nearby Y2, appears to be influenced by PWP. According to the individual characteristics of each one of them, they can be defined as dry soils of beach. They are distinguished as following:
7: Estuary beach soil = Dry with a sandy texture, saline, with high content of chlorine.
8: Beach San Carlos soil = Dry with a sandy texture, highly saline, with a high content of chlorine.
9: 'Mar Azul' Beach Soil = Dry with a sandy texture, of medium salinity, with a high content of chlorine.

As a global result, the majority of the soils evidence a capacity to support and sustain *Distichlis spicata*, under a hypersaline condition. This condition is an expression of soil, given the humidity available through the profile and the chemical exchange. The saline effect on the plants is the result of an action of a saline dissolution within the soil related to the content of the humidity in the soil. The dynamic effect of a concentration or compounds in dissolution is expressed by the electrical conductivity, which correlates the dissolved solids to the humidity at permanent wilting point or to the humidity of saturation in the different soils. On the basis of these results, we consider that the resultant variables as components of the lineal functions, expanded or reduced by the exhaustive mathematical analysis carried out, should be taken into account for future studies of the condition of the soils sustaining wild populations of *Distichlis spicata*. The numerical effect caused to every one of the variables, composing the expanded or reduced functions, have implications in the residuals, by the interrelations established in our analysis. Also, we consider that the soils of estuary have hypersaline and saline conditions, from the surface to the horizon C, presenting variable texture, with high content of chlorine, sodium, calcium and magnesium. In the center and edge of estuary, and also on the dry beach soils, it was observed a low humidity.

4. CONCLUSIONS

Salt grass wild plants growth and development under hyper-saline conditions are determinated by the permanent wilting point and the saturation soil humidity parameter. Two functions were obtained in order to express the hyper-saline conditions and to characterize the soils collected in the study sites. The hypersaline condition of the soils in the study sites was expressed by the high levels of electrical conductivity (30 to 100 dS/m), this range was observed in horizons estuary edge soils A-B-C, in the horizons estuary center soils A-B-C. A global EC average for all the sampled sites of 42 dS/m was found. This result contrasts with the extremely high salinity of 16 dS/cm reported by Seliskar and Gallagher (2000), and with the 4 dS/m, reported by Majerus (1996). Plant community presents a salinity average with a minimum saturation of water and another community of maximum salinity with a maximum hydromorphia. This suggest that the effects of salinity on the plants of *Distichlis spicata* depend on the content of water in the soil, which due to salinity can exert a stronger negative effect than the PWP value, during a period of drought, and a weaker effect during the period of rains because of the saturation of soils.

5. REFERENCES

Arkley, R.J. 1976. Statistical methods in soil classification research. Advances Agronomy 28: 37-70.

Arredondo, J.T., García-Moya, E. & Kobashi, J. 1991. Efecto de la temperatura, el fotoperíodo y la salinidad en el crecimiento y fisiología de Distichlis spicata. Agrociencia 25: 103-122.

Bresler, E., Dasberg, S., Russo, D. & Dagan, G. 1981. Spatial variability of cropyield as a stochastic process. Soil Science Society American Journal 45: 600-605.

Fielding, A. 2002. Biological Data Processing II: Multivariate Techniques, Biological Sciences, Manchester Metropolitan University, Manchester, M1 5GD, United Kingdom. 9 pp.

Figueras, M.S. 2000. Introducción al Análisis Multivariante. 5 campus.com, Estadística http: // www. 5camp us. com/leccion/anamul. Compiled edition, Spain: Universidad de Zaragoza. 15 pp.

Gould, F. & Moran R. 1981. The grasses of Baja California. México. San Diego, California: Society of Natural History. 97 pp.

Horvath, J. 2000. *Distichlis stricta*, saltgrass, desert saltgrass. Retrieved from http://www.usask.ca/agriculture/plantsci/classes/range/distichlis.html 2000.

Johnson, D.E. 1998. Applied multivariate methods for data analysis. 566 pp.Pacific Grove, California:Brooks Cole Publisher Leake, J., Barrett-Lennard, E., Sargeant, M., Yensen, N. & Prefumo, J. 2002. NyPa *Distichlis* Cultivars: Rehabilitation of Highly Saline Areas for Forage Turf and Grain. A Report for the Rural Industries Research and Development Corporation. Netherby, Australia: RIRDC Publisher. 39 pp.

Lewis, G.J. & Lisle T.A. 1998. Towards better canola yield; a principal components analysis approach. Proceedings of the 9th Australian Agronomy Conference. Wagga Wagga. School of Land and Food. The University of the Queensland: Australian Society of Agronomy. Lawea, Queensland. 395 pp.

Muñoz G., J.M. 1995. Técnicas de Análisis Multivariante. Banco Central de Costa Rica. División Económica. Departamento de investigaciones económicas. 11 pp.

Russo, D. & Bresler, E. 1981. Soil hydraulic properties as stochastic process: I. An analysis of field spatial variability. Soil Science. Society. American Journal 54: 1228-1233.

SAS. 1998. User's Guide (Relase 6.12). Statistics. Cary,North Carolina: SAS Isnt. Inc

Seliskar, D.M. & Gallagher, J.L. 2000. Exploiting wild population diversity and somaclonal variation in the salt marsh grass *Distichlis spicata* (Poaceae) for marsh creation and restoration. American Journal of Botany 87: 141-146.

Shaver, R.D., Undersander, D.J., Schwab, E.C., Hoffman, P.C., Lauer, J.G. & Combs, G. 2002. STATSOFT, Inc. Electronic Statistics Textbook. Tulsa, Oklahoma: StatSoft. 430 pp.

Shreve, F. & Wiggins, I.L. 1966. Vegetation and flora of the Sonoran Desert. Stanford. California:Stanford Press. 246 pp.

Uchytil, R.J. 1990. *Distichlis spicata*. U.S. Department of Agriculture, Forest Service, Rocky Mountain Research Station, Fire Sciences Laboratory. Fire Effects Information System. 118 pp. Velásquez, de R.M. & Colmenares, O. 1999. Análisis morfométrico de dos especies de Trichogramma (Hymenoptera: Trichogrammatidae) utilizando la metodología de componentes principales. Boletínde Entomología Venezolana 14: 191-200.

CHAPTER 17

COMPARATIVE SALT TOLERANCE OF PERENNIAL GRASSES

SALMAN GULZAR[2] AND M. AJMAL KHAN[1]

[1]*Department of Botany, University of Karachi, Karachi-75270, Pakistan*
[2]*Department of Botany, Government Superior Science College, Shah Faisal Colony, Karachi-75230, Pakistan.* [1]*Corresponding author:Ajmal@halophyte.org*

Abstract. Salt tolerance mechanisms of three perennial halophytic grasses (*Aeluropus lagopoides* (Linn.) Trin. ex Thw., *Sporobolus ioclados* (Trin.) C.E. Hubbard and *Urochondra setulosa* (Nees ex Trin.) Nees) were studied to determine if local species employ similar strategies to tolerate high salinity. We found different patterns of growth, water relations and ion uptake among the species tested. *Aeluropus lagopoides* and *U. setulosa* were grown in 0-1000 mM NaCl while *S. ioclados* in 0-500 mM NaCl under ambient conditions. Plants from non-saline controls had larger fresh and dry weights. Increasing concentrations of salinity from 600 - 1000 mM NaCl for *A. lagopodides* and *U. setulosa* and 500 mM NaCl for *S. ioclados* caused high salinity stress. Water and osmotic potential of the plants increased with increasing salinity and pressure potential decreased slightly in all species. Stomatal conductance in all grasses decreased substantially with the increase in salinity. Ash content remained low (~12%) in both shoot and root of all grasses and showed little change with the increase in salinity except for *S. ioclados*, where in root it increased up to 35%. Na^+ and Cl^- concentrations showed a small increase while Ca^{2+}, Mg^{2+} and K^+ remained constant with increasing salinity. Various ion ratios for shoot and root also showed variation between the species tested.

1. INTRODUCTION

Soil salinity is a major constraint to food production because it limits crop yield and restricts use of previously uncultivated land (Yokoi et al., 2002). Coastal areas of Pakistan have limited supply of good quality water, however, the coast of Balochistan, despite few seasonal rivers, does not have enough fresh water. Most of the sub-surface water is brackish with various levels of salinity and not fit for the cultivation of conventional crops. Therefore, use of brackish water is the only affordable alternative left to improve the ecology of the deserted lands and to produce cash-crops for alleviating the economic hardship of local populations. The coastal area of Pakistan is reported to have about 100 halophytes and a significant number of them are grasses (Khan & Gul, 2001). Salt tolerant perennial grasses could be an alternate source of forage and fodder because they can be grown with

M.A. Khan and D.J. Weber (eds.), Ecophysiology of High Salinity Tolerant Plants, 239-253.

brackish water irrigation (Khan, 2002). We selected three of the most common grasses viz. *Aeluropus lagopoides* (Linn.) Trin. ex Thw., *Sporobolus ioclados* (Nees ex Trin.) Nees, and *Urochondra setulosa* (Trin.) C.E. Hubbard to determine their comparative salt tolerance and suitability as forage crop for the region.

Perennial grasses are reported to be highly salt tolerant (Russell, 1976; Shannon, 1978; Venables & Wilkins, 1978; Marcum & Murdoch, 1990; Lissner & Shierup, 1997; Bajji et al., 2002; Bell & O'Leary, 2003; Shen et al., 2003; Alshammary et al., 2004; Debez et al., 2004). Similar reports are also available on the salt tolerance of some of the local grasses (Ashraf et al., 1986; Shamsi & Ahmed, 1986; Kumar, 1990; Ashraf & Naqvi, 1991; Bodla et al., 1995; Mahmood et al., 1996; Khan et al., 1999, Gulzar et al., 2003ab; Gulzar et al., 2005). These reports indicate that the salt tolerance of grasses could vary from 300 mM NaCl to 800 mM NaCl (Gulzar et al., 2005). Grasses like *Aeluropus lagopoides* and *Urochondra setulosa* could survive in up to 1000 mM NaCl (Bodla et al., 1995; Gulzar et al., 2003ab) while a number of them survived salinity (550 to 600 mM NaCl) approaching seawater (Glenn, 1987; Hester et al., 1996, 2001). Some grasses grew in soil salinity ranges between 300 to 500 mM NaCl (Mahmood et al., 1996; Bell & O'Leary, 2003; Peng et al., 2004) while others could not survive in salt concentrations above 300 mM NaCl (Khan et al., 1999; La Peyre & Row, 2003).

Osmotic adjustment under increased salinity occurred concurrent with increased shoot sodium and chloride concentrations, decreased shoot potassium concentration, and decreased shoot succulence (Marcum & Murdoch, 1990; Gulzar et al., 2003ab, 2005). Water and osmotic potential in *Halopyrum mucronatum* increased with increase in salinity while turgor decreased (Khan et al., 1999). Osmotic adjustment and maintenance of positive turgor under salt stress also occurred in *Paspalum vaginatum* (Peacock & Dudeck, 1985) and *Stenotaphrum secundatum* (Dudeck et al., 1993).

Halophytic grasses exclude salt effectively and use water loss to concentrate solutes for osmotic adjustment (Glenn, 1987; Munns, 2002; Parida & Das, 2005). Monocotyledonous halophytes generally have much lower water content, $Na^+ : K^+$ ratios and mineral content than dicotyledonous halophytes growing at the same location (Gorham et al., 1980; Glenn, 1987). Sodium exclusion method of salt tolerance appears less efficient than sodium accumulation particularly in the succulent xerophytes (Wang et al., 2004; Debez et al., 2004). At higher salinity, *Sporobolus arabicus* accumulated more Na^+ in comparison to other species studied (Mahmood et al., 1996). Khan et al. (1999) reported higher accumulation of Na^+ and Cl^- and corresponding decrease in K^+, Ca^{2+} and Mg^{2+} with increasing salinity in *Halopyrum mucronatum*.

High intracellular concentrations of Na^+ and/or Cl^- may inhibit the activity of many enzymatic systems and some cellular processes, such as protein synthesis or mRNA processing (Serrano, 1996; Yeo, 1998; Zhu, 2001; Forment et al., 2002). Sodium interferes with the uptake of essential cations, especially K^+ and Ca^{2+} and promotes oxidative stress through generation of ''reactive oxygen species'' (ROS)

(Serrano & Gaxiola, 1994; Yeo, 1998; Zhu, 2001). Substantial differences in Na^+ and K^+ accumulation between salt-resistant species may be due to differences in the selective ion transport capacities at root level (Hester et al., 2001; Wang et al., 2002). Salt secreting species would be expected to have the weakest selective transport capacity for K^+ over Na^+ as most of the salt would have to be transported up to the stem and excluded from the leaf via salt glands. *Aeluropus lagopoides* did show high selectivity for K^+ by retaining greater amounts of Cl^- and Mg^{2+} in roots than in shoots (Gulzar et al., 2003a), while *U. setulosa* shoots did not show high K^+ selectivity (Gulzar et al., 2003b) although countless salt crystals accumulate on its leaf surface. Salinity induced inhibition of plant growth may occur due to the effects of high Na^+, Cl^- or SO_4^{2-} by decreasing the uptake of essential elements such as P, K^+, NO_3^- and Ca^{2+}, ion toxicity or osmotic stress (Zhu, 2001, 2002). *Sporobolus spicatus* was found to secrete 93% NaCl by weight of salts secreted by plants from 4 different sites while K^+, Ca^{2+}, Mg^{2+} and SO_4^{2-} constituted only 5% of salts (Ramadan, 2001).

Ion ratios could be helpful in categorizing the physiological response of a plant (salt-excluding, salt-secreting or salt-diluting) in relation to ion selectivity under increasing substrate salt concentrations (Wang et al., 2002). However, the influence of various ion ratios on salt tolerance is quite complex and attempts to draw general conclusions have not been successful (Grieve et al., 2004). Sodium–potassium ion ratio is among the most important of these ion ratios and plants tend to maintain a low Na^+/K^+ ratio in the cytoplasm and low cytosolic Na^+ content below some crucial value (Greenway & Munns, 1980; Maathuis & Amtmann, 1999; Tyerman & Skerrett, 1999).

This research compares the effects of salinity on mechanisms of growth, water relations and ion uptake of three halophytic grasses (*Aeluropus lagopoides* (Linn.) Trin. ex Thw., *Sporobolus ioclados* (Nees ex Trin.) Nees, and *Urochondra setulosa* (Trin.) C.E. Hubbard from Pakistan.

2. MATERIALS AND METHODS

Propagules of *Aeluropus lagopoides* and *Sporobolus ioclados* were collected from University of Karachi campus and *Urochondra setulosa* from Hawkes Bay, at the Arabian Sea coast near Karachi. Seeds were separated from the hull and stored in a refrigerator at 4°C. Growth experiments were carried out in the University of Karachi under ambient atmospheric conditions. Seeds were germinated in 10 cm x 8 cm plastic pots filled three fourths with sandy soil. Plants were raised on half strength Hoagland and Arnon solution No. 2 (Moore, 1960) for two weeks until they achieved a height of about 5 cm. Plants were thinned to five similar sized plants in each pot. Pots were sub-irrigated and the water level was adjusted daily to correct for evaporation. Salt solutions were completely replaced once a week to avoid build-up of salinity in pots. Six salinity treatments with five replicates each were employed: 0, 200, 400, 600, 800 and 1000 mM NaCl for *A. lagopoides* and *U. setulosa* and 0, 100, 200, 300, 400 and 500 mM NaCl for *S. ioclados* after a preliminary test for salinity tolerance. Salinity levels were raised gradually at daily intervals. At the end of the experiment, plants were harvested and fresh and dry weights of stem and root were measured. Plants were oven-dried at 80°C for 48h before dry weight was determined. Water potential was measured on punched disks from randomly chosen leaves in a C-52 chamber with the help of a HR-33 dew point micro-voltmeter (Wagtech). Similarly, press sap technique on Whattman No. 1 filter paper disks was used for measuring leaf osmotic potential. Leaf turgor pressure was estimated from the difference of leaf osmotic and water potentials. Leaf stomatal conductance was determined with an AP-4 Porometer (Delta-T Devices). Chloride ion was measured with a Beckman specific ion electrode. Cation content of plant root and shoot parts were analyzed using a Perkin Elmer model 360 atomic absorption spectrophotometer. The Na^+ and K^+ levels of the plant were examined by flame emission spectrometry. Statistical analyses were performed with the help of SPSS (SPSS, 2002).

3. RESULTS

Best growth was recorded in the non-saline medium for all species (Figure 1). *Aeluropus lagopoides* had the highest fresh and dry weights for shoot and root as well as the highest number of tillers per plant. *Sporobolus ioclados* exhibited a progressive decrease in shoot fresh weight, root fresh weight, dry weight of shoot, dry weight of root, shoot length, root length, number of leaves and tillers with an increase in substrate salinity. However, *A. lagopoides* and *U. setulosa* showed little effect at 200 mM NaCl but a further increase in salinity level (200-1000 mM NaCl) substantially inhibited growth. *Urochondra setulosa* maintained greater root length at all salinities tested and shoot length at 400-1000 mM NaCl concentration as compared to *A. lagopoides* which had greater shoot length at 0 and 200 mM NaCl. In *A. lagopoides*, number of tillers per plant at 0 to 600 mM NaCl were twice those of *U. setulosa*, while the number of leaves was almost identical between these two species. Although *S. ioclados* had the highest shoot length of about 60 cm as compared to the other grasses in control, it had much lower root length at all salinity levels and both shoot and root length declined substantially with rising medium salinity (Figure 1). Shoot to root ratio on a dry weight basis increased with increase in salinity in *A. lagopoides* and *S. ioclados* in comparison to control while it decreased slightly in *U. setulosa* (Figure 2). Tissue water content on a dry weight basis declined at the higher NaCl concentrations in all the species but *A. lagopoides* and *S. ioclados* showed some increase at moderate salinities (Table 1).

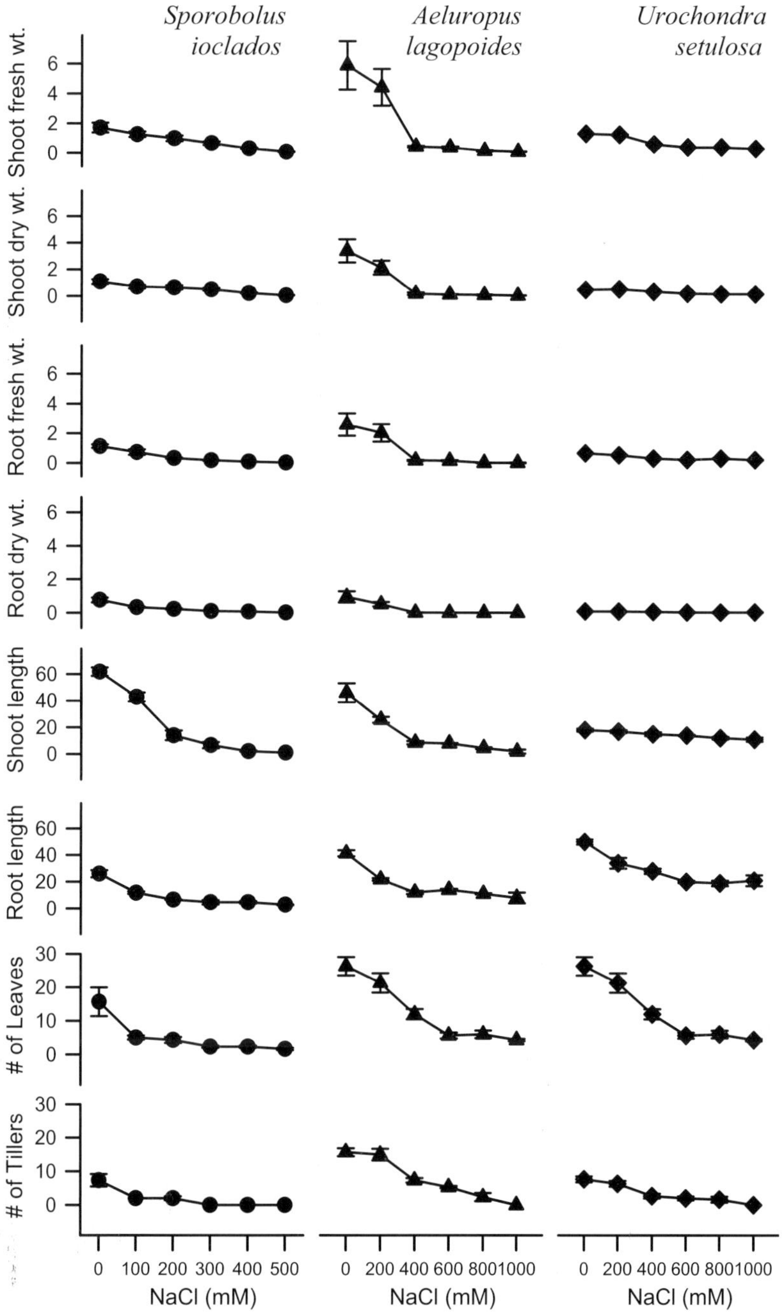

Figure 1. *Effect of increasing salinity on the growth parameters of* S. ioclados, A. lagopoides *and* U. setulosa. *Values for shoot and root weight are in g and of length in cm.*

A progressive increase in water potential and osmotic potential was observed with the increase in salinity in all species while stomatal conductance showed a progressive decrease (Figure 3). *Aeluropus lagopoides* showed the ability to maintain water potential, osmotic potential and stomatal conductance at salinities between 200-600 mM NaCl but salinities of 800 mM NaCl and above had inhibitory effects. All species maintained a higher osmotic potential than the water potential at each salinity concentration. Pressure potential did not vary much with the increase in salinity (Figure 3).

Table 1. *Effect of increasing salinity on the tissue water content of* Aeluropus lagopoides, Sporobolus ioclados *and* Urochondra setulosa. *Different letters in columns represent significant differences between mean tissue water content of each species.*

		Tissue water ($g\ g^{-1}$ dry wt.)			
NaCl (mM)	Sporobolus ioclados	*NaCl (mM)*	Aeluropus lagopoides	*NaCl (mM)*	Urochondra setulosa
0	$0.57^{a} \pm 0.23$	0	$0.73^{a} \pm 0.08$	0	$2.69^{a} \pm 0.14$
100	$0.78^{a} \pm 0.18$	200	$1.61^{b} \pm 0.53$	200	$2.04^{a} \pm 0.20$
200	$0.52^{a} \pm 0.11$	400	$1.26^{b} \pm 0.28$	400	$1.00^{b} \pm 0.09$
300	$0.31^{b} \pm 0.06$	600	$2.11^{c} \pm 0.43$	600	$1.00^{b} \pm 0.00$
400	$0.33^{b} \pm 0.12$	800	$0.58^{a} \pm 0.15$	800	$0.91^{b} \pm 0.16$
500	$0.14^{c} \pm 0.05$	1000	$0.40^{a} \pm 0.20$	1000	$0.69^{b} \pm 0.16$

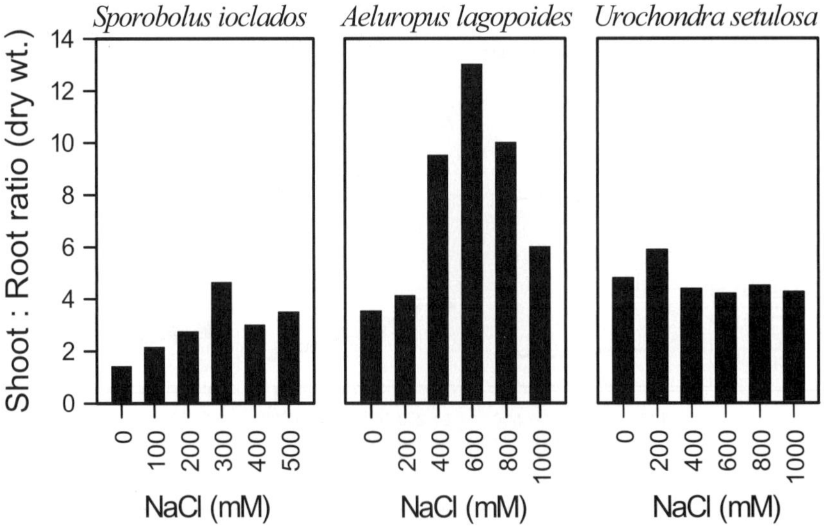

Figure 2. *Effect of increasing salinity on the shoot to root dry weight ratio of* A. lagopoides, S. ioclados *and* U. setulosa.

Ash content of shoots and roots of *A. lagopoides* and *U. setulosa* and shoots of *S. ioclados* was generally low (< 15% of dry weight) while *S. ioclados* roots had about 35% ash content (Figure 4). Ion content did not vary much in *A. lagopoides* and *U. setulosa* with an increase in salinity except for Cl^- which increased in roots of *A. lagopoides* with increase in salinity. Ion content was comparatively higher in

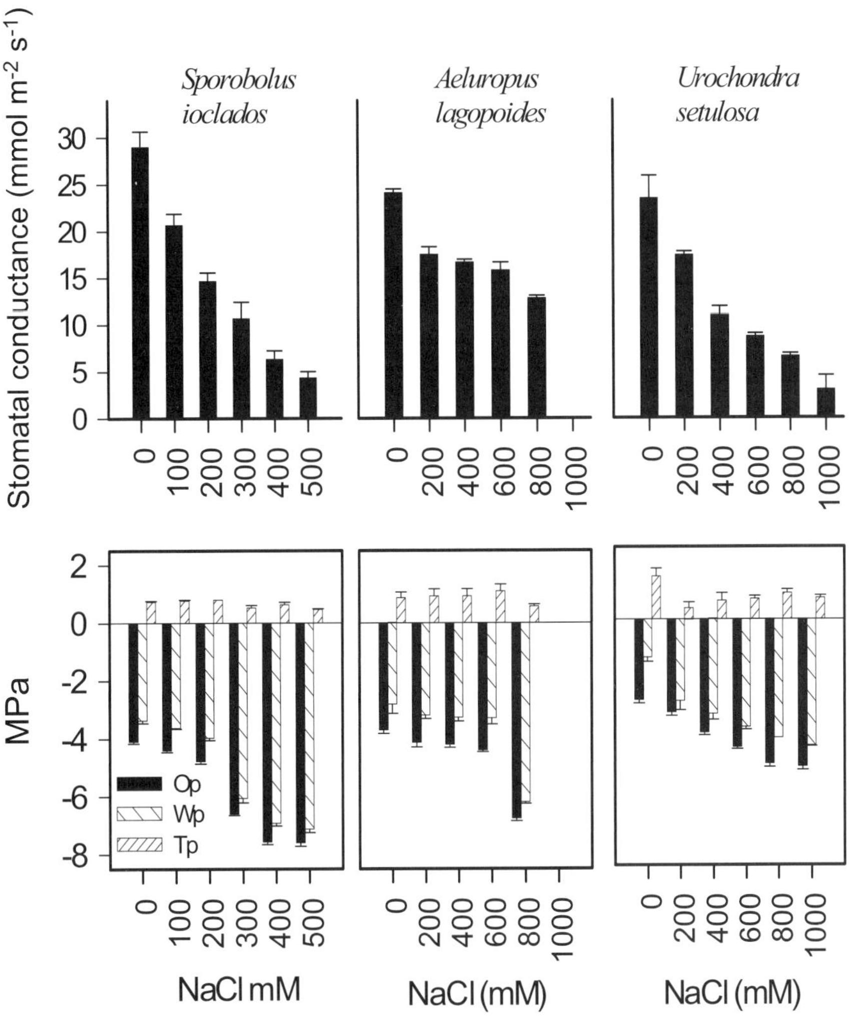

Figure 3. *Effect of increasing salinity on the water relations of* A. lagopoides, S. ioclados *and* U. setulosa.

roots of *A. lagopoides* but it was higher in shoot of *Urochondra setulosa*. In both the species mentioned Na^+ increased but K^+, Ca^{2+} and Mg^{2+} levels remained relatively unchanged. *Sporobolus ioclados* showed a different pattern of ion accumulation with rise in salinity. Sodium increased considerably while K^+, Ca^{2+}

and Mg^{2+} decreased concomitantly with increase in salinity concentrations (Figure 4).

These ratios were mostly higher in root than in shoot. At the highest salinities, *S. ioclados* roots had the highest Na^+/K^+, Na^+/Mg^{2+} and K^+/Ca^{2+} ratios. *Aeluropus lagopoides* showed comparatively lower ion ratios among the three species selected with little variations between salinity treatments.

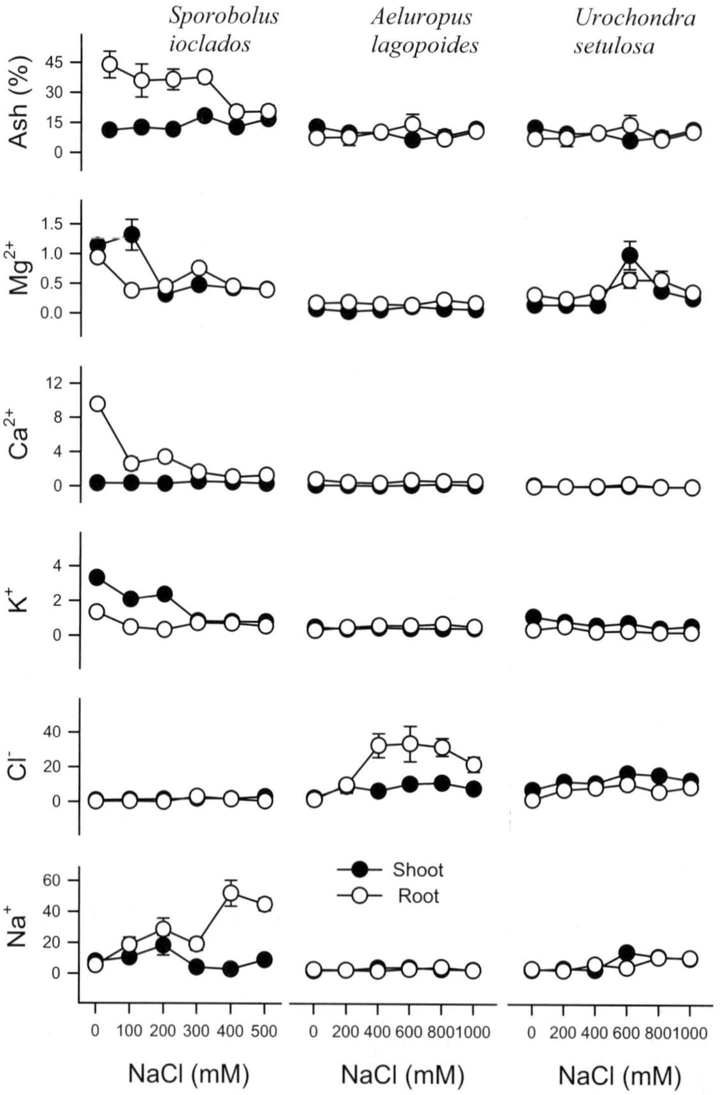

Figure 4. *Effect of increasing salinity on the shoot and root concentrations of Na^+, Cl^-, K^+, Ca^{2+}, Mg^{2+} and ash in* A. lagopoides, S. ioclados *and* U. setulosa*. Tissue ion concentrations are expressed in mmol g^{-1} dry weight.*

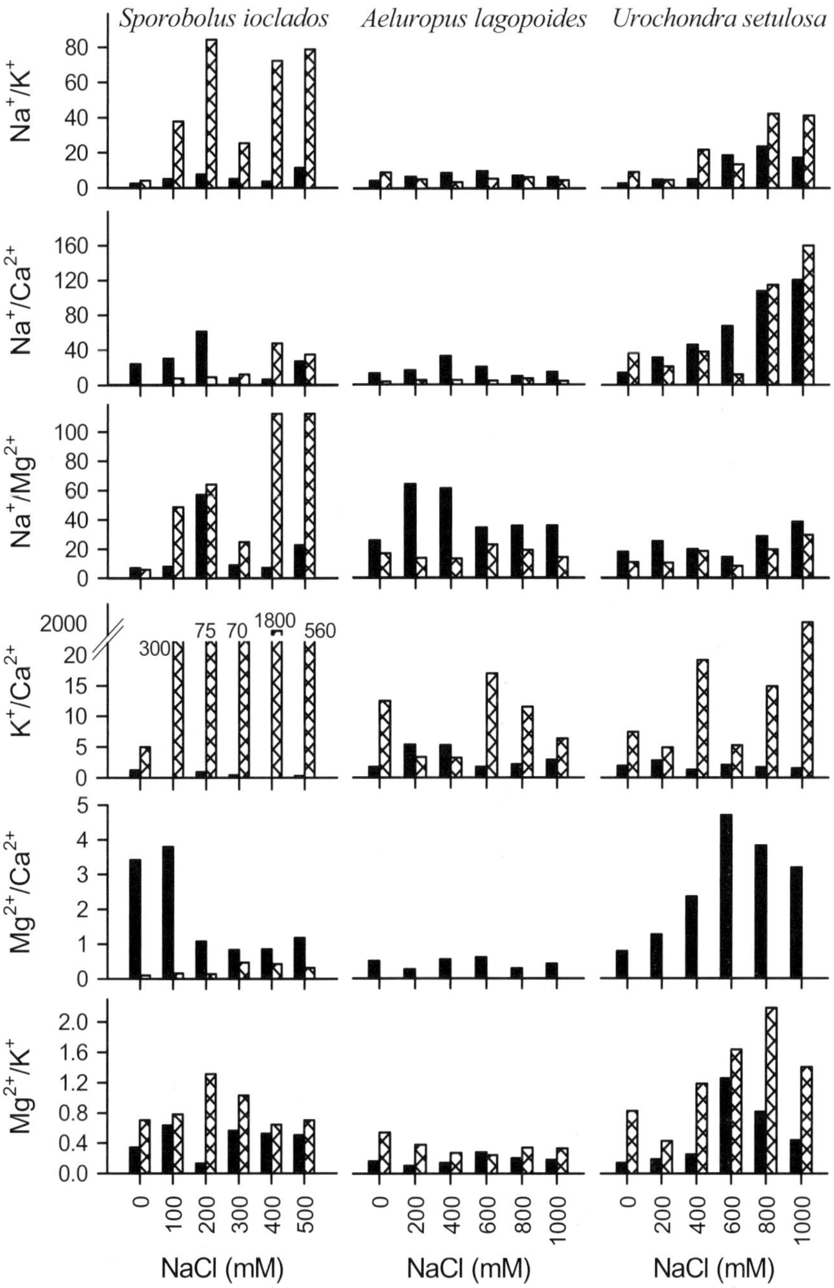

Figure 5. *Effect of increasing salinity on the Na/K, Na/Ca, Na/Mg, K/Ca, Mg/Ca and Mg/K ion ratios in shoot and root of* S. ioclados, A. lagopoides *and* U. setulosa.

4. DISCUSSION

Grasses are quite variable in their tolerance to salinity at growth (Khan et al., 1999; Hester et al., 2001; Muscolo et al., 2003; Joshi et al., 2004). Muscolo et al. (2003) reported that *Panicum clandestinum* could tolerate 200 mM NaCl while growth was little affected by 100 mM NaCl. Leaf and root growth, leaf number, root and leaf length also decreased with increase in salinity. Increasing salinity to 200 mM NaCl had no effect on growth parameters of *A. lagopoides* compared to control. Salt tolerance increased with the age of plants and after 30 days seedlings showed maximum growth inhibition in comparison to 90 days old plants (Debez et al., 2004). Lesser shoot length and fresh weight, but greater root length and dry weight were observed in 70 d plants of *A. lagopoides* and *Panicum hemitomon* grown in 16-48 dSm^{-1} seawater concentrations (Hester et al., 2001; Joshi et al., 2004). Khan et al. (1999) showed that growth of *Halopyrum mucronatum* was promoted at low salinity (90 mM NaCl) but decreased with increase in salinity. Bodla et al. (1995) showed that *A. lagopoides* could grow in up to 110 dS m^{-1} NaCl (1,500 mM NaCl). Mauchamp and Mésleard (2001) reported that growth of plants decreased by 50% in comparison to control at 0.75% (150 mM NaCl) NaCl and plants died at 2% (342 mM NaCl) NaCl. Other graminoid species like *Paspalum distichum* L. (Bodla et al., 1995), *Cyperus rotundus* L. (Shamsi & Ahmed, 1986), *Cynodon dactylon* (Kumar, 1990), *Cenchrus pennisetiformis* Hochst. and Steud; *Panicum turgidum* Forssk. (Ashraf & Naqvi, 1991) reported to survive at moderate salinities. Slower growth is a general adaptive feature for plant survival under stress, allowing to re-direct cell resources (e.g., energy and metabolic precursors) towards the defense reactions against stress (Zhu, 2001). Among the three species, we found that *U. setulosa* had the slowest growth, while *S. ioclados* showed robust growth in a short time and maximum growth was achieved in only 20 days (Gulzar et al., 2005) after salinity treatments in comparison to 45 and 60 days for *A. lagopoides* and *U. setulosa*, respectively (Gulzar et al., 2003ab).

Succulence is one of the mechanisms that halophytes utilize to deal with the high internal ion concentrations (Debez et al., 2004). In *Suaeda fruticosa* (L.) Forssk., shoot tissue water content at 200 mM external NaCl was higher than in non-treated controls, but it decreased at higher salt concentrations (Khan et al., 2000). Succulence expressed on a dry weight basis increased at 200 mM NaCl and then declined at the higher NaCl concentrations in *A. lagopoides* and *S. ioclados* but increased about two-folds in *U. setulosa*. Succulence plays a key role in the survival and maintenance of halophytes under saline conditions by maintaining positive turgor (Khan et al., 1999). These plants also maintained a high water and osmotic potential with progressively lower stomatal conductance with the increase in salinity. However, *A. lagopoides* maintained higher turgor, stomatal conductance and water and osmotic potential up to 600 mM NaCl as compared to the other two grasses. Salt inhibits growth due to stomatal closure subsequently reducing CO_2 uptake and limiting photosynthesis (Zhu, 2001). At high salinity water content of plants decreases and plants lose turgor causing growth inhibition and increase in mortality. Khan et al. (1999) reported that *H. mucronatum* adjusted osmotically, maintaining a more negative osmotic potential than that of the medium though the differences became less with increasing salinity. Antlfinger & Dunn (1983) found that species growing in higher soil salinities had more xylem pressure potential than plants growing in less saline areas. Xylem pressure potential for graminoids ranged from –2.2 to –2.9 MPa. Our data showed much higher ranges of water potential while growing under high salinity.

Salinity inhibits and delays growth due to primary (ion imbalance and hyper osmotic effects) as well as secondary (decrease in water uptake followed by an increase in ion uptake) stresses (Zhu, 2001). Delayed reserve mobilization from cotyledons (Gomes Filho et al., 1983; Prisco, 1987) and membrane stability due to increased leakage of materials from the embryo axis (Prisco, 1987). At the highest concentrations tested i.e., 200 mM NaCl, growth of *Cakile maritima* was inhibited but some plants appeared to maintain tissue hydration by Na^+ compartmentalization (Debez et al., 2004) for maintaining the osmotic potential (Blumwald et al., 2000). Evidently, this was not sufficient to overcome the Na^+ toxicity, particularly to the meristematic tissue involved in new leaf production (Debez et al., 2004). *Aeluropus lagopoides* and *U. setulosa* increased succulence for maintaining turgor by absorbing large quantities of Na^+. *Sporobolus ioclados*, however, lost tissue water with the increase in salinity and maintained higher K^+, Ca^{2+} and Mg^{2+} and lower chloride uptake in shoot in comparison to other two species, however, it was not successful in overcoming the Na^+ toxicity probably due to poor rate of secretion through leaf glands.

Chrysothamnus nauseosus appears to tolerate Na^+ toxicity equally well as compared to *Sarcobatus vermiculatus* but only in periodically wet times (Dodd & Donovan, 1999). However, decreasing soil moisture even under low saline conditions is inhibitory to seedling growth but not for *Sarcobatus vermiculatus* which takes up Na^+ to continue its growth (Dodd & Donovan, 1999). Sodium uptake by high salt tolerant plants promotes the water potential gradient between seedling and substrate and thus enhances the seedling ability to maintain turgor for growth. In other words under natural conditions osmotic and ionic effects act individually as well as interact with soil moisture content in affecting plant establishment in saline desert soils. Additionally, the proportion of various ions in the soil has been known to increase the chances of success for seedling establishment and hence the species distribution. The presence of Ca^{2+} for instance is known to improve the Na^+ and Mg^{2+} tolerance in saline soils to varying degrees in different halophytic species (Tobe et al., 2002).

A high cytosolic K^+/Na^+ ratio is important for maintaining cellular metabolism. Under salt stress, Na^+ competes with K^+ for uptake into roots (Zhu, 2003). The K^+/Na^+ ratios were considered to be an effective index for salt tolerance in wheat (Poustini & Siosemardeh, 2004). Potassium-sodium ratio in roots of *Puccinellia tenuiflora* (alkali grass) decreased from 1.64 at 3 mmol/L to about 0.6 at 100 mmol/L NaCl and remained at this level up to 300 mmol/L NaCl salinity in the culture solution and was higher than K^+ uptake of wheat at 0.37 (Peng et al., 2004). This higher uptake was found to be due to higher K^+/Na^+ selectivity of plasma membrane. A higher K^+ uptake under non-saline and saline conditions was observed in alkali grass, which indicated that a high K^+/Na^+ selectivity of potassium transport systems might be critical in plants salt tolerance. However, molecular evidence involving complex mechanisms at the cellular level would be required. Preference for K^+ over Na^+ did not appear to be a good predictor of plant salt tolerance in 10 forages tested in greenhouse cultures and each species had unique shoot ion concentrations (Grieve et al., 2004). The K^+/Na^+ ratio of the more salt tolerant grasses was lower than the less tolerant ones. In the present study, at all salinity levels, shoot and root of all the three grass species had high Na^+/K^+ that is K^+/Na^+ ratios of less than ~0.5.

Grieve et al. (2004) found that K^+/Mg^{2+} relations of grasses differed from legumes and did not change with the increase in either the external salinity from 15-25 dS m^{-1} or with an increase in Mg^{2+} from 14-28 mM. Similarly, a 50% decrease in Ca^{2+}/Mg^{2+} ratio in substrate from 0.8 to 0.4 did not affect grass species in contrast to alfalfa plants which showed subsequent decrease in Ca^{2+}/Mg^{2+}. The adequate range of Mg^{2+} between species varied mainly due to its competition with K^+ uptake. Magnesium is also strongly competitive with Ca^{2+} and may displace Ca^{2+} from extracellular binding sites within plant organs to further disrupt metabolic availability of Ca^{2+}. Plant growth may be reduced, when the external Mg^{2+}/Ca^{2+} increases above 1 (Carter et al., 1979). Complex ion ratios and balances in the saline irrigation waters did not adversely affect plant nutrient status and, with the exception of broadleaf trefoil (*Lotus ulginosus*), none of the forages showed symptoms of ion toxicities or deficiencies. Many researchers have suggested that the K^+/Na^+ ratio in tissues of glycophytes should be >1 to supply the K^+ necessary for the normal functioning of metabolic processes (Ashraf, 1994; Maathuis & Amtmann, 1999). The K^+/Na^+ and K^+/Na^+ selectivity ratios did not appear to be good indicators of plant salt tolerance (Grieve et al., 2004).

Aeluropus lagopoides, *S. ioclados* and *U. setulosa* varied in their growth, water relations and ion uptake responses with increasing salinity. Grass species studied showed various combinations of strategies for combating high salinity probably linked to their native soil and moisture conditions, the inherent salt and drought tolerating mechanisms as well the ability for intra- and inter-specific competition, unique for each species. Grasses like *A. lagopoides* and *U. setulosa* are highly salt tolerant and could be cultivated using seawater irrigation. *Sporobolus ioclados* is relatively less salt tolerant and suited for areas where brackish water is available. *Aeluropus lagopoides* and *U. setulosa* are good candidates for landscape development in highly saline areas. *Sporobolus ioclados* would be ideal for dune stabilization with good seepage where less saline water is available. These grasses could also be used for forage and fodder.

5. REFERENCES

Alshammary, S.F., Qian, Y.L. & Wallner, S.J. 2004. Growth response of four turfgrass species to salinity. Agricultural Water Management 66: 97–111.

Antlfinger, A. & Dunn, E.L. 1983. Water use and salt balance in three salt marsh succulents. American Journal of Botany 70: 561–567.

Ashraf, M. 1994. Breeding for salinity tolerance in plants. Critical Reviews in Plant Sciences 13: 17–42.

Ashraf, M. & Naqvi, M.I. 1991. Responses of three arid zone grass species to varying Na^+/Ca^{2+} ratios in saline sand culture. New Phytologist 119: 285–290.

Ashraf, M., McNeilly, T. & Bradshaw, A.D. 1986. Tolerance of sodium chloride and its genetic basis in natural populations of four grass species. New Phytologist 103: 725–734.

Bajji, M., Kinet, J.M. & Lutts, S. 2002. Osmotic and ionic effects of NaCl on germination, early seedling growth and ion contet of *Atriplex halimus* (Chenopodiaceae). Canadian Journal of Botany 80: 297–304.

Bell, H.L. & O'leary, J.W. 2003. Effects of salinity on growth and cation accumulation of *Sporobolus virginicus* (Poaccac). American Journal of Botany 90: 1416–1424.

Blumwald, E., Aharon, G.S. & Apse, M.P. 2000. Sodium transport in plant cells. Biochimica et Biophysica. Acta 1465: 140–151.

Bodla, M.A., Chaudhry, M.R., Shamsi, S.R.A. & Baig, M.S. 1995. Salt tolerance in some dominant grasses of Punjab. In: M.A. Khan and I.A. Ungar. (Eds.), Biology of Salt Tolerant Plants. Chelsea, Michigan: Book Crafters. 190–198 pp.

Carter, M.R., Webster, G.R. & Cairns, R.R. 1979. Calcium deficiency in some solonetzic soils of Alberta. Journal of Soil Science 30: 161–174.

Debez, A., Hamed, K.B., Grignon, C. & Abdelly, C. 2004. Salinity effects on germination, growth, and seed production of the halophyte *Cakile maritima*. Plant and Soil 262: 179–189.

Dodd, G.L & Donovan, L.A. 1999. Water potential and ionic effects on germination and seedling growth of two cold desert shrubs. American Journal of Botany 86: 1146–1153.

Dudeck, A.E., Peacock, C.H. & Wildmon, J.C. 1993. Physiological and growth responses of St. Angustine grass cultivars to salinity. Horticultural Science 28: 46–48.

Forment, J., Naranjo, M.A., Roldán, M., Serrano, R. & Vicente, O. 2002. Expression of Arabidopsis SR-like splicing proteins confers salt tolerance to yeast and transgenic plants. Plant Journal 30: 511–519.

Glenn, E.P. 1987. Relationship between cation accumulation and water content of salt-tolerant grasses and a sedge. Plant, Cell Environment 10: 205–212.

Gomes Filho, E., Prisco, J.T., Campos, F.A.P. & Eneas Filho, J. 1983. Effects of NaCl salinity in vivo and in vitro on ribonuclease activity of *Vigna unguiculata* cotyledons during germination. Physiologia Plantarum 59: 183–188.

Gorham, J., Hughes, L.L. & Wyn Jones, R.G. 1980. Chemical composition of salt marsh plants from Ynys-Mon (Angelsey): the Concept of Physiotypes. Plant, Cell Environment 3: 309–318.

Greenway, H. & Munns, R. 1980. Mechanisms of salt tolerance in non-halophytes. Annual Review of Plant Physiology 31: 149–190.

Grieve, C.M., Poss, J.A., Grattan, S.R., Suarez, D.L., Benes, S.E. & Robinson, P.H. 2004. Evaluation of salt-tolerant forages for sequential water reuse systems II. Plant–ion relations. Agricultural Water Management 70: 121–135

Gulzar, S., Khan, M.A. & Ungar, I.A. 2003a. Effects of Salinity on growth, ionic content and plant-water relations of *Aeluropus lagopoides*. Communications in Soil Science and Plant Analysis 34: 1657–1668.

Gulzar, S., Khan, M.A. & Ungar, I.A. 2003b. Salt tolerance of a coastal salt marsh grass. Communications in Soil Science and Plant Analysis 34: 2595–2605.

Gulzar, S., Khan, M.A., Ungar, I.A. & Liu, X. 2005. Influence of salinity on growth and osmotic relations of *Sporobolus ioclados*. Pakistan Journal of Botany 37: 119-129.

Hester, M.W., Mendelssohn, I.A. & Mckee, K.L. 1996. Intraspecific variation in salt tolerance and morphology in the coastal grass *Spartina patens*. American Journal of Botany 83: 1521–1527.

Hester, M.W., Mendelssohn, I.A. & Mckee, K.L. 2001. Species and population variation to salinity stress in *Panicum hemitomon*, *Spartina patens* and *Spartina alterniflora*: morphological and physiological constraints. Environmental and Experimental Botany 46: 277–297.

Joshi, A.J., Mali, B.S. & Hinglajia, H. 2004. Salt tolerance at germination and early growth of two forage grasses growing in marshy habitats Environmental and Experimental Botany. (In Press – published online).

Khan, M.A. 2002. An ecological overview of halophytes from Pakistan. In: H. Lieth, & M. Moschenko. (Eds.), Cash Crop Halophytes: Potentials, Pilot Projects, Basic and Applied Research on Halophytes and Saline Irrigation 10 years after Al-Ain Meeting (in press). Netherlands: Kluwer Academic Press. 167–187 pp.

Khan, M.A. & Gul, B. 2001. Salt tolerant plants of coastal sabkhas of Pakistan. In: B. Boer, & H, Barth. (Eds.), Sabkhas of Arabian Peninsula and adjacent regions. Dordrecht, Netherlands: Kluwer Academic Press. 123 – 140 pp.

Khan, M.A., Ungar, I.A. & Showalter, A.M. 1999. Effects of salinity on growth, ion content, and osmotic relations in *Halopyrum mucronatum* (L.) Stapf. Journal of Plant Nutrition 22: 191–204.

Khan, M.A., Ungar, I.A. & Showalter, A.M. 2000. The effect of salinity on the growth, water status, and ion content of a leaf of a succulent perennial halophyte, *Suaeda fruticosa* (L.) Forssk. Journal of Arid Environments 45: 73–84.

Kumar, A. 1990. Forage yield of grasses as affected by the degree of soil sodicity and soil amelioration caused by their growth. In: A. Kumar. (Ed.), Proceedings of Indo-Pak workshop on soil salinity and water management. pp. 434–445. Islamabad, Pakistan: PARC.

La Peyre, M.K. & Rowe, S. 2003. Effects of salinity changes on growth of *Ruppia maritima*. Aquatic Botany 77: 235–241

Lissner, J. & Schierup, H.H. 1997. Effects of salinity on the growth of *Phragmitis australis*. Aquatic Botany 55: 247–260.

Maathuis, F.J.M. & Amtmann, A. 1999. K^+ nutrition and Na^+ toxicity: the basis of cellular K^+/Na^+ ratios. Annals of Botany 84: 123–133.

Mahmood, K., Malik, K.A., Lodhi, M.A. K. & Sheikh, K.H. 1996. Seed germination and salinity tolerance in plant species growing on saline wastelands. Biologia Plantarum, 38: 309–315.

Marcum, K.B. & Murdoch, C.L. 1990. Growth responses, ion relations, and osmotic adaptations of eleven C_4 Turfgrasses to salinity. Agronomy Journal 82: 892–896.

Mauchamp, A. & Mésleard, F. 2001. Salt tolerance in *Phragmites australis* populations from coastal Mediterranean marshes. Aquatic Botany 70: 39–52.

Moore, R.H. 1960. Laboratory Guide for Elementary Plant Physiology; Minncapolis, Minnesota: Burgess Publishing Company.

Munns, R. 2002. Comparative physiology of salt and water stress. Plant, Cell & Environment 25: 239–250.

Muscolo, A., Panuccio, M.R. & Sidari, M. 2003. Effects of salinity on growth, carbohydrate metabolism and nutritive properties of kikuyu grass (*Pennisetum clandestinum* Hochst). Plant Science 164: 1103–1110.

Parida, A.K. & Das, A.B. 2005. Salt tolerance and salinity effects on plants: a review. Ecotoxicology and Environmental Safety 60: 324–349.

Peacock, C.H. & Dudeck, A.E. 1985. Physiological and growth response of seashore Paspalum to salinity. Horticultural Science 20: 111–112.

Peng, Y.H., Zhu, Y.F., Mao, V., Wang, S.M., Su, W.A. & Tang, Z.C. 2004. Alkali grass resists salt stress through high $[K^+]$ and an endodermis barrier to Na^+. Journal of Experimental Botany 55: 939–949.

Poustini, K. & Siosemardeh, A. 2004. Ion distribution in wheat cultivars in response to salinity stress. Field Crops Research 85: 125–133.

Prisco, J.T. 1987. Contribuicao ao estudo da fisiologia do estresse salino durante a germinacao e estabelecimento da plantula de uma glico′ fita (*Vigna unguiculata* (L.) Walp.). Tese de Professor Titular. Universidade do Ceara′, Fortaleza, Brasil.

Ramadan, T. 2001. Dynamics of Salt Secretion by *Sporobolus spicatus* (Vahl) Kunth from sites of differing salinity. Annals of Botany 87: 259–266.

Russell, J.S. 1976. Comparative salt tolerance of some tropical and temperate legumes and tropical grasses. Australian Journal of Experimental Agriculture and Animal Husbandry 16: 103–109.

Serrano, R., 1996. Salt tolerance in plants and microorganisms: toxicity targets and defence responses. International Review of Cytology 165: 1–52.

Serrano, R. & Gaxiola, R. 1994. Microbial models and salt stress tolerance in plants. Critical Reviews in Plant Sciences 13: 121–138.

Shamsi, S.R.A. & Ahmad, B. 1986. Studies on salt tolerance of purple nutsedge (*Cyperus rotundus*). Indian Journal of Experimental Biology 24: 499–504.

Shannon, M.C. 1978. Testing salt tolerance variability among tall wheatgrass lines. Agronomy Journal 70: 719 –722.

Shen, Y.Y., Li, Y. & Yan, S.G. 2003. Effects of salinity on germination of six salt-tolerant forage species and their recovery from saline conditions. New Zealand Journal of Agricultural Research 46: 263–269.

SPSS. 2002. SPSS 11 for Windows update. Chicago, Illinois: SPSS Inc.

Tobe, K., Li, X. & Omasa, K. 2002. Effects of sodium, magnesium and calcium salts on seed germination and radicle survival of a halophyte, *Kalidium capsicum* (Chenopodiaceae). Australian Journal of Botany 50: 163–169.

Tyerman, S.D. & Skerrett, M. 1999. Root ion channels and salinity. Scientia Horticulturae 78: 175–325.

Venables, V. & Wilkins, D.A. 1978. Salt tolerance in pasture grasses. New Phytologist 80: 613–622.

Wang, S., Zheng, W., Ren, J. & Zhang, C. 2002. Selectivity of various types of salt-resistant plants for K^+ over Na^+. Journal of Arid Environment 52: 457– 472.

Wang, S., Wan, C., Wang, Y., Chen, H., Zhou, Z., Fu, H. & Sosebee, R.E. 2004. The characteristics of Na^+, K^+ and free proline distribution in several drought-resistant plants of the Alxa Desert, China. Journal of Arid Environment 56: 525–539.

Yeo, A. 1998. Molecular biology of salt tolerance in the context of whole-plant physiology. Journal of Experimental Botany 49: 915–929.

Yokoi, S., Quintero, F.J., Cubero, B., Ruiz, M.T., Bressan, R.A., Hasegawa, P.M. & Pardo, J.M. 2002. Differential expression and function of *Arabidopsis thaliana* NHX Na^+/H^+ antiporters in salt stress response. Plant Journal 30: 529–539.

Zhu, J.K. 2001. Plant salt tolerance. Trends in Plant Science 6: 66–71.

Zhu, J.K. 2002. Salt and drought stress signal transduction in plants. Annual Review of Plant Biology 53: 247–273.

Zhu, J.K. 2003. Regulation of ion homeostasis under salt stress. Current Opinion in Plant Biology 6: 441–445.

CHAPTER 18

COMMERCIAL APPLICATION OF HALOPHYTIC TURFS FOR GOLF AND LANDSCAPE DEVELOPMENTS UTILIZING HYPER-SALINE IRRIGATION

M. W. DEPEW[1] AND P. H. TILLMAN[2]

[1]*Agronomist/Soil Scientist with Environmental Technical Services, LC.*
[2]*Engineer/Agronomist with Barbuda Farms International, Ltd.*

1. INTRODUCTION

1.1. Saline irrigation and turf management program developoment

Michael DePew and Paul Tillman were studying turfgrass as graduate students at Texas A&M University in College Station, Texas, USA in 1990. During this graduate study period we envisioned a saline irrigation and turf management program; in an effort to focus on a niche area of the turf industry with an expanding future potential. As we developed the idea, this program development process took on two main areas of focus. One in which the focus was on turf qualities and turf development. The second area was a focus on soil ecology issues, e.g. how soil processes occurring in saline soils affected the growth, development, aesthetics and utilitarian functions of salt tolerant grasses.

Initially our turf development program was primarily limited to making collections of turfs from the wild or from the fringe areas of landscapes that abutted adjoining saline areas. Our primary focus was on what we felt were two promising warm season turf species; Seashore paspalum (*Paspalum vaginatum*) and Seashore dropseed (*Sporobolus virginicus*).

Seashore paspalum has been used intermittently as a turfgrass cultivar in North and South America (including the US and Caribbean) from the early 20th century (Morton, 1973). In 1970, Hugh Whiting imported and introduced a named cultivar into the USA from Australia known as 'Adalayd'. While Adalayd was known to have adaptations for salinity and was a medium-textured turfgrass, this cultivar

M.A. Khan and D.J. Weber (eds.), Ecophysiology of High Salinity Tolerant Plants, 255-278.

failed to gain wide use in the USA. Our evaluations of Adalayd in the early 1990's showed that this cultivar only had the ability to sustain ongoing irrigation salinities of around 6,000 – 7,500 ppm TDS, but can withstand seawater spray and periodic seawater inundation. It was our desire to work at salinity ranges exceeding this range and to utilize turfgrass varieties that had better turf qualities than the Adalayd cultivar.

Seashore dropseed is known to flourish in many brackish and salt marsh areas in Texas and throughout the southern states of the USA. Due to the fine leaf texture of many of the native ecotypes of this species, potential use as a turf is readily indicated. Many Seashore dropseed ecotypes were collected for evaluation in 1990-1991 from Texas coastal locations. These early collections were later abandoned because they failed to thrive under fresh water irrigation regimes and we felt that they exhibited too much obligate halophyte characteristics. It was our desire to collect and develop grasses that could tolerate a salinity regime ranging from fresh water to high saline or even seawater.

Our initial studies involving soil ecology issue with saline soils began by reviewing the literature regarding processes that occur in coastal and marine soils and sediments. This included a review of processes relating to calcareous soils, sulfide-rich mine overburden and mine reclamation (acid-sulfate weathering), ironstone formation, beach hardness, sand podzolization, salt weathering, mineral transformations, etc. Integrating all of these processes and soil physics principles, we desired to understand the movement and fate of salt solutes in solution as they move into and through the soil.

2. HALOPHYTIC TURF DEVELOPMENT

2.1. Seashore Dropseed

After we abandoned our initial collection of Seashore dropseed ecotypes from Texas, we next began to evaluate the potential use of Seashore dropseed cultivars from the Caribbean. This aspect of the program coincided with Tillman's commercial activities with the K Club Resort Golf Course on the island of Barbuda, Antigua-Barbuda, West Indies. Tillman was hired (1993-1994) to manage the maintenance of the golf course, which is a seaside course that was at that time utilizing desalinated water (from reverse osmosis) to irrigate hybrid bermudagrass turfs (*Cynodon dactylon* x *C. transvaleensis*).

While the water provided for irrigation was from a desalinization plant, the quality of the water was not consistent and typically varied from a few hundred ppm TDS to a few thousand ppm TDS. Under these water quality variations, it was very difficult to manage the bermudagrass turf. Inconsistent water quality was also coupled with inadequate supply such that most available water was applied to tees and greens with little water available for fairways and roughs. In these circumstances, Tillman began to experiment with local collections of the Seashore dropseed on fairway areas of the golf course. These areas were then irrigated by direct pumping of salt water from lakes on the course with salinity ranging from 18,000 ppm TDS to seawater (~35,000 ppm TDS). Seashore dropseed ecotypes were evaluated under saline irrigation and turf management regimes (close mowing, high traffic, divoting) for utilitarian and aesthetic performance.

Once promising ecotypes were screened on the K Club golf course, they were then asexually propagated and placed on a commercial nursery facility that was developed in cooperation with a local businessman, Mr. David Shaw. At the nursery site, the turfs were then further evaluated for their commercial sod crop characteristics. From this process, an initial cultivar (BT-1) was imported as seed into the US and patented (US Plant Patent 13,652) and subsequently released as 'Saltfine®'.

2.2. Seashore Paspalum

Initially, we began to do some work utilizing Adalayd Seashore paspalum. Our initial screenings showed that a sustainable salinity range was somewhere around 6,000 – 7,500 ppm TDS. At higher salinity levels the Adalayd would exhibit increasing leaf firing, decreasing turf quality and substantial reductions in overall growth rate. However collections we had made of Seashore paspalum from the wild (Texas, Florida, California) had been of coarser ecotypes that while showing substantial increases in salinity tolerance compared to Adalayd, showed no improvement in overall aesthetic qualities.

While seeking for improved ecotypes to add to our collection, we traveled to many coastal communities and golf courses (1991-1999) within the US. This included areas such as Charleston, SC; San Diego, CA; Long Beach, CA; Savannah, GA; Galveston, TX; Corpus Christi, TX; Padre Island, TX; Key Largo, FL; Palm Beach, FL; Marco Island, FL; Sanibel, FL; Ft. Myers/Cape Coral, FL; Pine Island, FL; Charlotte, FL; Boca Raton, FL; Lover's Key, FL; Ruskin/Apollo Beach, FL; St. Petersburg Beach, FL; Indialantic, FL. This also included visits to golf courses using paspalum as turf such as;

Fairbanks Ranch, near San Diego, CA (≤ 2,000 ppm irrigation)
Kings Crossing, Corpus Christi, TX (≤ 8,000 ppm)
Alden Pines Country Club, Bokeelia, FL (≤5,000 ppm).

While many grass ecotypes were collected from these locations, most showed little to no improvement in a combination of turf quality and/or salinity tolerance compared to Adalayd. However, we found a different situation upon our visit to Alden Pines Country Club located on Pine Island, FL. Alden Pines had been planted with Adalayd around 1983 after the initial plantings of bermudagrass failed.

At Alden Pines (1997), we observed what appeared to be accessions of grass types from the coarse, native-type paspalum growing along the edges and into the golf course's salt water lakes to a progression of successively finer ecotypes moving outward from the salt lakes toward the fairway areas. We have attributed these various ecotypes to natural crosses that may have occurred between Adalayd and the native Seashore paspalum types.

Upon realizing the potential of some of these grasses at Alden Pines, we undertook efforts to secure some selections of these grasses. Mr. Stewart Bennett was the Superintendent of the golf course and a minority shareholder. We approached Mr. Bennett about our desire to obtain some of the turfs for evaluating the commercial potential. After hearing of our desire and enquiring as to our business objectives, Bennett approached us about becoming part of our venture and offered access to all the various ecotypes of paspalum turfs at Alden Pines as a contribution to the venture. Based upon this business proposal we formed Environmental Turf Solutions, Inc. (ETS) in 1997.

The remainder of 1997 and into 1998, we began a vigorous program to sort and evaluate the potential of the various turf ecotypes on the Alden Pines golf course. Having already been subjected to a high degree of selection pressure (low mow heights, soil salinity, high traffic, reduced soils, compaction, etc.) we began to evaluate various selections for use on other areas of the golf course to evaluate the adaptability of the grasses and to assure ourselves that the turfs remained true to type after asexual reproduction. From this process, two Seashore paspalum selections were chosen for initial commercialization and were patented and trademarked as Seagreen® (SGX-6, US Plant Patent 13,100) and Seaway (SFX-14, US Plant Patent 13,105).

Also during our involvement with Alden Pines, we initiated a process to improve conditions on the golf course by increasing the salinity of the irrigation water from about 5,000 ppm TDS to a range of 10,000 to 20,000 ppm TDS. This was done by the development of a brine water irrigation well (~45,000 ppm TDS) that could be blended with the brackish well water (~5,000 ppm TDS) to adjust the salinity range. Increasing the salinity of irrigation (along with improvement in drainage and soil conditions, and improved turf management protocols) resulted in less weed occurrence, reduced disease and insect pests and an improvement in turf conditions as the more saline tolerant ecotypes could then thrive.

In 1998, we established a turf nursery on Pine Island, FL separate from Alden Pines golf course. This nursery was irrigated exclusively with water ranging from 15,000 ppm TDS to 26,000 ppm TDS. This nursery was located on canals, which directly connected with the Gulf of Mexico. Water from the canals was blended with shallow brackish groundwater to modulate the salinity of the irrigation water.

At the Pine Island nursery, we manipulated collections of Seashore paspalum ecotypes (from Alden Pines and elsewhere) to produce additional crosses. We would then plant out these crosses for reselection based upon promising candidates. From this process we released the turf cultivar SeaDwarf® (SDX-1, US Plant Patent 13,294) for commercial production in 2001.

As these various cultivars became available for release, ETS executed various production license agreements including a Caribbean License Agreement with Barbuda Farms International, Ltd. (BFI) (formerly Barbuda Turf Company) on November 1, 1999, a US License Agreement with Environmental Turf Nurseries,

Inc. (ETN) on October 13, 2000, and an International License Agreement with Saline Resources International, Inc. (SRI) on March 5, 2001. Subsequently, ETS broke up and the ownership interests in the grasses were split between the shareholders. DePew and Tillman maintained rights to SeaDwarf and Saltfine while Bennett retained rights to Seagreen and Seaway. The respective parties warranted to each other that there were no international encumbrances to each of the other parties' respective turfs. DePew and Tillman formed EnviroTurf, LC as a holding company to retain rights in their grasses. As of this writing (June 2004), Bennett has not performed as to his warrant and indeed ETN claims international rights to EnviroTurf's grasses by a secretly negotiated and executed international license agreement that Bennett executed with ETN purportedly on behalf of ETS dated March 29, 2001. This issue is currently in litigation and remains to be resolved.

3. COMMERCIAL AVAILABILITY

Of the grasses the authors have patented and retain rights to through EnviroTurf (SeaDwarf and Saltfine), they are available commercially in the US through ETN. However, EnviroTurf is prosecuting a legal challenge to that license agreement on the basis of ETN's default of said license agreement for non-payment of license fees, royalties and other breach of terms. SeaDwarf and Saltfine are available in the Caribbean region (and for export to other world regions) through BFI (barbudaturf@yahoo.com). Grass is available internationally through SRF Holdings of Colorado, assignee to the SRI International License Agreement. EnviroTurf is seeking clarification through the courts as to the status of this issue and hopes to have the issue resolved by the date of publication. Please contact DePew at proturf@hotmail.com for additional details.

DePew and Tillman continue an ongoing turf development program with many crosses and reselections of Seashore paspalum and Seashore dropseed ecotypes. These include Seashore paspalums that have improved adaptations for shade and cool season color retention. These turfs will be commercially available by time of publication. The dwarf paspalum turfs developed by DePew and Tillman exhibit the following traits:

True dwarf paspalums (< 7.5 cm unmown mature height).
Saline tolerant up to 26,000 ppm TDS.
Low water consumption (Kc is ~0.33 - 0.40).
Deep rooted (> 2 feet, over 1 foot at greens cut).
Low N fertility.
Excellent cool weather color retention.
Can tolerate on-going mow heights of <0.10 inch (2.5 mm).
Rapid grow-in yet exhibits dwarf growth characteristics.
Low mowing frequency when grown for lawns or fairways.

Installations that include the author's turfs include:

Alden Pines, Bokeelia, FL (salt water)

Crown Colony, Ft. Myers, FL (effluent)
Hammock Bay, Naples, FL (effl/salt mix)
Parklands, West Palm Beach, FL (effluent)
Quail Ridge, Boynton Beach, FL (effluent)
Frigate Bay, St. Kitts (salt water)
San Felipe Beach Club, Baja California, Mexico (salt water)
Las Palomas Country Club, Puerto Penasco, Sonora, Mexico (saline-effluent blend)
Galveston Country Club, Galveston, TX (salt water)
K Club Resort, Barbuda, West Indies
Various resorts in Antigua, St. Kitts, Anguilla, St. Marteen, Puerto Rico.
Several other course currently under construction in Mexico, Florida and the Caribbean.

4. SALINITY MANAGEMENT

4.1. Salinity management protocols

Salinity management protocols were worked out in the US for the application of salt-affected water to irrigation tradition (salt sensitive) crops. These protocols were published by the US Salinity Laboratory (USSL) of Riverside, California as Handbook 60 in 1954. This program (and revisions to follow; Ayers & Westcot, 1976, 1985) primarily dealt with management of salinity (osmotic) effects in crop production and the management of sodium hazard (sodicity) as it impacts soil quality. Along with these two primary factors, considerable attention is also given to ion (nutrient) imbalances (or deficiencies) and/or toxicity from various elevated levels of essential and non-essential ions (such as boron, selenium and metals).

The salinity management protocols developed (and under development) by DePew and Tillman take a different approach. Our approach is to intentionally use salt water to irrigate salt-tolerant (halophytic) plants, especially turfgrasses. Having a high degree of salinity tolerance available to us through the inherent adaptations of the turf, our program focuses on landscape quality and the things that affect that quality and performance. These are primarily soil ecological issues and include such factors as:

4.1.1. Controlling and managing salt partitioning effects

4.1.2. Managing mineral transformations in the soil

4.1.3. Managing soil microbial reduction

4.1.4. Monitoring and managing soil redox conditions

4.1.5. Managing and reducing soil phytotoxic gaseous emission.

4.2. Controlling and Managing Salt Partitioning Effects

Controlling and managing salt partitioning effects requires an understanding of the movement and fate of solutes applied to the soil in salt-laden irrigation water. Salt partitioning controlled by the soil physics principle termed miscible displacement. Miscible displacement is regulated by two processes that include hydrodynamic dispersion and molecular diffusion. Of these two processes, hydrodynamic dispersion plays the dominant role. Hydrodynamic dispersion has also sometime referred to as mechanical dispersion.

In simple terms, hydrodynamic dispersion is the concept that when you apply salt water to a soil, the salts in the applied water do not move through the soil at the same rate as the water moving through the soil. Further, all the salts in solution do not move through soil at the same rate.

Molecular diffusion is a component of miscible displacement. However, a key to using salt water to irrigate halophytic turfs requires that the soils are not allowed to dry out excessively. Further, many of the soils required for a successful project are sandy, well-drained soils that will facilitate good aeration (gas exchange) and do not contain an abundant proportion of micropores where molecular diffusion primarily operates. In terms of soil water flux (under a high level of saturation), molecular diffusion is rather a static process. In other words, molecular diffusion of solutes primarily occurs when the soil is in unsaturated conditions and in soils with a high degree of microporosity and/or interconnectivity of micropores. Given these conditions, in many cases molecular diffusion does not play a large role in the movement of solutes except for short range movement under rapidly drying (unsaturated) conditions such as occur at/very near the soil surface in well-drained soils.

What should be known about molecular diffusion is that similar to hydrodynamic diffusion, all salt solutes do not diffuse at the same rate. Further, as soils dry all salts do not have the same solubility and will precipitate from solution leaving the higher solubility salts in solution and thus more prone to move within the soil profile through molecular diffusion. Molecular diffusion rates for various salt solutes is dependent upon the concentration of the solute (and its corresponding hydration energy/sphere), solute concentration gradient, temperature, the diffusion coefficient of the solute, soil moisture content (degree of saturation) and soil tortuosity. In general, the diffusion coefficient of a solute in a soil solution will be effectively reduced by a factor of 0.3 to 0.7 compared to diffusion in water due to the tortuosity of the soil.

Hydrodynamic dispersion is the dominating factor for miscible displacement (salt partitioning) in soils. This is a dynamic process in relation to the movement of water into (infiltration) and through (percolation) the soil at or near saturation. This process is dependent upon the solute size, solute charge (+ charge {cation} or – charge {anion}) valency (mono or polyvalent), pore geometry and tortuosity, soil charge and flow velocity.

Let's use mental imagery to envision how salts move through soil. First of all, keep in mind that the salt solutes are solids and not liquid. Salts dissolve and go into solution as discrete solid particles (as opposed to large salt crystals formed when they precipitate). As water moves through a soil pore, the various salt ions will move through at different rates. For example, as a general trend the larger solutes will move slower than the smaller solutes. The solutes in/near the center of the soil pore will move faster than those solutes near the "wall" of the soil pore. In that soil particles are negatively charged, the cations will be attracted to the pore "wall" while the anions will be repulsed and in higher concentration near the center of the pore where the water flow velocity is greater. This is termed anion exclusion.

Solutes also move in relation to how much the solute itself interacts with water molecules. This is referred to as the solute hydration energy and is related to the charge density of the solute. The charge density relates the overall charge of the ion in relation to its size. If a solute has a high charge density, then it will have a higher hydration sphere which then increases the solutes effective size. For example, sodium is a very small cation with a single charge. However since this single charge is contained within a very small ionic size, the charge density of sodium is quite high and therefore sodium has high hydration energy. This large hydration sphere associated with sodium is an important factor in understanding the movement (leaching) of sodium in the soil. As the concentration of sodium in solution increases, the hydration sphere of sodium decreases and the effective size of sodium is smaller and thus it will move through the soil more readily. The high hydration sphere of sodium is also the reason why sodium at low salinities (< ~2,600 ppm) disperses soil clay minerals. We will discuss this more detail later on in this article.

Other factors to keep in mind when thinking about solute movement through soil is the effective soil pore size. The larger the effective pore size (diameter) the larger the water flow velocity. Higher flow velocity can increase the amount of salt partitioning within a water stream. Also, the reactivity of the soil (soil charge or cation exchange capacity) effects how much attraction/interaction there is between soil particles and cations and how much repulsion of anions. Attraction of cations by the "walls" of the conduit (soil pores) reduces concentration of these ions in the center of the flow stream (higher flow velocity) while repulsion of anions puts a higher concentration of these ions into the center of the flow stream (anion exclusion). From Poiseulle's law we know that the quantity of water traveling through a conduit is related to the conduit radius by the simplified equation:

$$Q_{approx}=R^4.$$

What this demonstrates is that as the radius increases (doubles), the flow velocity increases by a factor of 16X.

Within a given pore, the difference between flow velocities in relation to the wall of the conduit is given by the relationship:

$$v=2v_a[1-(r^2/R^2)]$$

This equation demonstrates that at the center of the pore (r = 0) the flow velocity at the center is twice the average flow velocity. And the at wall, where r = R, thevelocity is zero.

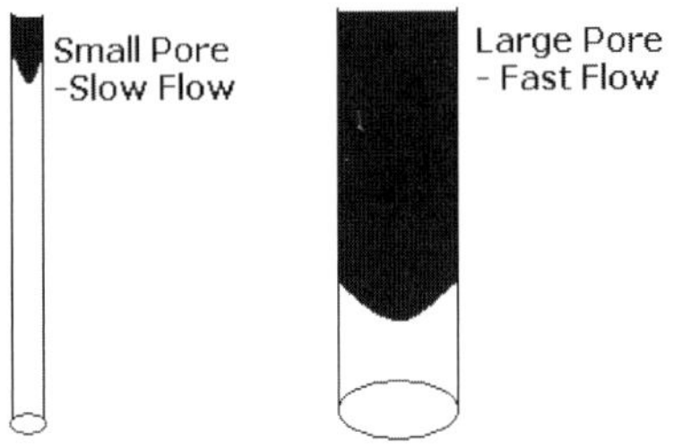

$$v= 2v_a \left[1-(r^2/R^2)\right]$$

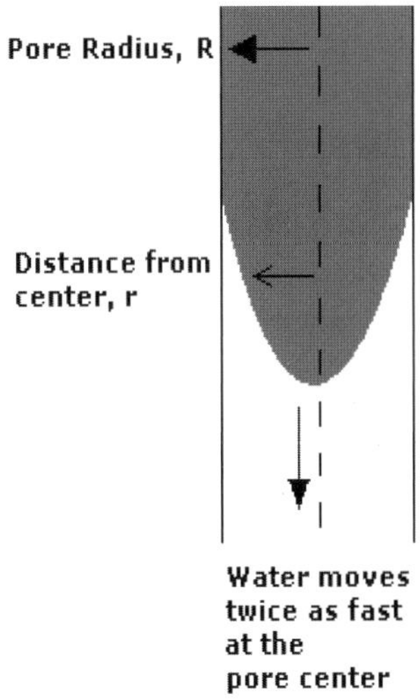

v is the velocity at a any given reference point, v_a is the average velocity in the pore, r = distance of the given reference point from the pore center, R = radius of pore.

In 1995, we undertook some in-house studies to determine the effects of salt partitioning within sand columns. We packed duplicate 7.5 cm PVC cores with medium sand to a depth of 30 cm. The columns were leached and then saturated with distilled water and then allowed to drain for 24 hours. Then we added a diluted synthetic seawater solution (~2,500 ppm TDS) to the columns in 1 pore volume increments. At the initiation of the pore volume additions, we collected the drainage water and analyzed it for ionic composition. This experiment was repeated several times after thoroughly leaching the sand columns (100 pore volumes) with distilled water. The average results from this study are given in (Figure 1) below. The line for a "non-ionic" solution is a theoretical line for comparative purposes only. Each pore volume was added to the surface of the column at one time such that the water movement into the core was at a high degree of saturation. Drainage water was collected at falling head increments representing partial pore volumes. Regression analysis was performed on the data to standardize the breakthrough curves and establish pore volume increments for standard concentration proportions (0, 0.25, 0.50, 0.75, and 1.0).

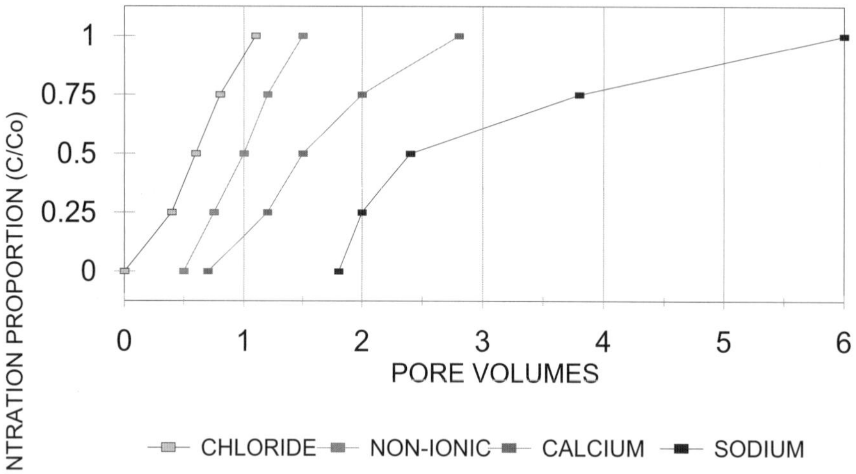

Figure1. *Miscible displacement or ion breakthrough curves for a medium sand and dilute seawater (~2,500 ppm TDS).*

What the Figure 1 graph shows is that while it takes about 2.75 pore volumes to leach through calcium salt ions; it takes 6+ pore volumes to leach through Na. Likewise it only takes about 1.3 pore volumes to see chloride leach through, illustrating the anion exclusion effect.

From a practical standpoint, what this type of data illustrates is that when you leach salts from the soil, sodium ions will be the last to leach. If you start with a dilute seawater solution and continuously leach the soil profile, you will have preferential accumulation of sodium within the upper portions of the soil profile.

Figure 1 is a simple illustration of data from one simplified study. Additional studies we have undertaken show that we can reduce the number of pore volumes for sodium breakthrough as we increase the overall strength of the salt solution, particularly as sodium concentration exceeds 2,500 ppm TDS. This data will be reported in later publications. What this means from a salinity management aspect is that if we want to reduce the effects of sodium (sodicity) hazard, then we should maintain solution concentrations of sodium greater than 2,500 ppm TDS. When using hyper-saline irrigation, subsequent leaching with water of low salinity or with fresh water only exacerbates the problems with sodium leaching and while it may serve to reduce the sodium and salt concentrations overall, it will result in a higher proportion of sodium remaining in the rootzone profile.

4.3. Mineral Transformations

Salts in solution in an open body of water are relatively free to dissociate and mix within the solution. During pumping, as water flows through an open pipe the salts move freely and readily as a response to pressure and velocity. So, the affects of salt partitioning in pipe systems is minimal. However when that same saline water is applied to a soil, the conduit for transmission is no longer a pipe but a set of tortuous conduits (soil pores) that vary in size and geometry, are anastomising and may be discontinuous. In these circumstances the effects of salt partitioning are exhibited as discussed previously and illustrated in (Figure 1).

Salts in soil solution are subject to precipitation upon soil drying. Soil drying can occur due to evaporation and/or root extraction. As soil solution decreases, the concentration of solutes within the solution will increase. When a critical concentration of solutes is reached, salt minerals will begin to precipitate from solution and form crystals within the soil void space. The ions (cation and anion) that pair up to form the salt crystal in the soil are dependent upon several factors. Primarily the sequence of precipitation is based upon the solubility of the various combinations of salts that can form from the ions within a given solution. To understand what salts will typically form from soil solution, it requires a working knowledge of solubility rules. Solubility rules are given below in (Table 1) and have been modified to only include those salts, which would commonly occur in soils.

Table1. *Solubility rules as they apply to common soil mineral salts that may form from soil solution.*

Soluble Salts
1. The Na^+, K^+, and NH_4^+ ions form soluble salts. Thus, NaCl, KCl, KNO_3, $(NH_4)_2SO_4$, Na_2S, and $(Na)_2CO_3$ are soluble.
2. The nitrate (NO_3^-) ion forms soluble salts. Thus, $Ca(NO_3)_2$ and $Fe(NO_3)_3$ NH_4NO_3, are soluble.
3. The chloride (Cl^-) ion generally forms soluble salts. Exceptions to this rule include salts of the Cu^+ ions.
4. The sulfate (SO_4^{2-}) ion generally forms soluble salts. Exceptions include $CaSO_4$, which is slightly soluble.
Insoluble Salts
1.Sulfides (S^{2-}) are usually insoluble. Exceptions include Na_2S, K_2S, $(NH_4)_2S$, MgS, and CaS.
2. Oxides (O^{2-}) are usually insoluble. Exceptions include Na_2O and K_2O, which are soluble, and CaO, which is slightly soluble.
3. Hydroxides (OH^-) are usually insoluble. Exceptions include NaOH, and KOH, which are soluble, and $Ca(OH)_2$, which is slightly soluble.
4. Phosphates (PO_4^{3-}) and carbonates (CO_3^{2-}) are usually insoluble. Exceptions include salts of the Na^+, K^+, and NH_4^+ ions.

(Table 2) shows the various solubility product constants for various salts (less soluble) that can/may form from a typical salt-water solution applied to a soil.

The more soluble salts are not listed because their solubility is complete (e.g. K_{sp} of 1) up to solution saturation. NaCl being very soluble, it has a saturated solution concentration of about 360,000 ppm or 36% solution. Keep in mind that ions will be present in the soil not only from the irrigation water solution but also from applied fertilizer salts and/or from ions inherent in the soil parent material. This could include ions of Fe (+2 or +3), PO_4^{3-}, NH_4^+, NO_3^-, SO_4^{2-}, Cu^{2+}, Mn^{2+}, Xn^{2+}, CO_3^{2-}. Carbonates and bicarbonates can also form in soil due to the presence of atmospheric CO_2 in the soil air or by direct release from the plant due to root respiration.

Table 2. *Solubility product of some soil salt/minerals.*

Name	*Formula*	*Solubility Product (K_{sp})*
Calcium sulfate	$CaSO_4$	4.9×10^{-5}
Calcium carbonate	$CaCO_3$	4.9×10^{-9}
Calcium phosphate	$Ca_3(PO_4)_2$	2.1×10^{-33}
Magnesium carbonate	$MgCO_3$	6.8×10^{-6}
Magnesium phosphate	$Mg_3(PO_4)_2$	1×10^{-24}
Magnesium hydroxide	$Mg(OH)_2$	3×10^{-14}
Ferrous carbonate	$FeCO_3$	3.1×10^{-11}
Ferric phosphate dihydrate	$FePO_4\times2H_2O$	9.9×10^{-16}
Ferrous sulfide	FeS	8×10^{-19}
Manganese sulfide	MnS	5.6×10^{-12}
Zinc sulfide	ZnS	2×10^{-25}
Copper sulfide	CuS	8×10^{-37}
Copper(II) phosphate	$Cu_3(PO_4)_2$	1.4×10^{-37}

It is important to understand that salts in solution will precipitate out of solution based upon the saturation percentage of a particular salt and/or the solubility product of the salt. When salt ions precipitate out to form a low solubility salt, this salt then is not likely to return to solution upon rewetting. For example, let's consider the case where we take the soluble salts of sodium sulfate and calcium nitrate and put them in a mixed salt solution (saturated). Then we evaporate the water to form solid precipitates. The ion pairs formed in the precipitate will include both the soluble salts (calcium nitrate, sodium sulfate and sodium nitrate) but will also form the less soluble salt of calcium sulfate. If we add water to the container to bring these salts back into solution, we will find that all of the salts will not go into solution, as the calcium sulfate will largely remain in the crystal form. The solubility of the calcium sulfate in this solution is governed by the solubility product constant (K_{sp}) of the salt.

It should be realized that salts precipitating from a soil solution do not always form pure mineral forms. In many instances the minerals formed are intergrade minerals with varying degrees of solubility. For example as carbonates precipitate from solution where calcium carbonate is predominant form, the solubility of the resulting precipitate may be reduced by the inclusion of small amounts of magnesium, iron or phosphate ions (Cole, et al., 1953; Griffin & Jurinak, 1973).

4.4. Mineral Transformation Processes

In addition to precipitation/dissolution processes, many other processes can occur in soils to facilitate mineral transformations. Mineral transformation and mineral transformation processes are of concern because of the results of such occurrences. Many mineral transformation processes can produce phytotoxic gaseous emissions or form cements (impervious or slowly permeable layers) and/or glues (plugging). Mineral transformation processes include various physiochemical and/or biological processes.
Some common soil transformation processes include:

4.4.1. Layer alkalization from Na_2CO_3 formation

4.4.2. Layer acidification from H_2SO_4 formation

4.4.3. Cementation processes from precipitation/formation of metallic sulfides, carbonates or hydrated silicates (opaline gels)

4.4.4. Biological reduction of solutes to produce phytotoxic emissions of H_2S, CH_4, R-CH_2S (thiols), NH_3, CN^{3-}, NO_2^-

4.4.5. Biofouling of reduced zones and of the soil surface from secondary algal and fungal growth resulting in soul plugging of mucilage

4.4.6. Hydrophobic soil conditions from the irreversible drying of biological mucilages

Layer alkalization occurs in soils due to the formation of sodium carbonate and/or sodium sulfate. The formation of these two salts occurs due to the ready leaching of chloride from soil. While chloride and sodium are the two most prevalent ions in solution of most salt-affected water (see table 3), the chloride ion readily moves in and through the soil profile. This is due anion exclusion effect and to the fact that there is little biotic or abiotic fixation of chloride in the soil system.

As indicated in (Figure 1), chloride leaches from the soil at a much higher rate than the second most predominant ion, sodium. The effect of this preferential leaching of chloride is that it results is an ion imbalance between cations and anions. Commonly ion pairing of sodium and sulfate occurs in partial response to the ionic imbalance. Sulfate is present because the large ionic size of the sulfate ion (even though an anion) does not lend to it being readily leached from the soil. However, sulfate is prone to leaving the soil solution from microbial reduction and the precipitation of low solubility metallic sulfide forms (biotic fixation) and/or the release as hydrogen sulfide gas. Additional charge imbalance initiates a hydrolysis reaction in which the end products result in the formation of sodium hydroxide (a strong base) and carbonic acid (a weak acid). Carbonates can readily be produced under such conditions in the soil due to atmospheric CO_2 and/or the production of CO_2 from root respiration. The pH of this reaction in equilibrium is about pH 10 (Whittig & Janitzky, 1963; Janitzky & Whittig, 1964). At soil pH increases above pH 9, the overall solubility of silica increases and in conjunction with the silica solubility as a function of particle size (smaller silicate minerals have higher solubility). The end result is that as pH increases toward 10, silica solubility increases and the higher the proportion of finer-grained silicate minerals (silts and clays) in the soil, the greater the total silica solubility. This alkalization-induced dissolution has also been referred to as "salt weathering". This process is also known in the concrete industry as alkali-aggregate reaction. This salt weathering process not only solubilizes fine-grained silicate minerals but also is also responsible for grain fracturing of larger grained silicates (Goudie et al., 1979).

Table 3. *Ionic composition of seawater (total ~ 35,000 ppm TDS).*

Element	*ppm*	*~ % of total*
Calcium (Ca)	419	1.2
Magnesium (Mg)	1,304	3.7
Sodium (Na)	10,710	30.5
Potassium (K)	390	1.1
Bicarbonate (HCO_3)	146	0.4
Sulfur (SO_4 or S_2)	2,690	7.7
Chlorine (Cl)	19,350	55.2
Bromine (Br)	70	0.2
Total	35,079	

Once silicates go into solution from the solid soil mineral phase they will upon reprecipitation form as hydrated silica gels (opaline gels) rather than discrete soil mineral grains (Flach et al., 1969; White, 1981; Wilding & Drees, 1971, 1973). As hydrated gels (viscous liquids) these gels tend to form at or near the surface within the pores and void spaces of the soil. As these gel precipitates become more extensive, they tend to plug the soil pore space and form surface seals. The sealing of the soil surface prevents for gas diffusion/exchange, which may lead to a build up of phytotoxic gases. Further, water infiltration and percolation through the soil profile is prevented. These are extremely hostile conditions for growing high quality turfgrass and often results in thinning and eventually loss of turf/plant material in these areas. It should be noted that silica solubility from gel phases is much higher than the solubility from crystalline soil minerals (such as quartz).

4.5. Microbial Reduction, Redox and Phytotoxic Gas Emmissions

Layer acidification is a condition that occurs as a result of the accumulation of metallic sulfides and/or the liberation of aluminum from layer alkalization. As soil alkalization processes solubilize silicates from layered-aluminosilicate minerals (clay minerals), mica or feldspar, the aluminum liberated from this process may then leach in the soil and accumulate at lower depths. Aluminum may then form/precipitate from solution as aluminum hydroxide in a hydrolysis reaction that will result in the liberation of H^+ ions. Acidic conditions may also be induced biologically by the reduction of sulfur to sulfide forms, which then precipitate out in insoluble metallic sulfide forms. During this process, phytotoxic H_2S gas is also produced. As the metallic sulfides later oxidize, the metallic sulfides transform in oxide or oxy-hydroxide mineral forms and in the process produce additional H_2S gas and sulfate ions, which will then induce a hydrolysis reaction to form sulfuric acid (localized pH often <3.0). Silica solubility increases as pH's approach 4.0 and lower (Beckwith & Reeve, 1964) in oxidized conditions (in the presence of oxides and oxyhydroxides).

In addition to the plugging or sealing of soils due to the processes involved in alkalization and acidification (and related processes such as metallic sulfide precipitation and hydrated silica gel formation), soil plugging may also occur due to the precipitation of other less soluble mineral forms. This could include gypsum, magnesium sulfate (bloedite or epsomite), phosphates, oxides/oxyhydroxides or more commonly carbonates. Magnesium sulfate often forms near the soil surface (often in conjunction with hydrated silica gels) due to its high solubility. Even though the various magnesium sulfate crystals have a high solubility, they may become somewhat less soluble upon precipitation due to their nature of forming overlapping crystals with low surface area. The overlapping crystal structure effectively reduces the solubility of these crystals (Driessen & Schoorl, 1973). The reduced solubility may also be a function of interaction with other mineral precipitates and/or some degree of intergrade mineral forms.

Carbonates commonly form layers within soils that restrict water movement and/or form layers with high microporosity, which retains excessive moisture. The alteration of hydraulic conditions within the soil profile by carbonates (or other cementation processes) may create zones within the profile that may then be subject to anaerobic conditions and the detrimental processes than can occur in soils under reduced conditions as previously discussed. The amount of soil carbonates (percentage of soil total) necessary to induce restricted layers/plugging in a soil is lower in coarser soils (aggregates) than in fine-textured soils.

As discussed previously with layer acidification processes, sulfuric acid forms from the oxidation of precipitated metallic sulfides. These metallic sulfide forms are often predominantly iron (ferrous) sulfide forms or pyritic materials (see figure 2, DePew & Pulley, 2004). As these metallic sulfides are deposited, they usually contribute a dark or black color to the soil. This can occur in concentrated zones (often termed black layer) or in nodules or streaks. These deposits can also occur in a more diffuse manner and impart a grayish or gleyed color to the soil. The conditions necessary for the formation of these metallic sulfide forms include anaerobic conditions that have become reduced (microsite redox potential of ~ 400 mV), the presence of sulfur-reducing bacteria (Desulfovibrio sp.), an energy source for the microorganisms (soil organic matter), and a source of iron (or other metallic ions) and sulfur (from salt water irrigation water and/or fertilizer) (Doner & Lynn, 1989; DePew et al., 2004). These sulfide deposits form in the voids and pore spaces and can become so extensive that they form a largely impermeable layer. Even upon oxidation following tillage, layer disruption or soil drying (and the formation of iron oxides/oxyhydroxides) the layers may still largely persist and begin the process of creating wet/reduced zones where the process can be repeated.

Other phytotoxic compounds can also be produced in the soil under reduced conditions and include such things as methane, thiol compounds, cyanides, ammonia and nitrites. Reduced soil conditions also favor the growth of organisms that are better adapted to those conditions than higher plants. This includes algal and fungal organisms. The mycelium from the growth of these organisms may be quite extensive and result in the formation of substantial quantities of organic mucilage, which further serve to plug the soil preventing infiltration and percolation of water and the release of phytotoxic gases. In turfgrass systems, it is not uncommon for these conditions to be countered by the use of chemicals in conjunction with mechanical aeration to physically disrupt the plugged soil layers and restore soil aeration. Often once reduced soil conditions are alleviated; the algal/fungal organisms die and leave behind the various mucilages. Upon drying, these mucilages then become resistant to rewetting and may lend hydrophobic properties to the soil profile.

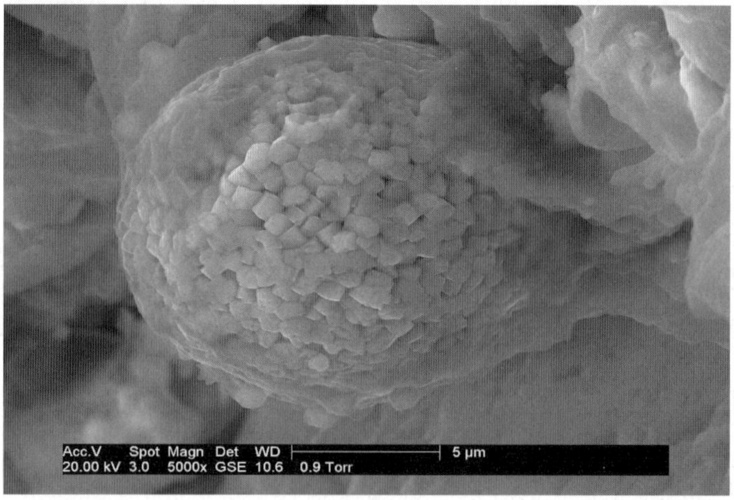

Figure 2. *Iron sulfide framboid from black layer formed in a sand-based turfgrass rootzone.*

5. SUSTAINABLE MANAGEMENT OF SALINE LANDSCAPE SYSTEMS

5.1. Irrigation

The most common error exhibited in using saline water to irrigated halophytes is one of over irrigation. Over irrigation should be avoided to reduce the impacts of salt partitioning on the soil. While leaching is an important factor to prevent the over accumulation of salts, it must be employed very judiciously. Only small leaching fractions should be employed on a regular basis as periodic flushes. Typically, we recommend that leaching fractions of one pore volume be applied and at repeating irrigation cycle frequencies so as to prevent surface runoff, when greater leaching is required.

5.2. Irrigation scheduling example

As an example, let's consider that one is managing a salinity profile (rooting depth) of a sandy soil to a depth of 8 inches. With a sandy soil your total porosity may be somewhere in the neighborhood of 40% depending upon bulk density (compaction). For our example, let's consider that our peak ET rates are around 2.25 inches per week (0.32 inches per day). Depending upon the salinity of our irrigation water and the cultivar of turf, our crop coefficient (K_c) will be somewhere in the range of 0.33 – 0.75. If we use a K_c value of 0.5, that will give us a consumptive use rate of 1.13 inches per week (2.25 x 0.50 = 1.13) or 0.16 inches per day. For our example only, we will consider efficiency of the irrigation system and of the irrigation application at 100% (although this never happens in reality). So our consumption of water is 0.16 inches per day and we desire to maintain that much available moisture in the top 8 inches. Let's consider that our sandy soil will retain (volumetric water holding capacity) 15% moisture content. In the top 8 inches of soil then, we will have about 1.2 inches of water. If we estimate that of that water is plant available, then our deficit water at which we would need to irrigate is 0.6 inches of water. At a consumption of 0.16 inches per day, that would give us an irrigation interval of 3.75 days. If this is a golf course, we would irrigate at an interval of 3 days to avoid irrigation during the hours of play. This adjustment would mean that our water requirements for 3 days would be 0.48 inches per irrigation.

To keep a net downward flux of salts (without overleaching), we would need to apply the consumptive water requirements plus a small leaching fraction. This can be done by a small over application of water at each irrigation event or by a larger application of irrigation water at set intervals. The use of a small leaching fraction employed with each irrigation is the preferred method. If we are using irrigation water with salinity in the range of 3,000 – 6,000 ppm TDS, we may want to use a leaching fraction of about 5% per irrigation cycle. This would increase our applied irrigation water requirements from 0.48 inches to 0.50 inches. In our example, we are considering an irrigation efficiency of 100% for simplification of the example. But, it should be noted that this is not achievable. Irrigation distribution/coverage efficiencies may vary from 60% to 90% and the efficiency of application depending upon such things as wind, potential runoff from sloped areas, etc. may reduce the overall efficiency another 10% - 20%.

Irrigation water should be applied in intervals such to avoid surface runoff. Water should be applied in small doses to avoid saturated or near-saturated conditions. Salt partitioning effects are lessened under lower (slower) flow velocities and when flow occurs in smaller pores. If your infiltration/percolation rate is 0.50 inches per hour (in the field, not a laboratory determined rate), you would need to run your system for 1 hour per station to meet your water requirements. To keep your infiltration flow velocities low, you should limit your irrigation to apply no more than 1/2 of your infiltration rate. For example if your irrigation system has a precipitation rate of 0.63 inches per hour, then you would expect to have runoff (0.50 inches/0.63 inches per hour) once you have run your system for 48 minutes (0.8 hours). To avoid reaching saturated flow, reduce this time to ~ 1/2 for each cycle and run your system for a maximum of only 24 minutes. Since your total time for irrigation is 60 minutes, this could be broken down into three 20-minute cycles.

Leaching fraction requirements for saline irrigation are site, soil and plant dependent. Some general guidelines are given in (Table 4). Values are given in the third column for average soil salinity values compared to the irrigation water salinity. You should note that this is approximation of the overall 'average' salinity while the salt profile in the soil will generally show lower salinity in the near surface and increasing salinity with depth in the profile. The cultivar of halophytic turf grass should be able to withstand the expected soil salinity levels. If not, you should consider using water of lower salinity and/or making a different choice of turf cultivar. As a note, we have found that the practical limits of salinity for golf courses to be about 20,000 ppm TDS irrigation salinity. Higher salinity levels are possible but the turf management becomes more extensive while the limits on play become more restrictive.

In most cases, the landscape or golf maintenance manager will have to make an initial best guess as to the required leaching fraction for the specific site, and then adjust the irrigation/leaching program to meet site conditions. Much of the guesswork can be reduced by using the services of an agronomist or soil scientist who is experienced in the use of hyper-saline irrigation. The amount or water applied for irrigation and the leaching factor are also dependent upon the drainage conditions of the soil. Where good drainage conditions exist, it is possible to accumulate salts in the soil at depths exceeding the rooting depth. For high value areas (such as greens and tees) or more salt-challenged areas (such as depressional basins), the use of artificial subsurface drainage (drain lines) is warranted. This allows for greater control over the overall salt accumulation and salinity profile of the soil. Salinity control is also given by applying the same amount of water over required over a given period of time in shorter, more frequent intervals. Shorter, more frequent intervals reduces the efficiency of your water use and reduces the overall leaching efficiency as you move deeper into the soil profile however, so you have a tradeoff between water efficiency and the management control of salinity and salt partitioning.

When considering the management of salts in a halophytic turf system when utilizing hyper-saline irrigation water, it is important to consider that salts applied to a soil are subject to biotic and abiotic fixation. This includes the various forms of precipitation and transformation processes, as well as plant uptake and plant salt excretion, which we have not discussed. Volatilization (or photovolatilization) of salts by halophytic grasses has also been proposed as a mechanism in saline systems (Yensen, 2004).

In any case, no matter what leaching fraction and water requirements, adjust your irrigation program to apply water only at precipitation rates/intensities at less than saturated conditions to reduce the effects of salt partitioning. In most cases, with many soils and with most irrigation systems, irrigation run times should not exceed 30-minute intervals. Also, when employing a hyper-saline irrigation program, it is critical that the soil is not allowed to dry out excessively. Excessive soil drying favors the precipitation of salt minerals, which may or may not retain solubility upon rewetting. Salt precipitation can result in mineral accumulations (low solubility salts) or the preferential accumulation of the more (problematic) soluble salts.

If it is found that smaller judicious leaching fractions have been inadequate, then you may desire to more heavily leach the soil to reduce the overall salinity. If the need arises to do this type of practice, there are several factors that should be kept in mind:

5.2.1. Leach only using saline water

Do not use fresh or fresher water as this will result in conditions favoring soil alkalization and soil degradation. High soil pH from alkalization processes also leaves many seashore paspalum cultivars subject to disease. For hyper-saline irrigated projects, NEVER use water for leaching that has a total salinity of less than 2,500 ppm TDS.

5.2.2. Do not over leach Use

Only about 1 pore volume for leaching (times the desired depth of leached soil profile) and apply the water in increments of 1/2 of the infiltration/percolation rate of the soil-turf system as measured in the field. Repeat the 1 pore volume leaching events on a daily basis until the desired soil salinity level is obtained. Remember, do not use water that has a salinity of less then 2,500 ppm TDS. If only fresher water is available, increase the salinity of the water by addition of salts (a blend of Ca and K salts work well for this purpose).

5.2.3. Do not use the one pore volume leaching fractions on a regular schedule

Use only as needed and then adjust your regular irrigation practices such to apply small judicious leaching fractions with each irrigation event.

5.2.4. Monitor and record your soil conditions.
Know your porosity, bulk density (compaction) and water retention rates. Know your actual (field) infiltration/percolation rates. Know your irrigation system precipitation rates and irrigation system efficiency. You should know such things as: When you run your irrigation system for x minutes, you apply y inches of water and it wets the soil to a depth of z inches.

5.2.5. If in a high rainfall season you may need to apply saline irrigation following heavy rain to restore/maintain soil salinity so as to prevent or reduce the effects of salt partitioning that could occur from the excessive leaching of salts by the rain water.

Table 4. *General recommended leaching fractions for saline irrigation of halophytic turfs.*

Salinity Range (ppm TDS)	*Leaching Fraction*	*Approximate Expected Average Soil Salinity Concentration Factor*
3,000 – 6,000	5%	3x - 5x
6,000 – 12,000	10%	2x - 4x
12,000 – 18,000	15%	1.5 x - 3x
>18,000	20% - 40%	≤ 2x

These leaching fractions are highly dependent upon site, soil, water (ionic composition) and cultivar conditions and must be used with professional judgment and adjusted for specific conditions and applications. Examples of commercial saline tolerant turf are shown in Figures 3, 4, and 5.

Figure 3. *One of the authors' dwarf Seashore* paspalum *cultivars being used on a putting green that is irrigated with reclaimed water Effluent.*

Figure 4. *A lawn-type* Seashore paspalum *used on a seaside overlook receiving heavy salt spray.*

Figure 5. *Seashore dropseed turf in a lawn application.*

5.3. Cultural Practices

Soil aeration (gas exchange) is a key to producing a high quality turf under a hyper-saline irrigation program. Mechanical aeration of golf, sports or high-traffic landscape areas will be required to maintain a good soil aeration state. Sands, particularly medium to coarse sands, have the best natural aeration states and are resistant to compaction. These types of sands will need less intensive aeration practices. Regular monitoring of soil conditions such as the appearance of layer features or the depression of soil redox states (reduced conditions) is an indication of the need for aeration. Soil redox levels can only be measured in the field and should be maintained at a level of $^{+}100$ mV or higher. When values reach $^{-}100$ mV, remedial action is indicated to increase the soil redox state and restore a more oxidized state.

The presence of sulfide gas odor (rotten egg smell) is also an indication of the need to take remedial action.

Fine-textured soils should be aerated with solid tines or with slit aerators and the aeration holes (or slits) back-filled with coarse sand topdressing to facilitate the formation of sand cores (channels) to serve as soil 'vents'. Additional practices of aeration/topdressing or topdressing alone should be undertaken using the same sand material. Using coarser material over finer material does not cause any hydraulic discontinuities. However, the use of finer textured material over the top of coarser-textured material does cause hydraulic discontinuities (interface problems) and must be avoided.

A proactive management approach is called for with hyper-saline irrigation programs. A reactive management program is likely to produce less than desirable results and may actually result in failure of the system. The golf/landscape manager should be prepared to observe soil conditions and measure changes in soil conditions. Of particular note is any conditions exhibited as layer features. These layers may only be as thin as a few mm but should be given attention. Soil redox should be measured regularly and in many locations. The soil pH should be monitored for discrete soil layers. Soil suction lysimeters should be placed in the soil of select greens and tees at various depths to facilitate the extraction of soil solution for direct salinity and pH monitoring and periodic laboratory analysis.

Other cultural practices to improve or maintain soil redox states include: Manage the irrigation reservoir so as to maintain high redox states (> $^{+}100$ mV). This is often best accomplished with an aggressive lake aeration system.

5.3.1. Adjust/treat irrigation water to remediate irrigation water and/or soil conditions

This could include peroxide or liquid oxygen injection, muriatic acid injection or even ozone generation/injection.

5.3.2. Use high rates of potassium fertility

To allow halophytes the proper amount of K to facilitate osmotic adjustment

5.3.3. A sand cap may be a necessary part

A sand cap may be a necessary part of the construction process if the project is challenged by soil quality and high salinity (> 10,000 ppm TDS).

6. SUMMARY

A commercial research and development program was begun by the authors in Texas (1990) for the purpose of developing a program to use saline water to irrigate salt-tolerant turfgrasses. This program developed with a two-prong approach, which included turf development and soil ecological processes. This program has led to the availability of improved types of halophytic grass species suitable for use as turfgrass on lawns, sports fields, and golf courses. Cultivars of warm-season turfs from the species *Paspalum vaginatum* and *Sporobolus virginicas* are currently available commercially. The authors continue to develop improved cultivars from these species and new cultivars of salt-tolerant turfs from other species (including cool season turfs).

The successful utilization of hyper-saline water for irrigation relies on careful application of such water using precise and judicious leaching fractions. Over application of hyper-saline water may lead to salt partitioning within the soil profile. Salt partitioning may induce certain biophysical processes within the soil that are detrimental to turf development and/or results in the release of phytotoxic substances. This may include such things as the production of black layer, sulfide gas production, hydrolysis-induced acidification and alkalization, silica solubilization, biofouling, and cementation.

In addition to careful irrigation management, maintaining the soil in a well-aerated state is critical for the maintenance of high-quality turf grass. Soil aeration is maintained by selective use of fertilizer carriers, water treatment (such as lake aeration), soil compaction control, artificial subsurface drainage and possibly soil amendment.

While the success enjoyed by the authors' program has been primarily a scientific/technical process, the legal/business aspects of bringing this program to the commercial market have proven difficult. It seems that once the commercial viability and potential is recognized by others, there is an overwhelming desire for others to take possession or control of the program and/or the resulting technologies. This has hampered the authors' ability to gain proper recognition in the industry for their contribution to this work, gain the credibility in the industry in order to profit from their work, and has resulted in the expenditure of much time, effort and resources in the process of litigation. It is our hope that we can steer through these legal entanglements and be able to offer viable saline solutions to the landscape and golf industry for the present and for the future as natural fresh water resources continue to diminish throughout the world.

7. REFERENCES

Ayers, R.S. & Westcot, D.W. 1976. Water Quality for Agriculture. FAO Irrigation and Drainage Paper No, 29 (Rev 1), Rome, Italy: Food and Agriculture Organization of the United Nations.

Ayers, R.S. & Westcot, D.W. 1985. Water Quality for Agriculture: FAO Irrigation and Drainage Paper 29 Rev. 1. Rome, Italy: Food and Agriculture Organization of the United Nations.

Beckwith, R.S. & Reeve, R. 1964. Studies on soluble silica in soils. II - The release of monosilicic acid from soils. Australian. Journal Soil Science 2: 33-45.

Cole, C.V., Olsen, S.R. & Scott, C.O. 1953. The nature of phosphate sorption by calcium carbonate. SSSA Proceedings 17: 352-356.

DePew, M.W., Nelson, S.D. & Pulley, G.E. 2004. Micromorphic observations of soil black layer formed in sand-based sport sturf root zone constructions. SSSA Journal, In: review.

Doner, H.E. & Lynn, W.C. 1989. Carbonate, halide, sulfate, and sulfide minerals. In: J.B Dixon & S.B. Weed. (Eds.), Minerals in soil environments. SSSA Book Series 1, 2nd edition. Madison, Wisconsin: SSSA. 302-309 pp.

Driessen, P.M. & Schoorl, R. 1973. Mineralogy and morphology of salt efflorescences on saline soils in the Great Konya Basin, Turkey. Journal. Soil Sciences 24: 436-442.

Flach, K.W., Nettleton, W.D., Gile, L.H. & Cady, J.G. 1969. Pedocementation: Induration by Silica, Carbonates, and Sesquioxides in the Quaternary. Soil Sciences 107: 442-453.

Goudie, A.S., Cooke, R.V. & Doornkamp, J.C. 1979. The formation of silt from quartz sand dune by salt-weathering processes in deserts. Journal Arid Environments 2: 105-112.

Griffin, R.A. & Jurinak, J.J. 1973. The interaction of phosphate with calcite. SSSA Journal 37: 847-850.

Janitsky, P. & Whittig, L.D. 1964. Mechanisms of formation of sodium carbonate in soils. Laboratory study of biogenesis. Journal Soil Sciences 15: 145-157.

Morton, J.F. 1973. Salt-tolerant silt grass (*Paspalum vaginatum* Sw.). Proceeding Florida. State Horticure Society 8: 482-490.

White, K.L. 1981. Sand grain micromorphology and soil age. SSSA Journal 45: 975-978.

Whittig, L.D. & Janitsky, P. 1963. Mechanism of formation of sodium carbonate in soils. Manifestation of biological conversion. Journal Soil Sciences 14: 322-333.

Wilding, L.P. & Drees, L.R. 1971. Biogenic opal in Ohio soils. SSSA Proceedings 35: 1004-1010.

Wilding, L.P. & Drees, L.R 1973. Scanning electron microscopy of opaque opaline forms isolated from forest soils in Ohio. SSSA Proceedings 37: 647-650.

Yensen, N.E. & Biel, K.Y. 2005. The hypotheses of halosynthesis, photoprotection and soil remediation via salt-conduction. Proceedings from The Symposium on High Saline Tolerant Plants, May 5-7, 2004. Brigham Young University, Provo, Utah: Springer Publisher.

CHAPTER 19

SALT TOLERANCE OF FLORICULTURE CROPS

CHRISTY T. CARTER[1] AND CATHERINE M. GRIEVE

George E. Brown, Jr., U. S. Salinity Laboratory, USDA/ARS, 450 W. Big Springs Road, Riverside, California 92507, USA
[1]Corresponding author:ccarter@ussl.ars.usda.gov

Abstract. The cut flower industry is an economically important industry in the United States, especially in the state of California. Growers have traditionally used the highest quality water to irrigate cut flower crops, but the need for alternative water sources for irrigation is increasing given the rising demand for quality water due to population growth and agronomic use. The reuse of saline wastewaters provides a viable option for the irrigation of salt tolerant floral crops. Investigations into the salinity tolerance of cut flowers have been initiated at the U. S. Salinity Laboratory to determine marketability based on stem length. Cultivars of Celosia, statice, stock, and sunflower were exposed to differing water ionic compositions and salinity levels. Most were found to be marketable at moderate salinities even though plant height tended to decline as salinity increased. Cultivars also showed differing responses based on the composition of ionic water treatments. Saline wastewaters and ground waters provide an alternative source for irrigation to produce marketable cut flower crops.

Species investigated: Celosia (*Celosia argentea* var. *cristata* (L.) Kuntze "Chief Rose" and "Chief Gold"); statice (*Limonium perezii* (Staph.) F. T. Hubb, "Blue Seas" and *L. sinuatum* (L.) Mill, "American Beauty"); stock (*Matthiola incana* (L.) R. Br. "Cheerful White" and "Frolic Carmine"); sunflower (*Helianthus annuus* L. "Moonbright" and "Sunbeam").

1. INTRODUCTION

The cut flower industry is an economically important industry in the United States. In 1998, over 2 million cut flower operations nationwide produced over \$512 million USD in total sales. Of these operations, approximately 550,000 were located in California which brought in nearly \$322 million USD (Census of Horticultural Specialties, 1998). These high cash crops, because of their sensitivity to salinity, have been irrigated with the highest quality water by growers. Yet in recent years, quality water has become in high demand due to population growth and as competition between agricultural users and municipalities has increased (Parsons, 2000). This is especially true in California where increasing concerns have prompted growers to incorporate water saving technologies into their operations (Parsons, 2000). Growing concerns of water usage have also prompted

M.A, Khan and D.J Weber (eds.), Ecophysiology of High Salinity Tolerant Plants, 279-287.

growers to reuse wastewaters or greenhouse effluents, which range in their salinity and ion composition, to irrigate crops. This becomes difficult given the sensitivity of many floral crops to salinity. There are, however, many floral crops that are tolerant to salinity. Our goal in these investigations was to determine the marketability (based on stem length) of cut flowers when grown under differing water ionic compositions and salinity levels.

1.1. Composition of Saline Substrates

The composition of salts in water varies widely across the globe. In most waters the dominant cations are Na^+, Ca^{2+} and Mg^{2+}, while the dominant anions are Cl^-, SO_4^{2-}, and HCO_3^- (Grattan & Grieve, 1999). Most horticultural crops are subjected to irrigation water or soil solutions with $Na^+/(Na^+ + Ca^{2+})$ in the range of 0.1 to 0.7, suggesting that the composition of saline water employed in experimental studies should reflect this ratio. Surprisingly a large percentage of salinity studies on horticultural crops use NaCl as the sole salinizing agent. This unrealistic salinizing composition may induce ion imbalances, which contribute to Ca-related physiological disorders in certain crops (Shear, 1975; Sonneveld, 1988). Furthermore, the use of sole-salt salinizing solutions may result in misleading and erroneous interpretations about plant response to salinity. Our investigations at the U. S. Salinity Laboratory have utilized irrigation water whose ion compositions mimic the ion compositions and ratios of different geographic locations across California. These waters differ markedly in ion composition. Saline "tail waters" in the southern inland valleys of Riverside and Imperial Counties (and include the Imperial and Coachella Valleys) are generally high in sodium, magnesium, chloride, and sulphate, whereas drainage waters in the San Joaquin Valley are dominated by sodium, sulphate, chloride, magnesium, and calcium, predominating in that order. A third area includes coastal regions where seawater (dominated by sodium and chloride) intrusion of ground water is an increasing problem.

Much of the information on the salt tolerance of floriculture crops relates to chloride-dominated salinity. However, many nursery and greenhouse operations also rely on high rates of fertilizer application to assure optimum and rapid growth. Fertilizers and other salts lost through leaching and the runoff may pose significant environmental risks by contamination of ground and surface waters. The capture and reuse of these effluents are environmentally sound options for pollution control (Alexander, 1993; Arnold et al., 2003). One of the production problems associated with this approach, however, is that high, and possibly unbalanced, concentrations of nutrients in the effluents may result in crop damage due to fertilizer-induced salinity (West et al., 1980). Optimal nutrient concentrations are obviously different for various floricultural crops, although requirements have been established for only a few species.

1.2. Salt Tolerance Screening Projects at the U. S. Salinity Laboratory

Tolerance information for conventional floriculture crops grown at high salinities has little practical importance due to their poor survival rates. On the other hand, certain plants do survive and complete their life cycles at high salinities (halophytes). Several plant families contain halophytic species, which are potentially useful as cut flowers, e.g. Asteraceae (*Inula* spp.), Gentianaceae (*Eustoma* spp.), Plumbaginaceae (*Armeria* spp., *Limonium* spp.), Portulacaceae (*Portulaca* spp.). Identification of species that are both salt tolerant and commercially important crops would permit the reuse of degraded, often saline, wastewaters for floriculture production. This approach would be expected to improve the sustainability of greenhouse and field floriculture operations and, at the same time, reduce the discharge of waste fertilizers and salt to the environment. Additionally, investigating the interactions of salinity with nutrients (nitrogen in particular) can provide insights as to the best combinations to reduce the amount of fertilizer while still producing a marketable product. Once salinity tolerance is determined for different cultivars, then investigations of the interactions of salinity and nutrients can be performed. The following cut flower species and varieties have been screened for salinity tolerance at the U. S. Salinity Laboratory.

2. METHODS AND RESULTS

2.1. Celosia

Celosia spp. is in the Amaranthaceae, a plant family closely associated with the Chenopodiaceae that contains many salt-tolerant species. Given this association, and its ability to withstand warm temperatures, *Celosia argentea* var. *cristata* (L.) Kuntze “Chief Rose” and “Chief Gold” were selected for their potential as salt tolerant cut flowers and were included in our salt screening program.

Seeds for the two cultivars were sown in greenhouse sand tanks. Water treatments simulated seawater and saline drainage waters from the Imperial and Coachella Valleys (ICV) of southern California. Electrical conductivities of the treatment waters included 2.5 (control) 4, 6, 8, 10, and 12 dSm^{-1}. Plant phenotypic measurements were recorded when plants were harvested.

As salinity increased, phenotypic measurements (including stem length and weight, inflorescence length and weight, stem diameter, and number of leaves) decreased for both cultivars. Based solely on a stem length marketability of 41 cm, “Chief Gold” could be produced with both water compositions up to 12 dSm^{-1} (Figure 1a), whereas “Chief Rose” could be produced in ICV waters up to 10 dSm^{-1} and in seawater up to 8 dSm^{-1} (Figure 1b). An additional benefit of salinity treatments was that higher salinities might be used in place of growth regulators to control for excessive stem lengths such as those found in the control. Either variety would be ideal for production in coastal areas where seawater intrusion may be problematic or in inland areas where wastewaters are typically used for irrigation.

2.2. Statice

In his compilation of salt tolerant plants of the world, Aronson (1989) lists a total of 52 *Limonium* species and notes that five, including *L. perezii*, and *L. sinuatum*, will survive at high salinity (irrigation waters with electrical conductivities as high as 56 dSm^{-1}). One would not expect, based on this information alone, that the flowers produced under hypersalinity would meet floriculture industry standards for these commercially important cut flowers. The question, then, was how the species would perform under more moderate saline environments. Therefore, we included the statice cultivars, *L. perezii* (Staph.) F. T. Hubb, "Blue Seas" and *L. sinuatum* (L.) Mill, "American Beauty" in our program designed to screen cut flower species for salt tolerance.

Plants were grown in greenhouse sand tanks irrigated with waters prepared to simulate saline wastewaters typical of those present in the inland valleys of California. Electrical conductivities of the irrigation waters ranged from 2 to 30 dSm^{-1}. Under field conditions, predicted average rootzone salinities of the soil waters would range from ~1 to 14 dSm^{-1}. Performance at flower harvest was rated on the yield components, stem length and weight, and quantified with the Maas Hoffman model (1977).

Both *Limonium* species were able to complete their life cycles when irrigated with saline solutions with EC = 30 dSm^{-1}, clearly a halophytic trait typical of the genus (Aronson, 1989). Growth response to salinity, however, more closely resembled that of glycophytes rather than halophytes. Maximum growth, as measured by dry weight and stem length, occurred at low salinity and decreased steadily as salt stress increased (Grieve et al., in press). In combination, these characters describe a class of halophytic plants (Flowers et al., 1986) termed "miohalophytes" (Salisbury, 1995). This nomenclature is undoubtedly useful in ecophysiological studies. However, neither cultivar examined in this study possessed a high degree of "salt tolerance" as understood by horticulturists and agronomists whose research focuses on crop yield response to salinity (Maas & Grattan, 1999). Evaluation of the marketable yield of *Limonium* suggests that *L. perezii* should be rated as salt sensitive; *L. sinuatum* as moderately tolerant. However, both species would be valuable as landscape plants in areas affected by salinity (Grieve et al., in press).

A second investigation focused on the germination stage of development. Using only *Limonium perezii*, seeds were exposed to two separate irrigation water treatments mimicking those of the San Joaquin Valley (SJV) and the Imperial and Coachella Valleys (ICV) of central and southern California, respectively, to determine the effects of salinity on seedling emergence. In comparison, San Joaquin Valley water is higher in sulphates than the Imperial and Coachella Valleys, which is higher in chloride. Seeds were sown directly into presalinized sand tanks under greenhouse conditions. Electrical conductivities of the irrigation waters ranged from 2 to 20 dSm^{-1}. Cumulative germination (emergence), survival, and ion uptake were measured.

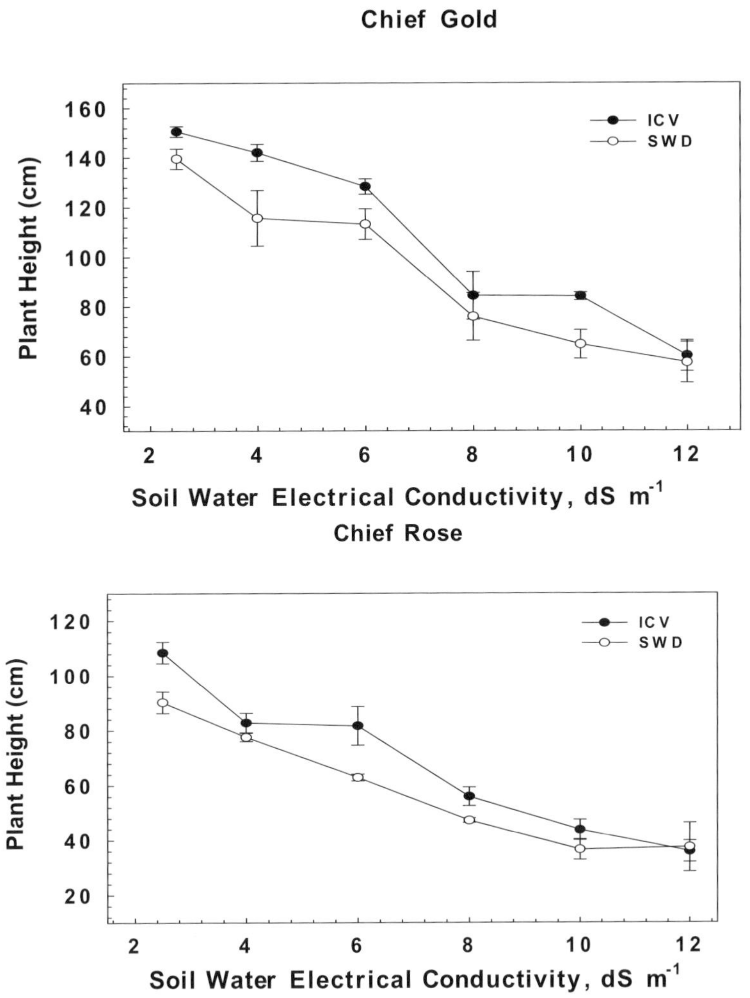

Figure 1. *Stem length (cm) (mean ±SE) of* Celosia argentea *(a) "Chief Gold" and (b) "Chief Rose" when exposed to two water ionic compositions Imperial/Coachella Valley (ICV) and Sea Water (SWD)) and six salinity treatments.*

Cumulative emergence declined markedly above 12 dSm^{-1} and was greater at 6 to 10 dS m^{-1} than in the control for both water treatments. Survival approximated 90% up to 8 dSm^{-1} (ICV) and 10 dSm^{-1} (SJV). At 12 dSm^{-1} and higher, survival decreased below 70% for both treatments. Stem length declined above the control (ICV) and 6 dSm^{-1} (SJV) (Carter et al., in press). Approximately 30% more marketable stems were produced under SJV treatments based on a 41 cm stem length for marketability (Barr, 1992). Carter et al. (in press) concluded that moderately saline wastewaters, especially those that are higher in sulphates representing water from the San Joaquin Valley, might be used to produce *L. perezii* commercially.

2.3. Stock

Two cultivars of *Matthiola incana* (L.) R. Br. were screened for salinity tolerance at the George E. Brown, Jr., Salinity Laboratory in Riverside, California. "Cheerful White" and "Frolic Carmine" were grown in greenhouse sand tanks and exposed to two types of irrigation waters designed to mimic the ionic composition of saline drainage waters from the San Joaquin Valley (SJV) and saline "tailwaters" of the Colorado River from the Imperial and Coachella Valleys (ICV). Electrical conductivities (EC) for each ionic water composition were 2, 5, 8, 11, and 14 dS m^{-1}. Three replicates were used for each treatment. At final harvest all morphometric data were collected.

Measurements tended to decrease overall for both cultivars and water treatments as salinity increased (height, weight, flower height, stem diameter, number of nodes, and number of axillary buds). For both cultivars and water treatments, stem length did not decrease significantly until EC exceeded 8 dS m^{-1}. Plant height for "Cheerful White" ranged from 60 cm in 11 dS m^{-1} to 75 cm in the control for ICV and from 58 cm in 14 dS m^{-1} to 71 cm in 8 dS m^{-1} for SJV. Plants grown in the SJV control approximated 68 cm in height (Figure 2a). "Frolic Carmine" plant height ranged from 58 cm in 11 dS m^{-1} to 69 cm in the control for ICV and from 53 cm in 14 dS m^{-1} to 66 cm in 5 dS m^{-1} for SJV. Plants grown in the control in SJV treatments approximated 62 cm in height (Figure 2b). Based on a minimum stem length of 41 cm for marketability, both varieties could be produced under saline conditions in inland areas where sulphate salts may be problematic.

Although not statistically significant at the 95% confidence interval, the number of flowers produced in ICV treatments was highest at 14 dS m^{-1} with "Cheerful White" producing 25 flowers per stem and "Frolic Carmine" producing 21 flowers per stem. The minimum number of flowers produced for "Cheerful White" was 19 in 11 dS m^{-1} and for "Frolic Carmine" was18 in 11 dS m^{-1} (unpubl. data, Salinity Laboratory). Statistical differences were found in the number of flowers produced in SJV treatments for "Frolic Carmine" with 18 flowers per stem in 14 dS m^{-1} and 22 flowers per stem in 5 dS m^{-1}. No statistical differences were found for "Cheerful White", but flower number ranged from 20 in 8 dS m^{-1} to 23 in 5 dS m^{-1}. There was an overall trend showing an increase in flower yield at moderate or even at high salinities for *Matthiola incana* (unpubl. data, Salinity Laboratory).

2.4. Sunflower

Two cultivars of *Helianthus annuus* L. were screened for salinity tolerance. "Moonbright" and "Sunbeam" were grown in greenhouse sand tanks irrigated with water that simulated drainage water from the San Joaquin Valley (SJV) and the Imperial and Coachella Valleys (ICV) of California. Electrical conductivities of the ionic water solutions were 2.5, 5, 10, 15, and 20 dS m^{-1}. All treatments were replicated three times. Water ionic composition treatments were applied after the appearance of first leaves. Plant height, inflorescence diameter, weight, and stem diameter were measured.

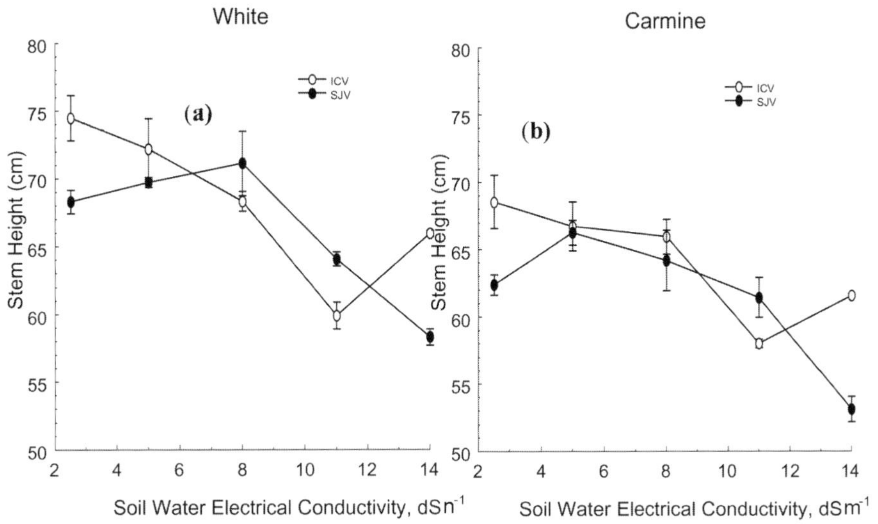

Figure 2. *Stem height (cm) (mean ± SE) of* Matthiola incana *(a) "Cheerful White" and (b) "Frolic Carmine" when exposed to two water ionic compositions (Imperial/Coachella Valley (ICV) and San Joaquin Valley (SJV)) and seven salinity treatments.*

Values for phenotypic features decreased as salinity increased for both cultivars. Statistically significant decreases in height were found for "Moonbright" and "Sunbeam". Height for "Moonbright" ranged from 170 cm in the control to 103 cm in 20 dSm^{-1} in ICV water treatments and from 162 cm in the control to 71 cm in 20 dSm^{-1} in SJV water (Figure 3a) . Height for "Sunbeam" also decreased significantly from 175 cm in the control to 73 cm in 20 dSm^{-1} in ICV water and from 168 cm in the control to 57 cm in 20 dSm^{-1} in SJV water (Figure 3b).

Even though inflorescence diameter decreased with increasing salinity, no statistical differences were found for either cultivar in ICV water treatments. "Moonbright" decreased from 10.9 cm in the control to 9.3 cm in 20 dSm^{-1} and "Sunbeam" decreased from 10.9 cm in the control to 9.4 cm in 20 dSm^{-1} (Figures 3a & b). Statistically significant decreases were found, however, in plants exposed to SJV water. Inflorescence diameter for "Moonbright" decreased from 10.7 cm in the control to 7.8 cm in 20 dSm^{-1} whereas "Sunbeam" decreased from 14.0 cm in the control to 7.5 cm in 20 dSm^{-1} (Figures 3a & b). These results suggested that stem length might be shortened without affecting inflorescence diameter, depending on the water composition. Differences in inflorescence diameter between the water compositions may be attributed to the higher sulphate concentration in the San Joaquin Valley water composition.

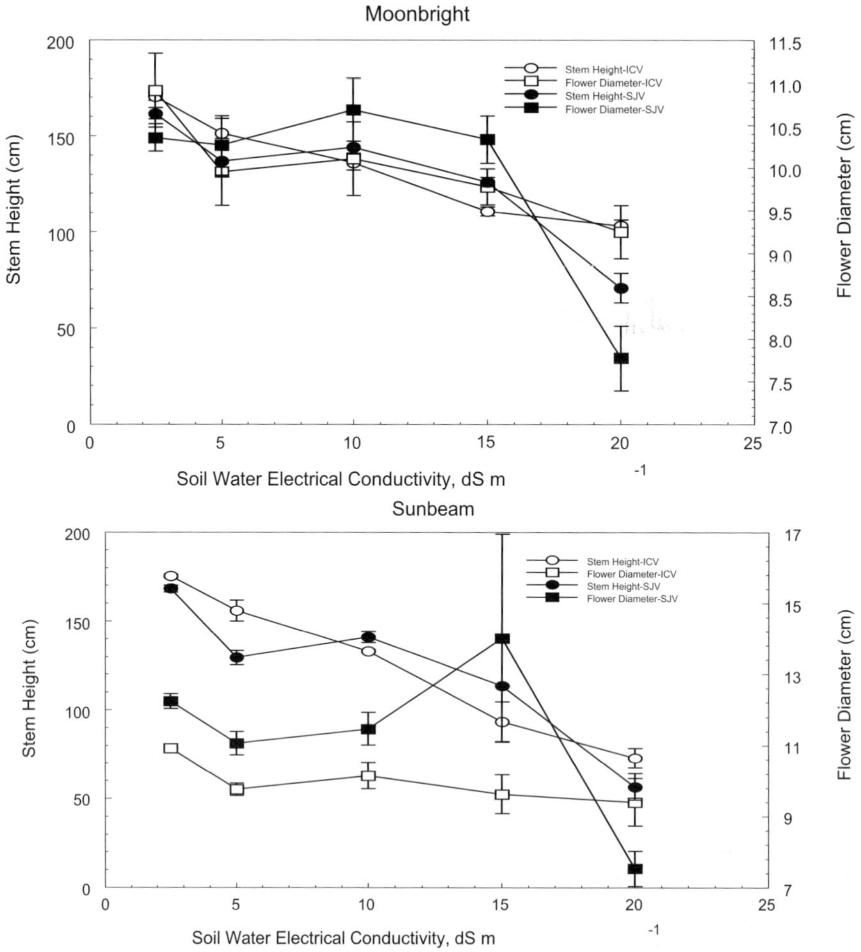

Figure 3. *Stem height (cm) and flower diameter (cm) (mean ± SE) and flower diameter of* Helianthus annuus *(a) "Moonbright" and (b) "Sunbeam" when exposed to two water ionic compositions (Imperial/Coachella Valley (ICV) and San Joaquin Valley (SJV)) and five salinity treatments.*

3. SUMMARY

Saline wastewaters and ground waters may be used to produce marketable cut flowers based on stem length, even though stem length tends to decrease as salinity increases. Species and cultivars respond differently to water ionic compositions and salinity. Some species would be better suited for field production in areas dominated by sulphate salts whereas others would be better produced in areas with chloride salinity. Further investigations will provide insight regarding the interactions of salinity and nutrients (nitrogen) on cut flower production. This is particularly important with regards to the recycling of greenhouse effluents for the purpose of irrigation.

Other species also being screened at the U. S. Salinity Laboratory: snapdragon (*Anthirrhunum majus* L. cultivars "Apollo Cinnamon" and "Monaco Rose"); zinnia (*Zinnia elegans* cultivars "Benary's Giant Golden Yellow" and "Benary's Giant Salmon Rose"); and lisianthus (*Eustoma grandiflorum* cultivars "Echo Pure White," "Echo Blue," and "Echo Blue Picotee").

4. ACKNOWLEDGEMENTS

These investigations were funded in part by a CAL-FED grant, administered by the California Department of Water Resources.

5. REFERENCES

Alexander, S.V. 1993. Pollution control and prevention at containerized nursery operations. Water Science and Technology 28: 509-516.

Arnold, M.A., Lesikear, B.J., McDonald, G.V., Bryan, D.L. & Gross, A. 2003. Irrigating landscape bedding plants and cut flowers with recycled nursery runoff and constructed wetland treated water. Journal of Environmental Horticulture 21: 89-98.

Aronson, J.A. 1989. Halophyte: A data base of salt tolerant plants of the world. Office of Arid Lands Studies. The Tucson, Arizona: University of Arizona Press. 1-114 pp.

Barr, C. 1992. The Kindest Cuts of All. Greenhouse Manager 11: 82-84.

Carter, C.T., Grieve, C.M. & Poss, J.A. 2005. Salinity effects on emergence, survival, and ion accumulation of Limonium perezii. Journal of Plant Nutrition, in press.

Census of Horticultural Specialties. 1998. Cut flowers sold by state. 167-181 pp. retrieved from. www.nass.usda.gov/census/census97/horticulture/horticulture.htm.

Flowers, T.J., Hajibagheri, M.A. & Clipson, N.J.W. 1986. Halophytes. The Quarterly Review of Biology 61: 313-337.

Grattan, S.R. & Grieve, C.M. 1999. Salinity-mineral nutrient relations in horticultural crops. Scientia Horticulturae 78: 127-157.

Grieve, C.M., Poss, J.A., Gratton, S.R., Shouse, P.J., Lieth, J.H. & Zeng, L. 2005. Productivity and mineral nutrition of Limonium species irrigated with saline wastewaters. HortScience in press.

Maas, E.V. & Grattan, S.R. 1999. Crop yields as affected by salinity. In: R. W Skaggs & J. Van Schilfgaarde. (Eds.), Agricultural Drainage. Agronomy Monograph No. 38. Madison, Wisconsin: American Society of Agronomy. 1328 pp.

Maas, E.V. & Hoffman, G.J. 1977. Crop salt tolerance 0-current assessment. Journal of the Irrigation and Drainage Division American Society of Civil Engineers 103: 115-134.

Parsons, L.R. 2000. Water management and water relations of horticultural crops: introduction to the colloquium. Introduction to Water Management and Water Relations of Horticultural Crops. HortScience 35: 1035-1036.

Salisbury, F.B. 1995. Units, Symbols, and Terminology for Plant Physiology. Logan, Utah: Utah State University Press. 1-256 pp.

Sear, C.B. 1975. Calcium-related disorders in fruits and vegetables. HortScience 10: 361-665.

Sonneveld, C. 1988. The salt tolerance of greenhouse crops. Netherlands Journal of Agricultural Science 36: 63-73.

West, D.W., Merrigan, I.F., Taylor, J.A. & Collings, G.M. 1980. Growth of ornamental plants irrigated with nutrient or polyethylene glycol solutions of different osmotic potentials. Plant and Soil 56: 99-111.

CHAPTER 20

UTILIZATION OF SALT-AFFECTED SOILS BY GROWING SOME ACACIA SPECIES

M. YASIN ASHRAF[1], M.U. SHIRAZI[2], M. ASHRAF[3], G. SARWAR[1] AND M. ATHAR KHAN[2]

[1] *Nuclear Institute for Agriculture and Biology (NIAB), P.O. Box 128, Jhang Road, Faisalabad, Pakistan. niabmyashraf@hotmail.com*
[2] *Nuclear Institutes of Agriculture, Tandojam, Pakistan*
[3] *Department of Botany, University of Agriculture, Faisalabad, Pakistan*

Abstract. In Pakistan most of the salt affected areas are located in the heart of its agriculturally important tract of the Indus plain. Growing conventional crops in these problem lands is not economical but these could be utilized gainfully by growing salt tolerant trees or shrubs. However, there is a need to identify suitable species for such conditions. The present study was, therefore, undertaken to evaluate the performance of some local and exotic trees, belonging to *Acacia* species, in salt affected soils of Pakistan. Five species of *Acacia*, i.e., *Acacia ampliceps*, *A. stenophylla*, *A. machonochieana*, *A.sclerosperma*, and *A.nilotica* were grown along with an Australian halophytic shrub *Atriplex lentiformis* in a field where salinity ranged from 4-25 dS m^{-1}. After three years of growth, *A. ampliceps* and *A. nilotica* showed markedly higher growth as compared with the other species examined. Although *A. ampliceps* grew well under saline environment, its maximum growth was observed under low to medium salinity patches (4-12 dS m^{-1}) showing survival percentage 80-90. However, at high salinity (12-16 dS m^{-1}) the percent survival of *A. ampliceps* was 50. In contrast, *Atriplex lentiformis* was mostly populated on medium to high salinity levels (8-16 dS m^{-1}), while *Acacia sclerosperma* and *A. machanochieana* were only populated on low salinity patches (4-8 dS m^{-1}). Analysis of plant leaves, for nutrient contents, showed that the concentrations of Na^+ and K^+ ions in the *Acacia* species were comparatively less than those in *Atriplex lentiformis*. However, *Acacia nilotica* had comparatively higher nitrogen and phosphorus than the other *Acacia* species examined. On the other hand, the concentrations of Na^+ and K^+ were higher in *Atriplex lentformis* while the concentrations of Ca^{2+} were higher in *Acacia sclerosperma* as compared to the other *Acacia* species and *Atriplex*. At the end of the experiment, complete soil analysis was carried out which showed that the soils on which *Acacia* species were grown, N, P, and K contents increased to some extent.

1. INTRODUCTION

Salinity is the presence of excessive amounts of soluble salts in the growth medium of plants. The salts responsible for causing salinity are mostly chlorides and sulfates of sodium, calcium and magnesium. Smaller quantities of potassium, ammonium and nitrate may also occur (Slavich, 1991; Qureshi et al., 1992). These salts affect the plant growth either by reducing the osmotic potential of the

M.A. Khan and D.J. Weber (eds.), Ecophysiology of High Salinity Tolerant Plants, 289-311.

medium or through specific ion effect. The presence of high concentration of ions in the rooting medium causes toxicity and nutrient imbalance by damaging the membrane and displacing other ions from the active absorption sites by mechanisms, not yet fully understood (Ramani & Joshua, 1993).

Plants have the ability to adjust osmotically by utilizing salts for maintaining turgor and thereby growth at high salinities (Yeo, 1994). Halophytes, for instance, are a class of plants capable of surviving high concentrations of electrolytes in their root environment without any injurious effects, through well established osmoregulatory systems. They absorb sodium and transport it to the leaves where it is either compartmentalized in the vacuole or excreted through salt glands. Special epidermal trichomes commonly called salt hairs are formed in *Atriplex* species, which accumulate salts from the adjacent leaf cells. These balloons like cells eliminate salts by bursting and thereby maintaining low salt concentration in leaf. Some trees may also maintain low foliar salt concentration in relatively high salinities as is the case with *Acacia amplicepes* (Neals & Sharkey, 1981; Grewal & Abrol, 1986; Marcar et al., 1990). They rely on avoidance mechanism, i.e., restricting the entry of salts into the roots and transport in the xylem and retaining most of the salts entering in old leaves (Yeo, 1994).

Most of the salt affected areas in Pakistan are located in the heart of its agriculturally important tract of the Indus plain. According to Rafiq (1990) out of 5.8 million hectares salt affected area, 3.16 Mha are within the canal commanded area and 2.64 Mha, outside it. In the Sindh province about half of the area is salt affected to varying degrees (Anonymous, 1981), of which about 51 percent is saline-sodic or sodic and the rest (49%) is saline (Muhammad, 1973).

Growing conventional crops in these problem lands is not economical, but they could be utilized gainfully for other purposes. For instance, according to Qureshi et al. (1992) about 4.1 million hectares have the potential of development through cultivation of salt tolerant forage and woody species. Apart from this coastal sea line area, sandy desert lands and coastal lands with possibilities of irrigation with saline water could also be considered for such a proposition.

Development of the above mentioned lands through tree cultivation is particularly important for Pakistan where annual domestic requirement of fuel wood by the end of 21st century, will be about 61.01 million m^3 (Anonymous, 2000). In addition to providing valuable wood and non-wood products, trees can have a significant positive impact by reducing water table and soil salinity in irrigated areas (Goodin, 1989; Khanzada et al., 1998).

In earlier studies conducted in Pakistan and India, several woody species have been identified as being either highly salt tolerant (*Prosopis juliflora*, *Prosopis chilensis*, *Prosopis alba*, *Tamarix aphylla*) or moderately tolerant (*Acacia tortilis*, *Eucalyptus camaldulensis*, *Casuarina equisetifolia*, *Azadirachta indica*, *Eucalyptus tereticornis*, *Eucalyptus microtheca*, *Acacia auriculiformis* and *Acacia nilotica*) (Jain et al., 1985; Sheikh, 1987; Singh, 1989; Yadav, 1980). However, there is still a need for further research to identify woody perennials and shrubs that can thrive well and produce reasonable wood or biomass on salt affected soils.

Therefore, the present study was undertaken to evaluate the relative performance of some local as well as exotic tree species on salt affected soils of Pakistan. It was also one of the objectives that whether these leguminous species could alter the nutritional status of the salt affected soils.

2. MATERIALS AND METHODS

The studies were conducted at the experimental farm of the Nuclear Institute of Agriculture (NIA), Tandojam. The farm is located at 25°, 24'North, 68°, 31' East, is within an intensively cultivated area, and irrigated with canal water from the Indus River. The area is a part of the lower Indus plain, with elevation of 28 m above the sea level. Soil of the experimental site, is medium in texture, varying from loamy clay to silty clay. The saline areas of the experimental farm were surveyed before the trials establishment, which showed wide heterogeneity in salinity. The estimated values of Electrical Conductivity (ECe) recorded with Electromagnetic Induction Instrument (EM-38), ranged from 7.31-26.45 (15.8) dS m^{-1}. The salts were mostly dominated at the upper surface and decreased with depth. The underground water is shallow (120-175 cm) but less saline. The other physico-chemical properties of the soil and water used are presented in (Table 1).

Table 1. *Physico-chemical properties of soil and underground water.*

Soil properties	*Depth (0-30 cm)*	*Depth (30-60 cm)*
(i) Clay %	23.5-37.0	19.0-34.5
(ii) Silt %	24.0-52.5	16.0-51.0
(iii)Sand %	12.0-45.5	16.0-65.0
Textural Class	Loamy Clay - Silty Clay	Sandy Clay -Silty Clay
Soil E C(1:2:5) dS m^{-1}	2.48-12.41	0.51 - 5.58
Soil Reaction (pH)	6.6 - 7.30	6.8 - 7.7
Soluble Cation & Anion		7.61 - 121.0
(i) Sodium (Na) (meq L^{-1})	48.0 -310.0	10.0 - 36.25
(ii) Potassium (K) (meq L^{-1})	42.5 - 75.0	0.4 - 6.57
(iii) Calcium (Ca) (meq L^{-1})	3.03-8.23	0.84 - 18.68
,(iv) Magnesium (Mg) (meq L^{-1})	4.17 -30.36	Absent
(v) Carbonates (CO_3) (meq L^{-1})	Absent	1.0 - 2.50
(vi) Bicarbonates (HCO_3) (meq L^{-1})	1.10 - 4.0	1.90 - 14.80
(vii) Chlorides (CI)(meq L^{-1})	7.10 -49.4	8.75 - 61.03
Sodium Adsorbtion Ratio (SAR)	18.14 - 86.00	10.43 - 46.65
Exchangeable Sodium Percentage (ESP)	20.31 – 100.0	—
Underground Water		
Underground Water Depth	1.30 - 1. 75m	—
ECw (dS m^{-1})	1.34	—
PH	7.5	—
(i) Na^+ (meq L^{-1})	4.78	—
(ii) Ca^{2+}(meq L^{-1})	3.38	—
(ii) K^+(meq L^{-1})	0.47	—

2.1. Nursery development

Seeds of different *Acacias* species, supplied by the CSIRO, Division of Forest and Forest Products, Canberra, Australia, were soaked in boiling water for five minutes and then left overnight in water at room temperature. The soaked seeds were sown in nursery bags (10 x 20 cm size), filled with river sand and perforated on sides for draining excess water. The seedlings were grown in nursery for six months before outplanting. *Atriplex lentiformis* (from Australia) was propagated through stem cuttings.

2.2. Experiment layout/design

Five species of *Acacia* (one native and rest Australian), which had given good performance during previous studies, were selected. *Atriplex lentiformis*, Australian halophytic shrub was also included in this study as a reference species. The botanical/local name and brief geographical description of their habitats is presented in (Table 2).

Table 2. *Geographical description of some exotic and native tree species.*

	Local name	*Origin*	*Latitudinal range*
Acacia ampliceps	Salt wattle	Western Australia	20-29
Acacia stenophylla	Rivercooba .	Victoria, N.S.Wales Queensland, N.T.	17-37
Acacia machanochieana	—	Northern Territory Western Australia	18-21
Acacia sclerosperma	—	Western Australia	20-29
Acacia nilotica	Babar,Babool	South Asia, Africa	15-60
Atriplex lentiformis	Salt Bush	Western Australia	—

After the preparation of land, soil was irrigated 2 days before transplantation. Seedlings were transplanted and sown at a distance of 2 m between rows/trees in auger holes (size 2.5 x 7 cm) along the sides of the ridges as was described by Gill and Abrol (1986), Dead plants were replaced 15-30 days after transplantation. The experiment was laid out in a Randomized Complete Block Design (RCBD) with three replicates. The other details of the experiment are: plot size = 32 m^2 (2 x 2 x 8 m), number of plots per replicate = six, number of trees per plot = 8 (4 in each row), row to row distance two meter, and plant-to-plant was 2 meter.

2.3. Growth parameters

The following observations were recorded at different intervals according to Field Trial Manual for Multipurpose Trees Species (Buford, 1990) and Standard Research Methods for Multipurpose Trees and Shrubs (MacDicken et al., 1991).

2.3.1. Survival percentage:

Number of trees at observation, divided by the total trees, planted x 100. Replants for initial mortality are not counted

2.3.2. Plant height (cm):

Vertical distance from the ground line to the apical bud of the main stem leaves and lateral branches were not considered

2.3.3. Stem diameter:

a) At 10 cm (D10): It was measured at the base i.e. 10 cm above the ground.
b) At breast height (DBH): It was measured at 1.3 m above the ground.

2.3.4. Plant canopy:

It was measured using non-clastic tape, which is stretched along an axis from one edge of the crown to its opposite edge, passing through the geographical center.

2.3.5. Biomass:

The total green foliage including leaves, small branches were measured in kg/tree.

2.3.6. Wood density:

Wood density was determined from small representative stem section (Majeed et al., 1994). Samples were air-dried and the density was calculated using the following formula:

Wood density = Air dry weight of wood/Apparent volume - volume of water

2.4. Physico-chemical analysis of soil, water and plants

Soil samples at two depths, i.e., 0-30 and 30-60 cm were collected at transplanting and 24 and 36 months after tree growth, from the center of the each plot. The soil texture was determined as described by Bouyuocos (1962). Similarly, EC and pH were also determined and the soluble cations and anions in soil and underground water were determined. The soil: water ratio for the extraction of solution was 1:2.5. Sodium, potassium and calcium were determined by flame photometer (Jenway, model PFP7), whereas, for the determination of magnesium, titration method was used. The total Ca^{2+} plus Mg^{2+} was determined by versinate solution (EDTA disodium) (Jackson, 1958). Carbonates (CO_3^{2-}) and bicarbonates (HCO_3^-), often termed as total alkalinity, were determined by titration with 0.5 N H_2SO_4 using phenolpthalein and methyl orange indicator. Chloride (Cl) was determined by titration with 0.05 N silver nitrate ($AgNO_3$) solutions.

To observe the subsequent effect of tree cover on the fertility status of soil, the soil sample, at the upper most layer (0-6 inches) were also collected for the determination of major nutrients and organic matter content. To observe the annual fluctuations of nutrient uptake, composite leaf samples (expanded) were collected around the canopy of the trees.

Total nitrogen was determined by Kjeldahl method (Jackson, 1958). Phosphorus in soil was determined by AB-DTPA extracting solution (Soltanpour & Schwab, 1977) and in plant by Barton's reagent. Potassium in soil and in plant was estimated by flame photometer (Jenway, PFP7). Organic matter percentage in soil was determined by Walkley and Black (1934).

3. RESULTS

3.1. Growth

Salinity under field condition varied from spot to spot. This heterogeneity in soil salinity was regularly monitored with the help of EM38 and the area was categorized into different salinity classes, i.e., low, medium, high and very high salinity having EC values ranging from 4-8, 8-12, 12-16 and >16 dS m^{-1}, respectively. All the species showed a decrease in seedling survival after planting in saline conditions (Table 3). *Acacia ampliceps* was least affected, while the maximum reduction due to salinity was observed in *Acacia machanochieana*, which could not survive under the salinity above 8.0 dS m^{-1}.

At 12 months after transplantation, survival generally became stable in all the species (*Acacia nilotica*, *Acacia sclerosperma*, and *Acacia machanochieana* had 45.8, 29.0, and 12.5 percent survival, respectively), except in *Acacia stenophylla*, which showed some mortality at every time of observation. However, *Acacia ampliceps* and *Atriplex lentiformis*, kept the population constant from early stage of out-planting, up to the last observation recorded at 36 months after transplantation, i.e., 75% and 50%, respectively.

Table 3. *Survival percentage recorded at different intervals.*

Tree species	*2.5 months*	*10 months*	*20 months*	*24 months*
Acacia ampliceps	75.0	75.0	75.0	75.0
Acacia stenophylla	41.6	37.5	20.8	16.0
Acacia mechanochieana	20.8	12.5	12.5	12.5
Acacia nilotica	33.0	29.16	29.16	29.16
Acacia sclerosperma	66.6	45.8	45.8	45.8
Atriplex lentiformis	50.0	50.0	50.0	50.0

The seedlings of *Acacia ampliceps* adapted well under saline environments, but were mostly populated under low to medium salinity patches, i.e., up to EC 12 dS m^{-1}. Soil salinity measured near the individual tree trunk showed that survival of *Acacia ampliceps* was only 50% at higher salinity patches, whereas under low to medium salinity patches it had performed extremely well (80-90% survival). Similar was the situation in case of *Acacia nilotica* and *Acacia stenophylla*. On the other hand, *Atriplex lentiformis* was mostly populated on medium to high salinities, whereas *Acacia sclerosperma*, and *Acacia machanochieana* were populated on low salinity patches.

All the species showed comparatively slow growth during the first 12 months after plantation except *Atriplex lentiformis* (Table 4), which grew vigorously during this period. Among the *Acacia* species examined, *Acacia ampliceps* produced maximum plant height (1.37 m) followed by *Acacia nilotica* (1.26 m), *Acacia stenophylla* (1.25 m), *Acacia machanochieana*, (0.98 m) and *Acacia sclerosperma* (0. 72 m).

Plant height increased with the passage of time. After 24 months of transplanting native *Acacia* (*Acacia nilotica*) outgrew *Acacia ampliceps* and the shrub (*Atriplex lentiformis*), while the other *acacias* grew in the same order, i.e. *Acacia stenophylla* > *Acacia machanochieana* > *Acacia sclerosperma*.

Table 4. *Effect of salinity on plant height (cm) of different species of* Acacia.

Tree species	*12 months*	*24 months*	*36 months*	*Mean*
Acacia ampliceps	137.50 b	467.6 b	535.6 b	380 B
Acacia stenophylla	125.6 bc	330.0 Cc	288.3 d	248 D
Acacia mechanochieana	98.0 bc	230.0 d	360.0 c	229 D
Acacia nilotica	126.0 bc	528.0 a	652.3 a	435 A
Acacia sclerosperma	71.6 c	221.6 d	288.6 d	194 E
Atriplex lentiformis	215.03 a	321.6 c	323.3 cd	286 C
Mean	128.99 C	349.83 B	408.06 A	

LS.D (0.05) Time = 22.01, Species = 31. 13028.
Means in the same column and same row sharing the same letters did not differ significantly according to Duncan's New Multiple Range test at 5% level.

During the first 12-24 months *Acacia nilotica* grew rapidly showing an increase of 168% over its first twelve months growth. The growth of *Atriplex lentiformis* was also fast during the early growth stages, but with the passage of time it slowed down due to its bushy nature. The relationship between plant height and estimated soil salinity (0-90 cm) showed a significant negative trend ($P \leq 0.05$) in *Acacia ampliceps,* and there was a decrease in plant height with increasing salinity. The coefficient of determination R^2 of salinity vs plant height was 0.33. This indicated that 33% of the variation in plant height could be attributed to the changes in soil salinity. This relationship in *Acacia stenophylla* was positive and significant ($P \leq 0.05$), the value of coefficient of determination was 0.256, indicating about 25% increase in plant height with the increase in soil salinity. *Acacia ampliceps* plants were, however, exposed to higher salinity than *Acacia stenophylla*. The coefficient of determination in *Acacia nilotica* and *Atriplex lentiformis* was also positive but non-significant.

There was a gradual increase in stem diameter (D 10) at the base with time (Table 5). The average values recorded 20, 24 and 36 months after transplantation were 5.45, 8.59 and 12.37 cm, respectively. *Acacia ampliceps* had a maximum diameter (15.40 cm) followed by *A. nilotica* (14.99 cm), *A. stenophylla* (5.13 cm), *A. sclerosperma* (4.9 cm) and *A. machanochieana* (3.57 cm). The increase in D 10 was more during 20-24 months as compared to that recorded after 36 months of growth.

Stem diameter at breast height (DBH) also showed similar trend (Table 6). The average diameter recorded at 36 months was 43% greater than that recorded 12 months before (i.e at 24 months after transplantation). Maximum DBH was recorded in *A. nilotica* (18.73 cm) followed by *A. ampliceps* (16.28 cm).

The correlation analysis between soil salinity and the diameter at base (D 10) and at breast height (DBH) in three *Acacia* species i.e. *A. ampliceps. A. nilotica* and *A. stenophylla* were non-significant. Negligible effect of soil salinity on the diameter of stem was recorded. The values for D 10 were positive in these species, but the situation was slightly different in case of DBH, where *A. ampliceps* and *A. nilotica* had negative values, while *A. stenophylla* was positively correlated.

Table 5. *Effect of salinity on stem diameter (D I0; cm) of different species of* Acacia.

Tree species	*20 months*	*24 months*	*36 months*	*Mean*
Acacia ampliceps	9.42 (0.425)	15.21 (4.36)	21.46(5.97)	15.40
Acacia stenophylla	3.83 (0.155)	5.33 (2.08)	6.23(2.55)	5.13
Acacia mechanochieana	1.87 (0.164)	4.02 (1.09)	4.83 (0.57)	3.57
Acacia nilotica	9.36 (0.310)	14.43 (2.22)	21.19 (3.05)	14.99
Acacia sclerosperma	2.78 (0.065)	3.98 (2.05)	7.95 (4.96)	4.9
Atriplex lentiformis	—	—	—	—
Mean	5.45	8.59	12.37	

Table 6. *Effect of salinity on stem diameter (DBH; cm) of different species of* Acacia

Tree species	*24 months*	*36 months*	*Mean*
Acacia ampliceps	11.27(3.29)	16.28(4.24)	13.78
Acacia stenophylla	2.57 (1.66)	6.23(2.95)	4.40
Acacia mechanochieana	1.78 (0.86)	3.60(0.41)	2.70
Acacia nilotica	11.01 (2.07)	18.73(3.31)	14.87
Acacia sclerosperma	1.34 (0.03)	4.18(1.013)	2.76
Atriplex lentiformis	—	—	—
Mean	5.59	9.80	—

Plant canopy (tree volume) of these species also varied with the passage of time (Table 7). At the period of 12 months after transplantation *Atriplex lentiformis* had the maximum volume followed by *Acacia ampliceps. A.nilotica. A. stenophylla, A. sclerosperma* and *A. machanochieana*, having plant canopy 5.82, 2.45, 1.99, 0.84 and 0.36 m, respectively. At 24 months *A. nilotica* showed a progressive improvement with the highest values of 13.43 m followed by *A. ampliceps*, (11.78 m), *Atriplex lentiformis* (10.15m), *Acacia sclerosperma* (6.3 m) *A. stenophylla* (5.54 m) and *A. machanochieana* (4.12 m). Similar was the trend at 36 months after transplantation, when the area covered by these species was 14.44 m each for *Acacia ampliceps* and *Acacia nilotica* followed by *Atriplex lentiformis* (11.59m), *A. sclerosperma* (6.3m), *A. stenophylla* (5.54m) and *A. machanochieana* (4.12m). The effect of salinity on plant canopy was non-significant, the R^2 values for *Acacia ampliceps, A. stenophylla* and *A. nilotica.* were 0.009, 0.15 and 0.06, respectively. However, the relationship was negative in case of the former species and positive for the latter two species. The relationship with the growth parameter like plant height, D 10, DBH, and number of stems showed that plant canopy is the function of these growth parameters.

***Table** 7. Effect of salinity on plant canopy (m) of different species of* Acacia.

Tree species	*12 months*	*24 months*	*36 months*	*Mean*
Acacia ampliceps	2.45 b	11.78 ab	13.6 ab	9.28 A
Acacia stenophylla	0.88 b	5.54 c	6.12 cd	4.10 BC
Acacia mechanochieana	0.36 b	4.12 c	4.53 c	3.00 C
Acacia nilotica	1.99 b	13.43 a	15.13 a	10.19 A
Acacia sclerosperma	0.84 b	6.3 c	6.96 c	4.70 B
Atriplex lentiformis	5.82 a	10.15 b	11.86 b	9.28 A
Mean	2.06 C	8.55 B	9.70 A	

LS.D (0.05) Time = 0.884, Species = 1.249

Means in the same column and same row sharing the same letters did not differ significantly according to Duncan's New Multiple Range test at 5% level.

Wood density was recorded at 24 and 36 months after planting the trees. There was a gradual increase in wood density with time, except in *Acacia sclerosperma* and *Atriplex lentiformis*. Both showed considerable reduction in wood density at 36 months after transplantation. The data recorded at 24 months, showed that *A. sclerosperma* had the maximum wood density followed by *Atriplex lentiformis. A. nilotica. A. stenophylla* and *A. ampliceps* (Table 8). At 36 months, *A. machanochieana* had the maximum wood density followed by *A. stenophylla, A. sclerospenna, A. ampliceps, A. nilotica* and *Atriplex lentiformis*.

Plant biomass above the ground was calculated on the basis of fresh weight (Table 8). The biomass of the individual species showed that *A. ampliceps* had maximum green foliage, i.e., 91.0 kg, and the minimum values (5.0 kg each) were recorded in *A. sclerosperma* and *A. stenophylla* (Table 8). The native *A. nilotica* also had high biomass (82.0 kg), while the halophyte *Atriplex lentiformis* had relatively lower values, i.e., 44.5 kg. The biomass data of *A. machanochieana* was not possible to be recorded because of its thin population and the surviving trees were left for recording data the following year.

Table 8. Wood density and plant biomass recorded at different intervals.

Tree species	Wood density ($g\ cm^{-3}$)		Plant biomass at 24 months (Kg/plant)
	24 months	36 months	
Acacia ampliceps	0.625	0.772	91.2
Acacia stenophylla	0.696	0.869	15.0
Acacia mechanochieana	—	0.870	—
Acacia nilotica	0.717	0.762	82.2
Acacia sclerosperma	0.920	0.824	5.0
Atriplex lentiformis	0.843	0.761	44.5
Mean	0.760 (±0.12)	0.81(±0.05)	45.58 (± 36.65)

3.2. Nutrients

Nutrient uptake by plants in saline medium is highly affected by several internal and external factors, such as level of the salinity of the growth medium, type and age of the trees, effect of toxic ions in and outside the plant environment. Analysis of the nutrients in plant leaf samples showed a wide variation within the species, with increasing plant age and with the increase in salinity.

Nitrogen contents were decreased due to increase in salt content of the medium with time (Table 9).

Table 9. Effect of salinity on Nitrogen (N)% in plant leaves of different species of Acacia.

Tree species	12 months	24 months	36 months	Mean
Acacia ampliceps	2.02 b	1.67b	1.53 b	1.74 BC
Acacia stenophylla	1.62 c	1.74 b	1.71 ab	1.69 C
Acacia mechanochieana	2.17 b	1.74 b	1.33 b	1.75 BC
Acacia nilotica	3.63 a	2.19 a	2.06 a	2.63 A
Acacia sclerosperma	1.32 c	1.89 ab	1.36 b	1.53 C
Atriplex lentiformis	2.13 b	1.63 b	2.10 a	1.95 B
Mean	2.15 A	1.81 B	1.69 B	

L.S.D (0.05) Time = 0.1529, Species = 0.2163

Means in the same column and same row sharing the same letters did not differ significantly according to Duncan's New Multiple Range test at 5% level.

Nitrogen concentration in leaves during the early 12 months ranged from 1.32-3.63%, with an average value of 2.15%. At 24 months, the average nitrogen content was 1.81% and further decreased to 1.69% at the age of 36 months. The maximum nitrogen content was recorded in *A. nilotica* (3.63%), whereas *A. sclerosperma* had the minimum nitrogen (1.32%). The remaining species *A. machanochieana*, *A. ampliceps*, *A. stenophylla*, and *Atriplex lentiformis* had more or less similar nitrogen concentrations (2.02-2.17%). At later growth stages, the contents of nitrogen in *A. stenophylla*, *A. sclerosperma*, and *Atriplex lentiformis* increased, whereas in *A. ampliceps*, *A. nilotica* and

A. machanochieana, there was decreasing trend. However, *A. nilotica* remained on the top over the period of three years, with maximum nitrogen concentration followed by *Atriplex lentiformis* (1. 95%), *A. stenophylla* (1. 71%), *A. ampliceps* (1.53%), *A. sclerosperma* (1.36%), and *A. machanochieana* (1.33%).

Phosphorus (P) content of all the tree species decreased (Table 10) with the increase in salinity. The P content among the species at 12 months after transplantation, ranged between 0.20-0.32% with an average of 0.28 percent. Annual variation in P content showed that there was a significant decrease in leaf P content at 24 months after transplantation compared to the values recorded at 12 months. The values at 2nd year, ranged between 0.09 - 0.14 with an average value of 0.12%. There was some improvement at the age of 36 months in leaf P content having average P value 0.18%. However, the mean values for P content at 12 months remained high. As in the case of nitrogen, *A. nilotica* also had higher P concentration followed by *A. machanochieana*, *A. sclerosperma*, *Atriplex lentiformis*, *A. stenophylla* and *A. ampliceps*. *Acacia ampliceps*, though had low phosphorus content, it maintained satisfactorily P level over a period of three years. On the other hand, *A. stenophylla* showed a consistent decline in P content with time. The mean values for individual species showed that *Atriplex lentiformis* and *A. nilotica* have similar values when compared statistically. Low phosphorus accumulation by plant might be due to antogonism between calcium and phosphorus in soil solution, as was reported by Grattan and Grieve (1993), that the reduction in P availability was not only because of strength that reduced the activity of phosphate, but also because P concentration in soil solution is tightly controlled by sorption process and by low solubility of Ca - P minerals. This was found true in *A. ampliceps*, *A. stenophylla*, *A. sclerosperma* and *Atriplex lentiformis*, where comparatively high calcium content was observed at later stages of growth.

Table 10. *Effect of salinity on Phosphorus (P) % in plant leaves of different species of* Acacia.

Tree species	*12 months*	*24 months*	*36 months*	*Mean*
Acacia ampliceps	0.20 b	0.12 a	0.19 a	0.17 BC
Acacia stenophylla	0.24 ab	0.11 a	0.09 b	0.15 C
Acacia mechanochieana	0.31a	0.11 a	0.18 a	0.20 AB
Acacia nilotica	0.32 a	0.13 a	0.20 a	0.22 A
Acacia sclerosperma	0.31 a	0.09 a	0.17 a	0.19 ABC
Atriplex lentiformis	0.29 a	0.14 a	0.24 a	0.22 A
Mean	0.28 A	0.12 C	0.18 B	

L.S.D (0.05) Time =3.029, Species = 4.284

Means in the same column and same row sharing the same letters did not differ significantly according to Duncan's New Multiple Range test at 5% level.

Potassium (K^+) content in plant was up to a reasonable level ranging from 0.96-1.32% over the period of three years (Table 11). The mean K^+ concentration after 12 months of transplantation was recorded 1.32%, after 24 months K^+ contents in leaves decreased, whereas at 36 months K^+ content again increased having similar concentration as in the first year after transplantation. The mean values for 24 and 36 months were 0.96 and 1.32%, respectively. The trend within the individual tree species was also similar, having relatively higher K^+ content during 1st and 3rd year than that during 2nd year after transplantation. All the *Acacia* species including native *Acacia* had lower K^+ content than that in *Atriplex lentiformis.*

Table 11. *Effect of salinity on Potassium (K^+) % in plant leaves of different species of* Acacia.

Tree species	*12 months*	*24 months*	*36 months*	*Mean*
Acacia ampliceps	0.99 c	0.83 b	1.20 cd	1.01 c
Acacia stenophylla	1.03 c	0.63 b	1.35 bc	1.01 c
Acacia mechanochieana	0.931c	0.78b	1.08 cd	0.93 c
Acacia nilotica	1.I2 bc	0.59b	0.95 d	0.89c
Acacia sclerosperma	1.46 b	1.42 a	1.58 ab	1.48 B
Atriplex lentiformis	2.42 a	1.49 a	1.77 a	1.89 A
Mean	1.32 A	0.96 B	1.32 A	

L.S.D (0.05) Time = 0.144, species = 0.203
Means in the same column and same row sharing the same letters did not differ significantly according to Duncan's New Multiple Range test at 5.% level.

Analysis of leaf samples after 12 months tree growth showed that all the species had more or less similar calcium (Ca^{2+}) content having non-significant difference among the species (Table 12). Calcium content after the first 12 months ranged between 0.83-1.75 percent with the average value of 1.29%. Calcium uptake by plants, increased with the age of the trees, ranging from 0.86-3.64% with the average value 2.80%, at 24 months. At 36 months, Ca^{2+} content was also high when compared with that at first 12 months after transplantation.

Table 12. *Effect of salinity on Calcium (Ca^{2+}) % in plant leaves of different species of* Acacia.

Tree species	*12 months*	*24 months*	*36 months*	*Mean*
Acacia ampliceps	1.75 a	2.24 c	2.54 a	2.17 BC
Acacia stenophylla	1.07 a	1.39 cd	2.20 a	I.55 CD
Acacia mechanochieana	0.85 a	0.86 d	1.50 a	1.07 D
Acacia nilotica	0.83 a	1.32 cd	2.13 a	1.42 CD
Acacia sclerosperma	1.57 a	7.38 a	2.67 a	3.87 A
Atriplex lentiformis	1.67 a	3.64 b	2.43 a	2.58 B
Mean	1.29 C	2.80 A	2.24 B	

L.S.D (0.05) Time = 0.506, Species = 0.716
Means in the same column and same row sharing the same letters did not differ significantly according to Duncan's New Multiple Range test at 5% level.

The trend for sodium (Na^+) content in plants was similar to potassium (K^+), i.e., low in *Acacia* species than *Atriplex lentiformis* (Table 13). Comparison of Na^+ concentration in *Acacia* species with *Atriplex lentiformis* showed wide differences. *Atriplex* species had the highest sodium concentration in leaves (6.62%), while the concentration of Na^+ in *Acacia* species ranged between 0.32% to 0.91%. The native species *A. nilotica* had the lowest sodium concentration during the whole growing period. Similar results for Na^+ were found by Grewal and Abrol (1986). They observed that *A. nilotica* accumulated low sodium and had the lowest Na/Ca and Na/K ratio.

Like other cations, there was a decrease in sodium in leaves at the age of 24 months in all the tree species. Sodium concentration at 36 months increased considerably in *Atriplex lentiformis* (8.25%) as compared to its concentrations at the age of 12 months, followed by *A. stenophylla* and *A. nilotica*. While in case of *A. ampliceps*, *A. machanochieana* and *A. sclerosperma*, the concentrations were very low.

Table 13. *Effect of salinity on Sodium (Na^+)% in plant leaves of different species of* Acacia.

Tree species	*12 months*	*24 months*	*36 months*	*Mean*
Acacia ampliceps	0.65bc	0.29b	0.45b	0.46 BC
Acacia stenophylla	0.91bc	0.77b	0.93 b	0.87 B
Acacia mechanochieana	0.98 bc	0.73b	0.79 b	0.83 B
Acacia nilotica	0.30 c	0.27b	0.39 b	0.32 C
Acacia sclerosperma	1.47 b	0.67b	0.60 b	0.91 B
Atriplex lentiformis	6.44 a	5.15 a	8.25 a	6.62 A
Mean	1.79 A	1.32 B	1.90 A	

L.S.D (0.05)Time = 0.321, Species= 0.454

Means in the same column and same row sharing the same letters did not differ significantly according to Duncan's New Multiple Range test at 5% level.

3.3. Changes in soil characteristics due to cultivation of Acacia species

Impact of tree planting was monitored at different intervals only in soil with planting of *Acacia ampliceps, A. nilotica* and *Atriplex lentiformis*, as these species had good survival followed by good growth. Planting effect on high salinity patches showed no regular trend of decrease or increase in soil salinity level. The estimated values of soil ECe during the first two months (June and July) showed a gradual decrease in soil salinity, under *Acacia nilotica* and *Atriplex lentiformis* planting, while under *A. ampliceps*, the ECe values increased. The salinity observed in March (20 months after planting) increased under all plantings with maximum values recorded under *Acacia ampliceps*, followed by *Atriplex lentiformis* and *Acacia nilotica*. Data recorded in August (24 months after planting) again showed a significant decrease in EC with maximum decrease in soil salinity under *A. nilotica* planting. Salinity fluctuations at 36 months after planting again showed a gradual increase in salinity of the soil under all the plantings but the values of salinity of the soil under *A. nilotica* remained lowest. Soil under *A. ampliceps* showed consistent decrease in salinity with time. However, the values for the soil under this planting remained higher when compared with soil under *A. nilotica*.

Changes in pH, soluble cations and anions were also monitored (Table 14). Soil pH increased marginally with time, and these values were relatively higher at 36 months after transplantation than the earlier observations. However, pH values determined at different intervals indicate that soil remained neutral alkaline in reaction throughout the study period, i.e., pH 7.0-8.3. Change of soil reaction was non-significant among plots occupied by different species so that pH values of individual plots ranged between 8.27-8.45 and 7.5-7.83 at 0-30 and 30-60 cm depth, respectively.

As the salinity was mostly dominated by sodium, a significant increase in sodium accumulation was observed with time. Concentration of Na^+ was higher at surface layer than at sub-surface. Analysis of individual plots showed that the only planting under which sodium concentration remained constant, was *A. nilotica*. On the other hand, *A. sclerosperma* planting showed a maximum increase in sodium, both at surface and subsurface. High concentration of Na salts under the planting of *A.* sclerosperma might have been due to very thin planting as the species had very low survival (20%) followed by poor growth.

The situation under *A. ampliceps* and *Atriplex lentiformis* plantings was quite different, because of both of them showed high sodium contents, although these two species had the thickest planting with high survival. However, the increase in sodium was relatively low in the period between 24 - 36 months than during transplanting and 24 months earlier. Accumulation of calcium was also increased with time. The highest Ca^{2+} content in soil was observed at 36 months after transplanting. Like Na^+, the concentration of Ca^{2+} was also high at 0-30 cm as compared to lower depth (30-60 cm). Analysis of individual plots having planting of *A. sclerosperma* from last two years had a maximum Ca^{2+} concentration, while at the end of 3rd year, *Atriplex lentiformis* planting had the highest Ca^{2+} contents. Concentration at lower depth also showed fluctuations, having similar trend, i.e., minimum Ca^{2+} at the end of 2nd year as compared to 3rd year. Concentration of potassium did not show any significant change with time, as well as within individual plots under different tree plantings. However, these concentrations were relatively high at the upper depth than the lower depth. Among the soluble anions, chlorides were dominant. Bicarbonates were also present in low concentration, but the carbonates were totally absent. Chloride concentrations decreased with increasing depth of the soil and were low during early establishment phase, their concentration increased with the passage of time and the values were comparatively high at 2nd and 3rd years after planting than initial values. Comparison between species showed that plots under *Atriplex lentiformis* accumulated maximum chlorides, while those under *A. nilotica* had minimum chloride contents. Sodium was also increased in plots left bare throughout the whole study period and this increase was more at surface soil layer and the values were relatively higher than the soil under planting. However, at subsurface, the concentration decreased and the values were lower than the soil under tree planting. Similarly, there was also an increase in calcium, but the values were more or less similar to the plots under plantings. The values for potassium were low at 36 months after planting than the values recorded at the time of transplanting. The difference between bare plots and the plots covered by trees also showed lower values of potassium in these plots. Among the anions the values of both bicarbonates and chlorides were higher than the initial values. Concentrations of bicarbonates were low at surface layer and were almost similar to those at subsurface when compared with the plots occupied by different trees. On the other hand, chlorides were relatively higher in bare soil than the soil planted with different species.

3.4. Changes in soil fertility

Soil samples were collected at 0-15 cm layer from the center of each plot at the time of transplanting, and then after 24 and 36 months of out-planting so as to observe the changes in fertility. There was an overall increase in organic matter and the major nutrients (NPK) in samples collected after 36 months of planting as compared to those collected at the time of transplanting (Table 15). The organic matter in 0-15 cm soil layer improved with planting. At 36 months, the impact of individual species showed that soil under *A. machonochieana* had the maximum organic matter content, followed by *A. ampliceps* and *A. nilotica* planting. The increase in organic matter was, however, statistically non-significant, both with time as well as within the individual species.

The soil under *A. nilotica* and *A. machonochieana* planting had maximum soil nitrogen values followed by *A. ampliceps* and *A. stenophylla* plantings. Increase in nitrogen was significant with time, while non-significant within the individual plantings. Phosphorus status was significantly improved with the tree cover, as was evident from the values observed at the time of last observation recorded. The average values were more than double, at three years after transplanting, having average values 13.51 mg/kg compared with initial values, i.e. 5.62 mg kg^{-1}. Maximum increase in phosphorus due to tree planting was observed in plots occupied by *Atriplex lentiformis* while the lowest increase was found in plots under *A. nilotica*. However, the differences in P values within individual species were not significant.

Like phosphorus, potassium content in soil also increased with time. Increase in potassium was more during after 3rd year of planting as compared to the values recorded at 24 months after transplanting. Among individual species, *A. machanochieana* was the highest in potassium content followed by *A. stenophylla*, *Atriplex lentiformis*, *A. ampliceps*, *A. sclerosperma* and *A. nilotica*. Differences among the individual species were non-significant.

Table 14. *Effect of tree plantation on soluble cations and anions content in soil.*

Species	Depth (cm)	EC (1:2.5)	pH	At the time of transplantation: Cations (meq L^{-1})			Anion (meq L^{-1})		
				Na^+	K^+	Ca^{2+}	CO_3^{2-}	HCO_3^-	Cl^-
Acacia ampliceps	0-30	8.00	7.10	18.3	1.46	5.80	A	1.30	25.5
	30-60	2.15	7.03	46.3	0.37	1.07	A	1.57	7.13
Acacia stenophylla	0-30	6.69	6.97	145	1.61	6.40	A	1.20	21.3
	30-60	3.96	6.90	86.5	0.70	3.00	A	1.47	11.37
Acacia mechanochana	0-30	8.50	6.87	206	1.46	5.80	A	1.60	25.3
	30-60	1.84	7.17	39.3	0.35	1.00	A	1.90	4.40
Acacia nilotica	0-30	6.90	6.97	176	1.30	5.73	A	2.20	18.87
	30-60	2.29	7.03	45.7	0.41	1.81	A	1.70	5.30
Acacia sclerosperma	0-30	8.36	7.00	208	1.50	4.40	A	1.43	27.53
	30-60	2.14	7.33	48.4	0.34	1.46	A	1.90	5.00
Atriplex lentiformis	0-30	6.54	6.97	137	1.30	5.00	A	1.40	24.8
	30-60	1.69	7.17	35.5	0.27	1.31	A	1.93	4.70
Bare plots	0-30	10.50	7.00	251	1.70	6.20	A	1.30	32.0
	30-60	3.64	7.2	82.3	0.48	1.17	A	1.16	10.3

Species	Depth (cm)	EC (1:2.5)	pH	At 24 months of transplantation: Cations (meq L^{-1})			Anion (meq L^{-1})		
				Na^+	K^+	Ca^{2+}	CO_3^{2-}	HCO_3^-	Cl^-
Acacia ampliceps	0-30	8.64	8.03	214.0	1.26	27.3	A	2.07	29.7
	30-60	4.11	7.87	89.0	0.48	16.43	A	2.07	11.33
Acacia stenophylla	0-30	6.72	8.00	113	1.67	29.0	A	2.07	32.0
	30-60	2.77	7.73	45.0	0.53	11.83	A	2.67	7.13
Acacia mechanochana	0-30	9.13	8.23	215.2	1.25	28.30	A	6.33	25.0
	30-60	3.96	7.80	100	0.46	12.80	A	2.60	10.13
Acacia nilotica	0-30	8.82	8.13	132	1.15	27.16	A	2.20	22.63
	30-60	2.72	7.73	57.0	0.36	5.30	A	2.20	5.73
Acacia sclerosperma	0-30	10.83	8.15	327	2.01	30.7	A	2.07	42.0
	30-60	2.78	7.63	33.0	0.45	9.63	A	2.07	7.30
Atriplex lentiformis	0-30	11.01	8.30	267	1.42	29	A	2.00	39.5
	30-60	5.88	7.87	1840	0.68	22.0	A	2.20	16.10
Bare plots	0-30	11.3	8.10	231.5	1.38	30.4	A	1.33	37.1
	30-60	4.02	7.60	81.2	0.43	13.5	A	2.20	10.33

Species	Depth (cm)	EC (1:2.5)	pH	At 36 months of transplantation: Cations (meq L^{-1})			Anion (meq L^{-1})		
				Na^+	K^+	Ca^{2+}	CO_3^{2-}	HCO_3^-	Cl^-
Acacia ampliceps	0-30	10.1	8.45	224	1.68	36.5	A	2.5	35.7
	30-60	2.97	7.8	60.0	0.34	10.1	A	2.0	7.93
Acacia stenophylla	0-30	9.13	8.35	196	2.21	35.3	A	2.7	29.7
	30-60	1.79	7.60	40.0	0.31	4.78	A	1.8	4.4
Acacia mechanochana	0-30	10.97	8.30	261	1.64	35.3	A	2.8	40.5
	30-60	3.05	7.67	66.3	0.37	10.9	A	1.8	8.13
Acacia nilotica	0-30	7.61	8.23	161	1.66	33.7	A	2.3	22.1
	30-60	2.71	7.7	53.7	0.41	11.3	A	2.2	5.60
Acacia sclerosperma	0-30	12.15	8.27	300	1.66	36.3	A	2.4	51.7
	30-60	4.18	7.50	79.0	0.45	10.9	A	2.2	9.2
Atriplex lentiformis	0-30	11.71	8.30	288	1.82	36.8	A	2.5	50.6
	30-60	2.90	7.83	68	0.33	9.5	A	1.8	9.66
Bare plots	0-30	12.53	8.40	315	1.45	36.8	A	2.1	52.4
	30-60	3.25	7.53	66	0.39	7.83	A	2.0	8.13

Table 15. *Effect of plantation table the fertility status of the soil.*

Tree species	*At the time of out planting*			
	Potassium (ppm)	*Phosphorus (ppm)*	*Total Nitrogen %*	*Organic matter %*
Acacia ampliceps	205	5.48	0.037	0.74
Acacia stenophylla	245	5.18	0.032	0.64
Acacia mechanochiena	194	6.2	0.031	0.61
Acacia nilotica	234	5.38	0.035	0.7
Acacia sclerosperma	214	5.4	0.037	0.73
Atriplex lentiformis	222	5.58	0.038	0.75
MeanBare Plots	216203	5.626.11	0.040.36	0.710.72
Tree species	*After 24 months*			
	Potassium (ppm)	*Phosphorus (ppm)*	*Total Nitrogen %*	*Organic matter %*
Acacia ampliceps	226	9.83	0.047	0.65
Acacia stenophylla	217	10.9	0.048	0.75
Acacia mechanochiena	256	7.2	0.04	0.62
Acacia nilotica	221	10.77	0.055	0.84
Acacia sclerosperma	217	9.4	0.042	0.63
Atriplex lentiformis	226	10.53	0.042	0.96
MeanBare Plots	218192	9.999.20	0.040.04	0.690.64
Tree species	*After 36 months*			
	Potassium (ppm)	*Phosphorus (ppm)*	*Total Nitrogen %*	*Organic matter %*
Acacia ampliceps	273	12.97	0.044	0.83
Acacia stenophylla	285	14.1	0.044	0.84
Acacia mechanochiena	394	13.72	0.047	0.75
Acacia nilotica	220	10.83	0.047	0.82
Acacia sclerosperma	270	14.2	0.038	0.72
Atriplex lentiformis	278	16.06	0.038	0.77
MeanBare Plots	271260	13.5113.10	0.0440.04	0.790.78

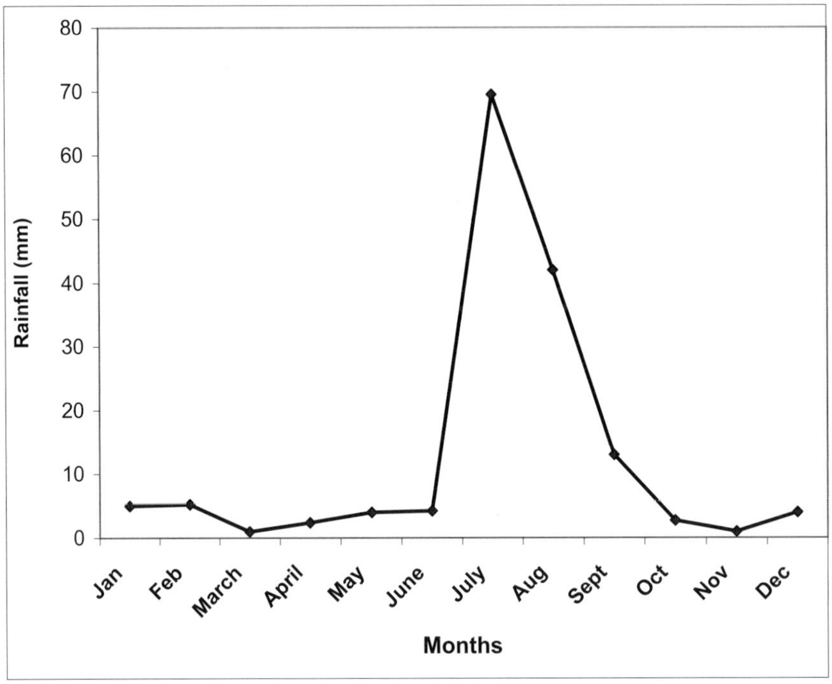

Figure 1. Ten years mean annual rainfall of experimental site.

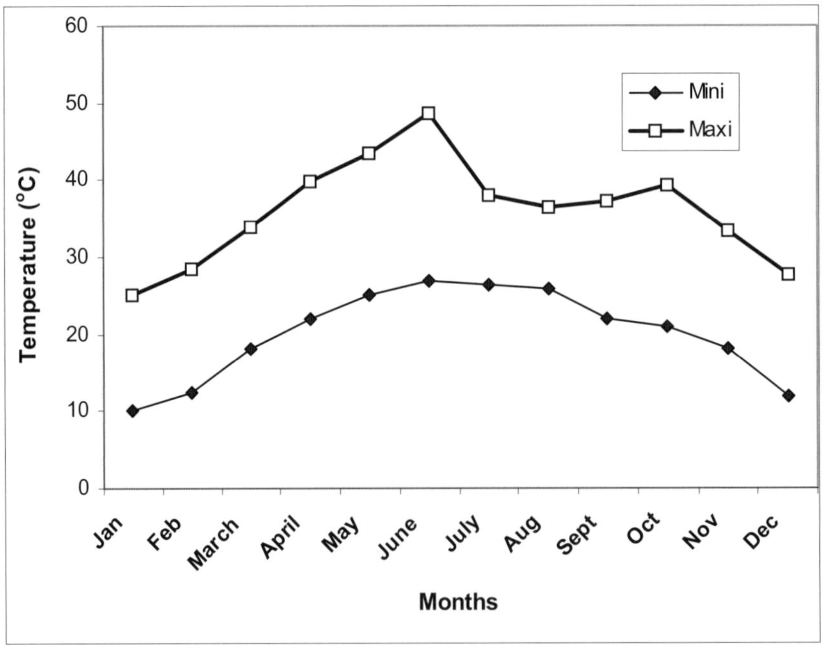

Figure 2. Ten years mean annual minimum and maximum temperature of experimental site.

4. DISCUSSIONS

Salt tolerance in tree plants is a complex phenomenon because many genetical, physiological and environmental factors are involved. The exotic *Acacia* (*A. ampliceps*) grew luxuriously like an indigenous species. The native *Acacia* (*A. nilotica*) showed difficulty during early establishment, but once it established, its growth was as good as in normal soils. This showed that *A. nilotica* can perform well under saline environment, provided that management practices during early stages are proper. Good performance of *A. nilotica* was also reported by Grewal and Abrol (1986) due to use of different management practices during the initial growth stages. Singh et al. (1986) also observed that there was no adverse effect on the growth of *A. nilotica* under high concentrations of NaCl at the top 28 cm of soil depth. Other exotic *Acacias* were relatively less salt tolerant in terms of seedling survival. High mortality of these *Acacias* might have been due to their lack of adaptability to the sudden shock of salinity. Pre-conditioning of seedlings might have improved it. The shock became more severe due to high temperature during summer, where the salt movement was mostly upward near the root zone due to capillary rise.

As far as the growth parameters are concerned, height and stem diameter are the main criteria, which affect the economic value of a particular tree species. Here again the main competition was between *A. ampliceps* and *A. nilotica*. The erectness of *A. nilotica* gave more and better quality wood. Periodical changes in growth also provided good information about the behavior of these species. The height of *A. nilotica* showed the fastest growth during the period from 12-24 months, which was not comparable with later growth stages. Similar growth of *A. nilotica* was also reported by Hussain et al. (1990) while studying the performance of different tree species under saline sodic soil, with or without gypsum ($CaSO_4$), indicating its better suitability for saline sodic soils. The mean annual increase in *A. nilotica* was much higher than the values observed earlier (Dass & Shunkarnarayan, 1984) under arid climate of Rayas, India. Among the *Acacia* species, salinity did not affect stem diameter. The decrease in stem diameter might have been due to reduction in vascular tissues as was observed by Hayward and Lang (1941). They reported the decrease in vascular bundle with increasing salinity and the reduction in the secondary xylem parallel to the reduction in vascular bundle. Maximum diameter of *A. nilotica* compared to other species further confirmed its better performance under adverse conditions. High stem diameter of *A. nilotica* was also reported by Khanduja and Goel (1986) while comparing the performance of 12 fast growing, multipurpose fuelwood tree species under alkali soils.

Accumulation of cations (Na^+, K^+, and Ca^{2+}) was generally associated with the concentration of these ions present in the growth medium. Concentration of these ions in all the *Acacias* examined showed a relatively low concentration as compared to that in *Atriplex lentiformis*. These results are in agreement with those of Neals and Sharkey (1981) who reported that *Acacia* species growing in saline environments do not accumulate the high foliar ion concentrations a phenomenon commonly observed in some halophytes such as *Atriplex*, *Suaeda*, and *Disphyma*. Similar were the conclusions of Marcar et al. (1990), Ansari et al. (1992) and Grewal and Abrol (1986) in *A. ampliceps* and *A. nilotica*. *Acacia* species had lower Na/Ca and Na/K ratios than those of *Atriplex lentiformis*. The differences in Na/Ca and Na/K ratios within the *Acacia* species were also evident. The tolerant species had low Na/Ca and Na/K ratio as compared to salt sensitive species. Low concentrations of sodium ions in tolerant species might have been due to ion regulation as was also observed by Khan et al. (1987) in *Prosopis juliflora*. They found that the species responded to salinity through K^+ selectively during ion influxes and exclusion of Na^+ from the leaves by retaining it in the roots. Similar were the conclusions of Marcar et al. (1990) in assessing the phenomenon in woody non-halophytic plants. They reported that these plants exclude salts (ions) but to differing degrees, the more salt tolerant species and provenances have greater ability to exclude salts. Exclusion refers to the procedure after reducing salt movement from root to leaves and to expanding leaves and the shoot apex in particular.

Nitrogen concentration in plant tissues was also affected by salinity. All the species showed lower N uptake, partly due to low concentration of nitrogen in soil and partly because of reduced uptake in the presence of salinity. Reduction in nitrogen content was more during later stages of growth, which may have been due to the dilution effect. The only species which maintained its nitrogen content in leaves, throughout the study period was *Atriplex lentiformis*. The possible decrease in N uptake by increasing salinity has also been attributed to a probable substitutions of Cl^- for NO_3^- ions as was suggested by Kadir and Paulsen (1982).

The accumulation pattern of phosphorus in plant leaves was quite different. The P concentration in leaves showed a decreasing trend with time, although soil P content at upper soil layer increased with increasing salinity of the soil. This showed that P uptake under saline medium is not only related with higher concentration of P present in soil, but under saline medium the activity of Ca^{2+} is increased which forms insoluble compounds of phosphorus rendering it unavailable for plants. Reduction of P concentration under salinity has also been reported by Grattan and Grieve (1993). They reported that reduction in P availability occurs, not only because of ionic strength effects that reduce the activity of phosphate, but also because P concentrations in soil solution are tightly controlled by sorption processes and by the solubility of Ca-P.

Tree planting brought good effects on the fertility of the soil. However, its effect was only at the upper soil layer. The situation for phosphorus and potassium was quite encouraging, especially in case of phosphorus, where the values were almost doubled from the values recorded at initial establishment of tree. Reynaldo and Cruz (1993) had defined two possible ways for the increased availability of phosphorus in soil under tree planting. One, as the mycorhizal association, which enables more efficient P uptake by creating oxalates, which bind with precipitating cations (aluminum, iron and manganese), thus releasing phosphate ions into the soil solution. Another possible mechanism is a greater activity of phosphate enzymes, which release organic P into available form. Potassium content in soil remained high throughout the whole study period. Potassium content further increased with time, which was not only due to tree planting, but its higher values are also the characteristics for the soils of arid climate where upward movement of salts is always high. The values for K content in soil throughout the study period remained in the limits as given by Wilde (1958), i.e. from 50 to 200 mg L^{-1}. Ameliorative effect of tree planting gave quite encouraging results. Organic matter content of soil, under the planting was increased, but values remained below one percent, which might have been due to high temperatures, resulting in fast decomposition of organic residues. Reduced organic matter content may decrease the soil aggregate stability, resulting in deteriorated soil physical properties and decreased infiltration. The resultant increased runoff and erosion may further accelerate the depletion of organic matter and nutrient status (Ohta, 1990).

The only and the largest single component of NPK content was litter-fall, which was mostly consisted of dead leaves. Nitrogen content in soil was also below the critical limits as given by Townsend (1972). Nitrogen content increased with the passage of time but the values remained within the critical limits (0.1-0.3%). The individual species had not given a clear-cut figure, however, the overall values were increased with time.

Appreciable amelioration in soil salinity did not occur in the present study, which is contrary to the general belief that tree planting may improve soil conditions. This, in this case, could be attributed to a number of reasons such as thin population stand, young age of the planting, and very shallow water table at the experimental site. This, however, does not rule out completely the possibility of using tree as ameliorative agent for use as "biological pumps" and improving soil conditions as has been in vogue since long in Pakistan "hurries" block planting of *A. nilotica.*

5. REFERENCES

Anonymous, 1981. Report of the National Commission on Agriculture, Ministry of Food and Agriculture. Islamabad, Pakistan: Government of Pakistan, 10 pp.

Anonymous, 2000. Manual of Salinity Research Methods, IWASRI.,Publication, Lahore, Pakistan: IWASRI.

Ansari, R., Khanzada, A.N. & Khan, M.A. 1992. Australian woody species for saline site of South Asia. Progress report, (ACIAR-8633) AEARC, Tandojam. Pakistan: AEARC. 3-11 pp.

Bouyoucos, G.J. 1962. Hydrometer method improved for making particle size analysis of soil. Agronomy Journal 54: 464-465.

Buford, C.B. 1990. Field Trial Manual for multipurpose Tree Species. Multipurpose Tree Species Network series. Thailand: Winrock International Institute for Agriculture Development.

Dass, H.C. & Shankarnarayan. 1984. Plant resources for wastelands of Rajisthan for bio-energy. In: R.N. Shanlla, O.P. Vimaland & P.D Tyagi. (Eds.), Proceedings of Bio-Energy Society First Convention and Syrup. Delhi, India: Bio-Energy Society of lndia. 84: 58-61.

Gill, H.S. &. Abrol, L.P. 1986. Grow Casuarina tree in sodic soils. Indian farming 35: 31-32.

Goodin, J.R. 1989. SCS Studies feasibilities of Eucalyptus plantation for salt affected farmland in Sanjoaquin valley. California Eucalyptus Grower 4: 1-10.

Grattan, S.R. & Grieve, C.M. 1993. Mineral Nutrient Acquision and Response by Plants in saline environment. In: M. Pessarakali. (Ed.), Handbook of plant and crop stress. New York, New York: Marcel Dekker, Inc. 203-266 pp.

Grewal, S.S. & AbroL, L.P. 1986. Agro-forestry on alkaline soils: Effect of some management practices on initial growth biomass accumulation and chemical composition of selected tree species. Agro-forestry System 4: 221-232

Hayward, I.E. & Lang, E.M. 1941. Anatomical and physiological response of tomato to varying concentration of sodium chloride, sodium sulphate and nutrient solution. Botanical Gazzette 102: 437-462.

Hussain, I., Mathur, A.N., & Ali, A. 1990. Effect of gypsum on survival and growth of tree species in saline sodic soil. Nitrogen Fixing Tree Research Reports.

Jackson, M .L. 1958. Soil Chemical Analysis. London, UK: Prentice Hall Publisher. 498 pp.

Jain, B.L., Muthana, K.D. & Goyal, R.S. 1985. Performance of tree species in salt affected soils in arid regions. Journal Indian Society Soil Science 33: 221-224.

Kadir, A. & Paulsen, G.M. 1982. Effect of salinity on nitrogen metabolism in wheat. Journal Plant Nutrition 5: 1141.

Khan, D., Ahamd, R. & Smail, S.I. 1987. Germination, growth and ion regulation in *Prosopis juliflora* (Swartz) DC. under saline conditions. Pakistan Journal Botany 19: 131-138.

Khanduja, S.D. & Goel, V.L. 1986. Pattern of variability in some fuel wood trees grown on sodic soil. Indian Forester 118-123.

Khanduja, S.D. 1987. Short rotation firewood forestry on sodic soils in Northern India research imperatives. Indian Journal Forest 102: 75-79.

Khanzada, A.N., Morris, J.D., Ansari, R., Siavich, P.G. & Collopy, J.l. 1998. Groundwater uptake and sustainability of *Acacia* and *Prosopis* plantations in Southern Pakistan. Agricultural Water Management 36: 121-139.

MacDicken, K.G., Wolf, G.V. & Briscoe, C.B. 1991. Standard research method for multipurpose trees and shrubs. Arlington, Virginia: Winrock International Institute for Agricultural Development. 92 pp.

Majeed, A., Yaqoob, S. & Qureshi, M.A. 1994. Distribution of wood density in a single stem *Populus euramericana*. Pakistan Journal Scientific Industrial Research 37: 436-438.

Marcar, N.E., Crawford, D., Ashwath, N. & Thomson, L.A.J. 1990. Salt and water-logging tolerance of subtropical leguminous Australian native trees: (A Review). In: S.S.M. Naqvi, R. Ansari, T.J Flower, & A.R. Aznii. (Eds.), International Conference on Current Development in Salinity and Drought Tolerance in Plants. Tandojam, Pakistan: AEARC. 375-387 pp.

Muhammad, S. 1973. Waterlogging, salinity and sodicity problems of Pakistan. Bulletin Irrigate. Drainage and Flood Control Research Council 3: 41-48.

Neals, T.F. & Sharkey, P.J. 1981. Effect of salinity on growth and on mineral and organic constituent of the halophyte, *Disphyma australe* (soland) J.M. Black. Australian. Journal Plant Physiology 8: 165-179.

Ohta, S. 1990. Influence of deforestation on the soil of the Pantabangan area, Central Luzon, the Philippines. Soil Science Plant Nutrition 36: 561-573.

Qayoom, M.A. & Malik, M.D. 1988. Farm production losses in salt affected soils. Proc. 1st National Congress Managing Soil Resources. Lahore, Pakistan: Soil Science Society Pakistan. 356-363 pp.

Qureshi, R.H., Shafqat, N. & Tariq, M. 1992. Performance of selected trees species under saline sodic field conditions in Pakistan. In: H. Lieth, &. A.M. Massoum. (Eds.), Towards rational use of high salinity. Dordrech, Netherlands: Kluwer Academic Publications. 1-11 pp.

Rafiq, M. 1990. Soil resources and soil related problems in Pakistan. In: M. Ahmed, M.E. Akhtar & M.L. Nizam. (Eds.), Soil Physics-Application under stress environments. Islambad. Pakistan: BARD, PARC. 16-23 pp.

Ramani, S. & Joshua, D.C. 1993. Studies on salt tolerance of plants mechanisms and physiological response. Annual Report. Nuclear Agriculture Division, Bombay, India: Bhabha Atomic Research Centre.

Reyrnaldo, E. & Cruz, D. 1993. *Acacia* for Rural Industrial and Environmental Development In: K. Awang & D.A Taylor. (Eds.), Proceeding of second meeting of the consultative group for Research and Development of *Acacias* (COGREDA). Udorn Thani, Thailand: COGREDA. 198-224 pp.

Sheikh, M.I. 1987. Energy plantation for marginal and problematic lands: Pakistan GCP/RAS/III/NET. Field Documents 81. Bangkok, Thailand: FAO.

Singh, K., Yadav, I.S.P. & Singh, B. 1986. Performance of *Acacia nilotica* on salt affected soils. Indian Journal of Forestry 9: 296-303.

Singh, P. 1989. Waste land, their problems and potential for fuel and fodder production in India. Report of the regional workshop on water and land development for fuel wood and other rural needs. Vadodora. India 19: 102-113.

Slavich, P. 1991. Properties of salt affected and waterlogged soil. In: W. Tahir & G.M. Black. (Eds.), A guide to tree planting on saline-sodic and waterlogged soils. Pakistan Forestry Planning and Development Project. Govt. of Pakistan. Arlington, Virginia: Winrock International Institute for Agricultural Development. 16-28 pp.

Soltanpour, P.N. & Schwab, A.P. 1977. A new soil test for simultaneous extraction of macro and micro nutrients in alkaline soils. Communications Soil Science Plant Analyist 8: 195-207.

Townsend, W.N. 1972. An introduction to scientific study of the soil. London,UK:Edward-Arnold Publishers Limited, 209 pp.

Walkey, A. &. Black, I.A. 1934. An examination of Degtjareff method for determining soil organic matter and proposed modification of the chromic acid titration method. Soil Science 37: 29-38.

Wilde, S.1. 1958. Forest Soils. New York: The Ronald Press Company. Yadav, I.S.P. 1980. Salt affected soils and their afforestation. Indian Forester 106: 259-272.

Yeo, A.R. 1994. Physiological criteria in screening and breeding. In: A.R. Yeo, & T.J. Flowers. (Eds.), Soil Mineral Stresses. Heidelberg, Germany:Springer Verlag, 37-60 pp.

CHAPTER 21

SOIL REMEDIATION VIA SALT-CONDUCTION AND THE HYPOTHESES OF HALOSYNTHESIS AND PHOTOPROTECTION

NICHOLAS P. YENSEN[1] AND KARL Y. BIEL[1, 2]

[1]NyPa International, Tucson AZ 85705, USA; [2]Institute of Basic Biological Problems of Russian Academy of Sciences, Pushchino Moscow Region 142290, Russia

Abstract. The present paper proposes that all terrestrial higher plants may be divided into three general categories based on their salt management: 1) Excluders, 2) Accumulators and 3) Conductors. Excluders exclude salt at the root level and are typically glycophytes, but there are a few euhalophytes as notable exceptions. Many miohalophytes and euhalophytes are salt accumulators, while virtually all crinohalophytes are salt conductors. Salt conductor plants move salt from the soil through the roots and plant tissues to salt glands on the plant surfaces.

We describe the hypothetical mechanisms and pathways of moving salt through the plant and predict that some plants with the ability to do this may acquire energetic benefits through a process described herein. This process we call Halosynthesis. Halosynthesis is any salt-mediated mechanism, by which exogenous energy is stored/utilized by organisms under appropriate conditions.

We predict in the present case, that the appropriate conditions for halosynthesis may be met in salt glands and possibly other structures. Halosynthesis utilizes environmental energy, in the face of otherwise stressful conditions, to optimize growth and other metabolic functions. The process of halosynthesis may have great potential to benefit conductor plants due to evapotranspirational energy and/or through the accumulation of a salt-mediated and/or salt-dependent, photon-electron, surface charge.

The first and best-known mechanism of energy extraction process is photosynthesis while the second discovered environmental energy source is that of thiobacteria which can extract heat energy from undersea hot-water vents to fix and reduce carbon. Conductor halophyte plants have been less well studied in this respect and this paper specifically examines the potential of the genus Distichlis as a good example of the quintessential halosynthesizing conductor plant. If our hypothesis of halosynthesis is proven correct it could represent the second known mechanism by which life can access energy from sunlight and the third mechanism by which organisms may extract energy from the environment.

1. INTRODUCTION

The estimated 1-10 billion hectares of salinized farm soils worldwide are an untapped resource opportunity for new halophilic (salt-loving) crops. Much of this "waste" cropland occurs in, north America, South America, the Middle East, Central Asia, Northern Africa and Australia. To meet the potential demand for

M.A. Khan and D.J. Weber (eds.), Ecophysiology of High Salinity Tolerant Plants, 313-344.

new halophilic crops, as water resources become scarce, many salt-tolerant plants and euhalophytes (plants that grow better under high salinities) are being studied by the emerging science of halophytology. Various new halotechnologies, such as the serial biological concentration of salts and biodrainage, incorporate halophytes for both practical (crops) and aesthetic (ornamental) reasons.

One of the first patents for a halophyte crop was issued in 1985 (Yensen, 1985). But the development of additional better-suited crops has been slow. We have just begun to develop new "crops" that can be used to eliminate toxic compounds and elements (e.g. Se, Pb, Cr, Cd, Zn, and U), petroleum products, asphalt, radioactive nucleides via phyto-remediating and bio-remediating halophytes. There is even working being done with vacuolar-salt-accumulating algae that can convert seawater to "brackish" water by sequestering salt within its tissues (H. Leith, per com.). And in new developments, ornamental halophytes could be irrigated with brackish or seawater, thus freeing up large amounts of fresh water for municipalities. Over 67 genera and hundreds of species are considered to have potential for development as cultivars (Yensen, 2002). Of the estimated 10,000 salt-tolerant species, there may be exist as many as 250 potential staple-crop halophytes plus ornamentals, dune stabilizers, pharmaceuticals and environment-improving halophytes.

The development of these needed crops is a major task and will be a significant milestone in the advancement of agriculture. The modicum of "embryonic halophyte crops," developed to date, herald from classical breeding of wild halophytes with minimal input from biotechnology, tissue culture, and/or plant screening techniques.

The incipient era of halotechnology must develop and carry forward a knowledge base. For this to happen, the support level will need to be at least on par with that of the US Salinity Lab by the Honourable George E. Brown and the early investigators: H.E. Hayward, L.A. Richards, F.M. Easton, A.D. Ayres, L. Bernstein, L.V. Wilcox, E.V. Mass, G.J. Hoffman and numerous other visionaries, who in 1928 established and/or carried out the early investigations in Riverside, California under the auspices of the USDA. Their efforts led the world in developing a knowledge base to understand how to manage "low salinity" and how it affects freshwater crops and soils. The US Salinity Lab still leads the world in how salt affects our fresh water crops. The new International Center for Biosaline Agriculture in Dubai, United Arab Emirates is now the world's pre-eminent center for high-salinity crop research.

Today, with the growing millions of acres of saline agricultural worldwide and the potential to export halophyte technology, we are entering a new era of halophytology. Some of the first observations of the increased productivity of a cereal grain on high salinity were noted in the genus *Distichlis* (Yensen et al., 1988). Fortunately, with regard to *Distichlis*, there is now a growing body of information on ecology, physiology, morphology, anatomy (including salt gland structure), etc. Still, basic knowledge of how "productivity can increase (reaching levels similar to or greater than freshwater crops) with an increase in salinity" still remains unexplained (Yensen, 2002).

2. THE HALOCONDUCTOR THEORY

2.1. The remediation of saline soils

World wide, saline soils reduce productivity by trillions of dollar annually. Thus, the remediation of salt-affected soils while growing productive salt-tolerant crops could be of great benefit … particularly in arid regions. The remediation of salt-affected soils by mechanical, chemical, and/or leaching, although effective, may be difficult, expensive and cause additional problems. Biological remediation may have the advantage of permanently restoring the land to its natural productive capacity. Certain plants may gradually remove salts from soil in beneficial ways.

The terms glycophyte, oligohalophyte, miohalophyte, and euhalophyte describe groups of plants with increasing tolerances to salt. Little attention, however, has been paid to classifying plants according to their ability to remove salt from soil. In Figure 1 it may be noted that we predict that both the glycophyte *Triticum* and the euhalophyte, *Salicornia*, do not effectively remove much salt from the soil. This is because *Triticum* excludes salt from entering its tissues, while *Salicornia* has limited removal ability once its tissues are full of salt.

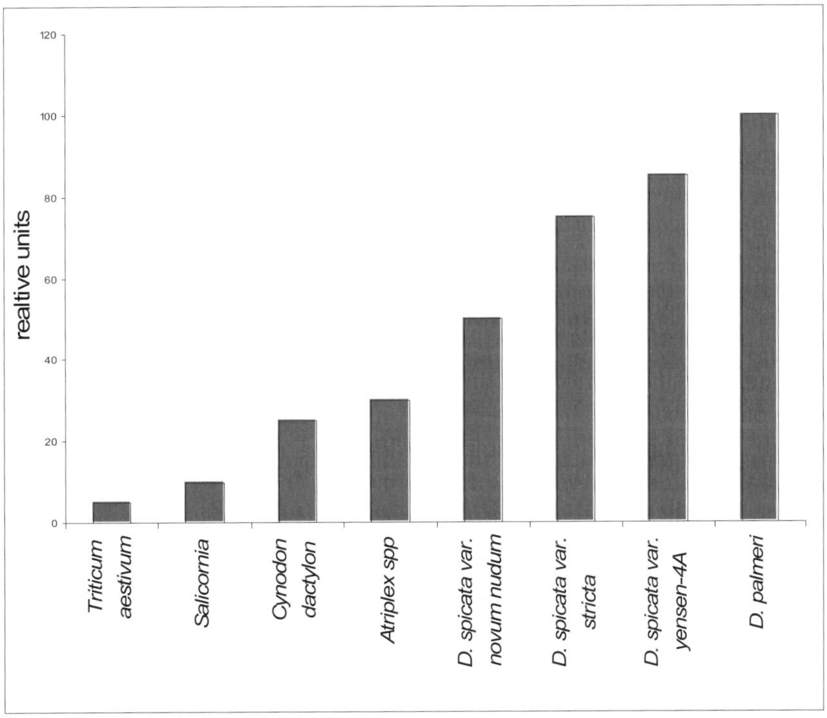

Figure 1. *Efficiency of soil-salt removal by different plants (from low to high removal ability):* Triticum aestivum *(wheat),* Salicornia spp. *(pickle weed),* Cynodon dactylon *(bermuda grass),* Atriplex spp. *(salt bushes),* Distichlis spp. *[4 NyPa lines] (saltgrass).*

In the past, it has been considered uneconomical to remediate saline soils using salt-accumulating halophytes to extract the salt from the soil for three reasons:

1. More salt is added to the soil via the water required to grow the halophyte than is extracted by the harvest of salt-laden plant material,
2. It is uneconomical to mechanically harvest and remove the salt and the otherwise useless plant material, and
3. The plant material is generally inedible and/or of little value due to the heavy load of salt in the plant tissue with 25-54% of dry weight as salt (Ungar, 1991).

The present concept considers salt-excreting halophytes that can be used in a method to potentially mobilize up to many tons of salt per hectare per annum with a certain percentage of this salt becoming airborne. This airborne salt may move to areas leached of salts. It has been long known that many crinohalophytes excrete salt onto their surfaces. The concept of using this ability to move salt from the soil through the plant, onto the plant surface, and potentially into the air, is a new concept and has not been described in detail in the literature. We term this process *haloconduction*.

The application of this process for the remediation of soils has not been described, but promises to be an environmentally friendly economical approach for the remediation of salt-contaminated soils. This process may be applied using the haloconducting principle with a number of species from various families and genera, such as, but not limited to: *Distichlis, Aeluropus, Spartina, Limonium, Avicennia,* and *Atriplex* (to a greater or lesser degree via bladder cells). Further, salt accumulators, such as *Climacoptera crassa* through senescence of leaves of accumulated salt may disperse salt if the leaves become wind blown or are other wise are dispersed.

The haloconducting approach and process may function in other plants with different genetic morphologies and physiological adaptations. The process may be applied to "salt-gland-bearing" species, which may range from being mildly salt-tolerant to highly euhalophytic.

2.2. New approach for soil remediation via salt conducting plants

A new classification system is necessary to understand salt balance and to remediate salinized lands. In the past, because some salt-accumulating halophytes may have up to 50% of their dry weight as salt, it is often suggested by neophytes and lay people that, "harvesting halophytes would remove salt from the soils." The typical response from halophyte and irrigation experts, however, has been that, "the water required to grow halophytes carries an influx of more salt than is removed by the harvested halophytes." We believe, however, that a re-evaluation of the different field situations, in conjunction with a new way to classify plants according to how much salt they can permanently remove from soil, may significantly alter this paradigm. A paradigm shift occurs when we realize that not all field situations require salt-water irrigation. Many areas of primary (nature caused) and secondary (human caused) salinization have influxes of water that are relatively low in salinity, such as in parts of Australia, western North America, northeastern China, northern Chile, Kara-Kum and Kyzyl-Kum regions of Asia and other areas salinized by salt-bearing "*produce*" water of petroleum and gas wells. In these situations, where the salinization has occurred over many years from relatively-fresh-water sources, or in the case of oil and gas wells where a one-time event has created a hyper-saline condition, it may be possible to remove salt from the soil using halophytes.

In the past there have been many ways to classify plants with regard to salinity and the various mechanisms that the plants employ to live in saline environments (see review Yensen, 1995). However, none of these classification systems have specifically considered, with respect to all plants, how a plant removes salt from the soil, how much salt is removed and where the removed salt goes.

We have therefore divided all plants into three groups with respect to their ability to remediate salty soils (Figure 2).

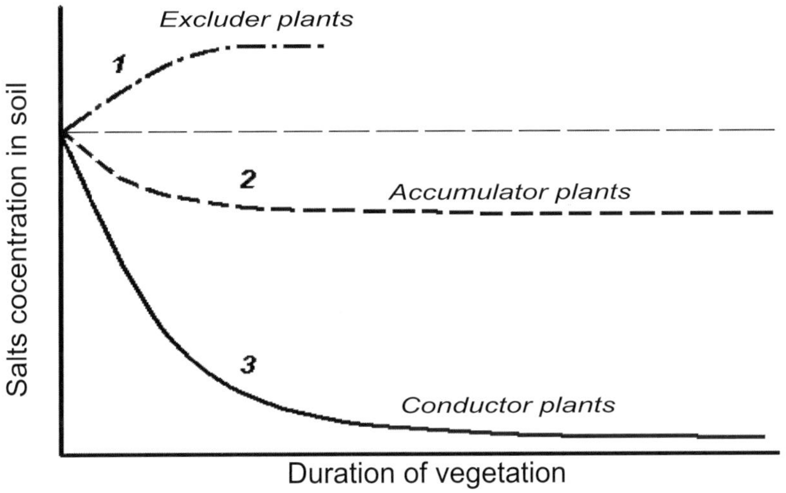

Figure 2. *Soil salt concentration over time with respect to (1) Excluder, (2) Accumulator, and (3) Conductor plants.*

2.2.1. *Excluders*

The first category of plants, exclude salts from entering their tissues at the roots. Most terrestrial plants fall into this category although many have limited abilities to exclude salts and as salt loads increase they die from excess penetration of salt. Some salt-excluding plants, however, are euhalophytes (e.g. the red mangrove, *Rhizophora spp.*). These plants, in general, have limited or no ability to remediate saline soils. And, because they selectively absorb water, but exclude salts at the root level, they tend to accumulate soil salts to the point that the salt tolerance of the plant is exceeded. This typically results in the demise of the plant with "plant-induced" salinization of the soil (Figure 2-*1*).

2.2.2. Accumulators

The second category, uptake salts and accumulate the salts in the vacuoles of the plant cells. These may be glycophytes, salt-tolerant or euhalophyte plants, with or without succulence. In general, these plants have remediation value only if they, or their parts, are removed from the site. With some exceptions, such salt-laden plants often do not have very valuable products, which typically do not cover the cost of removal. Also, they may become too salty to use as forage for grazing livestock. These plants also have a limited amount of salt that they may remove from the soil, i.e. 50% of their dry weight, due to the fact that their tissues become salt "saturated" (Figure 2-*2*).

2.2.3. Conductor

Plants, the third category (Figure 2-*3*), absorb salts from the soil, transport the salt through the plant to the shoot surface and, with the aid of wind, disperse the salt away from the site. These plants potentially have great ability to remove salt from the soil and widely distribute the salt via wind dispersal. It may be typical to have very wide dispersal of aeolian salt via air currents; for instance, it is well known that 80% of all inland salt arrived from the oceans via air currents carrying microscopic salt particles.

While this classification is phylogenetically artificial (and is not necessarily related to salt tolerance) it does allow the classification of plants with regard to their ability to remediate saline soils. Further, the application of the principals of this classification system may have additional economic advantages (such as for non-salt-laden forage). *Distichlis, Spartina, Aeluropus* and other conductor plants may serve as a biological tool to remove salts from the soil by placing the salts onto the shoot surfaces without accumulation of salts in the tissues. Moreover, a certain proportion of the excreted salt located on the surface may be released into the air and carried aloft with air currents, thus permanently eliminating the salt from the remediation area.

Putative advantages of salt conducting plants are that the salt is removed from the soil and released without accumulation in the plants. Our speculation is that in conductor plants the salts are essentially excluded from the cells. That is, we suggest that the salts are initially transported symplastically (xylem) from the roots and then apoplastically (and/or interstitially) to the basal cells of the salt glands (and to the papillae in some cases) and then symplastically through the basal cells to the cap cells where the salt is released externally on the cap cell (and/or papillae) surfaces and possibly in some cases (as of yet undocumented) as a "hollow" salt crystal filament.

Such a salt conducting system has not been fully described in the literature and at present is speculative. This speculation, although appearing simplistic, is important, with regard to the potential value to the plant and the environment, to uptake, transport, and eliminates salts from the immediate vicinity, as will be discussed later herein.

2.3. Advantages of conductor plants for soil remediation

(Figure 3) illustrates the advantage of conductor plants when compared to salt accumulator halophytes. Salt accumulating plants, although with great ability to tolerate high salinity, reach an absorption plateau, which limits their ability for continued uptake of additional salt. The maximum tolerance and initial rate of salt absorption varies between the many species of salt-accumulating plants, but in the long term they are highly restricted with regard to developing improved lines with greater ability to remove salt from the soil (2b, dotted line of Accumulator plants in Figure 3).

All salt conductors (even if limited in total flow of salt) have a continuous flow of salt from the soil to the shoot surface and potentially to the air. From this standpoint, it is important to understand the mechanisms and pathways of salt movement. It is likewise important to identify and/or develop plants with high capacity for an increased salt flow, cell-protection mechanisms, and potential for aeolian salt dispersal from the cap cell surface (3b, dotted line of Conductor plants in Figure 3).

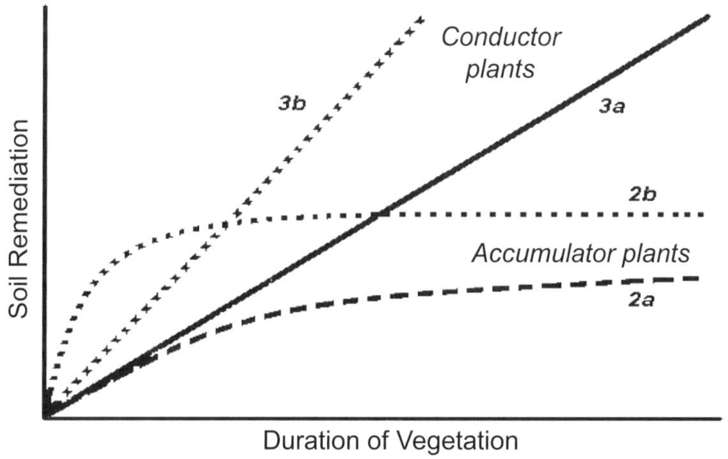

Figure 3. *Graph of generalized soil-salt remediation ability of existing and developed salt-accumulating plants (2a, 2b) versus conductor plants (3a, 3b). The 2a dashed line represents a wild-type accumulator plant. The 2b dotted line represents the theoretical limit of salt removal via accumulator plants through selection and breeding. The 3a solid line represents typical wild-type salt-conducting plants. The 3b dotted line represents the improved salt-conductor plants and their greater potential for soil remediation.*

2.4. Productivity considerations

With regard to productivity, higher plants are often classified into three types according to their sensitivity to salinity. For instance, *glycophytes* have productivity inversely related to salinity. The second, *miohalophyte* (salt-tolerant plants) maintain productivity with increasing salinity to some threshold, above which they have a linear decrease in productivity. The third, *euhalophytes*, have a convex curvilinear increase in productivity with salinity up unto some maximum level and then have a curvilinear decreased productivity with increased salinity. (Figure 4) considers the theoretical relative productivity limits for Excluder, Accumulator and Conductor plants.

Our new classification of plants according to how the plants manage the salts considers that most *glycophytes* are excluders, but not exclusively so. With regard to the *miohalophytes* and *euhalophytes* there can be major paradigm shifts with regard to whether a plant's ability to grow in highly saline substrates affects its ability to remove salts from the system. Conceptually, some *miohalophytes* may be able to conduct significant quantities of salt, over a number of years, while in contrast some *euhalophytes*, which tolerate very high salinities, may remove little or no salt from the system.

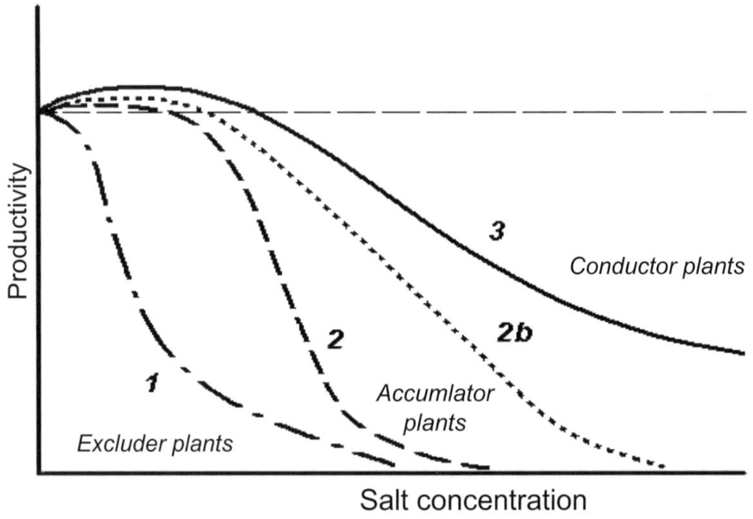

Figure 4. *Theoretical productivity of three types of plants with increasing salinity. Salt Excluder plants (1) typically perish in high salinity; Salt Accumulator plants (2, 2b) and Salt Conductor plants (3) often have the ability to survive in high salinity. The 2b shows the potential of some salt-accumulating plants to have high productivity at high salinity and have productivities similar to Conductor plants (3) even though they may accumulate up to 50% of dry weight as salt with insignificant release to the plant surface.*

Salt Excluder plants, such as beans (*Phaseolus spp.*), wheat (*Triticum spp.*), maize (*Zea spp.*), have a very rapid productivity drop rate with increasing salinity. Salt Accumulator plants, such as salt bush (*Atriplex spp.*), pickleweed (*Salicornia spp.*), seepweed (*Suaeda spp.*), are in general much more salt tolerant often due to their tissues and metabolic capacity functioning well under high salinity. Because salt-accumulator plants have evolved to tolerate high levels of tissue salt content, many are able to survive in remarkably high salinities, but still succumb to some threshold salt level above which their productivity drops precipitously even though they can survive the hypersaline conditions. From a practical consideration their accumulation of high levels of salts (some of which may be very toxic to animals) can make their use uneconomical and in some cases very negatively so.

In comparison to excluders and accumulators, salt-conductor plants, such as saltgrasses (*Distichlis spp.*), chord grass (*Spartina spp.*), black mangroves (*Avicennia spp.*), etc. have the advantage of maintaining low-salt levels in the shoot tissues (thus being palatable for animals) and some can grow under very high salinities to use this underutilized resource. Photosynthetic and non-photosynthetic (halosynthesis) energy of conductor plants gives the additional advantage of being able to function with either basal-cells mitochondrial energy and/or energy derived from salt glands (see below) to create a high flow rate from root to the shoot surface and maintain low volumes of salt solution within the plant.

2.5. *Intriguing productivity curves in a clonal conductor plant*

For nearly two decades we have conducted salinity trials with a fast-growing clone of *Distichlis spicata* var. *yensen-4a* (Yensen, 1985). Extensive experience with this euryhalic variety, which shows high productivities over a broad range of environmental conditions (Figure 5), demonstrated its advantages as a salt-conducting plant for soil-salt remediation work.

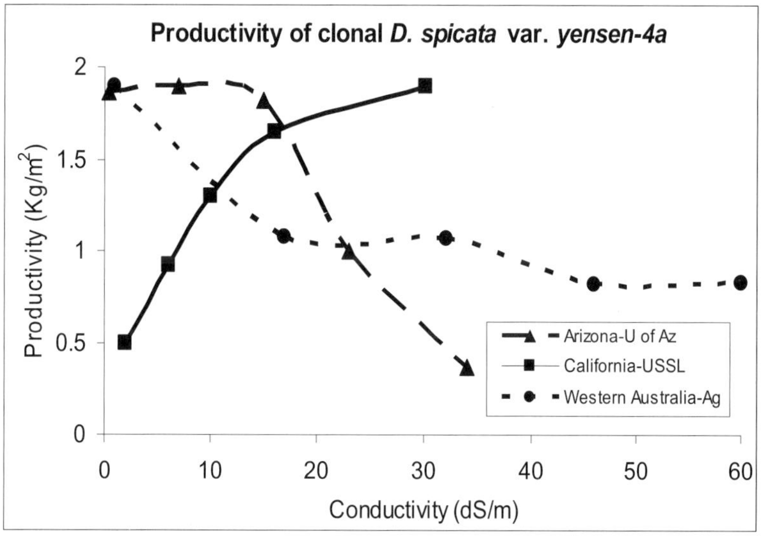

Figure 5. *Differing productivity-on-salinity behavior of clonal Distichlis spicata var. yensen-4a showing:* ▲ - ***threshold behavior*** *with high-light, warm, humid conditions (Arizona);* ■ - ***euhalophyte behavior*** *with high-light, hot, arid, sulfate-nutrient conditions (California); and* ● ***glycophyte-salt-tolerant behavior*** *with low-light, high-humidity, cool conditions (Western Australia) (Shannon, Greive, Barrett-Lenard, Perfumo, and Sergant, per. com.).*

Experimental work with this variety to better understand its full potential range of productivity under increasing salinity resulted in well-documented growth curves over salinity. The curves, however, under varying light, temperature and humidity conditions, showed differing productivity-on-salinity behaviors: i.e. similar to a *glycophyte*, a salt-tolerant *miohalophyte*, and a *euhalophyte* (Figure 5). Theses growth curves are so remarkably different that our disbelief prompted an investigation into the possible causes.

Consider the first trial in Tucson, Arizona (a two-month, open-air, high-light, hot-summer, humid/arid, rainy-season, modified-Hoagland's-nutrient solution study in the Sonoran Desert) a modest increase in productivity was found followed by a threshold point; where after a steep linear drop in productivity (Figure 5) resulted. This curve is typical of a threshold salt-tolerance type *halophyte*. This was not cause for concern.

A second trial (a one-month, closed-greenhouse, low-light, cool-winter, humid, modified-Hoagland's-nutrient solution study conducted in Perth by the Western Australia Agriculture Department) found an initial steep drop in productivity followed by a leveling out which extended into very high salinities (Figure 5). This is a curve more commonly associated with *glycophytic* C_3 plants that can tolerate high salinity by shutting down metabolic process to survive the higher salinities. Two other trials produced similar results under cool, low light; humid greenhouse conditions in Arkansas (Martin, 2003) and in humid, low-light growth chambers in Melbourne (Sargent per. com.). These results suggested that the first trial might have had some possible unexplained anomalies.

A fifth trial (a two-year, arid, open air, high-light, hot to cool, high-sulfate, drainage-water-analogue, nutrient-solution study conducted by the United States Department of Agriculture, Salinity Laboratory, in Riverside, California) found a dramatic increase in productivity with increasing salt levels up to salinities approaching ocean water (Figure 5). This growth curve is analogous to the productivities anecdotally observed for the plant in arid field situations around the world. This is a curve more commonly associated with obligate euhalophytic plants that optimize productivity in high salinities. This trial, conducted by highly reputable scientists (Grieve and Shannon) over a two-year period, was in stark contrast to other trials by conformed closely with our experience to the plant under field conditions.

The fact that material of the same clonal plant variety can have such varying productivities over increasing salinity appears astounding and suggested that we must be dealing with experimental error. Such a phenomenon has not been described in the literature. With regard to all greenhouse and seasonality trials calls into question the published results as a basis for suggesting potential productivity under different salt conditions.

Because these trials used different parameters for the productivity measurements, the absolute productivities may not be directly comparable. Still, the absolute productivities aside, the slopes of the curves clearly show how seriously the environmental conditions dramatically affect the productivity behavior of conductor plants to salinity.

These data infer that *Distichlis* may make efficient use of the available energy, particularly under arid, salty conditions. The mechanisms that allow saltgrasses to grow more rapidly in saline conditions are not understood.

We do know that most saltgrasses have bi-cellular salt glands on the shoot surface. There is an established correlation between salt-gland density and productivity under increasing-salt conditions (Marcum per. com.). It is also known that either higher temperatures or light levels can improve growth rates with increasing salinity (Kemp and Cunningham, 1981).

Our preliminary result with clonal *Distichlis spicata* trials under different environmental conditions, produced very different growth curves implies that an unknown mechanism is affected by environmental factors, salt-gland density, structure and physiological functions.

3. THE HALOSYNTHESIS HYPOTHESIS

Science advances through the process of observation followed by the developing and testing of hypotheses. Theories and laws are subject to change and revision, to wit, Newtonian physics changed dramatically by Arthur Eddington's measurement of light passing near the sun, heralding in the relativistic interpretation. Apropos to the present discussion, Einstein's Nobel-winning, corpuscular-photon, photoelectric effect changed our concept of light and paved the way for Feynmann's Nobel-winning electrochromodynamic.

The following hypothesis is not presented to presume such revolutionary changes to the paradigms of halophyte growth, but does advance hypotheses developed to satisfy an accumulation of contradictory observations regarding the growth of halophytes under stressful environments, and the discrepancy between observation and standard theory of halophytism. We would appreciate the reader's indulgence in considering that the following is presented to stimulate thought and discussion, and not necessarily to adequately describe the truth.

3.1. Concept description and terminology

The halosynthesis hypothesis is pure speculation, and to date is only supported by circumstantial evidence that may, or may not, potentially be explained through other known mechanisms. It is, however, an interesting hypothesis that could produce valuable results irrespective of its success as a correct hypothesis and may potentially lead to discoveries of new phenomenon related to the conversion of light energy without plant pigments and the protection of plants against destructive UV light (e.g. UV-C, $\lambda_{220\text{-}280\ nm}$).

Likewise, "halosynthesis-related" structures (epidermis cells, salt glands, trichomes, bladder cells, idioblasts, excreted salts or other compounds) may have important protective functions from certain destructive frequencies for light-sensitive metabolites and proteins, possibly mediated by Compton scattering (Compton, 1922, 1923) of destructive light-wave frequencies. For instance, the relationship of ultraviolet light and/or infrared wavelengths may have interactions (e.g. with shoots, in such ways as to utilize or store energies) that are not presently known. The potential destruction of organic compounds by certain light frequencies is analogous to how the frequency of a microwave oven quickly destroys flavonoid compounds in broccoli and other vegetables, (whereas steamed vegetables, when cooked to the same degree, lose few of their flavonoid compounds) (Garcia-Viguera, 2004).

Halosynthesis may utilize electromagnetic light energy to protect and/or benefit metabolic life functions. Certain aspects of this vision may potentially be applied to organisms other than plants and particularly in arid or non-electrical-conductive environments. Because the concept of halosynthesis (and associated concepts) is new and not yet tested, it is a field without extensive terminology. We have therefore taken some care in defining this term:

Halosynthesis is any salt-mediated (non-classical photosynthetic) mechanism (e.g. photon quanta or evapotranspiration related) by which energy from the environment is stored and/or utilized by organisms.

In the present case, described below, halosynthesis principally considers salt glands of plants and other relevant structures and organs. In salt-gland-bearing conductor *crinohalophytes* the process of halosynthesis may potentially benefit plants via one or more of the following processes: mechanical and osmotic pressure, evapotranspirational-hydraulic energy, photoelectric effect; Gaussian surface-charge distribution and photon/electron chain generation.

3.2. Hydraulic considerations and the salt gradient from soil to shoot surface

A generalized model of water and salt conduction from the soil to the salt glands, based upon the literature, considers that salt and water penetrate the root through the cells apoplastically, symplastically and/or transmembranally across temporary and permanent porous membranes (transvacuolar) (Newman, 1976; Zholkevich et al., 1989) to enter the xylem where the vascular tissue carries it upward. It is common knowledge that the vascular system of plants carries fluids from the roots to the shoots (xylem) and back to the roots (phloem). The mechanism of upward water flow in the xylem may be by passive (capillary and evapotranspirationally), active ("peristaltic" and hydraulic pumping) and/or by both (see Zholkevich et al., 1989). It is assumed that in a conductor plants the salt may be transported as salt water and via the same mechanisms within the xylem. Conductor-type plants may increase the hydrostatic pressure in the xylum by a peristaltic mechanism (per the "few-minute-contraction" form as opposed to the "diurnal" form) similar to that of *Zea mays* and *Helianthus annuus*.

From the xylum we suggest that the principal path of transport is apoplastically around the bundle sheath cells and then around the mesophyll cells to reach the epidermis, which may act as "salt collectors" and therefore as photoelectric collectors (see *Photoelectric effect* below).

Within the epidermal cells the salt and water may move apoplastically, symplastically and/or transmembranally to the basal cell of the salt glands. In the literature there is considerable information regarding different membrane tranporters for various salts in the roots, stems and leaf tissues. Because of the complexity of all the possible pathways for the various salts and fluids, we have created a simplified hydraulic model of fluid and salt transport in order to better understand the overall role of the salt glands with respect to the system as a whole.

A mechanistic theoretical model (Figure 6) of a conductor plants may further help to explain the relationship between sap pressure, salt conductance, water flow and volume. We may identify three hypothetical conductor plants: A. plants with an average number of salt glands, B. plants with below average number of salt glands and C. plants with above average number of salt glands. If we consider that the total conductive volume is equal for A, B, and C, we find that in:

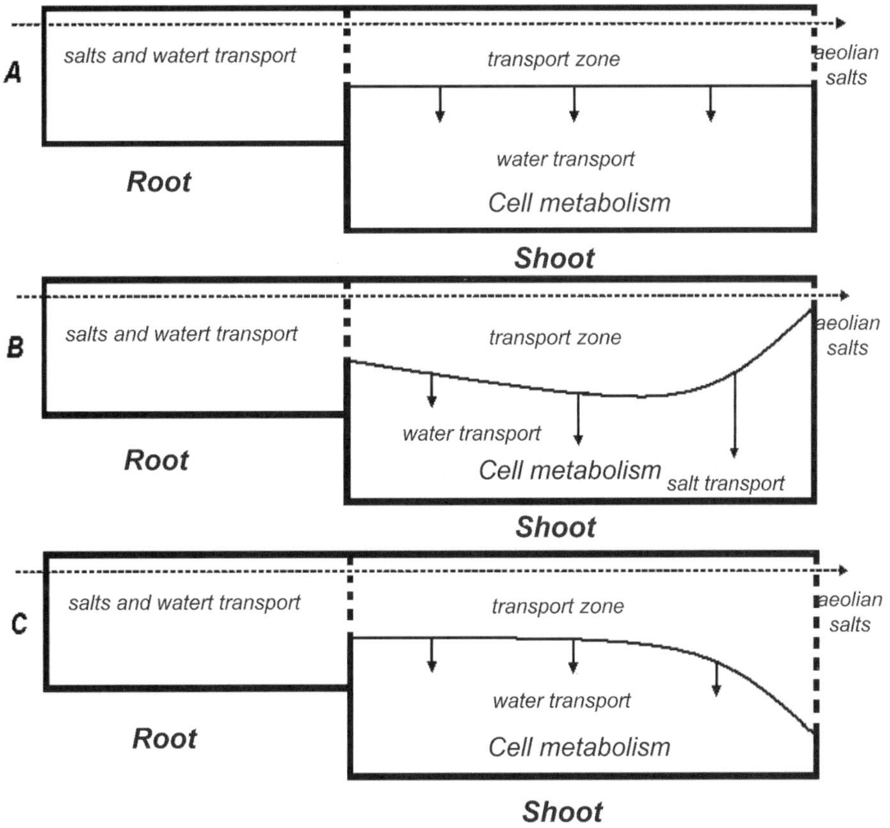

Figure 6. Diagrammatic representation of theoretical conductor plants: A. conductive rate of salt and water remain constant throughout the shoot; B. the conductive rate decreases at the salt gland potentially increasing salt entrance to the metabolic area; C. increased number of salt glands causing increased flow rate of salt and water.

A. The conduction flow rate is constant throughout the shoot resulting in average pressure in root (Figure 6A).
B. The conduction flow rate may be reduced throughout the shoot, resulting in an increased pressure in root and shoot, potentially resulting in reduced water and salt absorption in the roots and increased salt movement into the metabolic area (Figure 6B).
C. The conduction flow rate may be increased throughout the shoot, resulting in a decreased pressure in root and shoot and thereby a potential for an inverse salt pump.

From this simple diagram we can predict that plants with a greater number of salt glands may be better for both the goals of soil remediation and productivity due to reduced influx of salt into the metabolic area. The overall result is an increase in salt removal and remediation from the soil (3b in Figure 3).

The salinity in a conductor plant increases faster in a slower moving volume of water given the same stomatal conductance [note: conductor plants may also have a high or low volume with the same flow rate]. The flow rate may be limiting at the root uptake, entrance to the root-shoot conduction system, within the conducting zone, or at the salt glands. The salt-gland limitation may be dependent upon the number of glands and/or gland efficiency.

The conductance of any given volume through a conduction zone may be limited by a restriction at the entrance (Figure 6B), at the exit (Figure 6C) or at some place(s) within the conduction zone (not figured).

Both the relative volume of liquid entering or departing the conducting system may vary (as might the relative flow rates and associated energetic cost). In accord with (Figure 6B) a high transport-zone volume with restricted salt excretion may result in a relatively higher salt concentration in the transport zone. This in turn may either increase the salt level in the metabolic area or increase the cost to protect metabolism from salt intrusion. Such a cost may take several forms ranging from salt pumps and osmotica to a putative ammonia-ion protector (see below). A higher rate of salt-water departure from the transport zone (Figure 6C) may reduce the hydraulic pressure in both the transport zone. A reduced hydraulic pressure in the transport zone may result in a reduced pressure in the roots, which may facilitate the uptake of water from the soil. Also, the reduced hydraulic pressure in the shoot-transport zone may help to lower the cost for protection of the metabolic area and reduce the salt concentration within the shoot. Reduced shoot salt content has the potentially economic advantage of providing low-salt-content shoot tissue for animal forage.

The differential excretion of salt and the transpiration of water from the salt glands may affect the relative concentrations of salts and energetic costs in the above considerations. Likewise, an increased conductive volume may provide a larger potential volume available to the shoot-cell cytoplasm at a given salinity and flow rate. In order to provide an equivalent amount of water to shoot-cell cytoplasm under a reduced conductive volume, either a more rapid transport of salt water will be required or a higher energetic cost to move water against a greater osmotic gradient.

Some species and varieties of salt-conducting plants may have large underground root biomass, and significantly, may have an increasing salt-concentration gradient from soil to salt gland. This concentration gradient is determined by the existing soil water salt concentration versus the 100% solid salt crystals on the shoot surface. It should be clear that the salinity gradient would move water toward the increasing salt concentration, i.e. from root to shoot surface. It is likewise expected from this salt gradient that the diffusion of salt would be to move salt in the opposite direction, i.e. from shoot surface to the root, against the water movement.

At least two factors can prevent or reduce reverse salt movement from shoot surface to root: active fluid transport and unidirectional ports. The active fluid transport mechanism and pathways are mentioned and referenced (see Zholkevich et al., 1989) above. If the salt concentration in the immediate transport pathway between the roots and the epidermis and/or salt glands is isotonic the salt will move passively with the flow of the fluid (Figure 6A). Active transport in the roots and in the salt glands should also work synergistically at both ends of the transport zone to prevent the backflow of salt ions. The efficiency of preventing backflow could benefit from an increased number of salt glands at the shoot surface (Figure 6C). Unidirectional ports for salts may function at the root and salt gland salt-concentration gradient interfaces.

3.3. Salt glands and evapotranspirational halosynthesis

In the genus *Distichlis*, and other *crinohalophyte* grasses, their bi-cellular salt glands have a subdermal collector basal cells that takes up internal salts which move via symplastic transport to a hypodermal, excreting cap cells. The basal cell–cap cell interface is noted for having a high density of mitochondria closely associated with the area of plasmodesmata. Chloroplasts, however, are not known from either the salt gland cells or adjacent cells (see Liphschiz and Waisel, 1982).

It is generally assumed that the transport of salt from the basal collector cells to the excreting cells requires energy generated by the basal cell mitochondria. But the literature does not discuss how or what carbon substrate for Krebs cycle might reach the basal cells. Because the best potential source of a carbon substrate for the salt gland mitochondria is limited to only the mesophyll cells, we can predict that the products available to go to the salt glands are limited to the organic acids which are involved in the C_4 pump of the C_4-photosynthesis. In this case, the capacity of the C_4 pump and overall photosynthesis should be decreased. Therefore, we suggest that this transport of photosynthate to the basal cells during illumination of the plants may exist, but that it may be only temporary like a "starter motor" or that it is semi-catalytic to the overall process. Idealistically, salt glands should utilize and/or absorb solar energy from a source independent of mesophyll photosynthesis, if such a source existed.

We hypothesize that there is another energy source that may be tapped during, or before, the excretion process. For instance, in the excretory cap cell the evapotranspiration process itself may generate an energy gain; i.e., transpiration may result in an internal negative hydraulic pressure, resulting in useful potential energy at the plasmalemma between the basal collector cell and the excretory cap cell. Another possibility may be a result of the photoelectric effect.

3.4. The photoelectric effect

Due to the photoelectric effect, photons impinging on sodium ions (and other metals or compounds) will expel energized electrons. With sodium ions photons of sufficiently energetic light waves (λ = 220-550 nm) will expel electrons. These wavelengths are also potentially destructive to certain chemicals and pathways. Sodium in the epidermis will protect the underlying tissues from damaging UV-B and UV-C light (see Figure 10), both by converting the high energy waves into expelled electrons and by Compton scattering (Compton 1922, 1923) wherein the resulting wave lengths will be longer, less energetic and may be accepted by the pigment-protein light harvesting complexes of chloroplasts and/or others photoreceptors, such as phytochromes. At the same time the light can create electrons, which are critical to the photo-halosynthesis processes (see below).

3.5. Epidermal electro-halosynthesis

Sodium (and presumably other photoactive metals) is located in the shoot epidermis over virtually the entire light gathering surface (Koyro, 2003) potentially generating photoelectrically energized electrons (see above). These electrons, unable to move to the surface due to the non-conductive properties of the cuticular lipids, will move from cell to cell through the conductive cellular saline solution (see "hydraulic considerations" above) to the excretory cells and/or papillae. Upon reaching the surface they will assume a Gaussian electrostatic distribution. In this electron collection process, the current from the epidermal cells to the excretory cells may potentially be converting directly to stored energy when passing across the tonoplast and plasmalemma membranes (see below).

3.6. Salt-gland electro-halosynthesis

We suggest that the electrons produced by the photoelectric effect will move to the surface at the excretory cells, i.e. the salt glands and/or papillae. Because the lower and lateral surfaces of cap cells are typically covered with an insulating cuticle/lipid matrix, the electrons are further predicted to move to the non-cuticular distal surfaces (Figure 7) and to any externally excreted salt crystals.

This electrostatic negative charge distribution would ultimately be distributed in accord with Gauss' Law ($\Phi = EA \cos \theta$, where Φ is the electric flux through the surface, E is the electric field normal [perpendicular] to the surface, A is the area, and θ is the angle of the electric flux to the surface). Thus a negative electrostatic charge would be formed on the distal surfaces of the cap cells and papillae and would be more-or-less "equally" distributed on said "hemispherical" cap cells and papillae.

With a relatively strong electrostatic negative electron charge (albeit relatively low voltage) gathered from the surrounding leaf epidermis which will be concentrated at the tip of the salt glands (and/or at the tip of an attached crystal), and an electrostatic force may be exerted on the Na^+ ions (and other positive ions). This force may be calculated based upon the external voltage potential and the distance between the charged sodium ions within the surface and the distal surface of the salt gland/salt whisker where the negative electrostatic charge may reside, i.e. Force = $(k_e\ s\ c)/d$ (where $k_e = 8.987 \times 10^9\ N\text{-}m^2/C^2 = V^2$, s = surface electron charge, c = ion charge, d = distance). Simple calculations presuppose very strong forces at the small distances involved. Dimroth et al. (2000) indicate that as little as >50 mV potential is sufficient to drive the rotational ATP synthase structure. The photoelectric effect may produce electrons with up to 3000 mV.

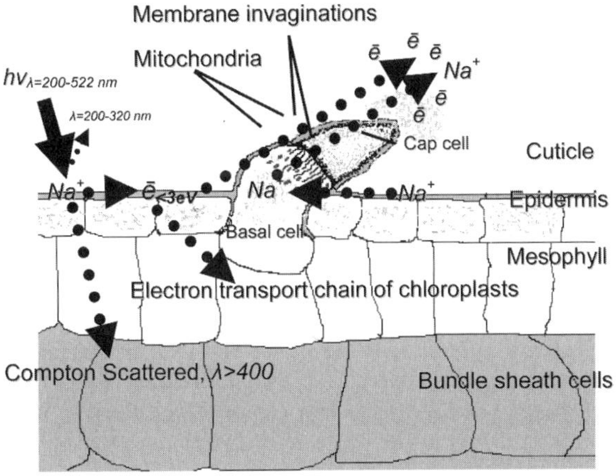

***Figure** 7. Diagrammatic representation of generalized salt gland and some of the hypothesized photo-halosynthesis and inverse sodium pump using ATP synthase molecules to generate ATP.*

Various forms of ATP synthase have been described since its remarkable structure and function was first identified and for which Paul Boyer, John Walker and Jens Skou were awarded a Nobel Prize (sees Boyer, 1993, 1997; Abrahams et al., 1994; Skou and Esmann, 1992). We propose a variant of ATP synthase that would have the stator and rotor assembly (F_1) on the basal-cell cytoplasm side of the plasmalemma membrane (Figure 8). This would allow production of ATP on the "plant-side" when the sodium or other cations move to the "excretory-side" of the membrane, thus keeping the useful ATP within the plant tissues.

Chlorine and other anions, due to their size and behavior are predicted to move toward the surface of a solution (Garrett, 2004), but may get an additional electrostatic boost from external cations during salt-crystal formation.

As the salt is excreted, as in the case of *Distichlis spp.* and other salt-gland-bearing species, the whiskers and crystals formed may facilitate or inhibit the electrostatic force depending on charge, direction of the whisker, crystal shape, humidity and other environmental factors.

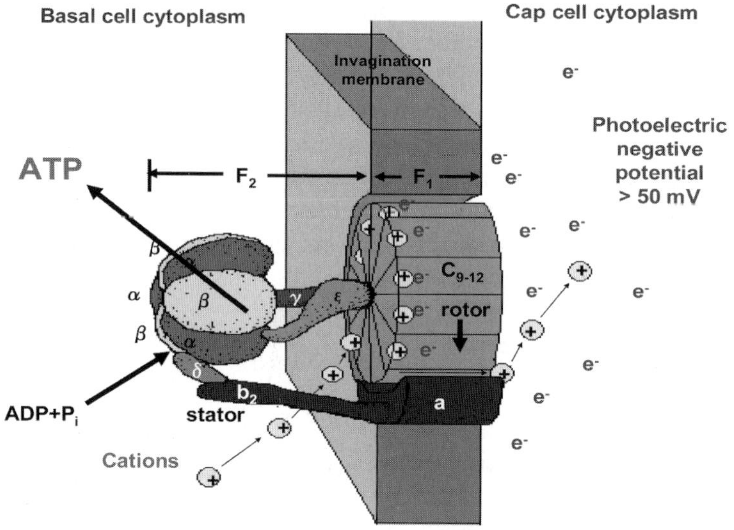

Figure 8. *Diagram of proposed "inverse sodium/cation pump." By arranging the ATP synthase* F_1 *unit (α, β, γ, ε, δ and* b_2 *stator) on the "plant-side" of the plasma membrane, the production of ATP will occur within the epidermal tissue. An external electrostatic charge > 50 mV is sufficient to drive the rotor for cation transport to the surface of papillae, cap cells, bladder cells or other charged structures. Note: the photoelectric effect on sodium ions should produce electrons with up to 3000 mV. The rotational torque of the rotor drives ATP off of the β subunit from where the ADP +* P_i *have been "stator-attached." (Modified from Boyer, 1993, 1997; Skou and Esmann, 1992; Abrahams et al., 1994; Dimroth et al., 2000)*

In appropriate conditions the salt whiskers (typically 2 μm to >500 μm long) have been observed to exceed several millimeters as crystals. It is speculated that, on occasion, the whiskers may be "tubular" analogous to the geological formation of hollow "soda straw" stalactites found in limestone caves wherein the salt water passes from the salt gland (and potentially thorough the salt-tube whisker) to the electrostatic charged tip. The salts would follow an osmotic and salt gradient maintained by the evaporation of water at the tip of the salt crystal where the highest salt concentration is expected to exist.

We predict that in dry air the evapotranspiration and electrostatic effects may be more pronounced, in part due to dry air's greater electrostatic capacitance and in part due to a greater evapotranspiration potential. Halosynthetically generated

ATP could help to explain the results of the different growth curves of the clonal *Distichlis* observed (Figure 5) under varying environmental conditions.

Under appropriate conditions, then, the electrostatic charge, and potentially in combination with hydrostatic pressure across a plasma membrane, may generate usefully energy. As the salt passes across the tonoplast, plasmalemma invaginations, and other membrane ports, in response to a >50 *mV* electropotential and/or hydrostatic potential, the energy may be converted to a biochemical energy storage product. As mentioned above (Liphschiz and Waisel, 1982) the centrifugal zone of the basal cells have a high density of mitochondria located near and surrounding in the area of the invagination of the plasmalemma and the plasmadesmata between the basal cell and the cap cell.

In conclusion, we predict the photoelectrically produced electrons from metallic salt cations in the epidermis will be intercepted by the basal cells before moving to the cap cells, papillae, bladder cells, and/or other surfaces, producing ATP. At that time, it is interesting to note that the electrons can reach the mitochondria located along the path to the cap cells. Some of these electrons may have the potential to reduce mitochondrial NAD^+ to NADH and/or convert directly into $\Delta\mu_{Na}^+$ and $\Delta\mu_H^+$; and indirectly into ATP (Figure 9). By these hypothetical schemes we present the possibilities of the conversion of photoelectric energy into biochemical storage products without classical photosynthesis.

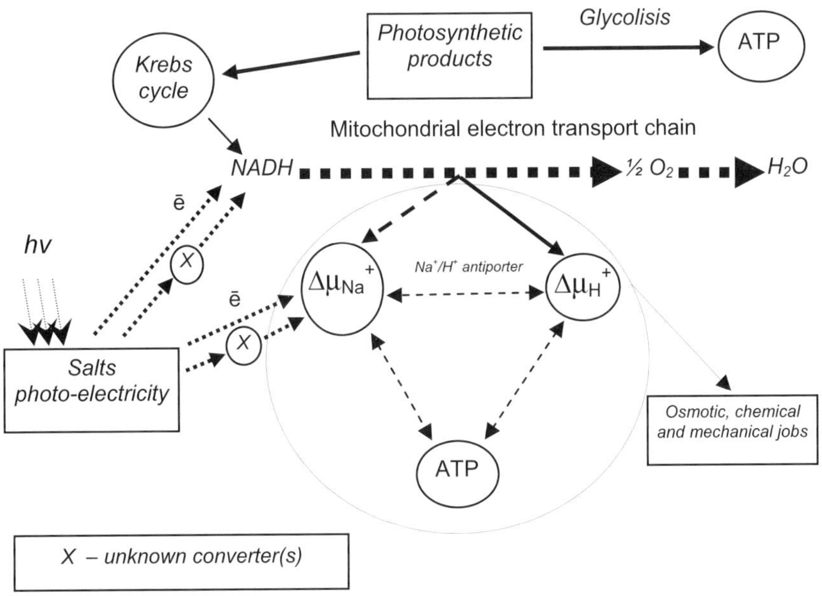

Figure 9. *Hypothetical scheme of photo-halosynthesis as a non-classical photosynthetic mechanism by which solar energy is utilized and/or stored by halophyte glands through the benefit saline conditions. (Note: the interconversion of $\Delta\mu_{Na}^+$, $\Delta\mu_H^+$; and ATP is well know in the literature, e.g. Skulachev, 1997; Semikhatova and Chirkova, 2001).*

3.7. co- and bio-chemical protection synergisms

We predict that all terrestrial plants may have similar protective mechanisms to defend themselves against high-energy photons. The epidermis cells are the primary protection layer against UV light. The protection may be provided by regular epidermal cells, trichomes, salt glands, cuticle, and other types of cells such as hypoderma, and hypodermal cells with calcium oxalate crystals (hypodermal idioblasts).

There is a steep gradient of decreasing light levels from the top to the bottom of the leaves (Cui et al., 1991; Vogelmann, 1993). Vogelmann (1993) found that the epidermal-cell-focusing of light might result in different light conditions for the cells and chloroplasts deep within a leaf. This may potentially result in destructive levels of light deep within the leaf and/or activate the production of oxygen radicals (in the electron transport chain of chloroplasts) that are likewise destructive (Bil' et al., 2003).

The physical protection mechanisms include, at least: reflection, Compton scattering, and absorption. And, we consider the absorption of UV light by metallic salts as perhaps the most important mechanism for the protection of plants against negative UV energy and possibly for the protection of animal skin as well (this question is beyond the scope of this paper, but is a very important consideration, and will be discussed by us in another article (Biel & Yensen, 2004, in preparation).

We thus assume that this may be a universal mechanism for many plants and animals, not only for protection, but also in the collection of photon energy (via conversion to electrons, Compton-generated lower-energy photons, surface charge and production of $\Delta\mu H^{+}$) in the photo-halosynthetic process.

In leaves, due to the sodium and other metallic salts, which may be sequestered in (or near) the epidermis, we find that the plants may receive protection against some of the destructive light (Figure 10). We would also predict that these metallic ions may serve to refocus and alter this potentially destructive UV light via the Compton effect resulting in longer, less energetic wavelengths, and more energized electrons.

Further the anatomical structure of the focusing apparatus should be such that they may take advantage of the reflective angle of the emissions in accord with $\lambda_f - \lambda_i = \Delta\lambda = h\,(1 - \cos\theta)/m_e c$ (where λ_f is the resultant wavelength, λ_i is the incident wavelength, h is Planck's constant = $6.626{\cdot}10^{-34}$ J-s, θ is the angle of divergence from the incident photon vector, m_e is electron rest mass and c is the speed of light in a vacuum) (Compton, 1922, 1923).

We propose these considerations because, as mentioned above, sodium and other metallic salts are primarily located in the epidermis (Koyro, 2003) and do three important things:

a) photo-electrically generate energized electrons via the halosynthesis processes (see above),
b) through Compton scattering may lengthen the wavelengths of destructive high-energy light to lower-energy non-destructive light more appropriate for

photosynthesis and photoregulation reactions, and, in the case of salt on animal surfaces, as non damaging to the animal body, and

c) via the preceding, protect the proteins, membranes, biochemical pathways and nuclear acids against destruction (Figure 10).

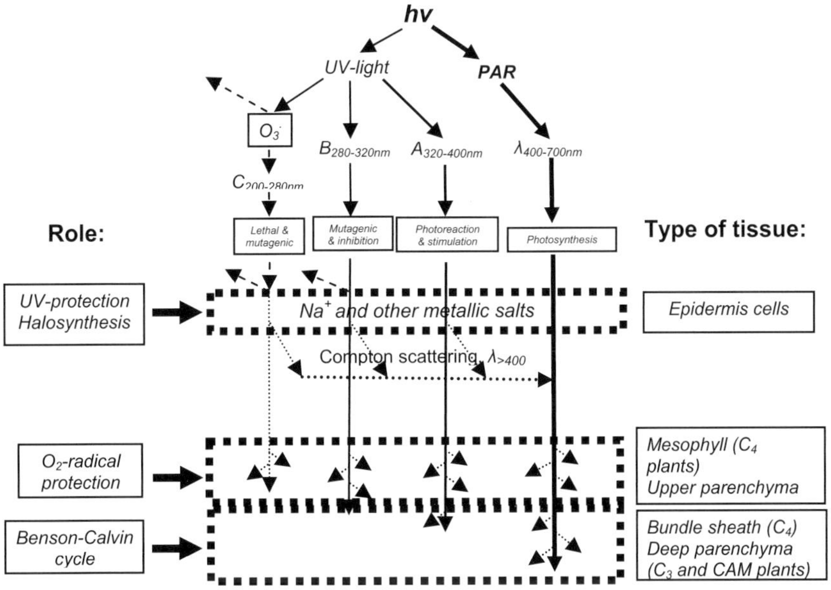

Figure 10. *Hypothetical scheme of UV- and excessive light resistance of photosynthetic organs of terrestrial plants (Modified from Bil', Fomina et al., 2003; Muzapharov et al., 1995). Abbreviations: PAR – photosynthetically active radiation.*

From this point of view it is easy to understand why C_4 plants should be less sensitive to UV destruction compared to mesophytic C_3 plants, as noted in the literature (Muzapharov et al., 1995), but not explained. Further, halophytes should have increased resistant to UV light when their epidermal tissues contain metallic salts. We would even expect this concept to apply to many non-halophytes, which have epidermal metallic salts. Also, some desert species (for example, *Haloxylon aphyllum, Suaeda acuminata*) have a vacuolized hypodermic layer and other species (for example, *Salsola dendroides*) may have calcium-oxalate crystals in this hypodermal layer of cells, which essentially are hypodermic idioblast layers (Biel, 1988). These hypodermal layers presumably function similar to the haloprotection mechanism of the epidermis cells.

3.8. Chemical protection against oxygen radicals

While photo-halosynthesis may utilize some of the incident light energy, there are many wavelengths that may penetrate further into the plant tissues irrespective of (and even due to) the halide elements, other metals, and other photoprotective compounds. Because these interceptive protections against destructive photons are not always sufficient, plants need "back-up" biochemical mechanisms to counter the potentially excessive light and effects of oxygen radicals formed in the chloroplasts. Some elements of this protection may include pigments (e.g. xanthrophyllic cycles), enzymes (e.g. super oxide dismutase [SOD], catalase, peroxidase), photorespiration pathway and some miscellaneous compounds (e.g. ascorbic acid, oxaloacetic acid [OAA], etc).

From our previous experience we have found that the upper part of mesophyll in the leaf of C_3 plants acts as a shield against excessive light for the deeper mesophyll. In particular, we have found that Rubisco activity near the surface of spinach leaves is minimal, but catalase activity is maximal. After flipping the leaf to an inverted position, both of these enzymes activity patterns also inverted (Soukhovolsky, Fomina et al., 2002; Bil', Fomina et al., 2003). It is logically assumed then that the low level of Rubisco near the upper surface of the leaf may be due to its destruction by active oxygen radicals.

Similar situations may exist in other and perhaps all terrestrial plant species. It would be clear then why Rubisco in C_4 plants is located in the bundle sheath cells. We speculate that the mesophyll of C_4 plants not only fixes atmospheric CO_2 and transports carbon via C_4-acids into the bundle sheath cells, but also function as a protector of the Benson-Calvin cycle and other important functions against the negative influence of oxygen radicals and excessive light (Figure 11).

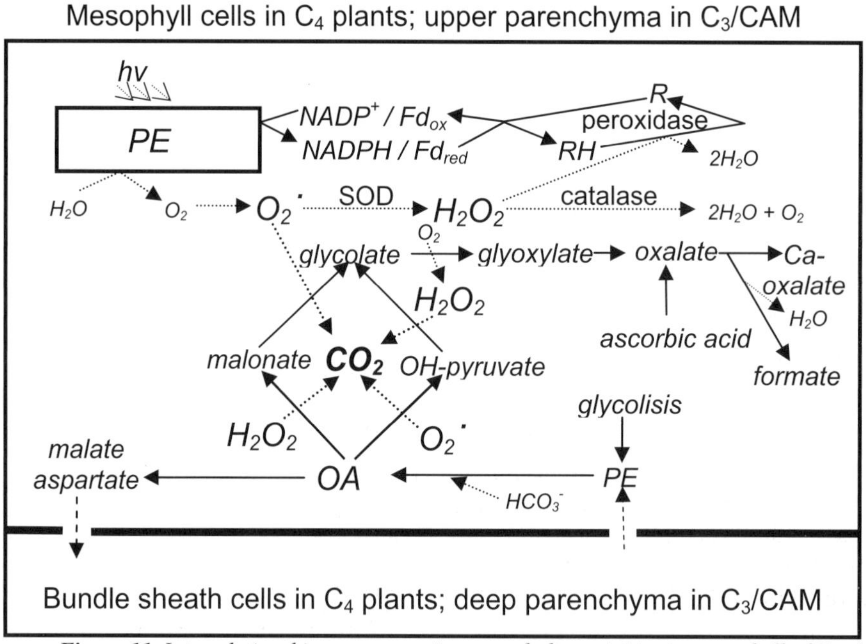

***Figure 11**. Interrelationships among various metabolic activities in mesophyll of C_4 plants and in upper mesophyll of C_3 and CAM plants related to stress tolerance of photosynthetic organs under high light illumination (Modified from Lyubimov et al., 1978; Bil' et al., 2003). Abbreviations: PET – photosynthetic electron transport chain; $F_{dox/red}$ – ferredoxin oxidized/reduced; PEP – phosphoenolpyruvate.*

We further speculate that this protection uses some portion of OAA, the first product of PEP-carboxylative reaction. It has long been shown (Luybimov et al., 1978) that, in the mesophyll cells of C_4 plants, up to 7% of the photosynthetically-formed OAA is oxidized by hydrogen peroxide and oxygen anion radicals to form glycolate, the first product in the photo-respiration pathway. We predict that the percent of OAA necessary to neutralize the oxygen radicals is not constant and may change with respect to various protective enzyme activities and with the light environment within the leaf.

Thus we have identified two protective mechanisms: 1) photo-haloprotection (against UV-light) located in the epidermis cells and/or in some other tissues, e.g. hypoderma, idioblasts, trichomes, scales, etc.; and 2) biochemical protection (against excessive light) located in the upper mesophyll in C_3 and CAM plants, and in the mesophyll of C_4 plants. These protective mechanisms allow optimization of the energy structure and function of the plant under varying light conditions.

4. A CASE STUDY, *DISTICHLIS*

Because conductor plants have been less well studied than excluders and accumulators, the following specifically examines what may be considered a quintessential halosynthesizing conductor plant, specifically plants of the genus *Distichlis*.

4.1. Ecophysiology

We have selected the genus *Distichlis*, a C_4 grass, as a plant that has increased productivity in saline environments and has the ability to conduct salts from soil to air. Preliminary studies indicate that *Distichlis* lines have varying capacities to conduct salt from the soil to the leaf surface for wind dispersal, halodispersion, a mechanism not well described or known.

Saltgrasses, such as *Distichlis,* may serve as productive crops under high salinities while removing salt from the upper soil horizons. How much salt is moved through the plant, and under what conditions, is not known, although estimates range from 2-150 tons/hectare/year. Anecdotal observations of salt whiskers and airborne salt dust suggest that, some percent of this salt becomes airborne. Aerially dispersed salt, to surrounding higher-elevation regions of salt-leached soils, may have the potential of stimulating productivity. It has been shown that many species, and in particular those having the C_4 photosynthesis, require sodium as a micronutrient (Brownell and Crossland, 1972).

4.2. Taxonomy and geographic distribution

Worldwide there are nine species in the genus *Distichlis* (*Poaceae*). The genus is placed by various authors in different subfamilies: *Festucoideae* (Hitchcock,

1950; Gould, 1951) or *Chloridoideae* (Texas A&M University herbarium; Marcum and Murdoch, 1994). Similarly, it has been placed in different tribes by various authors: *Festuceae* (Gould, 1951; Hitchcock, 1950) or *Aeluropodeae* (Texas A&M University herbarium).

Seven species occur in the Americas, one in Australia, and one in Sudan. The Australian species *Distichlis distichophylla*, however, is considered by Beetle (1955) a synonym of *Distichlis spicata* and *Distichlis sudanensis* (known from a single specimen) is most likely *Aeluropus massauensis*. Note, Yensen in agreement with Beetle (1943), treats *Distichlis stricta* as a subspecies of *Distichlis spicata* (NP Yensen per. obs.). Bermuda grass, *Cynodon dactylon*, is closely related and, like *Distichlis*, is in the subfamily *Festucoideae* (Hitchcock, 1950) or the subfamily *Chloridoideae* (Marcum and Murdoch, 1994), and like *Distichlis* it is in the tribe *Festuceae* (Hitchcock, 1950) or, unlike *Distichlis*, in the tribe *Chlorideae* (Avdulov, 1931 in Karpilov, 1970; Texas A&M University herbarium).

4.3. Root-soil restructuring capacity

Species of *Distichlis (Poaceae)* have strong penetrating rhizomes (see rhizocanicular percolation effect below) with the ability to open heavy-clay, deflocculated soils (Yensen, 2002; Leake, 2003). Further, they have the ability to grow aerobically within anaerobic sites due to the aerenchyma air passageways of the rhizomes and the roots (Leake, 2003). The oxygenation of black anaerobic mud is visible due to the brown iron oxide coloration that occurs around the roots. The rhizocanicular percolation effect of the rhizomes and roots physically opens up sealed soils such that water and gasses may penetrate more rapidly. The roots have air passageways (aerynchyma) and a ± biannual nature, which leaves the soil with holes full of organic matter. The organic material of the roots and rhizomes promotes biological activity on soil particle surfaces causing soil flocculation and complete restructuring of saline soils (Leake, 2003) to a depth of up to one meter or more in as little as three years (Yensen per. obs.).

In *Distichlis spp.* it is known that the rhizomes typically have a limited lifetime (annual to biannual) and when the internode regions atrophy the nodes produce new rhizomes. Over a number of years the opening of the soil, via new rhizomes and roots, results in channels of organic matter and can produce a dramatic increase in soil percolation rate, according to field observations (Yensen, 2002). Even with "cement-like" deflocculated soils the rhizocanicular effect facilitates percolation. For instance, in Mexico, Australia and USA (California) heavy-clay, deflocculated soils planted with *Distichlis spp.*, with water at 10-20 dS/m salinity, had increased percolation and salt leaching to the extent that the soils became amenable to fresh-water plants and earth worms (Yensen per. obs.). Other genera with rhizocanicular ability are: *Aeluropus spp., Agropyron spp., Agrostis spp., Arundo spp., Cynodon spp., Eragrostis spp., Jouvea spp., Monanthochloe spp., Odyssea spp., Pennisetum, Phragmites spp., Spartina spp.,* and *Sporobolus spp.* (Yensen per. obs.; Heuperman et al., 2002). But, perhaps the rhizocanicular effect phenomenon is better understood with *Distichlis* from which it was first described (Yensen and Bedell, 1995; Yensen et al., 1997; Leake, 2003).

4.4. Salt tolerance

Some populations of *Distichlis* are euhalophytes and have the ability to grow better in salty conditions than under fresh water conditions, while others do not have this ability, but are still salt tolerant. And, as noted above, within a single clonal line the environmental conditions can switch the plants from euhalophytic behaviour to a "salt-tolerant" behaviour, i.e. with reduced growth on increasing salinity (Yensen unpub.). It is not known how much of the salt tolerance of "salt-loving" varieties is due to whole plant structures and strategies and how much is due to the plant cell structures and metabolism. That is, it is not clear if the salt tolerance of *Distichlis* cells is due to its ability to keep salt out of the metabolic zones or if the cells are able to carry out total metabolism in the presence of salt ions.

4.5. Photosynthesis

Distichlis is a C_4 plant (Liphschitz and Waisel, 1982) with typical "Kranz" leaf anatomy. More than 30-35 years ago C_4 species were classified on basis of their first product, i.e. as a malic and/or aspartic type of C_4 plants (Karpilov, 1969; 1970; Black, 1973, *etc*), and, on basis of decarboxylating enzymes of C_4 acids, enzymes as NADP-ME (ME - malic enzyme), NAD-ME and PEP-carboxykinase in C_4 species (Edwards et al., 1971; Hatch et al., 1975; Edwards and Walker, 1983; etc.). It has also been found that the malate/aspartate ratio in the first photosynthetic products is not stable in C_4 plants and changes during leaf ontogenesis (Biel and Fomina, 1985; Biel, 1993), or may change due to environmental factors (Biel, 1988; 1993; Biel et al., 1990). It is important that, for example, even between different lines of corn there are some hybrids which have "aspartic-C_4 photosynthesis", and, as a result of that, these lines produce more protein in total biomass than the "malic-C_4 photosynthesis" lines (Biel, 1988; Biel et al., 1990).

A review of the literature (Liphschitz and Waisel, 1982) did not identify the first photosynthetic products for *Distichlis*. But, from our previous observations (Bil' and Gedemov, 1980; Bil' et al., 1982; Lyubimov et al., 1987; Biel, 1988; 1993) with some Kara-Kum desert plants, growing in saline soil, we hypothesize that, under high salinity, the C_4-photosynthesis of *Distichlis* will produce more aspartic acid than malic acid. And further, that protein production will increase over carbohydrate production. We believe that this phenomenon can correlate with the quality of the *Distichlis* grain and with total productivity of this genus and other C_4 desert plants.

4.6. Ammonia nutrition as a protector against salinity

In certain marine animals ammonia is used as osmotic protection against high salinity (Tam et al., 2003). It is also known that NH_3 is toxic to photosynthesis and other metabolic functions in plants. This toxicity can be eliminated if ammonia is maintained away from metabolic processes. For example, ammonia may be maintained in small quantities near the plasmalemma where its positive charge may serve to protect from Na^+ intrusion into the metabolic area and to maintain the sodium ion within the apoplastic area. We therefore speculate that, in *Distichlis* (and possibly other *halophilic* plants), ammonia may function as an electrostatic protector against toxic cations and a "binder" of anions. If verified, this would be a new approach to salt protection in plants.

Our preliminary data show that ammonia nutrition can change the rate of CO_2 photoassimilation and the quality of photosynthetic products and plant productivity compared to nitrate nutrition in C_4 species (Biel and Fomina, 1985; Bil' et al., 1987; Shabnova et al., 1997). That is, under ammonium nutrition a high aspartate/malate ratio is maintained. This is due to: 1) the stimulation of aspartate aminotransferase, 2) the inactivation of the enzymes that form and decarboxylate malate, 3) decreasing the C_4 pump activity, with the result of shifting the leaf metabolism to amino acid and protein synthesis (Biel and Fomina, 1985; Bil' et al., 1987; Biel, 1988, 1993).

Under nitrate nutrition, however, the aspartate/malate ratio decreases (Biel and Fomina, 1985; Bil' et al., 1985). Thus, ammonium nutrition has the overall effect of increasing the total nitrogen and protein content over that of nitrate nutrition in malic-C_4 species such as maize (Bil et al., 1987). Importantly, under ammonia nutrition the ammonium is evenly distributed throughout the plant parts. Under nitrate nutrition, however, virtually all the nitrate was concentrated in the stem (Shabnova, Fomina et al., 1993; Shabnova et al., 1997). Different genotypes of maize present varying responses to ammonia nutrition. Some responses are great (e.g. Early-ripening Sterling maize variety) and some responses are less (KVC-701 maize variety) (Biel et al., 1990).

Distichlis is taxonomically closely related to maize and, likewise, has shown to have varying responses to ammonia vs. nitrate nutrition (Yensen unpubl. field trials). We expect that the different *Distichlis* varieties, with their different salt tolerances will have different protein/carbohydrate ratios and productivities under ammonia vs. nitrate nutrition.

4.7. Soil salt removal and benefits to changes in soil properties

Successful remediation of salt-affected soils can be described as establishment of self-sustaining vegetation and the absence of migration of salts into objectionable locations (Yensen et al., 1999). It is further desired that the self-sustaining vegetation have beneficial value. Halophyte remediation trials in the Smackover Creek area of southwestern Arkansas have demonstrated that it is possible to take deflocculated, sealed soils with a pH ranging from 3.5 to 5.0, that have been heavily impacted with oil and brine to the extent that no plants have grown *in situ* for over 50 years and completely remediate the soil within a 3-5 year period using rhizocaniculating saltgrasses (Yensen et al., 1999). The trial involved 83 warm season and 33 cool season species, with less than 4-5 species showing remediation capabilities.

Certain rhizocaniculating saltgrass species, such as *Distichlis spp*., have penerhizomes allowing penetration through hard soils and extensive, deep roots with aerenchyma tissue, which contains spaces in the roots to permit gas exchange within anaerobic sites (Figure 12).

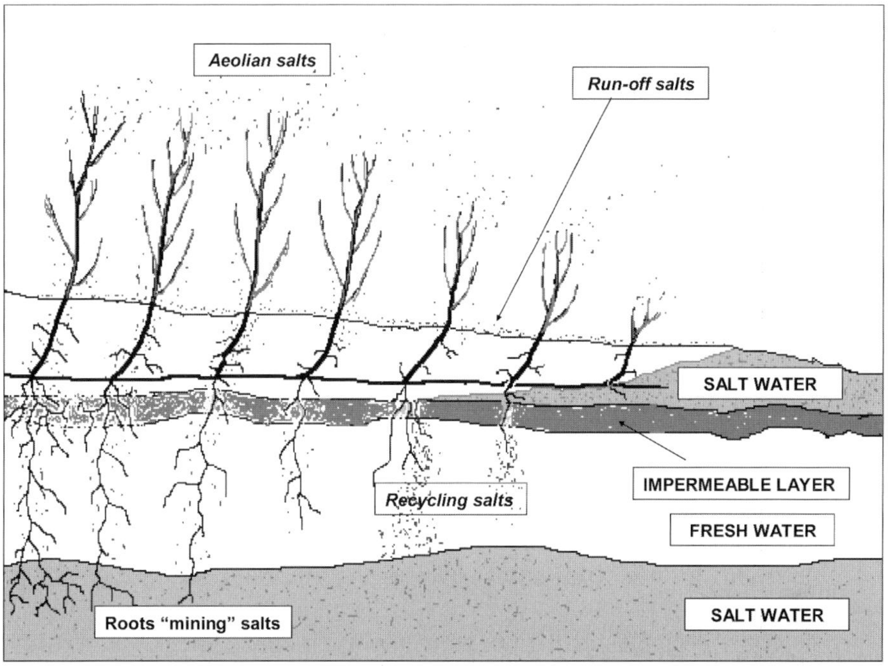

Figure 12. *Diagram of a generalized rhizocaniculating Distichlis plant. Sharp penerhizomes open heavy clay soils and even asphaltic soils in search of salt water. As the deeper aerenchymatous roots penetrate into anaerobic and impermeable aquacludes they facilitate the draining of pearched hypersaline water tables to be later recycled back to the plant. The roots may extend over 2 meters deep in active search of the deeper, heavier, hypersaline waters. The salt is then conducted to the leaf-blade surface where wind may disperse microscopic aeolian salt "whiskers" to the air. Some salt washes or falls back to the soil surface and is either washed away or, more commonly reabsorbed in situ by the vegetated surface.*

The rhizomes are biannual and leave the soil full of channels of organic matter, which allows bacteria to re-aggregate the clay particles (D. Beck, pers. comm.). The presence of organic matter will maintain aggregation of soil particles (I. Labron, pers. comm.).

5. CONCLUSIONS, PREDICTIONS AND POTENTIAL TESTS OF HYPOTHESES

5.1. Salt conduction

The convergent evolution of salt tolerance has allowed many species to persist, and in some cases thrive, in saline environments. The unique ability of certain of these plants to act as conductors of salts through their vascular system and to the shoot surface has given them significant advantages over other plants which can only survive by excluding the salt, or over those plants which accumulate salt in vacuoles until a limit of salt is reached.

Although conductor plants range from minimal to maximal ability to pass salts from roots to the shoot surface, we predict that:

- Conductor lines are the best candidates for soil remediation, forage and other products,
- Conductor lines have the ability to conduct "unlimited" amounts to salt from the soil to their shoot surfaces and into the air and they do not accumulate salt in their tissues,
- The potential presence of ammonia-nutrition-protection in the "root → transport → salt-gland" system protects against the toxic effects of salts on the cytoplasm and metabolism of cells.

The predictions of this concept will determine:

- the salt uptake from soils,
- quantification of the potential amount of salt removed from soil,
- salt movement through the plants, excretion and the aeolian dispersal,
- the specific environmental conditions (including protective nitrogen nutrition) for optimization of soil remediation, salt conductance and plant productivity,
- degree of soil re-flocculation, permeability, humic content, and fertility,
- the percentage of salt that becomes airborne, re-enters the soil, or is washed away, with regard to specific weather and surface conditions.

At this time we have the best-known lines of conductor grasses for the development and realization of evaluating the salt-conductor plant strategy. We recommend the testing of the above concepts using specific salt-excreting halophytes, such as four very closely related varieties, of *Distichlis spp.* or other specific conductor lines, which can verify the above predictions.

5.2. Halodispersion

Although this has not been studied, we predict that under certain conditions over 50% of the excreted salt may become airborne. In the case of certain lines of the saltgrass, *Distichlis*, we estimated that 5-50 tons/ha/year could be dispersed.

The concept of halodispersion is new and estimates of salt uptake and aerial dispersal is presently not available for any plant. Further application of this principal and associated knowledge may contribute to the remediation of soils in many parts of the world.

5.3. Metabolism

Work based upon these hypotheses may answer some fundamental questions about the effect of high salinity on metabolism and mechanisms important to saltgrass lines and/or other crop plants. The realized knowledge of these predictions may facilitate the selection of other commercial forages for practical application, e.g. role of the "ammonia protection" and optimization of protein for dairy vs. carbohydrate for meat production with minimal accumulated salts. Future cooperation with geneticists could create additional information useful for the development of more salt-tolerant glycophyte crops, such as maize, wheat, barley, oats, rye, rice, *etc.* and may provide information regarding nutritional and anti-nutritional characteristics of plants, seeds, and storage products in salty environments.

The study of the conductor phenomenon has led to predictions of basic physiological concepts and principals that were heretofore unknown. The ability of "salt conducting plants" with "salt membrane protection" to handle salt without negative effect on the growth and production of the plants can allow the dual benefits of remediating soils while achieving the production of photosynthetic products assisted by "halosynthesis."

5.4. Protection

From the above descriptions we can extrapolate that similar protection strategies may be utilized by other plants. For example, high altitude plants should resist high levels of UV light energy, and we predict that such plants use sodium and/or other elements and compounds in, or near, the epidermis for protection of the mesophyll against destruction. By contrast aquatic plants would not be expected to use or need similar protective strategies due to the near complete absorption of UV light in the first few centimeters of water. Protection against excessive light may also be manifested by biochemical processes located in the upper mesophyll in C_3 and CAM plants, in the mesophyll of C_4 plants, resulting in optimal function of the plant under varying light conditions.

5.5. Halosynthesis

If our hypothesis of halosynthesis is proven correct it would represent the second mechanism by which life can access energy from sunlight and the third from the environment. The first and best known mechanism is photosynthesis while the second discovered environmental energy source being that of thiobacteria which can extract heat energy from undersea hot-water vents to fix and reduce carbon.

These new concepts of haloconduction, haloaspersion, halosynthesis, destructive light protection and protection of metabolism from salts, and their predictions should provide value to researchers, farmers and politicians involved in a myriad of global salinization and other anthropogenic problems, and potential opportunities.

6. REFERENCES

Abrahams, J.P., Leslie, A.G., Lutter, R. & Walker, J.E. 1994 Structure at 2.8 Å resolution of F_1 - ATPase from bovine heart mitochondria. Nature 370: 621-628

Badger, M.R. & Price, G.D. 2003 CO_2 concentrating mechanisms in cyanobacteria: molecular components, the diversity and evolution. Journal of Experimental Botany 54(383): 609-622

Beetle, AA. 1943 North American variations of *Distichlis spicata*. Bull. Torr. Bot. Club 70(6): 638-650

Beetle, A.A. 1955 The grass genus *Distichlis*. Revista Argentina de Agronomía 22(2): 86-94

Biel, K.Y. 1988 Phototrophic carbon metabolism in plants with differential organization of photosynthetic apparatus (*Dr. Sc. Thesis*), Pushchino [in Russian] 543 pp.

Biel, K.Y. 1993 Ecology of photosynthesis. Moscow "Nauka" [in Russian], 221 p.

Biel, K.Y. & Yensen, N.P. 2004 The hypothesis of photo-halosynthesis and its application for some agricultural and medical problems (in preparation)

Bil', K.Y. & Gedemov, T. 1980 The Structural and functional properties of assimilation apparatus in *Suaeda arcuata* Bunge (*Chenopodiaceae*) related to cooperative photosynthesis. Proceedings of the Academy of Sciences of the USSR 250(4): 9-13

Bil', K.Y., Fomina, I.R. & Nishio, J.N. 2003 Stress tolerance and light dependent organization of carbon metabolism within spinach leaves. In: New and Nontraditional Plants and Prospects of Their Utilization. *Materials of 5thInternational Symposium*, Moscow-Pushchino, Russian University of People Friendship, Moscow 3: 24-27

Bil', K.Y., Fomina, I.R., Nazarova, G.N. & Nishio, J.N. 2003 Light dependent organization of carbon metabolism within spinach leaf. Physiology and Biochemistry of Cultivated Plants 35(5): 392-402

Bil', K.Y., Fomina, I.R. & Tsenova, E.N. 1985 Effects of nitrogen nutrition on photosynthetic enzymes activities, type of photosynthates, and photosystem 2 activity in maize leaves. Photosynthetica 19(2): 216-220

Bil', K.Y., Fomina, I.R. 1985 Chlorophyll-protein complexes from green algae and higher plants. On the role of metabolic regulation of the photosystem 2 complex formation in bundle sheath cell chloroplasts of C_4 plants. Biochemistry-USSR 50(2): 219-224

Bil', K.Y., Fomina, I.R. & Shabnova, N.I. 1987 Content of nitrogen compounds in Maize shoots, structural and functional features of photosynthetic apparatus under ammonium and nitrate nutrition. Photosynthetica 21(4): 525-534

Bil', K.Y., Fomina, I.R., Shabnova, N.I., Terekhova, I.V. & Tsenova, E.N. 1990 Influence of ammonium nitrogen nutrition and conditions of illumination on activity of enzymes of photosynthetic carbon metabolism in corn of different genotypes. Biochemistry-USSR 54(7): 920-925

Bil', K.Y., Lyubimov, V.Y., Demidova, R.N. & Gedemov, T. 1982 Assimilation of CO_2 by plants of the family *Chenopodiaceae* with three types of autotrophic tissues in the leaves. Soviet Plant Physiol 28(6): 808-815

Black, C.C. 1973 Photosynthetic carbon fixation in relation to net CO_2 uptake. American Review on Plant Physiology 24: 253-287

Boyer, P.D. 1993. The binding change mechanism for ATP synthase - some probabilities and possibilities. Biochimica et Biophysica Acta 1140:215-250

Boyer, P.D. 1997. The ATP synthase - a splendid molecular machine. Annals Review on Biochemistry 66: 717-749

Brownell, F.P. & Crossland, C.J. 1972. The requirement for sodium as micronutrient by species having the C_4 dicarboxylic photosynthesis pathway. Plant Physiology 49: 794-797

Compton, A.H. 1922. Total reflection of X-rays from glass and silver. Physics Review 20: 84

Compton, A.H. 1923. A quantum theory of the scattering of x-rays by light elements. Physics Review 21(5): 483-502

Cui, M., Vogelmann, T.C. & Smith, W.K. 1991. Chlorophyll and light gradients in sun and shade leaves of *Spinacia oleracea*. Plant Cell Environment 14: 493-500

Dimroth, P., Kaim, G. & Matthey, U. 2000 Critical role of the membrane potential for ATP synthesis by F_1F_0 ATP-synthases. Journal of Experimental Biology 203: 51-59

Edwards, G.E., Kanai, R. & Black, C.C. 1971. Phosphoenolpyruvate carboxykinase in bows of certain plants which fix CO_2 by the C_4–dicarboxylic acid cycle of photosynthesis. Biochem. Biophys. Res. Commun 45(2): 278-285

Edwards, G.E. & Walker, D. 1983. C_3, C_4: mechanisms, and cellular and environmental regulation, of photosynthesis. Blackwell Scientific Publications. Oxford-London-Edinburgh-Boston-Melbourne

Garrett, B.C. 2004. Ions at the air/water interface. Science 303:1146-1147

Gould, F.W. 1951. Grasses of southwestern United States. University of Arizona Press, Tucson, 352 pp.

Hatch, M.D., Kagawa, T. & Craig, S. 1975. Subdivision of C_4 pathway species based on differing C_4 acid decarboxylating systems and ultrastructural features. Australian Journal of Plant Physiology 2: 111-128

Heuperman, A.F., Kapoor, A.S. & Denecke, H.W. 2002. Biodrainage: principals, experiences and applications. FAO, Rome, 79 pp.

Hitchcock, A.S. 1950. Manual of the grasses of the United States. USDA publ. 20, Washington, D.C., 1051 pp.

Karpilov, Y.S. 1969. Specificity of the structure and functions of the photosynthetic apparatus in some tropical plant species. Proceedings of the Moldavian Institute for Research in Irrigation Farming and Vegetable Growing, Kishinev (Special Issue) [in Russian] 11: 1-35

Karpilov, Y.S. (editor), 1970. Cooperative photosynthesis of xerophytes. Proceedings of the Moldavian Institute for Research in Irrigation Farming and Vegetable Growing, Kishinev [in Russian] 11(3): 67

Kemp, P.R. & Cunningham, G.L, 1981, Light, temperature and salinity effects on growth, leaf anatomy, and photosynthesis of *Distichlis spicata* (L.) Greene. American Journal of Botany 68(4): 507-516

Koyro, H.W. 2003. Study of potential cash crop halophytes by a quick check system: Determination of the threshold of salinity tolerance and the ecophysiological demands. In: Cash Crop Halophytes: Resent Studies. 10 Years after the Al Ain Meeting, Eds: Helmut Lieth and Marina Mochtchenko, Kluwer Academic Publishers (Dordrecht-Boston-London): 5-17

Leake, J.E. 2003. The role of NyPa *Distichlis spp.* cultivars in altering groundwater and soil conditions – a work in progress. *Salinity Seminar,* Hydrological Society of South Australia, Australian Geomechanics Society, South Australian Chapter, and International Association of Hydrogeologists, Australian National Chapter, North Adelaine, 14 July, 1-8

Liphschiz, N. & Waisel, Y. 1982. Adaptation of plants to saline environments: salt excretion and glandular structure. In: Tasks for vegetation science, Contribution to the ecology of halophytes, Junk Publ., The Hague 2: 197-214

Lyubimov, V.Y., Atakhanov, B.O. & Bil', K.Y. 1987. Adaptation of photosynthetic apparatus to extreme conditions in plants of the Kara-Kum desert. Soviet Plant Physiology 33(5): 680-686

Lyubimov, V.Y., Biel, K.Y. & Karpilov, Y.S. 1978. Peculiarities of the glycolate formation in assimilation tissues of C_4 plants. In: Plant Physiology. Materials of the 6th National Conferences of Bulgarian Academy of Sciences, Sofia 4(1): 41-56

Marcum, K.B. & Murdoch, C.L. 1994. Salinity tolerance mechanisms of six C_4 Turfgrasses. Journal Amer. Soc. Hort. Sci. 119(4): 779-784

Martin, R. 2003. The salinity tolerance of six bermudagrasses (*Cynodon dactylon* (L.) Pers., one dropseed (*Sporobolus virginicus* (L.) Kunth) and one saltgrass (*Distichlis spicata* var. yensen-4A) for phytoremediation of salt-affected soils in oil, gas and brine fields. M.S. thesis, Louisiana Tech Univ., Ruston, 66 pp.

Muzapharov, E.N., Kreslavsky, V.D. & Nazarova, G.N. 1995. Light- and hormone-dependent regulation of photosynthesis and plants growth. Pushchino, ONTI PSC RAS [in Russian] 140 pp.

Newman, E.I. 1976 Water movement through root systems. Philos. Trans. Roy. Soc., London 273(927): 463-478

Semikhatova, O.A. & Chirkova, T.V. 2001. Physiology of plants respiration. Issued by S-Petersburg's University [in Russian] 220 pp.

Shabnova, N.I., Fomina, I.R. & Biel, K.Y. 1997. The productive process and qualitative features of overhead mass of corn plants with regard to a source of nitrogen nutrition. Journal of Agricultural Biology [in Russian] 1: 57-63

Shabnova, N.I., Fomina, I.R., Biel, K.Y. & Stakhova, L.N. 1993. The amino acid composition of the maize green mass in connection with the conditions of nitrogen nutrition. Physiology and Biochemistry of Cultivated Plants [in Russian] 25(4): 346-352

Skou, J.C. & Esmann, M. 1992. The Na, K-ATPase. Jour. Bioenergetics and Biomembranes 24: 249-261

Skulachev, V.P. 1997. Laws of bioenergetics. Soros Educational Journal [in Russian] 1: 9-14

Soukhovolsky, V.G., Fomina, I.R., Bil', K.Y., Nishio, J.N. & Khlebopros, R.G. 2002. An optimization model of the photosynthetic leaf: the model of optimal photosynthetic CO_2 fixation within leaves of mesophytic C_3 plants. Doklady (Proceedings) Biological Sciences of Russian Academy of Sciences 382: 28-30

Tam, W.L., Wong, W.P., Loong, A.M., Hiong, K.C., Chew, S.F., Ballantyne, J.S. & Ip, Y.K. 2003. The osmotic response of the Asian freshwater stingray (*Himantura signifier*) to increased salinity: a comparison with marine (*Taeniura lymma*) and Amazonian freshwater (*Potamotrygon motoro*) stingrays. The Journal of Experimental Biology 206(17): 2931-2940

Ungar, I. 1991. Ecophysiology of vascular plants. CRC Press, Boca Raton, Florida 209 pp.

Vallejo, F., Tomás-Barberán, F.A. & Garcia-Viguera, C. 2003. Phenolic compounds content in edible parts of broccoli inflorescences after domestic cooking. Jour. of the Science of Food and Agriculture 83(14)

Vogelmann, T.C. 1993. Plant tissue optics. Annals Review on Plant Physiology and Plant Molecular Biology 44: 233-251

Yensen, N.P. 1985. Saline agriculture and new halophyte crops. Proc. of the 61st annual meeting of the Southwestern and Rocky Mountain Division, AAAS, 19-23 March, 27

Yensen, N.P. 1995. International Symposium on high salinity tolerant plants summary of papers presented. In: Biology of Salt Tolerant Plants (M.A Khan & I.A Unger eds.), Ohio University, Athens: 1-12

Yensen, N.P. 2002. New developments in the world of saline agriculture. In: Proceeding of Prospects for saline agriculture, an international seminar, Islamabad, Pakistan; Pakistan Academy of Science, Pakistan Agricultural Research Council and COMSTECH, 10-12 April, 2000

Yensen, N.P. & Bedell, J.L. 1995. The use of saline water for forage production in coastal ecosystems. Abstracts of International Symposium on Salt-Affected Lagoon Ecosystems, Valencia, Spain, 18-25 September, 221-222

Yensen, N.P., Bedell, J.L. & Yensen, S.B. 1997. The use of saline water for forage production in coastal ecosystems. In: International Symposium on Salt-Affected Lagoon Ecosystems, Jorge Batlle Sales (ed.), Universitat de Valéncia and International Soil Science Society, Spain, 18-25 September, 297-307

Yensen, N.P., Hinchman, R.R., Negri, M.C., Mollock, G.N., Settle, T., Keiffer, C.S., Carty, D.J., Rodgers, B., Martin, R. & Erickson, R. 1999. Using halophytes to manage oilfield saltwater: disposal by irrigation/evaporation and remediation of spills. *In:* Proc. Sixth international petroleum environmental conference: environmental issues and solutions in petroleum exploration, production and refining, K.L Sublette (ed.), 16-18 November, Houston

Yensen, N.P., Yensen, S. & Weber, C.W. 1988. A review of *Distichlis* spp. for production and nutritional values. In: Arid Lands: Today and Tomorrow, E.E. Whitehead, C.F. Hutchinson, B.N. Timmermann & R.B. Varady (eds.), Proceedings of an international research and development conference, 20-25 October, 1985, West view, Bolder, Colorado; Belhaven, London, 809-822

Zholkevich, V.N., Gusev, N.A., Kaplya, A.V., Pakhomova, G.E., Pilshchikova, N.V., Samuilov, F.D., Slavnii, P.C. & Shmatko, I.G. 1989. Water Exchange in plants. Moscow "Nauka" [in Russian], 256 pp.

CHAPTER 22

MECHANISMS OF CASH CROP HALOPHYTES TO MAINTAIN YIELDS AND RECLAIM SALINE SOILS IN ARID AREAS

HANS-WERNER KOYRO[1, 3], NICOLE GEISSLER[1], SAYED HUSSIN[1] AND BERNHARD HUCHZERMEYER[2]

[1]*Institute for Plant Ecology, Justus-Liebig-University of Giessen, D-35392 Giessen,*
[2]*Botany Institute, Plant Developmental Physiology and Bioenergetics, Hannover University, D-30419 Hannover*
[3]*Corresponding author: Hans-Werner.Koyro@bot2.bio.uni-giessen.de*

1. INTRODUCTION

About 7% of the world's total land area is affected by salt, as is a similar percentage of its arable land (Ghassemi et al., 1995; Szabolcs, 1994) when soils in arid regions of the world are irrigated, solutes from the irrigation water can accumulate and eventually reach levels that have an adverse affect on plant growth. Of the current 230 million ha of irrigated land, 45 million ha are salt-affected (19.5 percent) and of the 1,500 million ha under dryland agriculture, 32 million are salt-affected to varying degrees (2.1 percent). There are often not sufficient reservoirs of freshwater available and most of the agronomically used irrigation systems are leading to a permanent increase in the soil-salinity and step by step to growth conditions unacceptable for most of the conventional crops. Significant areas are becoming unusable each year. It is a worldwide problem, but most acute in Australasia (3.1 million hectars), the Near East (1.802 million hectars) and Africa (1.899 million hectars), North and Latin America (3.963 million hectars) and to an increasing degree also in Europe (2.011 million hectares of salt-affected soils; FAO Land and Plant Nutrition Management Service). Although careful water management practices can avoid, or even reclaim damaged land, crop varieties (such as cash crop halophytes) that can maintain yields in

M.A. Khan and D.J. Weber (eds.), Ecophysiology of High Salinity Tolerant Plants, 345-366.

saline soils or allow the more effective use of poor quality irrigation water will have an increasing role in agricultural land use in near future.

In contrast to crop plants, there exist specialists that thrive in the saline environments along the seashore, in estuaries and saline deserts. These plants, called halophytes, have distinct physiological and anatomical adaptations to counter the dual hazards of water deficit and ion toxicity. Salinity can affect any process in the plant's life cycle, so that tolerance will involve a complex interplay of characters. Research projects investigated details of the physiology and biochemistry of salt tolerance and also searched for methods to screen overall plant performance that could be used in breeding programs.

The sustainable use of halophytic plants is a promising approach to valorize strongly salinised zones unsuitable for conventional agriculture and mediocre waters (Boer & Gliddon, 1998; Lieth et al., 1999). There are already many halophytic species used for economic interests (human food, fodder) or ecological reasons (soil desalinisation, dune fixation, CO_2-sequestration). However, the wide span of halophyte utilisation is not jet explored even to a small degree.

2. DEFINITIONS OF THE TERMS HALOPHYTE AND SALINITY TOLERANCE

Halophytes are plants, able to complete their life cycle in a substrate rich in NaCl (Schimper, 1891). One of the most important properties of halophytes is their salinity tolerance (Lieth, 1999). This substrate offers for obligate halophytes advantages for the competition with salt sensitive plants (glycophytes). There is a wide range of tolerance among the 2600 known halophytes (Pasternak, 1990; Lieth & Menzel, 1999). However, information about these halophytes needs partially careful checking. A precondition for a sustainable utilisation of suitable halophytes is the precise knowledge about their salinity tolerance and the various mechanisms enabling a plant to grow at (their natural) saline habitats (Marcum, 1999, Warne et al., 1999; Weber & Dántonio, 1999; Winter et al., 1999). The many available definitions especially for salinity tolerance or threshold of salinity tolerance impede a uniform description and complicate the comparison between species:

a) Phytosociologists are using this term only for plants growing natural in saline habitats. In order to get first information on salinity tolerance phytosociological vegetation analysis is very helpful and salinity tolerance numbers are widely applied for qualitative approximations (Ellenberg, 1974; Landolt, 1977).
b) Another group of scientists describes salinity tolerance with polygonal diagrams of the mineral composition in the plants (see literature in Kinzel, 1982; Marschner, 1995).
c) The threshold level of salinity tolerance is described in a further definition as the point (salt concentration) when the ability of plants to survive and to

reproduce is no longer given (Pasternak, 1990). This definition is a modification of the definition of halophytes presented above. Survival and reproduction of a plant are not always impeded at the same salinity level (Tazuke, 1997). However, the definition of Pasternak (1990) is still important for the interpretation of the ecological dissemination and can be used as a solid basis for physiological studies concerning the survival strategies of plants.

d) Generally, classification of the salinity tolerance (or sensitivity) of crop species is based on the threshold EC (electrical conductivity) and the percentage of yield decrease beyond threshold (Greenway & Munns, 1980; Marschner, 1995). Salinity tolerance is usually assessed as the percentage biomass production in saline versus control conditions over a prolonged period of time (Munns, 2002). The substrate-concentration leading to a growth depression of 50 % (refer to fresh weight, in comparison to plants without salinty) is widely used by ecophysiologists as a definition for the threshold of salinity-tolerance (Kinzel, 1982). This definition is based on the same background as the Michaelis/Menten factor, because it is as difficult to fix the upper limit of salinity tolerance, as it is to determine the lowest substrate concentration for an optimal enzymatic activity. The agreement to the above-mentioned growth depression is comparatively arbitrary, but it leads to a precise specification of a comparative value for halophytic species and is especially expressive for applied aspects such as economic potentials of suitable halophytes.

e) It is worth mentioning that there is another definition of salinity tolerance in use for glycophytic species. Especially in agriculture it is very common to speak of salinity tolerance if a variety of a glycophyte such as *Hordeum vulgare* survives at a slightly higher salinity level than another variety of the same species. However, the tolerated NaCl-substrate concentrations are in both variieties far beyond seawater salinity (Amzallag, 1994; Jeschke et al., 1995).

This paper concentrates on eco-physiological mechanisms. We will carry on using the term (threshold of) salinity tolerance for unambiguous understanding as defined in (d).

3. QUICK CHECK SYSTEM FOR THE SELECTION OF USEFUL PLANTS AND THE PHYSIOLOGICAL CHARACTERIZATION OF SALINITY TOLERANCE

It is - without doubt - necessary to develop sustainable biological production systems, which can tolerate higher water salinity because freshwater resources will become limited in near future (Lieth, 1999). A precondition is the identification and/or development of salinity tolerant crops. An interesting system approach lines out that after halophytes are studied in their natural habitat and a determination of all environmental demands has been completed, the selection of potentially useful plants should be started (Lieth, 1999). The first step of this identification list contains the characterisation and classification of the soil and climate, under which potentially useful halophytes grow.

The measurement of the EC (electrical conductivity in [$\mu S * cm^{-1}$] offers a simple method for characterizing the salt content (Osmotic potential MPa = EC * -0,036, Koyro & Lieth, 1998). A saline soil has an EC greater than 4 mmho * cm^{-1} (equivalent to 40 mol * m^{-3} NaCl; U.S. Salinity Staff, 1954; Koyro & Lieth, 1998) and is widely used for this purpose. Spatial variability in salt-affected fields is normally very high. Since the habitats are often complex and the concentrations vary with water content, the EC of saturation extract only is an insufficient indicator for salinity tolerance. Plant growth in saline soils can be influenced although the EC indicates no changes, because the actual salt concentration at the root surface can differ as compared to the bulk soil. The EC characterises only the total salt content but not changes in its spatial composition. The importance of micro-heterogenity of salinity and fertility for maintenance of the plant diversity was shown for example from several authors (Igartua, 1995; Abdelly et al., 1999). The study in the natural habitat represents a mean behaviour but the major constraints can vary this much that a precise definition of the salinity tolerance of a species (and a selection of useful plants) is not possible. These arguments open a conflict of interests: Do we need results of local importance or is it necessary to perform universally valid studies? We think a useful database for people interested in the use of cash crop halophytes should contain both information!

However as a first step, only artificial conditions in seawater irrigation systems in a growth cabinet under photoperiodic conditions offer the possibility to study potentially useful halophytes under reproducible experimental growth and substrate conditions. The supply of different degrees of sea water salinity [0%, 25%, 50%, 75%, 100% (and if necessary higher) sea water salinity] to the roots in separate systems under otherwise identical or/and close to natural conditions gives the necessary preconditions for a comparative study in a quick check system (QCS) for potential cash crop halophytes. Former studies have shown that hydroponic cultures (soil-free) and soil cultures were not as reliable as a gravel/hydroponic system with drip (sprinkler, ditch, slack water or intertidal) irrigation (Koyro & Huchzermeyer, 1999a). Only the latter system had the potential to work under (nearly) completely artificial test conditions with high reproducibility and under close to natural conditions.

The experiments of the QCS started off at steady state conditions in a gravel/hydroponic system imitating the climatic conditions of subtropical dry regions. (Koyro & Huchzermeyer, 1999a). It is well known that salinity tolerance depends on the stage of development and period of time over which the plants have grown in saline conditions (Munns, 2002). Plants were exposed to salinity in the juvenile state of development and were studied until achieving the steady state of adult plants.

It is worth mentioning that the reliability of this cultivation system depended beside the climatic conditions (light intensity, relative humidity, air temperature) directly on a constant periodical irrigation (15 min per 4 hrs), a sufficient O_2 supply to the roots and constant nutrient conditions (pH, nutrient composition, temperature).

The screening of potential crop halophytes (Quick-check-system, QCS) comprises the following eco-physiological tests:

1) A collection of general scientific data and some special physiological examinations of adult plants. General scientific data contain informations about factors such as germination rate, growth development, yield, reproductivity, survival (perennial species), salt induced morphological changes, photosynthesis (such as gas exchange), water relations, mineral content, content of osmotically active organic substances (such as carbohydrates and amino acids).
2) Special physiological examinations are mainly on cellular level. They include the study of the relations inside single cells such as the compartmentation between cytoplasm and vacuole, the distribution of elements in different cell types or along a diffusion zone in a root apoplast and ultrastructural changes. However, especially in the last decade studies about gene-expression (proteomics) or genomics are getting increasingly more important.

This list is not unchangeable and can (or has to) be extended if the general and special scientific data do not allow uncovering the individual mechanisms for salinity tolerance. This variable applicable QCS seems to be valuable for the selection of useful plants and it suggest itself as a first step for the controlled establishment of cash crop halophytes because it provides detailed information about the three major goals:

1) The threshold of salinity tolerance at idealized growth conditions
2) To uncover the individual mechanisms for salt tolerance
3) The potential of utilization for the pre-selected species (=> cash crop halophytes).

The next chapters will demonstrate the red thread of the screening procedure and the achievement of the quick check system. Mainly, the threshold of salinity tolerance is determined as one of the first steps.

4. THRESHOLD OF SALINITY TOLERANCE

In correspondence with the definition for the threshold of salinity tolerance according to Kinzel (1982), the growth reaction and the gas exchange are used during the screening of halophytes as objective parameters for the description of the actual condition of a plant (Ashraf & O'Leary, 1996). There are now reliable informations available about studies with several halophytic species from different families such as *Aster tripolium* (figure 1a), *Inula crithmoides*, *Plantago cf.*

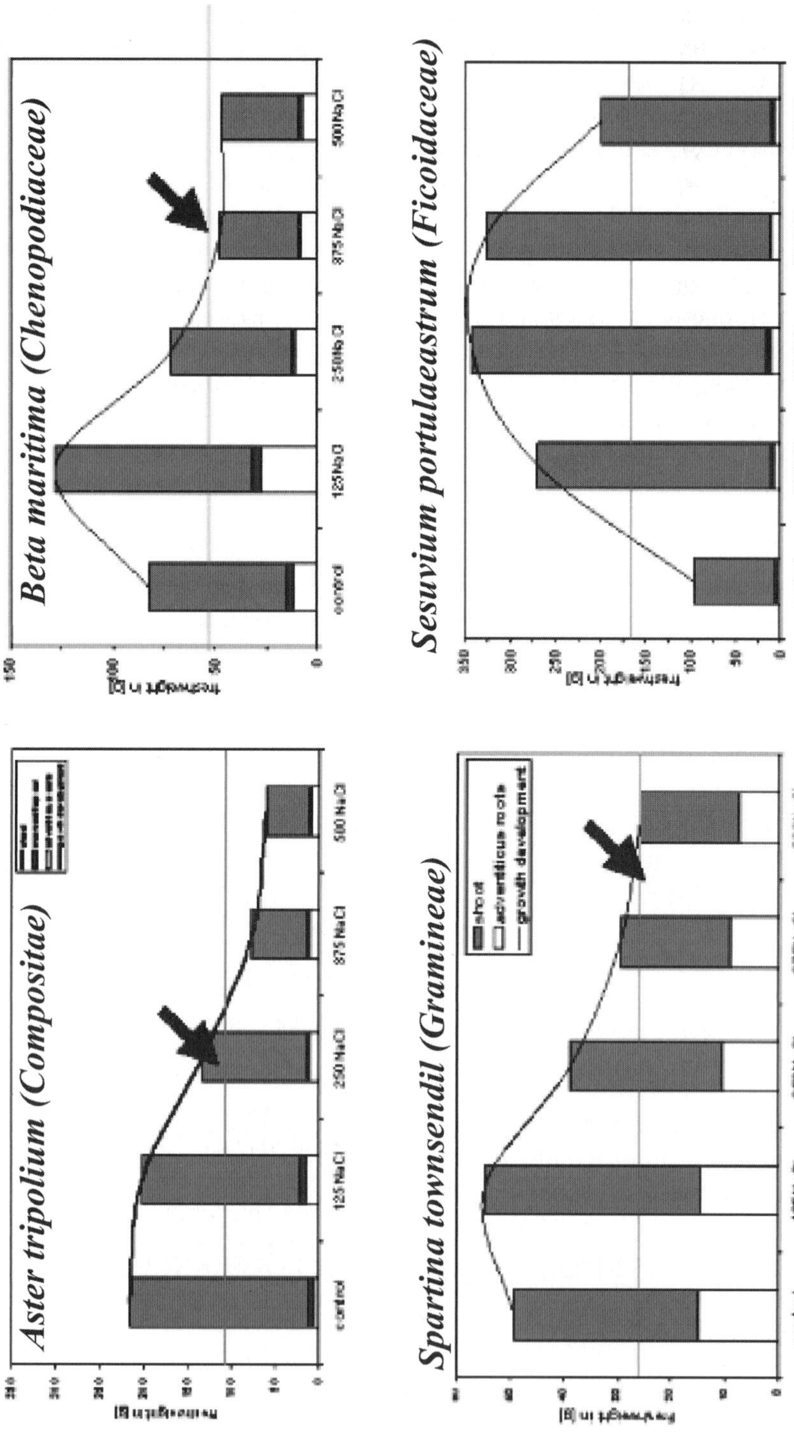

Figure 1. Threshold of salinity tolerance

coronopus, *Laguncularia racemosa, Limoneastrum articulatum*, *Beta vulgaris* ssp. *maritima* (Figure 1b), *Atriplex nummularia, Atriplex leucoclada, Atriplex halimus, Chenopodium quinoa, Batis maritima, Puccinellia maritima*, *Spartina townsendii* (Figure 1c) and *Sesuvium portulacastrum* (Figure 1d) (Pasternak, 1990; Koyro & Huchzermeyer, 1997, 1999a; Koyro et al., 1999; Lieth & Menzel, 1999; Koyro, 2000; Koyro & Huchzermeyer, 2003). The substrate-concentration leading to a growth depression of 50% (refer to freshweight, in comparison to plants without salinty) is easy to calculate with the QCS (by extrapolation of the data) and it leads to a precise specification of a comparative value for the threshold of salinity tolerance (Figure 1a-d). Dramatic differences are found between halophytic plant species. The threshold of salinity tolerance amounts to 300 mol * m^{-3} NaCl in *Aster tripolium*, 375 mol*m^{-3} in *Beta vulgaris* ssp. *maritima*, 500 mol*m^{-3} in *Spartina townsendii* and 750 mol*m^{-3} in *Sesuvium portulacastrum* (Figure 1). These results prove that it is essential to quantify differences in salinity tolerance between halophytic species as one basis for assessment of their potential of utilization.

5. BALANCE BETWEEN WATER LOSS AND CO_2- UPTAKE

Terrestric plants at saline habitats are often surrounded by low water potentials in the soil solution and atmosphere. Plant water loss has to be minimized under these circumstances, since biomass production depends mainly on the ability to keep a high net photosynthesis by low water loss rates. In this field of tension, biomass production of a plant has to be seen always in connection to the energy consumption and gas exchange [for example water use efficiency (WUE)]. A critical point for the plant is reached if the CO_2-fixation falls below the CO_2-production (compensation point). Therefore, one crucial aspect of the screening procedure is the study of growth reduction and net photosynthesis especially at the threshold of salinity tolerance (Figure 2).

Many plants such as *Beta vulgaris* ssp. *maritima*, *Spartina townsendii* or *Plantago* cf *coronopus* reveal at their threshold salinity tolerance a combination of low (but positive) net photosynthesis, minimum transpiration, high stomatal resistance and minimum internal CO_2-concentration (Koyro, 2000; Koyro & Huchzermeyer, 2003). However there is a big bandwidth between halophytes. Especially succulent halophytes such as *Sesuvium portulacastrum* or *Avicennia marina* have alternatives if the water balance is still positive (water uptake minus water loss) and not a limiting factor for photosynthesis. In case of Sesuvium net photosynthesis and WUE increase but stomatal resistance decrease. These results show that it is quite important to describe the regulation of gas-exchange at high salinity in strong reliance with other parameters (such as water relations). Water deficit is one major constraint at high salinity and can lead to a restriction of CO_2-uptake. The balance between water loss and CO_2-uptake is another basis for assessment of their potential of utilization. Additionally it helps to find weak spot in the mechanisms of adjustment (of photosynthesis) to high salinity.

Sesuvium portulacastrum	A (µmol*m^{-2}s^{-1})	E (mmol*m^{-2}s^{-1})	WUE (A/E)	rs (m^2*s* mol^{-1})
control	5,47	1,46	3,76	28,45
	+1,06	+0,20	+0,22	+4,88
125NaCl	6,53	1,67	3,90	26,40
	+0,25	+0,34	+0,34	+4,28
375NaCl	6,95	1,42	4,87	24,36
	+0,57	+0,02	+0,50	+0,28
500NaCl	7,44	1,51	4,98	22,60
	+0,81	+0,26	+0,36	+2,32

Aster tripolium	A (µmol*m^{-2}s^{-1})	E (mmol*m^{-2}s^{-1})	WUE (A/E)	rs (m^2*s*mol^{-1})
control	24,30	3,81	6,58	3,21
	+3,14	+0,37	+1,25	+0,29
125NaCl	17,20	3,89	4,38	5,29
	+3,69	+0,58	+0,35	+0,55
375NaCl	7,88	2,10	3,76	42,33
	+2,22	+0,63	+0,50	+2,55
500NaCl	4,26	1,22	3,49	156,74
	+0,66	+0,47	+1,68	+13,64

Figure 2. *Net photosynthesis rate [µmol * cm^{-2} * s^{-1}, figure 2a and b] and water use efficiency (of the photosynthesis, [µmol CO_2 * mmol^{-1} H_2O, figure 2a and b] of leaves of.* Aster tripolium *(a) and* Sesuvium portulacastrum *(b) at different NaCl-salinities. Many excluder-species such as* Beta vulgaris ssp. maritima *show at their threshold salt tolerance low (but positive) net photosynthesis as well as decreasing transpiration. Succulent, highly salt tolerant species show only minor changes (sometimes even an increase!) of net photosynthesis.*
0% sea water salinity = control, 25% = 125NaCl, 50% = 250NaCl, 75% = 375NaCl and 100% = 500NaCl

6. MORPHOLOGICAL STRUCTURES TO REDUCE SALT CONCENTRATIONS

In many cases various mechanisms and special morphological structures are advantageous for halophytes since they help to reduce the salt concentrations especially in photosynthetic or storage tissue and seeds. Salt glands may eliminate large quantities of salt by secretion to the leaf surface. This secretion appears in complex multicellular organs, for example in *Avicennia marina* or by simple two cellular salt glands, for example in *Spartina townsendii* (Sutherland & Eastwood, 1916; Walsh, 1974; Koyro & Stelzer, 1988; Marcum et al., 1998). Several halophytes can reduce the salt concentrations in vital organs by accumulation in bladder hairs (*Atriplex halimus, Leptochloa fusca* (L.), *Halimione portulacoides*), enhancing the LMA (leaf mass to area ratio, e.g. by *Suaeda fruticosa, Salicornia europaea, Salsola kali, Sesuvium portulacastrum*), establishing apoplastic barriers

(Freitas & Breckle, 1992, 1993ab; Hose et al., 2001), translocating NaCl into special organs (z.B. *Kandelia candel* L.), using of ultrafiltration at the root level to exclude salt. (*Avicennia marina, Sonneratia alba*) or shedding of old leaves (*Beta vulgaris* ssp. *maritima*, see literature in Marschner, 1995; Schröder, 1998; Glaubrecht, 1999; Koyro, 2002).

7. SCREENING OF MECHANISMS TO AVOID SALT INJURY IN INDIVIDUAL SPECIES

7.1. Major constraints for plant growth on saline habitats

Many halophytic species can tolerate high sea water salinity without possessing special morphological structures (see section 6). The salinity tolerance of halophytic plants is in most cases multigenic and there is often a strong reliance between various mechanisms. It is the exception, that a single parameter is of major importance for the ability to survive at high NaCl-salinity (as already shown in section 5). A comprehensive study with the analysis of at least a combination of several parameters is a necessity to get a survey about mechanisms constitution leading at the end to the salinity tolerance of individual species. These mechanisms are connected to the four major constraints of plant growth on saline substrates:

(a) Water deficit
(b) Restriction of CO_2 uptake
(c) Ion toxicity
(d) Nutrient imbalance

Plants growing in saline habitats face the problem of having low water potential in the soil solution and high concentrations of potentially toxic ions such as chloride and sodium. Salt exclusion minimizes ion toxicity but accelerates water deficit and deminishes indirectly the CO_2-uptake. Salt absorption facilitates osmotic adjustment but can lead to toxicity and nutritional imbalance. However it should not be overlooked that beside the major constraints of high salinity less universally valid factors could be also of importance and should be kept in mind before expressing general statements:

The vegetation cycle (annual, biennial or perennial),
Ion buffer capacity of the substrate,
Duration of exposure,
Stage of plant development,
Plant organ
Environmental conditions (such as tides)

7.2. *Major plant responses to high NaCl-salinity*

In general the presence of soluble salts can affect growth in several ways (Mengel & Kirkby, 2001). In the first place plants may suffer from water stress, secondly high concentrations of specific ions can be toxic and induce physiological disorders and thirdly intracellular imbalances can be caused by high salt concentration. In principle, salinity tolerance can be achieved by salt exclusion or salt inclusion. The following physiological mechanisms to avoid salt injury (and to protect the symplast) are known as major plant responses to high NaCl-salinity (Marschner, 1995; Mengel & Kirkby, 2001; Munns, 2002; Koyro & Huchzermeyer, 2003):

(a) Adjustment of the water potential, decrease of the osmotic and matric potential, enhanced synthesis of organic solutes
(b) Regulation of the gas exchange (H_2O and CO_2), high water use efficiency (H_2O- loss per net CO_2-uptake) or/and switch to CAM-type of photosynthesis,
(c) Ion-selectivity to maintain homeostasis especially in the cytoplasm of vital organs
  - Selective uptake or exclusion (e.g. salt glands)
  - Selective ion-transport in the shoot, in storage organs, to the growing parts and to the flowering parts of the plants, retranslocation in the phloem
  - Compartmentation of Na and Cl in the vacuole
(d) High storage capacity for NaCl in the entirety of all vacuoles of a plant organ, generally in old and drying parts (e.g. in leaves supposed to be dropped later) or in special structures such as hairs. The dilution of a high NaCl content can be reached in parallel by an increase in tissue water content (and a decrease of the surface area, succulence)
(e) Avoidance of ionic imbalance
(f) Endurance of high NaCl-concentrations in the symplast
(g) Compatibility of whole plant metabolism with high NaCl-concentrations (synthesis of NaCl-tolerant enzymes, protecting agents such as proline and glycine-betaine).
(h) Restricted diffusion of NaCl in the (root-) apoplast

7.3. *Benefical scientific data*

Useful parameters for screening halophytes should be based on the major plant responses to high NaCl-salinity (Volkmar et al., 1998). It seems to be essential that such a screening system should include salt induced morphological changes (see section 6, succulence, LAR: leaf mass to area ratio), growth (see section 4), water relations, gas exchange (see section 5) and composition of minerals (and compatible solutes) at different parts of the root system and in younger and older leaf tissues. The measurement of such general scientific data at plant-, organ- or tissue level reveals general trends – but since these represent a mean behaviour of several cell types, much information on single cell adjustment are lost. They cannot give sufficient information about the compartmentation inside a cell or along a diffusion zone in a root apoplast or about ultrastructural changes such as apoplastic barriers (Hose et al., 2001). The collection of scientific data should be completed if necessary (to uncover the individual mechanisms for salt tolerance) by a special physiological research at single cell level supplemented optionally by methods such as the analysis of the gene-expression and its genetic basis (genomics & proteomics, Winicov & Bastola, 1997, 1999; Winicov, 1998).

7.4. Collection of general scientific data

Information about salt induced morphological changes; gas exchange and growth were already presented in the sections 4, 5 and 6 of this article. The list of general scientific data can be completed by information on water relations (for example leaf water potential see section 7.4.1) and composition of organic and inorganic solutes (see section 7.4.2).

7.4.1. Leaf water potential

Recognition of the importance of time frame led to the concept of a two-phase growth response to salinity (Mengel & Kirkby, 2001; Munns, 1993, 2002; Munns et al., 2002). The first phase of growth reduction is essentially a water stress or osmotic phase and presumably regulated by hormonal signals coming from the roots.

Data of the leaf water potentials (measured by dew point depression with a WESCOR HR-33T) demonstrated clearly that leaf water potential of halophytes does not correlate alone as a single factor with salinity tolerance. *Aster tripolium* (Figure 3a), *Beta vulgaris* ssp. *maritima* (Figure 3b), *Spartina townsendii* (Figure 3c), and *Sesuvium portulacastrum* (Figure 3d), have a sufficient adjustment mechanism even at high salinity treatment. The osmotic potentials were for all four halophytes (and many other) at all salinity levels sufficiently low to explain the full turgescence of the leaves (results not shown).

Assuming there is no interruption of the water supply water can flow passively from the root to the shoot and there seems to be no reason for growth reduction by water deficit for any of the studied species. However, by regulating the extent of apoplastic barriers and their chemical composition, plants can effectively regulate the uptake or loss of water and solutes (by structures such as barriers in the hypo- or exodermis). This appears to be an additional or compensatory strategy of plants to acquire water and solutes (Hose et al., 2001) and at the extremes of growth under conditions of drought and high salinity make the exodermis an absolute barrier for water and ions in the strict sense (North & Nobel, 1991; Azaizeh & Steudle, 1991) as shown for *Spartina* in figure 4.

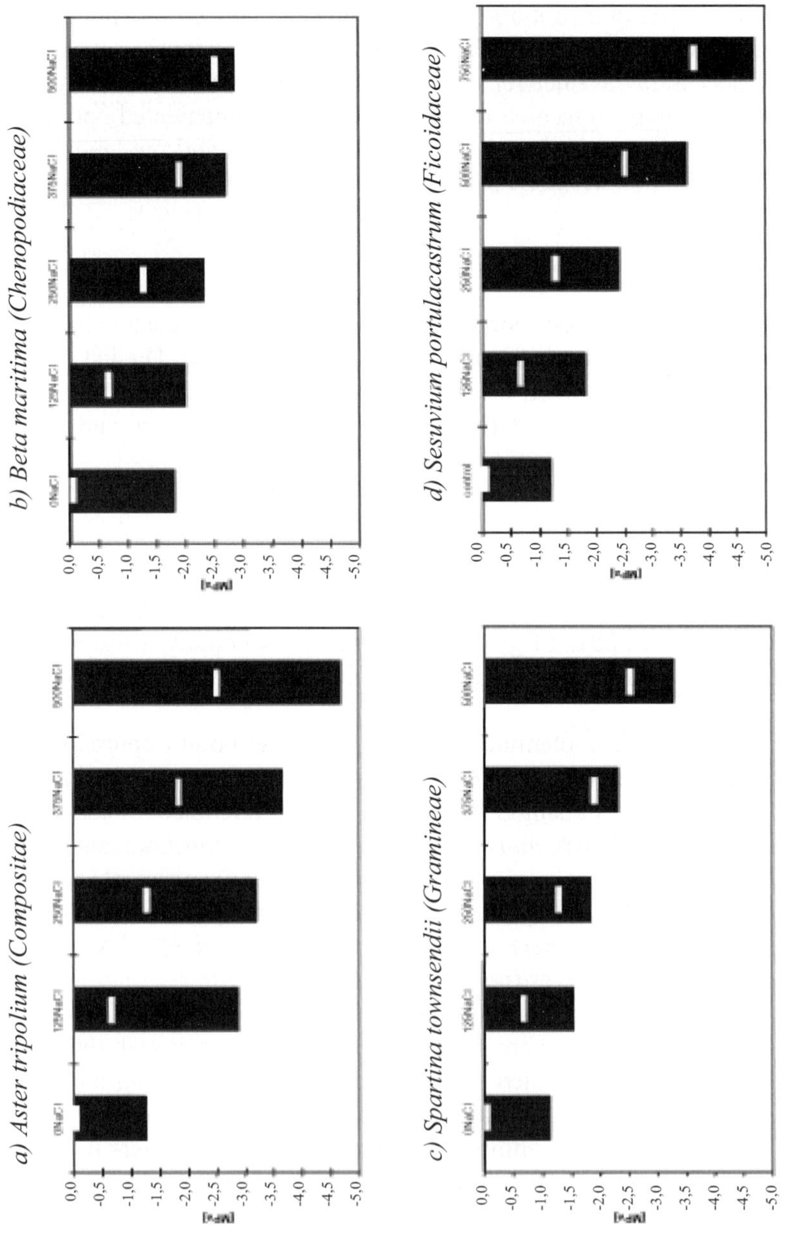

Figure 3. *Leaf water potential*

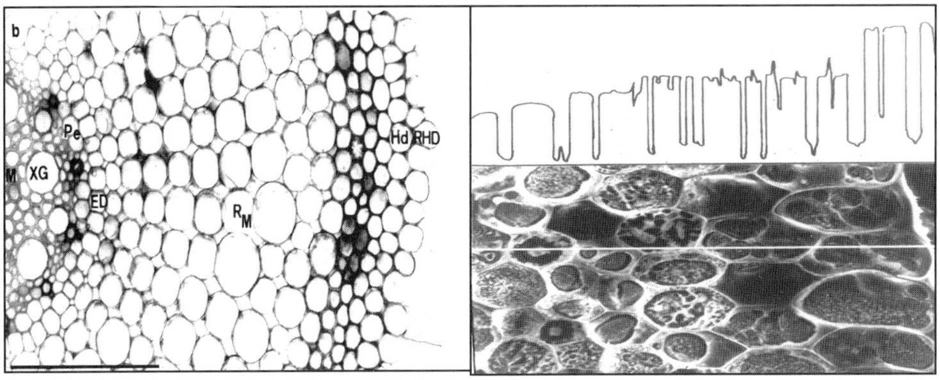

Figure 4. Radial cross sections in adventitious roots of Spartina townsendii.
a) Light microscopical proof of suberin (dye Sudan III) in the cell walls of hypodermal root cells (arrow)
b) EDXA-chlorine-specific line scans in a frozen section of a root. The white line in the electron micrograph marks the course of the beam.

Thus, the rate of supply of water to the shoot can be restricted due to the coupling between the flows of water and solutes (Na and Cl) even if the leaf water potential is low. Therefore, the balance between water flow (sum of water accumulation and transpiration) and the decrease in the amounts of nutrients or unfavorable nutrient ratios (e.g. Na^+/K^+) are important factors for impaired leaf elongation (Lynch et al., 1988; Munns et al., 1989) and plant growth. This is summarised a good example for the strong reliance between various mechanisms as already mentioned in section 7.1.

7.4.2. Organic and inorganic solutes

There is a second phase of growth response to salinity, which takes time to develop, and results from internal injury (Mengel & Kirkby, 2001; Munns, 1993, 2002). It is due to salts accumulating in transpiring leaves to excessive levels. Ion toxicity and nutrient imbalance are two major constraints of growth (see section 7.1) at saline habitats and therefore of special importance for the salt tolerance of halophytes. Data of additional scientific studies have shown that halophytes exhibit very different ways of adjustment to high NaCl-salinty. Generally, salt tolerant plants differ from salt-sensitive ones in having a low rate of Na^+ and Cl^- transport to leaves (Munns, 2002). However, some halophytes (so-called salt includers) even need an excess of salts for maximum growth and for attaining low solute potentials (Flowers et al., 1977; Greenway & Munns, 1980). Alternatively, high concentrations can be avoided by filtering out most of the salt. These halophytes so-called salt excluders adapt to saline conditions by ion exclusion so that osmotically active solutes have to be synthesized within the plant to meet turgor pressure demands (Mengel & Kirkby, 2001). This adaptive feature can be of importance even in species that have salt glands or bladders. However, NaCl-

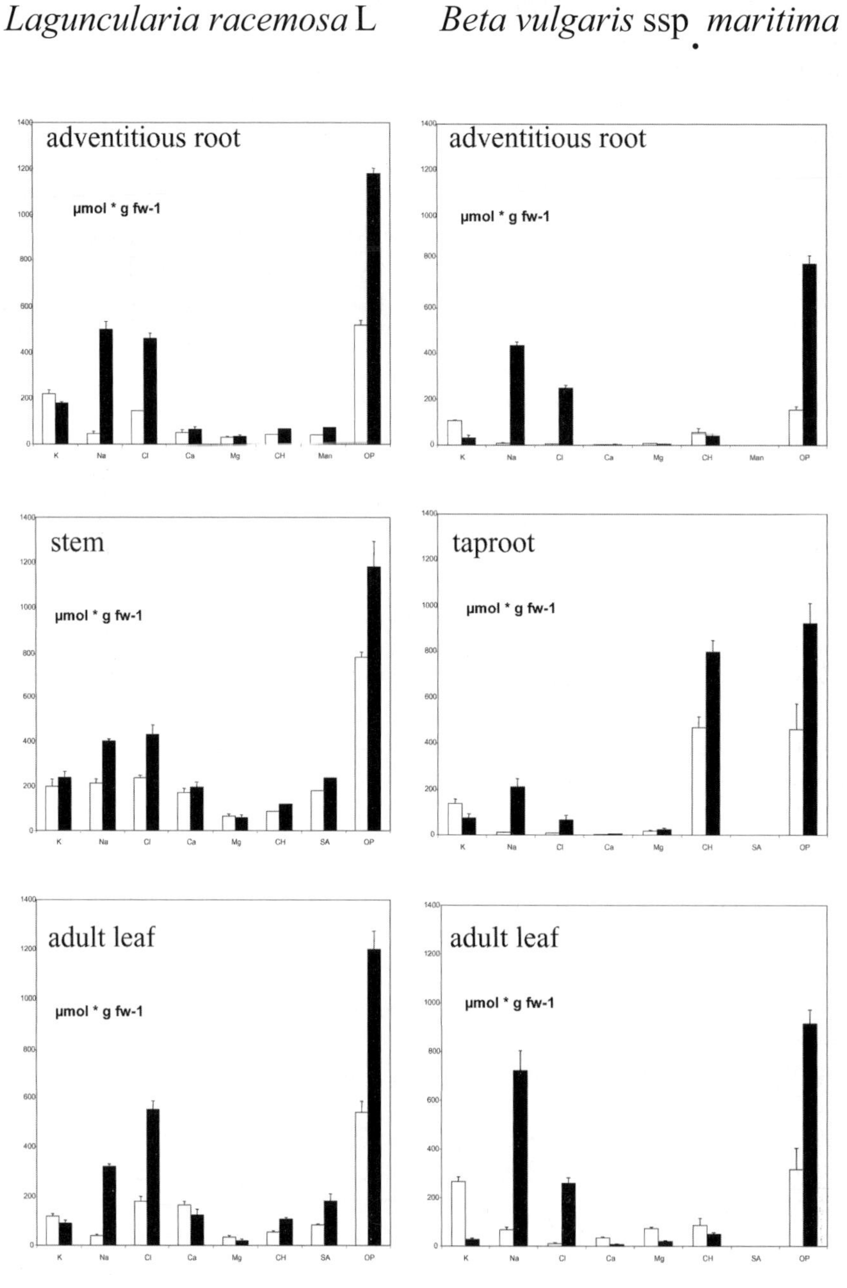

Figure 5. *Potassium-, sodium-, chlorine-, calcium, magnesium, carbohydrate and sugar alcohol-concentrations in mol * m^{-3} and osmotic values (in mOsmol) in different tissues of* Laguncularia racemosa *and* Beta vulgaris ssp. maritima.

salinity is discussed in literature mainly as if a common reaction of both ions (Na^+ and Cl^-) is leading to a salt injury. This is not always the case! For example in maize, Schubert and Läuchli (1986) did not find a positive correlation between salt tolerance and Na^+ exclusion. It is quite important to distinguish between both ions to uncover the individual mechanisms for salt tolerance.

Halophytes are able to distinguish precisely between the metabolic effects of both ions Cl^- and Na^+:

a) Some halophytes such as *Scirpus americanus*, *Avicennia marina* (with salt glands) or *Rhizophora mangle* are able to exclude Na and Cl (see literature in Kinzel, 1982) from the leaves,
b) *Laguncularia racemosa* (with salt glands) is a typical Na-excluder but with high Cl-accumulation in the leaves (figure 5a; Koyro et al., 1997),
c) *Beta vulgaris* ssp. *maritima, Suaeda brevifolia, Suaeda vera, Limoneastrum monopetalum, Allenrolfea occidentalis*, or *Spartina townsendii* are typical Cl-excluder with high Na-accumulation in the leaves (figure 5b; see literature in Kinzel, 1982; Koyro & Huchzermeyer, 1999a),
d) *Salicornia rubra, Salicornia utahensis, Suaeda occidentalis, Atriplex vesicaria, Atriplex nummularia, Atriplex papula, Atriplex rosea* or *Inula crithmoides* accumulate Na and Cl in the leaves in a range above the saline environment (salt-includers). Typical halophytic adaptation includes in this case leaf succulence in order to dilute toxic ion concentrations (Kinzel, 1982; Mengel & Kirkby, 2001).

In Na^+ and/or Cl^- excluding species (a-c), however, a lack of solutes may result in adverse effects on water balance, so that water deficiency rather than salt toxicity may be the growth-limiting factor (Greenway & Munns, 1980; Mengel & Kirkby, 2001). To achieve a low water potential and/or a charge balance the solute potential in these species is decreased by the synthesis of organic solutes (figure 5a & b) such as sugar alcohol (e.g. mannitol in leaves of *Laguncularia racemosa*; see also figure 5a), soluble carbohydrates (e.g. sucrose in taproots of *Beta maritima* ssp. *maritima*; see also Figure 5b), organic acids (incl. amino acids) or by reducing the matrical potential (e.g. with soluble proteins in leaves of *Beta vulgaris* ssp. *maritima*; results not shown). However, the synthesis of organic solutes is energy demanding and the formation of these solutes decreases the

energy status of the plant. Thus for plant survival, growth depression is a necessary compromise in Na^+ and /or Cl^- excluding species and not a sign of toxicity or nutrient imbalance.

7.4.3. Division of labor between leaves

The comparison between ion relations in different plant organs (root, stem, leaf) can be insufficient to uncover the individual mechanisms for salt tolerance. It can be necessary to distinguish also between different phases of growth development of one organ especially in leaves.

The comparison of the K- and Na-concentrations in juvenile and adult leaves reveals obviously that *Beta vulgaris* ssp. *maritima* uses the older parts for internal detoxification or better internal exclusion of sodium (Koyro, 2000). The steep inverse Na/K gradient between juvenile and adult leaves of the sea beet is a typical reaction of many halophytes to high NaCl-salinity (Wolf et al., 1991; Koyro & Huchzermeyer, 1999a). There is scientific debate whether high potassium concentrations in young leaves and reproductive organs can be achieved by low xylem import of both potassium and sodium, and/or high phloem imports from mature leaves (Wolf et al., 1991).

The chlorine just like the sodium concentrations was much higher in adult than in juvenile leaves. An effective restriction of sodium and chlorine import into young leaves as compared to old leaves is also typical for *Agrostis stolonifera* (Robertson & Wainwright, 1987). Another mechanism of sea beet to reduce internal NaCl-concentrations was shedding of old leaves (see literature in Koyro, 2000). This system successfully deminishes and controls the accumulation of Na and Cl in younger or metabolically active parts of the plant and buffers against an imbalance of nutrients such as K.

7.5. Special physiological examination

The general scientific data give an impression of various mechanisms of adaptation to high NaCl-salinity. Beside water stress and ion specific toxic physiological disorders on tissue level (see section 7.4.1 and 7.4.2) intracellular ionic imbalances (K^+, Ca^{2+} and Mg^{2+}) can be caused by high salt concentrations (Mengel & Kirkby, 2001). The capacity of plants to maintain K^+ homeostasis and low Na^+ concentrations in the cytoplasm appears to be one important determinant of plant salt tolerance (Yeo, 1998; Läuchli, 1999). A possibility to find such limiting factors is the study of the relations inside single cells such as the compartmentation between cytoplasm and vacuole, the distribution of elements in different cell types or along a diffusion zone in a root apoplast and ultrastructural changes. However, the analysis of the distribution of elements (compartmentation) in a distinct cell type can be used to answer several questions such as:

(a) Is there a compartmentation and a division of labour in cytoplasm and vacuole?
(b) Is it necessary to look more in detail to toxic effects caused by high cytoplasmic sodium or chlorine concentrations?
(c) Is there a compatibility of the metabolism with high NaCl-concentrations in the symplast (cytoplasm)?
(d) Is there a high ionic imballance in the symplast (cytoplasma)?

Beta vulgaris ssp. *maritima* and *Spartina townsendii* are typical Cl-excluder with high Na-accumulation in the leaves (see section 7.4.1). Both species seem to react similar to salinity with changes of leaf water potential, gas-exchange and nutrients. However, this number of congruence does not allow concluding analogical intracellular relations. The comparison of their intracellular ionic balance will be used to demonstrate the necessity of special physiological investigations.

In contrast to water stress effects, occurring in the meristematic region of younger leaves, salt toxicity predominantly occurs in mature leaves (Mengel & Kirkby, 2001). This is because Na and Cl are stored mainly in the shoot of halophytes such as *Beta vulgaris* ssp *maritima* and *Spartina townsendii* leading to a growth reduction of the above ground parts much higher than of the root (Koyro, 2000; Koyro & Huchzermeyer, 2003). These changes can be interpreted as signs of a critical load. Therefore, to distingiuish between the individual mechanisms of salinity tolerance further investigations of the intracellular ionic balance were performed first of all at epidermal leaf cells (the end of the transpiration stream) of these both species.

The single cell data of the vacuolar and cytolasmatic composition in cells of the upper leaf-epidermis are summarized for the controls and the high-salinity treatments (at seawater salinity) in figure 6. The intracellular composition of the leaf epidermal cytoplasm and vacuoles of controls of *Beta vulgaris* ssp. *maritima* and *Spartina townsendii* show some more congruities of both species. The epidermal vacuoles of controls of both species contains most of the elements (with the exception of P) in higher concentrations as the cytoplasm indicating the overall picture of a vacuolar buffer. The leaf-vacuoles in its entirety can be described as a voluminous potassium-pool with high storage capacity for sodium and chloride. This pool is needed in case of high NaCl-salinity for the maintenance of the K-homeostasis in the cytoplasm. The dominant elements in the cytoplasm were P and K. The K-concentrations were in the epidermal cytoplasm of control plants in an ideal range for enzymatic reactions (Wyn Jones et al., 1979; Wyn Jones & Polard, 1983; Koyro & Stelzer, 1988).

It is obvious that seawater salinity leads to a decrease of P, S, Mg and K in the epidermal vacuoles of both species. The remaining K, S and Mg concentrations were only in Spartina two-digit and especially for K much higher as in Beta. The vacuolar buffer of the latter one seems to be exhausted.

NaCl salinity led to a significant decrease of the K and P concentrations especially in the cytoplasm of Beta and to a breakdown of the homeostasis. This result points at a deficiency for both elements in the cytoplasm. Additionally the concentrations of sodium and chlorine were at high NaCl-salinity below 5 mol * m^{-3} in the cytoplasm of the epidermal cytoplasm and the gradients between cytoplasm and vacuole were higher in comparison with the results of *Spartina* In summary these results support the hypothesis that the sea beet does not sustain ion-toxicity but ion-deficiency! It is hypothesized that such low K^+ levels in the cytoplasm can lead to a reduction of protein synthesis, which is of utmost importance in the process of leaf expansion (Mengel & Kirkby, 2001). One possible consequence is the supply of sufficient fertilizers (especially K and P) at high NaCl-salinity to reduce the symptoms of K- and P-deficiency in Beta.

Beta vulgaris ssp. *maritima*

adaxial	vacuole				cytoplasm			
leafepidermis	control		480 NaCl		control		480 NaCl	
Cl	22.6	±4.7	654.3	±54.8	0.0		<5	
P	22.2	±4.6	6.4	±4.2	58.2	±8.4	28.1	±6.7
S	40.4	±7.9	2.7	±1.9	10.5	±4.0	<5	
Na	12.2	±3.1	724.9	±65.1	<5		<5	
Mg	24.0	±1.2	4.3	±1.7	<5		<5	
K	282.5	±14.7	9.3	±6.1	88.9	±9.5	66.7	±7.2
Ca	<5		<5		<5		<5	

Spartina townsendii

adaxial	vacuole				cytoplasm			
leafepidermis	control		480 NaCl		control		480 NaCl	
Cl	21.2	±3.2	324.3	±64.8	<5		<5	
P	11.1	±1.1	5.3	±2.1	81.5	±6.8	71.6	±10.8
S	24.4	±6.9	20.8	±2.6	8.8	±3.0	<5	
Na	16.4	±3.1	521.0	±54.9	<5		15.3	±3.2
Mg	29.6	±2.1	18.8	±1.6	<5		<5	
K	212.5	±34.8	71.5	±6.9	92.7	±12.7	78.4	±6.9
Ca	<5		<5	±2.06	<5		<5	

Figure 6. *Sodium- magnesium-, phosphor-, sulphur-, chlorine- and potassium-concentrations in mol * m^{-3} (measured with EDX-analysis in bulk frozen tissues) in the vacuoles and in the cytoplasm of adaxial epidermis cells of* Beta vulgaris ssp. maritima *and* Spartina townsendii.

The salt induced reductions of the cytoplasmic K and P concentrations were much less pronounced in *Spartina* as in *Beta*. The results of *Spartina* point at a working system to keep ionic homeostasis. However there was one important exception: The sodium concentration increased significantly in the epidermal cytoplasm. Sodium could (try to) substitute potassium in its cytoplasmic functions or it could be the first sign of intoxication.

The results and interpretations are in agreement with the hypothesis that plant growth is affected by ion imbalance and toxicity and probably lead to the long-term growth differences between the salt-tolerant and sensitive species.

However, Beta and Spartina are also two excellent examples how important it can be to validate intracellular ionic imbalances (K^+, Ca^{2+} and Mg^{2+}) at high salt concentrations to uncover the individual mechanisms for salt tolerance and to understand the threshold levels of individual species.

7.6. Evaluation of the screening procedure

The results presented in this paper contain a lot of information about the essential eco-physiological needs of several halophytes at high salinity. The very variable screening of individual species enables to study the characteristic combination of mechanisms against salt injury and the threshold of salinity tolerance. The so-called QCS can be modified to the special characteristics and needs of other species and is therefore useful to study a wide range of suitable halophytes. This screening procedure is a practical first step on the selection of economically important cash crop halophytes.

8. SELECTION OF SALINITY TOLERANT SPECIES WITH PROMISING TOLERANCE AND YIELD CHARACTERISTICS

For future studies on utilisation potentials of halophytes precise data about the ecological demands of halophytic species are required. Comparative physiological studies about salinity tolerance are essential. A precondition for this demand is a precise specification of a comparative value for halophytic species as shown in this paper. Four steps are prerequisites for the selection of appropriate salinity tolerant plants with promising tolerance (and yield) characteristics.

(a) The literature has to be screened prior to the selection of priority species (potentially useful species) in order to get first order information about their natural occurrence in dry or saline habitats, existing utilisation (because of their structure, chemical content or other useful properties), natural climatic and substrate conditions, water requirement and salinity tolerance.
(b) Soon after the selection of a priority species, the threshold of salinity should be determined according to Kinzel (1982) and Munns (2002). The characteristic major plant responses have to be evaluated for precise informations of ecophysiological demands. The data can build up a well-founded basis for the improvement of the utilisation potential.

9. DEVELOPMENT OF CASH CROP HALOPHYTES

The physiological studies with the sea water irrigation system have the potential to provide highly valuable means of detecting individual mechanisms of species against NaCl stress, and may also provide opportunities for the comparison and screening of different varieties for their adaptation to salinity (QCS for cash crop halophytes). However, it can be only the first step for the development of cash crops or other usable plants from existing halophytes. After the selection of halophytic species suited for a particular climate and for a particular utilisation a gradual realization of the following topics could be one way to establish potentially useful cash crop halophytes:

(a) Green house experiments at the local substrates (and climatic conditions) to select and propagate promising sites (Isla et al., 1997).
(b) Studies with Lysimeters on field site to study the water consumption and ion movements.

(c) Design of a sustainable production system in plantations at coastal areas or at inland sites (for example for economical use).
(d) Testing yield and (economic) acceptance of the product.

10. REFERENCES

Abdelly, C., Lachaal, M. & Grignon, C. 1999. Importance of micro-heterogenity of salinity and fertility for maintenance of the plant diversity. In: H. Lieth, M. Moschenko, M. Lohmann, H. W. Koyro & A. Hamdy (Eds.), Progress in Biometeorology. Leiden, Netherlands: Backhuys Publishers, 65-76 pp.

Amzallag, G.N. 1994. Influence of parental NaCl treatment on salinity tolerance of offspring in *Sorghum bicolor* (L) Moench. New Phytologist 128: 715-723.

Ashraf, M. & O'Leary, J.W. 1996. Effect of drought stress on growth, water relations, and gas exchange of two lines of sunflower differing in degree of salt tolerance. International Journal of Plant Sciences 157: 729-732.

Azaizeh, H. & Steudle, E. 1991. Effects of salinity on water transport of excised maize (*Zea mays* L.) roots. Plant Physiology 97: 1136-1145.

Boer, B. & Gliddon, D. 1998. Mapping of coastal ecosystems and halophytes (case study of Abu Dhabi, United Arab Emirates). Marine and Freshwater research 49: 297-301.

Ellenberg, H. 1974. Zeigerwerte der Gefäßpflanzen Mitteleuropas. Scripta Geobotanica 9: 97.

Flowers, T.J., Troke, P.F. & Yeo, A.R. 1977. The mechanisms of salt tolerance in halophytes. Annual Review Plant Physiology 28: 89-121.

Freitas, H. & Breckle, S.W. 1992. Importance of bladder hairs for salt tolerance of field-grown *Atriplex*-species from a Portuguese salt marsh. Flora 18: 283-297.

Freitas, H. & Breckle, S.W. 1993a. Progressive cutinization in *Atriplex* bladder stalk cells. Flora 188: 287-290.

Freitas, H. & Breckle, S.W. 1993b. Accumulaton of nitrate in bladder hairs of *Atriplex* species. Plant Physiological Biochemistry 31: 887-892.

Ghassemi, F., Jakeman, A.J. & Nix, H.A. 1995. Salinisation of land and water resources: Human causes, extent, management and case studies. Sydney, Australia: USNW Press. 540 pp.

Glaubrecht, M. 1999. Mangrove der tropischen Gezeitenwälder; Naturw. Rdsch 52.

Greenway, H. & Munns, R. 1980. Mechanisms of salt tolerance in nonhalophytes. Annal Review Plant Physiology 31: 149-190.

Hose, E., Clarkson, D.T., Steudle, E., Schreiber, l. & Hartung, W. 2001. The exodermis: a variable apoplastic barrier. Journal of Experimental Botany 52: 2245-2264.

Igartua, E. 1995. Choice of selection environment for improving crop yields in saline areas. Theoretical and Applied Science 91: 1016-1021.

Jeschke, W.D., Klagges, S., Hilpert, A., Bhatti, A.S. & Sarwar, G. 1995. Partitioning and flows of ions and nutrients in salt-treated plants of *Leptochloa fusca* L Kunth .1. Cations and chloride. New Phytologist 130: 23-35.

Isla, R., Royo, A. & Aragues, R. 1997. Field screening of barley cultivars to soil salinity using a sprinkler and a drip irrigation. Plant and Soil 197: 105-117.

Kinzel, H. 1982. Pflanzenökologie und Mineralstoffwechsel. Stuttgart, Germany: Eugen Ulmer Publisher.

Koyro, H.-W. & Stelzer, R. 1988. Ion concentrations in the cytoplasm and vacuoles of rhizodermal cells from NaCl treated Sorghum, *Spartina* and *Puccinellia* plants. Journal of Plant Physiology 133: 441-446.

Koyro, H.-W. & Lieth, H. 1998. Salinity conversion table. 2nd enlarged Edition, © H.Lieth ISSN 09336-3114, Osnabrück.

Koyro, H.-W. & Huchzermeyer, B. 1997. The physiological response of *Beta vulgaris* ssp. *maritima* to seawater irrigation. In: H. Lieth, A. Hamdy & H.-W. Koyro, (Eds.), Water management, salinity and pollution control towards sustainable irrigation in the mediterranean region. Salinity problems and halophyte use. Bari, Italy: Tecnomack Publications. 29-50 pp.

Koyro, H.-W., Wegmann, L., Lehmann, H. & Lieth, H. 1997. Physiological mechanisms and morphological adaptation of *Laguncularia racemosa* to high salinity. In: H. Lieth, A. Hamdy & H.-W. Koyro, (Eds.), Water management, salinity and pollution control towards sustainable irrigation in the mediterranean region: Salinity problems and halophyte use. Bari, Italy: Tecnomack Publications, 51-78 pp.

Koyro, H.-W. & Huchzermeyer, B. 1999a. Influence of high NaCl-salinity on growth, water and osmotic relations of the halophyte *Beta vulgaris* ssp. *maritima*. Development of a quick check In: H. Lieth, M. Moschenko, M. Lohmann, H.-W. Koyro & A. Hamdy, (Eds.), Progress in Biometeorology, Leiden, Netherlands: Backhuys Publishers. 87-101 pp.

Koyro, H.-W. & Huchzermeyer, B. 1999b. Salt and drought stress effects on metabolic regulation in maize. In: M. Pessarakli, (Ed.), Handbook of plant and crop stress 2nd Ed. New York, New York: Marcel Dekker Inc. 843-878 pp.

Koyro, H.-W., Wegmann, L., Lehmann, H. & Lieth, H. 1999. Adaptation of the mangrove Laguncularia racemosa to high NaCl salinity. In: H. Lieth, M. Moschenko, M. Lohmann, H.-W Koyro, & A. Hamdy. (Eds.) Progress in Biometeorology, Leiden, Netherlands: Backhuys Publishers. 41-62 pp.

Koyro, H.-W. 2000. Untersuchungen zur Anpassung der Wildrübe (*Beta vulgaris* ssp. *maritima*) an Trockenstreß oder NaCl-Salinität. Habilitation. Giessen, Germany: Justus-Liebig-University.

Koyro, H.-W. 2002. Ultrastructural effects of salinity in higher plants. In: A. Läuchli & U. Lüttge. (Eds.), Salinity: Environment – Plants – Molecules. Dordrecht, Netherlands: Kluwer Academic Publication. 139-158 pp.

Koyro, H.-W. & Huchzermeyer, B. 2003. Ecophysiological needs of the potential biomass crop Spartina townsendii Grov.. Journal of Tropical Ecology, (in press).

Landolt, E. 1977. Ökologsche Zeigerwerte zur Schweizer Flora. Veröffentlichungen des geobotanischen Instituts der Eidgenössischen Technischen Hochschule in Zürich 64, Stiftung Ruebel, 208 pp.

Läuchli, A. 1999. Potassium interactions in crop plants. In: D.M. Oosterhuis, & G.A. Berkowitz. (Eds.), Frontiers in Potassium Nutrition. New perspectives on the effects of potassium on physiology of plants. New York, New York: Marcel Dekker. 71-76 pp.

Lieth, H. 1999. Development of crops and other useful plants from halophytes, In: H. Lieth, M. Moschenko, M. Lohmann, H.-W Koyro, & A. Hamdy (Eds.), Halophytes Uses in different Climates, Ecological and Ecophysiological Studies. Leiden, Netherlands:Backhuys Publishers, 1-18 pp.

Lieth, U. & Menzel, U. 1999. Halophyte Database Vers. 2, In: H. Lieth, M. Moschenko, M. Lohmann, H.-W Koyro, & A. Hamdy. (Eds.), Halophytes Uses in different Climates, Ecological and Ecophysiological Studies, Leiden, Netherlands:Backhuys Publishers. 159-258 pp.

Lieth, H., Moschenko, M., Lohmann, M., Koyro, H.-W. & Hamdy, A. 1999. Halophyte uses in different climates I. Ecological and ecophysiological studies. In: H. Lieth, (Ed.), Progress in Biometeoroogy, Leiden, Netherlands:Backhuys Publishers. 1-258 pp.

Lynch, J., Thiel, G. & Läuchli, A. 1988. Effects of salinity on the extensibility and Ca availability in the expanding region of growing barley leaves. Botanica Acta 101: 355-361.

Marcum, K.B., Anderson, S.J. & Engelke, M.C. 1998. Salt gland ion secretion: A salinity tolerance mechanism among five zoysiagrass species. Crop Science 38: 806-810.

Marcum, K.B. 1999. Salinity tolerance mechanisms of grasses in the subfamily Chloridoideae. Crop Science 39: 1153-1160.

Marschner, H. 1995. Mineral nutrition of higher plants. New York, New York: Academic Press. 1-889 pp.

Mengel, K. & Kirkby, E.A. 2001, Principles of Plant Nutrition. Dordrecht, Netherlands: Kluwer Academic Publisher. 1-849 pp.

Munns, R., Gardner, P.A., Tonnet, M.L. & Rawson, H.M. 1989. Growth and development in NaCl-treated plants. II Do Na+ or Cl- concentrations in deviding or expanding tissues determine growth in barley. Australian Journal Plant Physiology 15: 529-540.

Munns, R. 1993. Physiological processes limiting plant growth in saline soils: some dogmas and hypotheses. Plant Cell and Environment 16: 15-24.

Munns, R. 2002. Comparative physiology of salt and water stress. Plant Cell and Environment 25: 239-250.

Munns, R., Husain, S., Rivelli, A.R., James, R.A., Condon, A.G., Lindsay, M.P., Lagudah, E.S., Schachtman, D.P. & Hare, R.A. 2002. Avenues for increasing salt tolerance of crops, and the role of physiologically based selection traits. Plant and Soil 247: 93-105.

North, G.B. & Nobel, P.S. 1991. Changes in hydraulic conductivity and anatomy caused by drying and rewetting roots of Agave desertii, Agavaceae. American Journal of Botany 78: 906-915.

Pasternak, D. 1990. Fodder production with saline water. The institute for applied research, Beer-Sheva/Israel: Ben Gurion University of the Negev. 173 pp.

Robertson, K.P. & Wainwright, J.J. 1987. Photosynthetic responses to salinity in two clones of Agrostis stolonifera. Plant Cell and Environment 10: 45-52.

Schimper, A.F.W. 1891. Pflanzengeographie auf physiologischer Grundlage. Jena:Fischer Publication.

Schroeder, F.G. 1998. Lehrbuch der Pflanzengeographie; Wiesbaden: Quelle & Meyer.

Schubert, A. & Läuchli, A. 1986. Na^+ exclusion, H^+ release and growth of two different maize cultivars under NaCl salinity. Journal Plant Physiology 61: 145-154.

Sutherland, G.K. & Eastwood, A. 1916. The physiological anatomy of *Spartina townsendii.* Annuals Botany 30: 333-351.

Szabolcs, I. 1994. Soils and salinisation. In: M. Pessarakli. (Ed.), Handbook of Plant and Crop Stress. New York: Marcel Dekker. 3-11 pp.

U.S. Salinity Laboratory Staff 1954 Diagnosis and improvement of saline and alkali soils. In: L. A. Richards. (Ed.). Agricultural Handbook of the U.S. Department of Agriculture, Washington D.C.:Government Printing. 157 pp.

Tazuke, A. 1997. Growth of cucumber fruit as affected by the addition of NaCl to nutrient solution. Journal of the Japanese Society for Horticultural Science 66: 519-526.

Volkmar, K.M., Hu, Y. & Steppuhn, H. 1998. Physiological responses of plants to salinity: A review. Canadian Journal of Plant Science 78: 19-27.

Walsh, G.E. 1974. Mangroves. A review. In: R. J. Reimold, & W. H. Queen. (Eds.), Ecology of halophytes. New York, New York: Academic Press. 51-174 pp.

Warne, T.R., Hickok, L.G., Sams, C.E. & Vogelien, D.L. 1999. Sodium/potassium selectivity and pleiotropy in stl2, a highly salt-tolerant mutation of *Ceratopteris richardii*. Plant Cell and Environment 22: 1027-1034.

Weber, E. & D'Antonio, C.M. 1999. Germination and growth responses of hybridizing *Carpobrotus* species (Aizoaceae) from coastal California to soil salinity. American Journal of Botany 86: 1257-1263.

Winicov, I. & Bastola, D.R. 1997. Salt tolerance in crop plants: New approaches through tissue culture and gene regulation. Acta Physiologiae Plantarum 19: 435-449.

Winicov, I. 1998. New molecular approaches to improving salt tolerance in crop plants. Annals of Botany 82: 703-710.

Winicov, I. & Bastola, D.R. 1999. Transgenic overexpression of the transcription factor Alfin1 enhances expression of the endogenous MsPRP2 gene in alfalfa and improves salinity tolerance of the plants. Plant Physiology 120: 473-480.

Winter, U., Kirst, G.O., Grabowski, V., Heinemann, U., Plettner, I. & Wiese, S. 1999. Salinity tolerance in *Nitellopsis obtusa*. Australian Journal of Botany 47: 337-346.

Wolf, O., Munns, R., Tonnet, M.L. & Jeschke, W.D. 1991. The role of the stem in the partitioning of Na+ and K+ in salt treated barley. Journal of Experimental Botany 42: 697-704.

Wyn Jones, R.G., Brady, C.J. & Speirs, J. 1979. Ionic and osmotic relations in plant cells. In: D.L. Laidman, & R.G. Wyn Jones. (Eds.), Recent Advances in the Biochemistry of Cereals. New York, New York: Academic Press. 1-391 pp.

Wyn Jones, R.G. & Pollard, A. 1983. Proteins, enzymes and inorganic ions. In: A. Läuchli, & R. L. Bieleski. (Eds.). Inorganic Plant Nutrition. Encyclopedia of Plant Physiology 15b, Hiedelberg, Germany: Springer Verlag. 528-555 pp.

Yeo, A. 1998. Molecular biology of salt tolerance in the context of whole-plant physiology. Journal of Experimental Botany 49: 915-929.

CHAPTER 23

HALOPHYTE USES FOR THE TWENTY-FIRST CENTURY

NICHOLAS P. YENSEN

NyPa International and Centro de Investigación en Alimentación y Desarrollo (CIAD), 727 North Ninth Ave., Tucson, Arizona, 85705 USA
nypa@aol.com

Abstract. There are about a billion hectares of salt-affected land world wide, which may be resource opportunities for halotechnologies, such as halophyte crops and landscape plants, which grow better under high salinities. While much of this land occurs in the Middle East, Central Asia, Northern Africa and Australia it seems that no country is free of salinity issues. The first patent for a halophyte crop was issued less than 20 years ago and at present, crops are being developed by classical breeding, biotechnology, tissue culture and plant exploration. An extensive spectrum of salt-tolerant and halophilic crops is being developed by the new science of halophytology, with over a hundred genera already being studied. Of the estimated 10,000 salt-tolerant species, a potential may exist to develop as many as 250 halophyte staple crops, not to mention many ornamentals, dune stabilizers and environment improving halophytes. New "crops" are being developed that can be used to eliminate toxic compounds and elements (e.g. Se, Pb, Cr, Cd, Zn), petroleum products, asphalt, or radionuclides via halophyte phytoremediation and bio-remediation. Desalinization can also now be done with algae, which convert seawater to "brackish" water and Capacitive Deionization Technology utilizing a carbon aerogel capable of efficiently removing salts via electrostatic charge. In landscaping, ornamental halophytes can use brackish or seawater thereby freeing up large amounts of fresh water for municipal and domestic use. These new halotechnologies will carry their own knowledge bases and terms yet to be invented, much as the US Salinity Lab developed a glycophyte knowledge base for freshwater crops growing in brackish waters and soils. Today, we are entering a new era where the science of halophytology is being developed to understand euhalophytes and how we may benefit from their abilities. Much practical work remains to be done, as well as developing the basic science and concepts necessary to properly understand this new information base. An adequate hypothesis for why some halophytes have increased productivity with increasing in salt levels is needed.

1. INTRODUCTION

The 21st Century will likely be the century of halophyte agriculture expansion, as diminishing fresh water resources put pressure on civilization to utilize the vast saline soils and aquifers. Even today, trillions of dollars are lost annually in mildly saline fields that could be billions of dollars gained from new salt-loving crops. Additional benefit may come from extensive saline areas suitable for halophyte crops and landscape plants. This may include many areas of the world, but particularly Australia, Central Asia, the Middle East and Northern Africa.

M.A. Khan and D.J. Weber (eds.), Ecophysiology of High Salinity Tolerant Plants, 367–396.

To better understand the extent of the "problem" (think "opportunity"), the extent of salinization in the world is herewith briefly described. This area enumeration is followed by an extensive listing of the numbers and kinds of plant species that can survive in salty conditions and represent potential for halophyte crops. From this information it is possible to predict the potential and the extent of developing new salt-tolerant and salt-loving (halophilic) crops, potential for remediation and ornamentals. New developments and discoveries may be expected to provide fresh drinking water for animals and people. While the author has a 30 year bias toward certain crops that have been developed, in some part through his effort, these crops represent only a sample of what is being done and can be done in the future. The brief sampling of halophyte crops presented are taken from some of those that the author knows best and there are now many halophyte crop developers who deserve respect for their efforts. The Biosalinity Awareness Project website covers many of these other developments and should serve as an excellent additional source of information on the other work. The site was created by Sasha Alexander and may be found at http://www.biosalinity.org

1.1. Terminology

The concept of Halotechnologies or salt-related technologies is gradually changing the ways that we think about our salty resources for sustainable applications. Drainage and Biodrainage, the consumptive water use by plants, (Heuperman et al., 2002) may be included in the proper management of excess water, and when carefully applied, have saved hundreds of thousands of hectares world wide. Amendments also provide some degree of soil management.

Halophytology is the study of plants in saline environments. This is a relatively new science and will be key to the commercial application of high salinity cultivation or haloculture, the cultivation of plants and animals in highly saline environments.

Halophytes (halo=salt + phyte=plant), are plants capable of completing their life cycle under highly saline conditions. Glycophytes (glyco=sweet + phyte=plant), are conventional "sweet-water plants" which do well on fresh water and have decreasing productivity with increasing salt levels. Significant work has been done in developing Oligohalophytes (oligo=few) or "salt-resisting glycophytes" which resist salty conditions but do not grow well until returned to fresh water cultivation. Miohalophytes (mio=partly) are "salt-tolerant plants" which can maintain their productivity up to some thresh-hold salt level and then have decreasing productivity with increasing salt levels. Miohalophyte crop application (such as sugar beet, dates, Lucerne, olives, etc.) has helped to slow the salt-induced, productive-land loss in some marginal areas. Contemporarily, there has been research into the development of "salt-loving plants" i.e. halophilic (philic=loving) halophytes which are also called euhalophytes (eu=true) or "true halophytes" because they have increased productivity with increasing salt levels and actually grow better under salty conditions than under fresh-water conditions. The advantage of halophilic crops is that high salt levels, impermeable soils, and water logging can actually improve their productivity.

In this paper we will examine the potential uses for halophytes, haloculture and halotechnologies for more productive usage of resources for the benefit of humanity and for the complimentary protection of natural habitats and ecosystems.

2. SALT-AFFECTED AREAS OF THE WORLD

The principle source of the salt in salt-affected land is from microscopic aeolian salt carried inland from the oceans (13 kg/ha/year), followed by human-caused (secondary salinization) (2.3 kg/ha/year). Volcanoes contribute 0.2 kg/ha/year and, contrary to the popular belief, the weathering of rocks contributes the least (0.04 kg/ha/year) (Waisel, 1972). Salt accumulates in many inland areas through this aeolian process (Figure 1).

Figure 1. *Salt-affected inland areas of the world (after Szabolcs, 1994b).*

By tabulation, there are approximately one billion hectares of saline land on this planet (Table 1) (Yensen, 1985). Another billion acres of desert overlie saline aquifers (Neary, 1981). According to Szabolcs (1992a) account, 10 % of the Earth's land surface is salt affected. Of the irrigated land, it is estimated that 10-50 % (1.5 to 7 billion additional hectares) has reduced productivity due to salinity. Of the approximately 14.6 billion hectares of land, the area available for irrigation is only 1 % (Espenshade, 1960; Israelsen & Hansen, 1962; Yensen et al., 1981). Given then that 10% of the Earth's land surface is salt affected, and by tabulation we may calculate that 1.0 - 1.5 billion hectares of land are salt-affected. The majority [~90%] of salt-affected land is naturally saline (primary salinization)(Szabolcs, 1992a, 1992b; Yensen, 2002). Each year, approximately 5 million acres still go out of productivity due to salinity.

One of the reasons for the variation in the estimated salt-affected areas is because there are a number of ways to define and classify salt-affected soils. Because there is an inconsistent usage of classification around the world, a precise across-the-board tabulation and comparison is not possible. The terms solonetz, solonchaks, and saline phase are all commonly used categories of salt-affected soils; and, which one is used (or used in combination) can significantly alter the number of salt-affected hectares listed.

For instance, Canadian and the United States, by some accounts have similar numbers of salt-affected soils, (7,238,000 and 8,517,000 ha. respectively). Canada, however, has only a few hectares (264,000) in a saline phase and is mostly considered solonetz (6,974,000 hectares), while the United States is mostly saline phase (5,927,000 ha) and less than half as many hectares of solonetz (2,590,000 ha). In contrast, Mexico (with a total of 1,649,000 salt-affected ha) has only 242,000 hectares cited as solonchaks, compared to 1,407,000 hectares in a saline phase and none listed as solonetz.

Another example is Argentina where 85,612,000 hectares are salt-affected. Kovda (1977) and White (1978) indicate that nearly half of the salt-affected land (41,321,000 ha) is in an different catagory, an alkaline phase, with only 1,905,000 hectares in solonchaks, 30,568,000 hectares in a saline phase, 11,818,000 hectares in solonetz. Some estimates are by a few soil profiles, while others consider only surface salts and/or satellite imagery. Also, some salt-affected areas become remediated while others are becoming increasingly salt-affected resulting in differing estimates over time.

Table 1. *Salt-affected area (hectares) of the regions of the world.*

Region	*Area salinized (Ha)*	*Potentially affected*	*% irrigated area affected*	*Ref.*
North & Central	15 755 000			7
America	2 300 000			5
Canada	7 238 000			2, 23
	2 187 000			16
Manitoba	364 000			16
Saskatchewan	1 623 000			16
Alberta	200 000			16
United States	8 517 000		20-25	2, 4, 23
	30 269 500		5*	16
Alabama	26 200			16
Arizona	2 222 600			16
Arkansas	13 500			16
California	3 572 200			16
Colorado	845 800			16
Connecticut	13 400			16

Delaware	2 000	16
Florida	1 275 200	16
Georgia	105 700	16
Hawaii	21 000	16
Idaho	701 300	16
Illinois	159 200	16
Indiana	28 100	16
Iowa	343 800	16
Kansas	1 059 100	16
Kentucky	1 000	16
Louisiana	301 100	16
Maine	24 100	16
Maryland	10 800	16
Massachusetts	18 300	16
Michigan	60 500	16
Minnesota	1 571 000	16
Mississippi	24 400	16
Missouri	340 500	16
Montana	7 071 600	16
Nebraska	2 040 500	16
Nevada	1 335 100	16
New Hampshire	3 500	16
New Jersey	29 500	16
New Mexico	3 361 200	16
New York	55 800	16
North Carolina	12 700	16
North Dakota	5 227 000	16
Ohio	10 600	16
Oklahoma	546 300	16
Oregon	273 200	16
Pennsylvania	14 700	16
Rhode Island	4 700	16
South Carolina	133 100	16
South Dakota	5 252 900	16
Tennessee	2 500	16
Texas	4 749 900	16
Utah	671 000	16
Vermont	200	16
Virginia	101 800	16
Washington	190 900	16
West Virginia	3 600	16
Wisconsin	40 300	16
Wyoming	1 780 600	16
Puerto Rico	7 400	16
Mexico	1 649 000	2, 23
Mexico & Central America	1 965 000	7
Cuba	316 000	2, 23

South America	129 000 000		7
	2 100 000		5
Argentina	32 000 000	11*	9
	85 610 000		2, 23
Bolivia	5 949 000		23
Brazil	4 503 000		23
Chile	8 642 000		23
Columbia	907 000	20	4, 23
Ecuador	387 000		23
Paraguay	21 902 000		23
Peru	21 000	12	4, 23
	> 339 633		26
Venezuela	1 240 000		23
Europe	50 804 000		7
	3 800 000		5
Africa	80 538 000		7
	14 800 000		5
Algeria	3 150 000		8, 23
		10-15	4
	7 200 000	3*	22
Angola	526 000		23
Botswana	5 679 000		23
Cameroon	671 000		23
Chad	8 267 000	43	10, 23
Djibouti	1 741 000		23
Egypt		30-40	4
	7 360 000	7*	8, 21, 23
	8 700 000	9*	22
		33	25
Ethiopia	20 000 000	16*	11
	11 033 000		23
Gambia	150 000		23
Ghana	318 000		23
Guinea	525 000		23
Guinea-Bissau	194 000		23
Kenya	9 000 000	15*	13
	4 858 000		23
Liberia	406 000		23
Libya	4 000 000	2*	22
	2 457 000		23, 8
Madagascar	1 324 000		23
Mali	2 770 000		23
Mauritania	640 000		8, 23
Morocco	1 148 000		8, 23
	2 300 000	5*	22
Namibia	2 313 000		23
Niger	1 489 000		23
Nigeria	6 502 000		23
Senegal	765 000	10-15	23, 4

Sierra Leone	307 000		23
Somalia	5 602 000		8, 23
Sudan	4 874 000	< 20	8, 23, 4
Sudano-Sahelian	50 000 000	4*	10
Tanzania	3 537 000		23
Tunisia	1 300 000	8*	22
	990 000		8, 23
Zaire	53 000		23
Zambia	863 000		23
Zimbabwe	26 000		23
Western Sahara	0	0	22
Middle East			
Iran		<30	4
	23 800 000	15*	22
	27 085 000		23
		15	25
Iraq	6 726 000	50	8, 4, 23
	8 800 000	20*	12
	6 100 000	14*	22
		70	25
Jordan	180 000	16	8, 4, 21,
	300 000	3*	23
		3.5	22
			25
Kuwait	209 000	85	8, 23, 25
	200 000	8*	22
Lebanon	0	0	22
Oman (& Muscat)	290 000		8, 23
	> 127 400		20
	2 000 000	7*	22
Palestine/Israel		13	4
	28 000		23
Qatar	225 000		8, 23
	200 000	18*	22
Saudi Arabia	6 002 000		8, 21, 23
	9 300 000	4*	22
Syria	532 000	30-35	8, 4, 23
	321 000	30	15
	500 000	3*	22
		50	25
Turkey	> 351 025		18
	2 500 000		23
United Arab	1 089 000	25	8, 23, 25
Emirates	1 000 000	13*	22
Asia	52 700 000		5
North & Central Asia	211 686 000	50	7, 25
South Asia	87 000 000		7
South-East Asia	19 983 000		7

Afghanistan	3 700 000		6*	22
	3 101 000			23
Bangladesh	< 571 508			19
	3 017 000			23
Cambodia	1 291 000			23
China	36 658 000			23
India	7 000 000	12 000 000	27	3, 4
	8 500 000			17
	23 796 000			23
Indo-Gangetic plains	3 046 000			17
Arid West	1 450 000			17
Peninsular region	9 932 000			17
Coastal region	2 173 000			17
Kazakhstan	21 500 000		8*	22
Kyrgyzstan	100 000		0.5*	22
Malaysia	3 040 000			23
Peninsular (by difference)	1 502 000			-
Sarawak	1 538 000			23
Mongolia	4 070 000			23
Myanmar	634 000			23
Pakistan	5 700 000			14
	600 000		>40	4, 25
	10 456 000			23
Russia & Soviet Union	170 720 000			23
Sri Lanka	200 000		13	23, 4
Tajikistan	700 000		5*	22
Turkmenistan	7 300 000		15*	22
Thailand	3 430 000			6
	1 456 000			23
Uzbekistan	2 400 000		5*	21
	6 300 000		14*	22
Viet Nam	983 000			23
Australasia	357 330 000			7
	900 000			5
Oceania				
Australia	2 476 000	>11 783 000		1
	357 240 000			23
Western Australia	1 804 000	6 109 000		1
South Australia	402 000	600 000		1
Victoria	120 000	Unknown		1
New South Wales	120 000	5 000 000		1
Tasmania	20 000	Unknown		1

Queensland	10 000	74 000	1
Northern Territory	minor	Unknown	1
Fiji	90 000		23
Indonesia	13 213 000		23
New Zealand	8 800		24
Solomon Islands	238 000		23

**of total region land surface. Italics indicate secondary human-caused salinization.1-Bennett, 1998 after LWRRDC 1997 (Hayes, 1997); 2- Kovda, 1977; White, 1978:35; 3- Girdhar,1993; 4-World Resources, 1987; 5-Ghassemi, et al., 1995; 6-Arunin, 1992; 7-Szabolcs, 1992b; 8-Batanouny, 1993; 9-Maddaloni, 1986; 10-Kraiem, 1986; 11-Sissay, 1986; 12-Abdul-Halim, 1986; 13-Kanani & Torres, 1986; 14-Sandhu & Qureshi, 1986; 15-Sankary, 1986; 16-McKell et al., 1986; 17-Dagar, 1995; Öztürk, et al., 1995; Nazrul-Islam, 1995; 20-Ghazanfar, 1995; El-Betagy, 2004; 22-De Pauw, 2004; 23-Kodva & Szabolcs, 1979; 24-Cromarty & Scott, 1995; 25-Al-Attar, 2002; 26-Zavaleta, 1969.*

In Australia a number of estimates have been made for various reasons, some political. There is no question, however, that the amount of salinization in Australia is increasing. It currently has 2,476,000 hectares affected by dryland salinity (Bennett, 1998 citing Hayes, 1997). But, there are 11,783,000 or more hectares potentially affected by dryland salinity (Bennett, 1998). The expansion rate of the area of dryland salinity is estimated to be 3-5% per year. It is also estimated that this is an annual loss of $270 million Australian dollars (Bennett, 1998).

World wide, billions of dollars are lost each year due to the effect of salt (Yensen et al., 1995). As noted above approximately 5 million acres a year are still going out of productivity due to salinity. In the last 30 years we have nearly pushed to the limit the salinity envelope for conventional crops and the predicted global climate change threatens to increase salinity. We should therefore look at these salty areas not as problems, but as opportunities.

Because salt-affected soils and ground waters cross national boundaries, international collaboration and coordination become essential in the development of regional and global salinity strategies. It is critical to initially involve politicians, institutions, select farmers and other potential beneficiaries from the onset. Thus, when the critical time for expansion comes, all parties will be familiar with their role.

Worldwide, there is a huge opportunity for salt-loving and salt-tolerant crops and technologies. It may be possible to produce twice as much food on this planet with halophyte crops than is currently being produced. Remarkably, it was not until the latter part of the 20th century that serious efforts were begun to develop euhalophyte land crops. And, surprisingly, the first euhalophyte crop patent [*Distichlis palmeri* var. y. 1a] issued as late as 1985.

To better understand the potential of developing halophyte crops it is necessary to examine the base of wild halophyte species from which such crops may be derived.

3. THE NUMBER OF HALOPHYTE SPECIES

Worldwide the diversity of salt-tolerant and halophyte plants is very great. Choukr-Allah et al. (1996) estimated that there are about 6000 species of terrestrial halophytes. Menzel and Lieth (1999), based on Aronson's (1989) listing of 1560 halophytes species, have estimated that 0.6% of the" >250,000 plant species" are halophytes. The present author (Yensen et al., 1988) had previously estimated that approximately 10,000 species (4%) of higher plants were salt-tolerant or halophytic, based on a study of Sonoran Desert halophytes. The present paper, however, estimates that 11.1 ± 0.9% of the world's terrestrial plant flora is salt tolerant or halophytic (Table 2).

If this is multiplied by Raven et al.'s (1986) estimate of the number of plant species in the world, 235,000 species, then we find that there are 26,000 ± 2,000 species of salt-tolerant and halophyte plants in the world (Table 3). Estimating the number of halophytes species is, at present, more difficult. Because distinctions between salt-tolerant and halophytic species were not made in the lists of (Table 2) the less-accurate percentage (14%) of Sonoran Desert halophytes (Table 4) (to salt-tolerant plants) suggests that there are approximately 3640 ± >280 species of euhalophytes in the world.

3.1. The Number of Halophyte and Salt Tolerant Families and Genera

In order to develop new crops suitable for many different climates, soils, and needs of humanity, it is valuable to have a diverse "Library" of germplasm to draw upon. Fortunately, the occurrence of halophytism is wide spread throughout the plant kingdom and appears to have evolved a number of times lending support to the concept that many fresh-water tolerant glycophytes may have halophyte genes permanently switched off, but under mutagenesis may revert to halophytism. (Table 5) based on a listing of only 1861 halophytes and salt-tolerant species identifies 139 families and 636.

3.2. The Salt-Management Classification of Halophytes

There is basically three ways that halophytes manage their salt load: 1) exclusion, 2) excretion and/or 3) accumulation.Excluder plants (e.g. *Hordeum* spp., *Melilotus* spp., *Rhiozophora* spp. and most glycophytes, glyco = sweet + phyte = plant) exclude the salts from entering the vascular system at the root level. The use of excluder halophytes is best at the lower salinities as the exclusion of salt is often an energy-expensive process and because when the excluders extract water the result is the concentration of salt in the soil, which may induce the plants to expire.

Table 2. *Regional percentages of floras as "halophytes" (and/or salt-tolerant) as potential source material for developing halophyte crops.*

Region	Total flora (# species)	"Halophytes" (# species)	Percent halophytic	Ref.
Australia	17590	1800	10.2%	1
Algeria	3500	360	10.3%	2
Egypt	2100	250	11.9%	2
Europe	6550	303	4.6%	4
France	4500	265	5.9%	2
Indo-P (New Guinea)	12000	900	7.5%	3, 5
Iran	6170	354	5.7%	4
Iraq (lowland)	1200	135	11.3%	2
Jordan	2100	260	12.4%	2
Kuwait	450	80	17.8%	2
Libya	1800	295	16.4%	2
Morocco	4200	380	9.0%	2
N. Africa steppes	2200	390	17.7%	2
Palaestina flora*	2800	300	10.7%	2
Qatar	435	70	16.1%	2
Romania	3350	294	8.8%	4
Sahara	2800	400	14.3%	2
Saudi Arabia	2200	250	11.4%	2
Sonoran Desert	2891	150	5.2%	6
Syria	3460	260	7.5%	2
Tunisia	2250	303	13.5%	2
UAE	450	70	15.6%	2
Estimated % of world flora as "halophytes"		mean	11.1%	
(estimate is un-weighted and desert biased)		SEM	± 0.9%	

** Palestine/Israel, Jordan and Sinai. 1 - NP Yensen unpubl. list of 600 spp. estimated to be approximately 1/3 of the "halophytic" flora. 2 - Le Houerou, 1993. 3 - http://www.eurekalert.org/pub releases/2002-04/uom-tsf042302.php; George Weiblen, plant biology department, (612) 624-3461 gweiblen@umn.edu. 4 – Breckle, 1986. 5 - Mepham & Mepham 1985. 6 - Yensen et al., 1988. [also see Moghaddam & Koocheki, 2004]*

Table 3. *Number of potential halophyte crops in the world.**

*Number of Plant species In the World ***	*Number of Halophytes* in the World****	*Number of Crops (per Sturtevant) in the World*	*Number of Potential Halophyte Crop Species*
235 000	26 000 ± 2 000	32 000	3 550

**Includes both miohalophytes and euhalophytes* ***Raven, et al., 1986:720.*
****235000 x 0.11 = Number of plant species x halophyte %, from Table 1.*

Table 4. *Potential halophyte crops from the Sonoran Desert (After Yensen et al., 1988)*

	Total	*Edible*	*"Staple"*
Higher plant species	2 891	596	47
Salt tolerant	150 (86%)	38 (78%)	9 (75%)
Euhalophyte	25 (14%)	11 (22%)	3 (25%)

Table 5. *Taxonomic distribution of approximately 10% of the possible salt tolerant species (1861 spp.) in descending order from the most speciose family and indicating the number of genera in a family. Based on species accounts by author, which may be found at www.nypa.net.*

Family	*Species*	*Genera*	*Family*	*Species*	*Genera*
Chenopodiaceae	382	63	Asclepidaceae	7	6
Poaceae	188	71	Bignoniaceae	7	6
Asteraceae	118	56	Celastraceae	7	3
Fabaceae	89	37	Juncaginaceae	7	1
Cyperaceae	74	12	Liliaceae	7	5
Aizoaceae	69	23	Nyctaginaceae	7	3
Plumbaginaceae	58	5	Zannichelliaceae	7	1
Tamaricaceae	39	2	Lythraceae	6	4
Solanaceae	30	8	Meliaceae	6	2
Zygophyllaceae	28	12	Najadaceae	6	1
Euphorbiaceae	27	13	Primulaceae	6	3
Apiaceae	25	15	Ranunculaceae	6	2
Myrtaceae	25	7	Sonneratiaceae	6	1
Polygonaceae	25	4	Sterculiaceae	6	3
Rhizophoraceae	25	8	Typhaceae	6	1
Scrophulariaceae	25	12	Amaryllidaceae	5	1
Brassicaceae	24	7	Anacardiaceae	5	4
Arecaceae	23	15	Capparaceae	5	3
Caryophyllaceae	22	9	Phoenicaceae	5	2
Malvaceae	21	12	Portulacaceae	5	2
Zosteraceae	21	4	Resedaceae	5	3
Juncaceae	18	1	Sapotaceae	5	5
Plantaginaceae	18	1	Tiliaceae	5	3
Convolvulaceae	15	6	Barringtoniaceae	4	1
Frankeniaceae	15	1	Curcubitaceae	4	3
Myoporaceae	15	2	Onagraceae	4	2
Phytolacaceae	14	4	Rubiaceae	4	4
Amaranthaceae	13	8	Rutaceae	4	4
Verbenaceae	13	7	Salvadoraceae	4	2
Avicenniaceae	12	2	Acanthaceae	3	1
Cymodoceaceae	12	4	Basellaceae	3	2
Pandanaceae	12	1	Iridaceae	3	2
Potamogetonaceae	12	1	Lamiaceae	3	3
Boraginaceae	11	5	Moraceae	3	2
Combretaceae	11	7	Myristicaceae	3	2
Gentianaceae	10	4	Myrsinaceae	3	2
Casuarinaceae	9	1	Olacaceae	3	2

Goodeniaceae	9	3	Polypodiaceae	3	1
Hydrocharitaceae	9	3	Posidoniaceae	3	1
Cactaceae	8	7	Rosaceae	3	2
Alismataceae	7	2	Ruppiaceae	3	1
Apocynaceae	7	4			

Families with 2 species: *Annonaceae, Batidaceae, Bombacaceae, Bromeliaceae, Chrysobalanaceae, Crassulaceae, Cupressaceae, Ebenaceae, Elatinaceae, Ephedraceae, Flacourtiaceae, Lauraceae, Loasaceae, Melastromaceae, Orchidaceae, Pinaceae, Pontederiaceae, Portaliaceae, Pteridaceae, Thymelaceae.*
Families with only one species: *Aquifoliaceae, Blechnaceae, Brexiaceae, Callitrichaceae, Calyceraceae, Campanulaceae, Cistanchaceae, Clusiaceae, Commelinaceae, Cuscutaceae, Cycadaceae, Dilleniaceae, Epacridaceae, Eriocaulaceae, Fagaceae, Geraniaceae, Haloragaceae, Hippuridaceae, Hydrocaryaceae, Leitneraiaceae, Linaceae, Myricaceae, Nymphaeaceae, Oleaceae, Osmundaceae, Podocarpaceae, Proteaceae, Rhamnaceae, Salicaceae, Sapindaceae, Saururaceae, Simaroubaceae, Steraceae, Surianaceae, Theaceae, Vitaceae.*

TOTAL FAMILIES	TOTAL GENERA	TOTAL SPECIES
139	636	1861

Accumulator plants (e.g. *Atriplex* spp., *Salicornia* spp. etc.) sequester salts in the cell vacuoles as an osmoregulation mechanism and presumably to avoid toxic effects. In accumulator plants salt may account for as much as half the dry weight. For this reason ash-free dry weight (ash = salt) determines the actual productivity.

Accumulator plants usually do not make good forages due to the accumulated salt. Under rangeland conditions, however, saltbush (*Atriplex* spp.) and others, with deep roots, can grow well on the fresh rainwater and do not accumulate high levels of salt. Also, if grown on salt water but foraged during and after the rainy season they may be good forages. In *Salicornia* spp. the salt-free seed and the low-salt growing tips may be used.

Some accumulated salts can be toxic. In Mexico I have observed *Atriplex canescens*, "happily" growing on such highly toxic saline soil that the plants had to be fenced off to reduce cattle mortality. Under other circumstances, however, many *Atriplex* spp. (low in oxalates, malates, etc.) can serve as an excellent browse. The selection and development of saltbushes without these liabilities would be an excellent program. (see Malcolm, 1969, 1971, 1979, 1980, 1982; Barrett-Lennard & Malcolm, 1995 for discussions of saltbushes). It is important to understand the plants, the soil and the ground water on which accumulator plants are grown.

Excreter plants (e.g. *Distichlis* spp., *Avicennia* spp., etc.) excrete salts through salt glands. Excreter plants (Crinohalophytes) can have low levels of tissue salt. The *Distichlis* spp. salt gland, like other salt-loving grasses, is bi-cellular having a basal cell embedded in the leaf's epidermal layer and modified trichome (hair) cell extending above the surface of the leaf. Some salt accumulator halophytes (such as *Atriplex* spp.) do "excrete" salt with above-the-surface bladder cells, which may burst to release salts. Thus, as noted, their use is best after rains. Whereas plants that directly excrete salts without salt accumulation (cf. *Distichlis* spp.) have their biomass immediately available as forage.

3.3. Future Halophyte Cultivars

Over a hundred genera are now being trialed for commercial applications. Iran, for example, currently has 66 research projects on halophytes (Cheraghi, 2004). Local abundances of halophyte species may determine the opportunities for a specific need or application. For instance, of the 2981 species of higher plants in the Sonoran Desert, there are approximately 150 species that are salt-tolerant (~miohalophyte) and 25 species that are euhalophytes (Table 4). There are beginning to be a few halophyte flora listings that can assist specific regions in identifying potential halophytes to investigate, such as A. Kahn's (per. com.) remarkable list of approximately 400 salt-tolerant and halophytes species for Pakistan.

Patented cultivars are still few in number, but are a harbinger of what the future could hold. Clearly a large body of information on halophytes is accumulating and the effort to develop useful varieties is only beginning. Due to the large number of potential halophyte species the identification of one for development is actually a daunting proposition. It is the reduction of the potential option that is difficult, not the identification of a potential halophyte to develop.

A short listing from Yensen and Bedell (1993) illustrates some genera with salt-tolerant species that may be developed into useful cultivars: *Abronia, Acacia, Acrostichum, Anemopsis, Arenaria, Arthrocnemum, Batis, Brahea, Bruguiera, Cakile, Carex, Chloris, Coccoloba, Convolvulus, Cotula, Cressa, Crithmum, Dactyloctenium, Disphyma, Eleusine, Frankenia, Galenia, Grindelia, Halimone, Halocharis, Halosarcia, Halostachys, Hedysarum, Heliotropium, Hyphaene, Ipomea, Jaumea, Kunzea, Laguncularia, Leptochloa, Maytenus, Melaleuca, Monanthochloe, Myoporum, Myristica, Nitraria, Odyssea, Oncosperma, Parapholis, Pelliciera, Pemphis, Phyla, Phylospadix, Physalis, Pinus, Pluchea, Potamogeton, Rhizophora, Rumex, Ruppia, Sarcocornia, Scirpus, Scolopia, Sesuvium, Sonneratia, Spergularia, Syringodium, Tabeuia, Tetragonia, Thespesia, Typha, Zizania.*

Halophytic genera with varieties already at some stage of use: *Agropyron, Allenrolfea, Atriplex, Avicennia, Casuarina, Cenchrus, Diplachne, Distichlis, Eucalyptus, Juncus, Kochia, Kosteletzkya, Lycium, Maireana, Nypa* [no relation to NyPa Inc.], *Pandanus, Panicum, Paspalum, Plantago, Puccinellia, Salicornia, Salsola, Spartina, Sporobolus, Suaeda, Tamarix and Zostera.*

Salt-tolerant/halophyte genera, that could be crossed to increase the salt-tolerance of our present glycophyte cultivars: *Apium, Annona, Achillea, Acacia, Aster, Asparagus, Beta, Bougainvillea, Capparis, Carpobrotus, Cassia, Castilleja, Ceriops, Chenopodium, Chloris, Chrysanthemum, Cocos, Combretum, Conocarpus, Cynodon, Daucus, Ficus, Helianthus, Hibiscus, Hordeum, Limonium, Lumnitzera, Medicago,*

Mesembryanthemum, Mora, Olea, Phoenix, Phragmites, Portulaca, Prosopis, Rubus, and *Trifolium,* to name a few.

These incomplete lists illustrate the considerable untapped potential.

3.4. Patenting

Wild halophytes and/or domesticated plants of the public domain are not patentable; that is, in order to obtain a patent a potential inventor must demonstrate: 1) the creation of a unique cultivar, 2) through significant effort and 3) which results from at least one step that was unobvious to others of the discipline.

The actual research costs to develop a new cultivar are minor compared to the patenting costs, which are minor compared to the agricultural development costs; and which are minor to the business establishment, marketing and distribution costs. Because of these exponentially increasing costs, and without serious funding, only a few of the new cultivars may ever become successfully distributed, irrespective of their intrinsic value for society.

Obtaining Intellectual Property (IP) rights, however, is the next step after the development of a new cultivar. Ironically, this is not so much for protection, as it is for credibility. In many countries of the world where halophyte crops are needed, IP is not respected (even if governments have signed the international patent treaty and a world patent is registered in that country). Legal protection is only at great expense (note: the US, for instance, in spite of being a developed country is not part of the World Court making any international lawsuits difficult). A patent does, however, give credibility to potential funding organizations, who must then realize that at least the inventors are serious. And secondly, a patent provides assurances to farmers and the non-cognoscenti who must now believe that halophyte crops can grow better on salt-ruined land than in fertile soil irrigated with fresh water, i.e., a patent indicates that at least some official body accepts the fact that these new halophyte cultivars can grow well in salty conditions. Our first patent for a halophyte crops was issued on 9 August 1988, US Patent 4762964, (Yensen, 1988), but now totals seven patents with worldwide registry at a cost of approximately $150,000 USD.

For countries that respect patent rights, the use of new halophyte crops can quickly result in a better standard of living, because it will be these countries that inventors and developers will want to market other protected cultivars. For the all peoples, as virtually all countries have a salt problem, there is long-term potential to raise the standard of living. This is especially true in the 21st Century, as it is expected that many other inventors and breeders will be patenting new halophyte crops; not to mention ornamentals, turf grasses, dune stabilizers and oil-well remediation halophytes for economic benefit, all of which are ideal plant-types that may benefit from the patent process.

4. THE FIRST HALOPHYTE PATENTS

While selection and use of halophytes extends from pre-historical times, the first new halophyte crop uniquely bred, selected and designed with unique agronomic characteristics sufficient to satisfy the US Patent Office has only recently been developed as noted above. After six years of working with over a million individual plants the author first filed for a halophyte crop patent in 1985. The patent issued on 9 August 1988. It is possible that local patents in the Soviet Union may predate this patent, but patent searches have not yet located them. Information on earlier patented halophytes is being sought. In any case US Patent 4762964 may be recognized as a patented euhalophyte cereal grain.

This cereal grain development and patenting was funded, in part, by Engineering Research Associates (SEBRA) and carried out by NyPa Incorporated of Tucson, Arizona. Following the direction of SEBRA, the new taxon was given the name *Distichlis palmeri* var. *yensen-1a* and trademarked "NyPa® grain" and "WildWheat® grain." This first invention was followed by 5 other halophyte patents of cereal grains and in 1994 a patent was issued for a halophyte forage crop, *Distichlis spicata* var. *yensen-4a*, "NyPa® Forage," again with the varietal nomenclature by the funding institution. This patent, however, was completed pro se - i.e. without a patent attorney, wherein the government patent examiner is required to act as the inventor's attorney as well as the examiner. Pro se patenting is allowed in some countries for small companies or un-funded individuals and we were a small, under-funded company at that time.

There was a movement by the Environmental Research Laboratory (ERL) at the University of Arizona to patent a variety of *Salicornia bigelovii* var. "*SOS-7*." While ERL did not finish the patent process, it appears that a proprietary cultivar, which is "leased" only under a use license, has more protection than a patent. This may even be true for a patent registered worldwide. This is due primarily to new international laws, originally designed to provide protection of computer software. NyPa, for instance, currently has far more proprietary plants than patented plants, yet continues to patent certain cultivars for various reasons, such as increased capital worth to a company, credibility of claims about a cultivar, reduced incentive for illegal propagation, and for industries that are accustomed to patented plants as opposed to "leased plants," such as the turf-grass industry.

While 16 years has passed since our first patented halophyte (and there are still few other patented halophytes), we optimistically expect that the 21st century will find many other inventors and breeders patenting and declaring proprietary rights over new halophyte cultivars.

By the time this goes to press the first halophyte patent will expire without having covered its patent costs. By contrast, the more recently patented forage will have paid for its itself many times over from the standpoint of establishing projects. As time goes on the use of halophyte crops will become much more understood and accepted.

Halophytology is a field that has been barely recognized much less utilized. As awareness of the potential of halophytic crops grows, humanity will come to recognize its economic and social worth. Hopefully, other inventors will find it easier to initiate and implement projects. To date, lack of understanding of halophytes and their application has been the greatest deterrent.

4.1. Cereal Grain Halophytes

Compared to the extensive "salt-tolerance" breeding programs of glycophyte cereals, there has been little work with of euhalophyte cereals. For instance, the first euhalophyte bread made in recent times was made from the grain of *Zostera marina* (Felger & Moser, 1973).

The following discussion, however, will deal with the euhalophyte cereal of, due the 20+ years of experience with this euhalophyte cereal grain, which, through a unique out-crossing-selection program, has had an increased yield from 1 kg/hectare to 4+ metric tons/hectare (in small test plots).

The development program had many difficulties. One of the first *D. palmeri* fields (in an alkaline/sodic, "heavy" clay soil, irrigated with ~7-10 g/L saline drainage water) was lost when the plants eliminated the salt from the rhizosphere in less than 3 years. Fields established near Hermosillo, Mexico were again in heavy clay but irrigated with a saline well. The well would pump 10-20 g/L salt water for 24 hours to (one to four times a month) before the cone of depression dropped sufficiently to obtain fresh water. Ironically, we lost these fields when the host farm went broke due to pumping costs and high salinity before the program could be scale up sufficiently to help them. The next trials (in Namibia) succumbed to a native common root rot fungus. In Arizona and California the fields and water were not sufficiently salty. Outstanding growth occurred in a Sahara Desert site (at 10-20 g/L salt) and a number of new cultivars were produced, but appropriate seed cleaning equipment was needed for this program in order to scale up. Although slowly progressing, Australia now has the world's pre-eminent halophyte grain development program under the direction of John Leake and Mark Sargeant.

While the agronomic development has been a somewhat step-wise progression, the marketing and product development has been highly variable, albeit an with an interesting history. Shortly after the first filing for patents in 1985, the NyPa WildWheat® grain of *Distichlis palmeri* var. *yensen-2* was initially wholesaled at $36 US/pound and retailed at $72/pound through Nieman Marcus, a gourmet food outlet (Anonymous, 1985). For this sale a hundred pounds of the grain was harvested (the second largest quantity ever amassed, excepting prehistory harvest by the Cocopah Indians. The demand, however, for this nutty-flavored, sweet grain far exceeded (and still exceeds) our ability to supply. So, in order to avoid "loosing" a market from an inconsistent supply, we have opted to wait until there is a stable supply capable of rapid expansion to meet the market.

For interest and the record, the largest quantity of *D. palmeri* grain we ever amassed (circa 200 kg ~500 lbs.) was sent to CPC International for an "unbiased" food industry evaluation. The agreement was that the evaluators would not have any information about the grain, which could have included a dissertation, two theses and our 10-years-worth of food science experience and information. Coarse-ground, gritty "flour" was sent off ("they would have milling capabilities"). The, London-based, third party food scientists, however, believed they had a final milled product and did not mill or sift the flour. Using only a couple of kilograms they made one "cake" and some biscuits, which turned out "unsatisfactory" and the largest mass of *Distichlis palmeri* flour ever produced (worth $36,000 circa 1990) was summarily tossed out without contacting us. In retrospect, it is easy to realize that they thought that they just had a different variety of coarse ground wheat flour. Also, wheat gluten normally requires 30 minutes of kneading to develop elasticity, whereas the gluten-free *D. palmeri* flour completely loses elasticity after only 10 minutes of kneading.

4.2. Forage

Animal forage such as the NyPa Forage, *Distichlis spicata* var. *yensen-4*, has just begun to contribute to economic worth, particularly in Australia. NyPa Forage is the result of over 20 years of research, breeding and selection of thousands of varieties.

Recent trials and economical studies in various countries suggest that, under proper management and cultivation practices, the forage variety can profitably produce forage suitable for goats, sheep, beef-cattle and form the bulk in formulated diets for dairy, ostrich, poultry, aquaculture and horses. Yearling dairy calves grew virtually as well (1 kg/day/4 kg feed) as with a conventional formulated diet (Yensen, 1997). The protein content is typically 8-12% with an unknown max-min nitrogen fertilizer requirement, but has a good energy value, 1200 KJ/100 g, making it particularly suitable for sheep, cattle forage and meat production.

The stem and leaf chemical composition ranges are: 6-19% protein; 8% gross fat; 47% non-nitrogenous matter (by difference); 20-37% gross fiber/cellulose; 5-14% ash; 1245 KJ/100 g; 5-7 % % lignin; 0.2-0.9% Ca; 0.1-1.0% P; 0.7-1.2% K; 0.14-0.33% Mg; 0.18-0.27% S; 0.2-2.8% Na; 0.8-1.1% Cl; 2-5 ppm B; 4-200 ppm Cu; 250-600 ppm Fe, 39-75 ppm Mn, and 30-67 ppm Zn (Yensen, 1997; Shannon, 1999 per. com.).

NyPa Forage does very well under a wide range of conditions, has high palatability for a wide variety of livestock, and can survive fresh water, unlike the grain. It has produced 13.9 ± 3.5 MT/ha dry weight in a 4 month period (5.7 T/A/4-mo.) compared to 1.6 ± 0.4, 1.3 ± 0.3, 0.6 ± 0.1, 0.7 ± 0.2, 0.5 ± 0.1 for other saltgrasses in the trial, i.e. 2-5 times that of other saltgrasses (Yensen et al., 1988).

Proper management is critical to maintaining suitable forage. Frequent harvests, adequate nitrogen, and salt levels 10-35 g/L (~15-45 dS/m, i.e. up to ocean water) in the irrigation water are essential. It prefers highlight under hot, dry conditions with a good supply of salty water. Productivity trials by University of Arizona (Yensen et al., 1988), the US Salinity Lab (M. Shannon & K. Grieve, 1999 unpubl.), Western Australia Dept. of Agriculture (J. Prefumo & E. Barrett-Lennard, 1999 unpubl.) and Latrobe University (M. Sargeant, 1999 unpubl.) Louisiana Tech University (Martin,

2003) observed very different growth curves over salinity levels. This difference may be due to photo-halosynthesis (see article in this volume by Yensen & Biel).

5. BIO-REMEDIATION AND SALT-CONDUCTOR HALOPHYTES REMEDIATION

Bio-remediation, the removal of toxic salts and compounds by organisms, is a new rapidly developing area. Phyto-remediation refers to plant-activated bioremediation and typically occurs by elemental up-take, volatilization, sequesterization and/or precipitation to reclaim toxic waste sites, saline petroleum sludge and mining areas.

While some plants can accumulate the toxic salts directly from the soil, other applications rely on volatilization and/or precipitation of toxins into harmless compounds by microbes that use halophytes as a carbon energy source. Also, some macrophytes can directly bio-remediate certain toxins. Toxic elements, such as chlorine (Cl) and selenium (Se), can be remediated by volatilization (D. Davis per. com.; B. Frankenberger per. com.). There is some highly controversial discussion, at present, about the presently dubious possibility to volatilize and/or sequester sodium (Na) via microbes (Maton et al., 1986; Nelidov, 1981).

Halophyte crops, especially accumulator and excretor types, are being developed that can eliminate toxic compounds and elements (especially Se, Pb, Cr, Cd, Zn, As and others), petroleum products, asphalt, or radio nucleides via halophyte phytoremediation and bio-remediation. Halophytes have adapted to many kinds of salts. For example, some gypsophilous halophytes have become specialized in growing on gypsum salts ($CaSO_4$: $2H_2O$) and gypsiferous soils. Some are obligate gypsophilous halophytes. Closely related halophyte populations may have widely differing tolerances to the same salts.

5.1. Rhizocanicular Effect

Saltgrasses and some other plants have extensive rhizomes that open up the soil for water percolation even in highly deflocculated sealed soils. New understanding is being gained from this old phenomenon, termed the rhizocanicular effect (rhizo = root + caniculi = channels). The rhizocanicular effect occurs when the rhizomes, principally of the saltgrasses, penetrate and open sealed soils and upon dying leaving extensive organic channels through the soil permitting the percolation of water in otherwise impermeable soils (Yensen et al., 1995). Thus, the old adage that, "adding salt water to a clay soil will ruin the soil" is not always true. We have been able to irrigate heavy, clay, alkaline soils with salty water (from 15-46 dS/m depending on site) to grow halophyte crops in heavy clay soils.

Rhizome-bearing halophytes with rhizocanicular action typically have sharply-pointed rhizomes, called pene-rhizomes for their penetration ability. It is even possible to cut one's fingers on sharply pointed pene-rhizomes, which can penetrate heavy deflocculated clay soils, hardpan, caliche, asphalt, and brick-like clays sufficiently hard that a knife blade had difficulty scratching the surface.

Salty soils lose their clumped (flocculated) structure because the salt ions cause the clay soil particles to repel each other (deflocculate) and form a thick colloid, resulting in the soil pores becoming clogged. Rhizocaniculated soils may improved soil structure by providing organic matter by which bacteria may draw the clay plates together forming clumps and re-establishing soil structure (Llebron, US Salinty Lab, per. com.; D. Beck per. com.). While mulch is ephemeral, and oxidizes into CO_2, H_2O and byproducts, halophytes with a rhizocanicular effect provide a persistent sustainable "ever-growing improvement" to the soil.

5.2. Halodecking

A strong rhizocanicular halophyte will also allow downward water drainage and movement of salts down and out of the upper soil horizons. The salts become sequestered or halodecked at the bottom or below of the root zone. Halophytic trees, grasses, etc. are capable of growing over salt beds after moving the salts down and/or building up the surface soil matrix. This halodecking process causes salty soil layers to form well below the upper soil layers, which often become very high in organic matter. This "bedding down" of the salts moves the salts down to an "inactive zone" with regard to the local rhizosphere and virtually frees the ecosystem of hypersaline interactions (J. Darling per. com., and per. obs.). Halodecked salt in the Atacama Desert can form meter-thick rock salt layers below a surface rhizosphere of fertile soil where vegetable gardens could be grown. "Halodecking" by *Distichlis* sp. mats in this area may form vertical columns up to 2 meters high (Yensen et al., 1981). Vernadsky (1998) first described some of these biogeological processes in 1926, but has been little recognized in the West until recently.

5.3. Sustainable Haloculture and Serial Biological Concentration (SBC)

Except for certain coastal areas and low impact riparian systems where drainage waters can return to the sea without environmental impacts, the practice of serial haloculture of the different crops may be necessary to eliminate and/or reduce salt impacts on adjacent farms and ecosystems. Halophytes, combined with serial biological concentration systems, can provide sustainable haloculture. There are enough salt-tolerant/halophyte species known that an entire community's food, shelter, fuel and ornamentals could be generated by salt-water cultivation.

In SBC haloculture systems the effluent/drainage waters from each crop is collected and reapplied to the next, more-halophilic, crop. Each crop in the series must be increasingly more salt-tolerant than the last due to evapotranspiration and salt concentration in such hydrologically "open" systems. This process was suggested by Dr. Jim Rhodes (former Director of the US Salinity Lab) in the 1970's for application in the San Joaquin Valley. Vashek Cervinka and Doug Davis have been active in the San Joaquin Valley working out the details of SBC systems. These systems in California have produced substantial financial benefit for its operators and expansion may include dairies in the near future (Davis, per. com.). In Mexico a SBC system at ex-Lago Texcoco has been in operation since the 1970's (Llerena et al., 1980). While SBC is a concept that has developed into reality in only the last few decades, we feel privileged to have worked with virtually all of the systems.

In a SBC system the last crops are brine-loving species (*Artemia* spp., *Spirolina salina, Dunniella* spp. *Halobacter* spp., *Halomonas* spp, etc.). SBC projects are present or planned for Mexico, California, Australia, Italy, Central America, and Jordan and can utilize a new re-inoculation system developed in Italy (FAO) by Dr. Maurizio Giannotti to efficiently treat water. Sewage treatment using these systems can be very effective in using salt water for toilets and plumbing which also includes non-corroding water delivery and effluent piping.

A scalable model of a small-pilot SBC system might utilize: 50-100 cubic meters/ha/day of water per day at 10,000-15,000 ppm (15-20 dS/m) on a miohalophyte. The second level might utilize 25-50 cubic meters/half-hectare/day at 30,000 ppm (nearly ocean water). The third level might utilize 10-25 cubic meters/quarter-hectare/day at 60,000 ppm (hypersaline) on a euhalophyte crop. While the number of potential hypersaline crops is limited, there are more than enough options to satisfy virtually any SBC system.

5.4. Desalinization by Algae and Capacitive Deionization Technology (CDT)

While the production of forage, fodder, food, construction materials, clothing, etc. can all be derived from halophytes, the production of fresh water for drinking and/or bathing is still a limiting factor for any community dependent solely on salt water. Desalinization can now be done with algae and/or with a new technology Capacitive Deionization Technology (CDT) that electronically removes salts using sheets of inert carbon aerogel.

The alga, cf. *Pheridia tenuis*, if left to grow in a bucket of sea water, will sequester salt within the vacuoles of its tissues until the water in the bucket is relatively salt free (H. Leith, per. com.). The growth curve of this alga, on increasing salt levels, should be most interesting. This represents a condition wherein the salt must be accumulated against an increasing osmotic gradient. Other ways to obtain fresh drinking water has been by distillation, electrodialysis, ultra filtration, freezing point separation, methyl hydrate separation, reverse osmosis, and flash distillation. Flash distillation with a "free" heat source, is generally considered the most efficient.

A new technology is now being developed also has the potential to help the farmers obtain drinking water for themselves and their animals. Capacitive deionization technology (CDT) is a new technology first developed with carbon aerogel at Lawrence Livermore National Laboratory (LLNL) in the USA. The initial work was conducted by Joseph C. Farmer, Jeffrey H. Richardson, David V. Fix, Gregory V. Mack, Richard W. Pakala, and John F. Poco of the LLNL. CDT can electronically remove dissolved salts and toxic elements. With brackish water, the process is potentially more efficiently than Reverse Osmosis and Flash

Distillation, and has many other niche applications where it is far more effective due to using inert carbon aerogel which is neither affected by heat, acids, alkali nor bacteria. CDT is still in the developmental stages, but someday soon may provide pure drinking water for farmers and animals.

The key ingredient of CDT is carbon aerogel, a new material. It is a carbon composite made by coating a sheet of carbon felt with formaldehyde and a resin. The dried polymer is then pyrolized at high temperature. The resultant carbon aerogel has a density of 0.3-1.0 g/cm^3 (based on .005 inch thick material) and has an internal ion-collecting surface area that can be greater than 60,000 times the "plane form" surface area. That is, a piece of aerogel the size of piece of typing paper would have a collecting-surface area the size of a football playing field. The ion-collecting surface area varies from 100-700 m^2/cm^3. Aerogel is a form of carbon with a reticulated structure whose pore size is measured in tens of nanometers. It acts like a "microscopic sponge!" (Sheppard et al., 1998)

Aerogel removes salt ions when an electrostatic field (1-2 volts) is established on alternating charged sheets of aerogel, as in a capacitor. The electrode sheets are stacked and electrically bussed in the shape of a flow-through cell. Cells are connected in series to meet water quality requirements and in parallel to meet water volume requirements. When salt water passes between the aerogel sheets, the ions are attracted to their opposite electro-potential and are electro-adsorbed to the aerogel. Any particle with a net charge (positive or negative) will be held to its respective oppositely charged aerogel surface of the capacitor and later released into a waste steam by removing or reversing the electrostatic charge. Because the ion-collecting surface area of the aerogel is very large, it is capable of removing relatively large quantities of salt from a fluid passing across the electrodes.

CDT is ideally suited for Ultra Pure water production because little energy is lost through the "non-conductive" water and large quantities of water may pass through the cells before the collected salt needs to be flushed. Models predict that less than 10 Watt-hours/gallon are required to purify water with a TDS of a few hundred ppm (parts per million) to less than 100 ppb (parts per billion) (deionization only) and the pumping energy is minimal (no pressurized membranes or filters). Ocean water and brine desalinization is the most difficult for CDT as the electrodes need frequent flushing. Never-the-less estimated energy requirements are low compared to other technologies. A comparison of the estimated energy requirements for purification of seawater using various technologies is:

Thermal Evaporation:	2720 Watt-hours/gallon
Electro dialysis:	75 Watt-hours/gallon
Reverse Osmosis:	35 Watt-hours/gallon
CDT Process:	25 Watt-hours/gallon

Ideal applications for CDT are: Rural and Municipal Water Suppliers, Electronics Manufacturing Facilities - low cost ultra pure water, Nuclear Process

Water Cleanup - removal of radioactive ions, EPA Site Cleanup - underground water supplies, e.g. Hexavalent chromium and Ammonium Perchlorate removal, Medical Research and Care Providers, and Boiler Operations - "At temperature" water purification

6. A PHILOSOPHY OF SALINE AGRICULTURE

A few million dollars of halophyte developmental costs, over a number of years, can provide people with a value worth many times the development costs. Ironically, the just the increase in the value of the land from having a productive halophyte crop can easily pay for the development costs. Yet, salinization continues to increase in spite of some success stories. Humankind has had the habit of salinizing an area and then moving on and on to salinize other areas leaving a trail of "troubled" land.

Today the technology exists that will allow us to take the salinized land and make it productive again. It may need a different crop and a landowner with a different mind-set, but we no longer have to move on to destroy other lands. We have an opportunity. There are many options for going forward to use our salty resources. There is, however, an inherent logical order of priorities in the development of saline agriculture, which flows intrinsically from our fresh-water past:

Arguments, discussion and debates over whether to invest in: drainage, genetically-induced/selected salt-tolerant conventional crops, basic science on halophyte physiology or soil management, mapping the salinity, if planting halophytes - which ones salt bush/grass/trees/etc, pumping the ground water, trees planted in the recharge zones, adding amendments, magnetizing the water, mulching the soil, etc. usually follow the interests of the particular individuals in the discussion. In order to help us to establish our priorities the author has identified four "laws" of saline agriculture priorities:

"The Four Laws of Saline Agriculture":

6.1. First priority law-protect fresh-water crops from salt

The sustainable maintenance of the population's current fresh water crops is critical and commands the highest priority. Fresh water crops provide the bulk of the food and agriculture of our current civilization and we must protect this resource via drainage, leaching, contours, ripping, chemical amendments, adequate management and engineering, dams, canals, crop rotation, etc.

6.2. Second priority law- develops salt tolerant varieties

The development on new salt-tolerant varieties of our current fresh-water crops is next in priority because our infrastructure is designed for these same crops. Thus, the existing agro-socio infrastructure does not need to change and will help to maintain or augment production on marginal farms, while avoiding the necessity to re-educate farmers and consumers unaccustomed to a non-traditional food or product.

6.3. Third priority law- develop local halophyte species

Useful wild, or partially domesticated, halophytes already exist in virtually all-regional ecosystems. Here the local plants are already adapted to the climate and environment. Local knowledge of how to use them may already exist. The domestication of these new/ancient crops can be an important addition to the agricultural base of the area and to the world's food bio-diversity. Consideration and caution should be taken that the rich diversity of the natural floral and faunal are not destroyed by the planting of these new locally developed halophyte crops.

6.4. Fourth priority law-introduce domesticated halophytes

Domesticated halophyte crops may be introduced to the region after careful study and evaluation programs. In situ trials under varying conditions will help determine where the introduced crops should be grown. Although introductions are fourth in priority, their importance cannot be underestimated. Thomas Jefferson, arguably the most intelligent US President, wrote, "the greatest thing that any one individual can do for their country is to introduce a new crop." witness our introduced fresh-water crops: tomato, wheat, corn, rice, potato, sugar cane, sugar beets, alfalfa, actually, most of our crops!

Fortunately, the introduction of salt-loving crops to species-rich natural areas is not often economical. It is well known from our and others trials that ocean-water irrigation gives substantially lower yields than the brackish water readily available in salinized farmland. This higher productivity on brackish water is true for almost all halophytes and aquaculture. How halophyte crops grow better with increasing salinities (the curvilinear increase in productivity with increase in salt) is still unexplained. The photo-halosynthesis concept (see this volume) is an unproven hypothesis that suggests that the photoelectric effect and Gaussian charge distribution may drive inverse sodium pumps to generate stored energy without chlorophyll. The economics of utilizing this new concept in its natural form and/or in synthetic membranes, however, are still unknown.

Never-the-less, there are long-term economic reasons for preserving our species-rich coastal and inland salt marshes and salt flats. Apart from their natural beauty and aesthetics, these areas have a rich diversity that can provide future generations with the raw genetic material to develop new halophyte crops. The habitat destruction and subsequent extinction of our diverse halophyte flora would be tantamount to the burning of the library of Alexandria, which was perhaps the single greatest loss in the history of humankind. The establishment of protected halophyte areas should go hand-in-hand with the development of new halophyte crops. (Yensen, 1988)

7. CONCLUSION

The economics and cost benefit ratios for food, products, turf, landscaping and ornamental halophytes that can use brackish (or in certain cases sea water) is similar to that of fresh water plants and has the added benefit of freeing up large amounts of fresh water for municipal and domestic use. The selection of the appropriate technology and halophyte plants is not always easy. To assist in the selection of crops for different regions a prototype computer program has also been developed by Edwin Ongley, Sarah Dorner and the author with the assistance of the Food and Agriculture Organization (FAO) of the United Nations.

The development and use of halophilic crops is only now beginning. Ancient halophyte crops are being re-examined, bred and selected like our glycophyte crops were centuries and millennia ago. With the obvious benefit of salinity information and potential new technologies and crops there is an increasing need for the establishment of salinity centers and the opening of communication between such centers to generate and distribute these advances. We are pleased to recognize the International Center for Biosaline Agriculture (ICBA), the Halophyte Laboratory, and other new centers that will make the Twenty-first Century an exciting time to be working with halophytes and identifying their uses.

8. REFERENCE VOLUMES

A good halophytology library going into the 21st century might hold the following references. The author apologizes to those authors not listed, as only these works were at hand.

8.1. Halophyte lists and uses

Saline Agriculture: Salt-tolerant Plants for Developing Countries (Shay, 1990) is an excellent book by the US National Research Council on the uses of 215 species. It has been translated into Chinese.

Halophyte Uses in Different Climates, I, II, and III. (Lieth et al., 1999; Hamdy et al., 1999; Güth, 2001, respectively): Volume I has some 2500 species listed and a salinity conversion table and considers many aspects of ecology and physiology of halophytes; Volume II describes pilot and commercial aspects of haloculture (e.g. *Atriplex, Aster, Brassica, Avicennia, Batis, Nitraria, Limoneastrum, Spartina, Prosopis, Eucalyptus, Casurina*); Volume III considers the economics and commercializing halophyte crops and includes a CD on Cash crop halophytes for future halophyte growers by Leith and Lohmann (2000).

Haloph (Aronson, 1989) lists some 1500 species of halophytes.

The flora of tidal forests, a rationalization of the use of the term Mangrove (Mepham, & Mepham, 1985) lists 927 species from the Indo-West Pacific, 424 arborescent and 503 non-arborescent.

Halophytes of the Gulf of California and their Uses (Yensen, 2001) Spanish and English.

8.2. Forage halophytes and trees

Bibliography of Forage Halophytes and Trees for Salt-Affected Land: Their Use, Culture and Physiology (Ismail, 1990) A fine compilation/survey of halophyte literature.

8.3. Soils and Water

Alkali Soils: their formation, properties and reclamation (Kelley, 1951) covers the basics of soil salts even though it is a bit out of date.

Diagnostic and Improvement of Saline and Alkali Soils (Richards, 1954) more commonly known as "Handbook 60" by the US Salinity Lab was edited by Richards and is a classic compilation of the basics.

All India Symposium on Soil Salinity (U.P. Institute of Agri. Sciences, 1971) again, although mostly about India, it covers many different aspects of agriculture using saline soils.

Irrigation, Drainage and Salinity (Kovda et al., 1973) is an excellent resource for irrigation and drainage.

Modeling of Soil Salinization and Alkalization (Kodva & Szabolcs, 1979) is mathematically oriented, but does include basic soil chemistry.

Management of Saline Soils and Waters (Gupta & Gupta, 1987), although oriented toward practices used in India, includes a lot of basic information.

Water Quality for Agriculture (Ayers & Westcot, 1989) is another classic with information on both water, irrigation and soils.

Remediation of Salt-Affected Soils at Oil and Gas Production Facilities (Carty et al., 1997).

Agriculture Salinity and Drainage (Hanson et al., 1999) considers toxicity to crops, leaching fractions, amendments and drainage calculations in an understandable format.

8.4. Proceedings

Proceedings of the Symposium on Sodic Soils (Szabolcs, 1964)

Symposium on the Reclamation of Sodic and Soda-Saline Soils (Szabolcs, 1969)

Ecology of Halophytes (Reimold & Queen, 1974)

Managing Saline Water for Irrigation (Dregne, 1977)

The Biosaline Concept (Hollaender et al., 1979)

Proc. of the Hungaro-Indian Seminar on Salt Affected Soils. (Szabolcs & Yadav, 1981)

Biosaline Research (San Pietro, 1982)

Contributions to the Ecology of Halophytes (Sen & Rajpurohit, 1982)

Prospects for Biosaline Research (Ahmad & San Pietro, 1986)

Strategies for Utilizing Salt Affected Lands (Moncharoen et al., 1992)

Toward the Rational Use of High Salinity Tolerant Plants-1, -2 (Lieth & Masoom, 1993)

Halophyte Utilization in Agriculture (Choukr-Allah, 1993)

Hungarian Contributions to the 15th Int'l Congress of Soil Science (Szabolcs, 1994a)

Biology of Salt Tolerant Plants (Khan & Ungar, 1995)

Halophytes and Biosaline Agriculture (Choukr-Allah et al., 1996)

International Symposium on Sustainable Management of Salt-Affected Soils in the Arid Ecosystem (Hanna & Elgala, 1997) is a collection of over 200 abstracts.

International Symposium on Salt-Affected Lagoon Ecosystems (Batlle-Sales, 1997) deals primarily with saline soils in coastal areas.

9. REFERENCES

Abdul-Halim, R.K. 1986. Soils salinization and the use of halophytes for forage production in Iraq. In: E.G..Barrett-Lennard, C.V. Malcolm, W. R. Stern, & S.M Wilkins, (Eds.), Reclamation and Revegetation Research. Amsterdam, Netherlands: Elsevier. 75-82 pp.

Ahmad, R. & San Pietro, A. (Eds.), 1986. Prospects for Biosaline Research. Karachi, Pakistan: University of Karachi. 587 pp.

Al-Attar, M. 2002. Role of biosaline agriculture in managing freshwater shortages and improving water security. World Food Prize Symposium, Des Moines, Iowa,

Anonymous. 1985. Nieman-Marcus proudly introduces NYPA™ Indian Wild Wheat™. N-M BY POST; Holiday Gourmet, 1985, 11.

Aronson, J.A. 1989. Haloph, a data base of salt tolerant plants of the world. Tucson, Arizona: Office of Arid Land Studies, Univ. of Arizona. 77 pp.

Arunin, S. 1992. Strategies for utilizing salt affected lands in Thailand. In: Moncharoen et al., (Eds.), Strategies for Utilizing Salt Affected Lands, Proc. Int'l Symposium, Bankok, Thailand,17-25 February, Bangkok,Thailand: Funny Publications. 26-37 pp.

Ayers, R.S. & Westcot, D.W. 1985. Water Quality for Agriculture. FAO #29, Rome, reprinted 1989. Rome, Italy: FAO. 174 pp.

Batanouny, K.H. 1993. Ecophysiology of halophytes and their traditional use in the Arab World. In: R.,Choukr-Allah & A. Hamdy, (Eds.), Halophyte Utilization in Agriculture, Advanced Course on, Proceeding of a symposium in Agadir, Morocco 12-26 September 1993. Agadir, Morocco:Symposium Proceedings. 37-70 pp.

Batlle-Sales, J. (Ed.), 1997. International Symposium on Salt-Affected Lagoon Ecosystems. Proceedings of symposium, 18-25 Sept. 1995, Valencia, Spain; Univ. de Valéncia and International Soil Science Society, Depto. de Biología Vegetal, 46100 Burjasot, Valencia. 463 pp.

Barnett, H., Holmes, J. & Newman, K. 1998. Microbial Remediation of Saline Soils- EnviRx (un-reviewed manuscript).

Barrett-Lennard, E.G. & Malcolm, C.V. 1995. Saltland Pastures in Australia: a practical guide. Western Australia, South Perth: Dept. of Agriculture, Bulletin 4312: 112 pp.

Bennett, B. 1998. Rising salt, a test of tactics and techniques. Dealing with Dryland Salinity; Ecos 96 July-September (1998), 28 pp. publ. of Western Australia Department of Conservation and Land Management, CSIRO.

Breckle, S.W. 1986. Studies on halophytes from Iran and Afghanistan. In: II. Ecology of Halophytes along Salt Gradients, Proceedings. Royal Society of Edinburgh 89B: 203-215.

Carty, D.J., Swetish, S.M., Priebe, W.F. & Crawley, W. 1997. Remediation of Salt-Affected Soils at Oil and Gas Production Facilities. American Petroleum Institute. Washington DC: API Publication. 247 pp.

Cheraghi, S.A.M. 2004. Institutional and scientific profiles of organization working on saline agriculture in Iran. In: F.K. Taha, S. Ismail, & A Jaradat. (Eds.), Prospects of Saline Agriculture in the Arabian Peninsula, Amherst, Mass:Amherst Scientific Publishers. 399-412 pp.

Choukr-Allah, R. (Ed.), 1993. Advanced course on "Halophyte Utilization in Agriculture." Proceeding of symposium, 12-26 Sept. 1993, Agadir, Morocco, R. Choukr-Allah. (Ed.), Salinity and Plant Nutrition Laboratory, I.A. V. Hassan II, Rabat, Morocco:Symposium Proceedings. 538 pp.

Choukr-Allah, R., Malcolm, C.V. & Hamdy, A. (Eds.), 1996. Halophytes and Biosaline Agriculture. New York,New York: Marcel Dekker. 400 pp.

Cromarty, P. & Scott, D.A. 1995. A directory of Wetlands in New Zealand. Wellington, New Zealand: Dept. of Conservation. 46 pp.

Dagar, J.C. 1995. Characteristics of halophytic vegetation in India. In: M.A. Khan & I.A Ungar. (Eds.), Biology of Salt Tolerant Plants. Chelsea, Michigan: Book Crafters. 255-276 pp.

De Pauw, E. 2004. Agroecological characterization for salinity assessment in the dry areas. In: K.T. Faisal, S. Ismail, & A. Jaradat, (Eds.), Prospects of Saline Agriculture in the Arabian Peninsula, Amherst, Massachusrtts:Amherst Science.Publications. 37-52 pp.

Dregne, H.E. (Ed.), 1977. Managing Saline Water for Irrigation. Proceedings of the International Conference on Managing Saline Water for Irrigation: Planning for the Future. Lubbock, Texas: Texas Tech University. 618 pp.

El-Beltagy, A. 2004. Strategies to meet the challenges of salinity in Central and West Asia and North Africa. In: K.T. Faisal, S. Ismail, & A. Jaradat, (Eds.), Prospects of Saline Agriculture in the Arabian Peninsula, Amherst, Massachusrtts:Amherst Science.Publications. 1-14 pp.

Espenshade, E.B. (Ed.), 1960. Goode's World Atlas. 11th edition, Rand McNally. 285 pp.

Felger, R.S. & Moser, M.B. 1973. Eelgrass (*Zostera marina* L.) in the Gulf of California: discovery of its nutritional value by the Seri Indians. Science 131: 355-356.

Ghassemi, F., Jakeman, A.J. & Nix, H.A. 1995. Salinisation of Land and Water Resources: human causes, extent, management, and case studies. Sydney, Australia: University South Wales Press, 526 pp.

Ghazanfar, S.A. 1995. Vegetation of coastal sabhkas: an analysis of the vegetation of Barr Al Hikman, Sultanate of Oman. In: M.A. Khan, & I.A Ungar. (Eds.), Biology of Salt Tolerant Plants. Chelsea, Michigan: Book Crafters. 277-283 pp.

Girdhar, I.K. 1993. Occurrence and properties of salt-affected soils in India. In: P. Lal, B.R. Chhipa & A. Kumar, (Eds.), Salt Affected Soils and Crop Production: A modern synthesis. Bikaner, India:Agro Botanical Publishers. 7-30 pp.

Gupta, S.K. & Gupta, I.C. 1987. Management of Saline Soils and Waters. New Delhi, India: Oxford & IBH Publisher. 339 pp.

Güth, M. 2001. Halophytes uses in Different Climates III: Computer-Aided Analysis of Socio-Economic Aspects of the Sustainable Utilization of Halophytes. Progress in Biometeorology, Leiden, Netherlands: Backhuys Publisher. 99 pp.

Hamdy, A., Leith, H., Todorović, M. & Moschenko, M. (Eds.), 1999. Halophytes uses in Different Climates II: Halophyte Crop Development: Pilot Studies. Progress in Biometeorology, Leiden, Netherlands: Backhuys Publisher. 144 pp.

Hanna, F., & Elgala, A.M. (Eds.), 1997. International Symposium on Sustainable Management of Salt-Affected Soils in the Arid Ecosystem. University of Ain Shams and ISSS, Cairo, Egypt, 22-26 Sept. Cario, Egypt: University of Ain Shams.

Hanson, B., Grattan, S.R. & Fulton, A. 1999. Agriculture Salinity and Drainage. Division Agriculture and Natural Resources Publication. 3375, Univ. California, Davis, California: Natural Resources Publisher. 160 pp.

Hayes, G. 1997. An assessment of the National Dryland Salinity Research Development and Extension Program. LWRRDC Occasional Paper, 16/97.

Heuperman, A.F., Kapoor, A.S. & Denecke, H.W. 2002. Biodrainage: Principals, experiences and applications. IPTRID, Rome, Italy: FAO. 79 pp.

Hollaender, A. (Ed.), 1979. The Biosaline Concept: an approach to the utilization of underexploited resources. New York,New York:Plenum Press. 391 pp.

Ismail, S. 1990. Bibliography of Forage Halophytes and Trees for Salt-affected Land: their use, culture and physiology. Dept. of Botany, University of Karachi, Karachi-75270, Karachi, Pakistan: University of Karachi. 258 pp.

Israelsen, O.W. & Hansen, V.E. 1962. Irrigation Principles and Practices. New York, New York: Wiley.

Kanani, S.S. & Torres, F. 1986. The extent of salinization and use of salt tolerant plants in Kenya. In: E.G. Barrett-Lennard, C.V. Malcolm, W.R. Stern & S.M. Wilkins (Eds.), Reclamation and Revegetation Research: Amsterdam, Neetherlands: Elsevier. 97-103 pp.

Kelley, W.P. 1951. Alkali Soils: Their Formation, Properties and Reclamation. New York, New York: Reinhold Publisher. 176 pp.

Khan, M.A. & Ungar, I.A. (Eds.), 1995. Biology of Salt Tolerant Plants. Dept. of Botany, University of Karachi, Karachi-75270, Karachi, Pakistan: University of Karachi. 419 pp.

Kovda, V.A. 1977. Arid land irrigation and soil fertility: problems of salinity, alkalinity, compaction. In: E.B. Worthington (Ed.), Arid land irrigation in developing countries. Environmental problems and effects. Oxford, UK: Pergamon Press. 211-236 pp.

Kovda, V.A., Berg, C. van den & Hagan, R.M. (Eds.), 1973. Irrigation, Drainage and Salinity. An international source book, FAO/UNESCO, London,UK:Hutchison Publisher. 510 pp.

Kovda, V.A. & Szabolcs, I. 1979. Modeling of Soil Salinization and Alkalization. Agrochemistry and Soil Science, Budapest 28(suppl.): 1-208.

Kraiem, H. 1986. Vegetation of salt affected land in Chad and Senegal. In: E.G. Barrett-Lennard, C.V. Malcolm, W.R. Stern & S.M. Wilkins (Eds.), Reclamation and Revegetation Research: Amsterdam, Netherlands: Elsevier. 31-39 pp.

Le Houerou, H.N. 1993. Salt-tolerant plants for the arid regions of the Mediterranean isoclimatic sone. In: H. Lieth, & A.A. Al Masoom (Eds.), Towards the rational use of high salinity tolerant plants, Dordrecht,Netherlands: Kluwer Academic Publication. 403-422 pp.

Lieth, H. & Al Masoom, A.A. (Eds.), 1993. Toward the rational use of high salinity tolerant plants, Agriculture and forestry under marginal soil water conditions. Proceedings of the first ASWAS conference, 8-15 Dec. 1990, UAE University, Al Ain. Dordrecht, Netherlands: Kluwer Academic Publication. 447 pp.

Leith, H. & Lohmann, M. 2000. Cash Crop Halophytes for Future Halophyte Growers. Institute of Environmental Systems Research, Univ. of Osnabrück, Germany, 32 pp.

Leith, H., Moschenko, M., Lohmann, M., Koyro, H.W. & Hamdy, A. (Eds.), 1999. Halophytes uses in Different Climates I: Ecological and Ecophysiological Studies. Progress in Biometeorology, Leiden,Netherlands: Backhuys Publisher. 258 pp.

Llerena, A. & Tarin, M. 1980. Una alternativa para integrar a la productividad algunas de las áreas altamente salino-sódico del País. In: Proceedings of The Second Inter-American Conference on Salinity and Water Management Technology, Juarez, Mexico, 11-12 December 1980. Juarez, Mexico: Conference Proceedings.1-27 pp.

Maddaloni, J. 1986. Forage Production on saline and alkaline soils in the humid region of Argentina. In: E.G. Barrett-Lennard, C.V. Malcolm, W.R. Stern, & S.M. Wilkins (Eds.), Reclamation and Revegetation Research: Amsterdam, Netherlands: Elsevier. 11-16 pp.

Malcolm, C.V. 1969. Use of halophytes for forage production on saline wetlands. Journal of the Australian Institute for Agricultural Sciences 35: 38-49.

Malcolm, C.V. 1971. Plant collection for pasture improvement in saline and arid environments. Western Australian Department of Agriculture Technical Bulletin No. 6.

Malcolm, C.V. 1979. Selection of shrubs for forge production from saline soils under natural rainfall. Paper presented at the International Conference on Indian Ocean Studies, Perth, Australia.

Malcolm, C.V. 1980. Production from salt-affected soils. Report of the senior research officer on a visit to India and Pakistan, Soil Research and Survey Section, Western Australian Department of Agriculture.

Malcolm, C.V. 1982. Samphire for waterlogged saltland. Western Australian Department of Agriculture Farmnote No. 4/82.

Martin, R.K.T. 2003. The Salinity Tolerance of Six Bermudagrasses [Cynodon dactylon (L.) Pers.], one Dropseed [*Sporobolus virginicus* (L.) Kunth] and Saltgrass [Distichlis spicata var. yensen-4a] for Phytoremediation of Salt-affected Soils in Oil, Gas and Brine Fields. MS thesis, Ruston, Lousiana: Lousiana Tech University. 66 pp.

Maton, T., Kairusmee, P. & Takahashi, E. 1986. Salt-induced Damage to Rice Plants and Alleviation Effect of Silicate. Soil Science and Plant Nutrition 32: 295-304.

McKell, C.M., Goodin, J.R. & Jefferies, R.L. 1986. Saline land in the United States of America and Canada. In: E.G. Barrett-Lennard, Malcolm, C.V. Stern, W.R & Wilkins, S.M. (Eds.), Amsterdam, Netherlands: Elsevier. 159-165 pp.

Mepham, R.H. & Mepham, J.S. 1985. The flora of tidal forests, a rationalization of the use of the term Mangrove. South African Journal. Botany 51: 77-99.

Moghaddam, P.R. & Koocheki, A. 2004. History of research on salt-affected lands of Iran. Present status and future prospects: Halophytic ecosystems. In: F.K. Taha, S. Ismail, & A. Jarada, (Eds.), Prospects of Saline Agriculture in the Arabian Peninsula, Amherst, Mass: Amherst Scientific Publishers. 83-95 pp.

Moncharoen, L., Jantawat, S., Kheoruenromne, I., Vearasilp, T. & Piyapun, B. (Eds.), 1992. Strategies for Utilizing Salt Affected Lands. Proceedings of the International Symposium, 17-25 February 1992, Bankok; 549/1 Soi Senanikom 1 Phaholyothin 32. Bankok, Thailand: Funny Publishing. 586 pp.

Nazrul-Islam, A.K.M. 1995. Ecological conditions and species diversity in Sundarban mangrove forest community, Bangaladesh. In: M.A Khan & I.A Ungar. (Eds.), Biology of Salt Tolerant Plants. Chelsea, Michigan: Book Crafters. 294-305 pp.

Neary, J. 1981. Pickleweed, Palmer's Grass and Saltwort: Can we grow tomorrow's food with today's salt water. Science 81: 38-43.

Nelidov, S.N. 1981. Effect of Straw on Microbial Activity of Soil and Rice Crop Production. Dissertation, Candidate of Biological Science, Alma-Ata, 181 pp.

Öztürk, M., Ozcelik, H., Behcet, L., Güvensen, A. & Özdemir, F. 1995. Halophytic flora of Van Lake Basin-Turkey. In: M.A Khan & I.A Ungar. (Eds.), Biology of Salt Tolerant Plants. Chelsea, Michigan: Book Crafters. 306-315 pp.

Reimold, R.J. & Queen, W.H. (Eds.), 1974. Ecology of Halophytes. 605 pp. New York, New York: Academic Press.

Sandhu, G.R. & Qureshi, R.H. 1986. Salt affected soils in Pakistan and their utilization. In: E.G. Barrett-Lennard, C.V Malcolm, W.R Stern, & S.M. Wilkins, (Eds.), Reclamation and Revegetation Research. Amsterdam,Netherlands: Elsevier. 105-113 pp.

San Pietro, A. (Ed.), 1982. Biosaline Research: a look to the future. New York, New York: Plenum Press. 578 pp.

Sankary, M.N. 1986. Species distribution and groth on salt affected land in Syria. In: E.G. Barrett-Lennard, C.V. Malcolm, W.R. Stern & S.M Wilkins. (Eds.), Reclamation and Revegetation Research: Amsterdam,Netherlands: Elsevier. 125-141 pp.

Sargeant, M. 1999. Personal communication and BS. thesis. Latrobe University, Melbourne, Victoria.

Sen, D.N. & Rajpurohit, K.S. 1982. Contributions to the Ecology of Halophytes. The Hague.Netherlands: W. Junk. 272 pp.

Shay, G. (Ed.), 1990. Saline Agriculture: Salt-tolerant Plants for Developing Countries. Washington, DC: National Academy Press. 143 pp.

Sheppard, C., Yensen, N.P. & Vaught, C. 1998. Capacitive Deionization (CDI) with Carbon Aerogel: A New Technology for Purifying Contaminated Waters of Salts and Pollutants. Taller Internacional sobre Gestión de la Calidad del Agua y Control de la Contaminación en América Latina [International Workshop regarding Water Quality and Control of Contamination in Latin America]. 30 Sept. - 4 Oct. 1998, Arica, Chile: Food and Agriculture Organization of the United Nations.

Sissay, B. 1986. Salt affected wasteland in Ethiopia: Potential for production of forage and fuel. In: E.G. Barrett-Lennard, C.V. Malcolm, W.R. Stern, & S.M. Wilkins (Eds.), Reclamation and Revegetation Research: Amsterdam, Netherlands: Elsevier. 59-64 pp.

Szabolcs, I. (Ed.), 1964. Proceedings of the Symposium on Sodic Soils. Agrochemistry and Soil Science, Budapest 14(Suppl.): 1-480.

Szabolcs, I. (Ed.), 1969. Symposium on the Reclamation of Sodic and Soda-Saline Soils. Yerevan, Agrochemistry and Soil Science, Budapest, 18(Suppl.), 1-480.

Szabolcs, I. 1992a. An overview of the salt affected lands in the world. In: Moncharoen et al. (Eds.), Strategies for Utilizing Salt Affected Lands, Proc. Int'l Symposium, Bankok, Thailand, 17-25 February, Bankok,Thailand: Funny Publications. 19-25 pp.

Szabolcs, I. 1992b. Salt affected soils as the ecosystem for halophytes. In: V.R. Squires, & A.T. Ayoub, (Eds.), Halophytes as a Resource for Livestock and for Rehabilitation of Degraded Lands, Proc. of an Int'l Workshop on Halophytes for reclamation of saline wastelands and as a resource for livestock: problems and prospects, London,UK: Kluwer. 19-24 pp.

Szabolcs, I. (Ed.), 1994a. Hungarian Contributions to the 15th International Congress of Soil Science. Agrochemistry and Soil Science, Budapest, 43(1-2), 1-240.

Szabolcs, I. 1994b. Prospects of soil salinity for the 21st century. Agrochemistry and Soil Science, Budapest 43(1-2): 5-24.

Szabolcs, I. & Yadav, J.S.P. (Eds.), 1981. Proceedings of the Hungaro-Indian Seminar on Salt Affected Soils. Agrochemistry and Soil Science, Budapest 30(Suppl.); 1-256.

Raven, P.H., Evert, R.F. & Eichhorn, S.E. 1986. Biology of Plants. New York,New York: Worthen Publications. 775 pp.

U.P. Institute of Agriculture Sciences. 1971. All India Symposium on Soil Salinity. Kampur, India: U.P. Institute of Agriculture Sciences. 239 pp.

U.S. Salinity laboratory Staff. 1954. Diagnosis and improvement of saline and alkali soils. In: L.A. Richards. (Ed.), Agricultural Handbook of the U. S. Department of Agriculture. Washington DC: Government Publisher. 157 pp.

Vernadsky, V.I. 1998. The Biosphere. New York, New York: Springer-Verlag. 192 pp.

Waisel, Y. 1972. Biology of Halophytes. New York, New York: Academic Press. 393 pp.

White, G.F. (Ed.), 1978. Environmental effects of arid land irrigation in developing countries. UNESCO, MAB Technical Notes 8, Paris, France: UNESCO. 67 pp.

World Resources. 1987. A report by the International Institute of Environmental Development and World Resources Institute. New York, New York: Basic Books. 375 pp.

Yensen, N.P. 1988. Plants for salty soils. Arid lands newsletter 27: 3-10.

Yensen, N.P. 1985. Saline Agriculture and New Halophyte Crops. Proceedings of the 61st Annual Meeting of the Southwestern and Rocky Mountain Division, March 19-23 American Association for the Advancement of Science.

Yensen, N.P. 1995. International Symposium on High Salinity Tolerant Plants Summary of Papers Presented. In: M.A. Khan, & I.A. Ungar. (Eds.), Biology of Salt Tolerant Plants, Chelsea, Michigan: Book Crafters. 1-12 pp.

Yensen, N.P. 1997. The Agronomic Production and Nutritional Characteristics of NyPa Forage, *Distichlis spicata* var. *yensen-4*, (Poaceae) when grown on highly Saline Water in Arid Regions of the World with respect to Economic Worth. In: A.M. Elgala, (Ed.), International Symposium on Sustainable Management of Salt Affected Soils in the Arid Ecosystem Ain Shams University, 22-26 Sept. 1997. Cairo, Egypt: Symposium Proceedings. 363 pp.

Yensen, N.P. 1988. Halophytes of Latin America and the World: their use with saline & waste waters and marginal soils. International Workshop regarding Water Quality and Control of Contamination in Latin America, 30 Sept.-4 Oct. 1998, Arica, Chile: Food and Agriculture Organization, UN, (Spanish in press). 48 pp.

Yensen, N.P. 2005 Salt-tolerant Plants of the World and their Uses. Hermosillo, Mexico: CIAD (in prep.) 383 pp.

Yensen, N.P. 2001. Halófitas del Golfo de California y sus Usos. Halophytes of the Gulf of California and their Uses. Hermosillo, México: University of Sonora Press (Spanish and English). 296 pp.

Yensen, N.P. 2002. New developments in the world of saline agriculture. In: R. Ahmad, & K.A. Malik (Eds.), Prospects for Saline Agriculture, Dordrecht, Netherlands: Kluwer Publications. 321-332 pp.

Yensen, N.P. & Bedell, J.L. 1993. Consideration for the selection, adaptation, and application of halophyte crops to highly saline desert environments as exemplified by the long-term development of cereal and forage cultivars of *Distichlis* spp. (POACEAE). In: H. Lieth, & A. Al Masoom (Eds.), Toward the rational use of highly salinity tolerant plants, Dordrecht, Netherlands: Kluwer Academic Publishers, 305-313 pp.

Yensen, N.P., Bedell, J.L. & Yensen, S.B. 1995. Domestication of *Distichlis* as a Grain and Forage. In: M.A. Khan, & I.A. Ungar (Eds.), Biology of Salt Tolerant Plants, Chelsea, Michigan: Book Crafters. 388-392 pp.

Yensen, N.P., Bedell, J.L. & Yensen, S.B. 1995. The Use of Agricultural Drain Water for Forage Production in Coastal Ecosystems. In: J. Battle-Sales. (Ed.), Proceeding of the International Symposium on Salt-Affected Lagoon Ecosystems, Universidad de Valencia, Valencia, Spain: Universidad de Valencia.

Yensen, N.P., Fontes, M.R., Glenn, E.P. & Felger, R.S. 1981. New salt-tolerant crops for the Sonoran Desert. Desert Plants 3: 111-117.

Yensen, N.P., Yensen, S.B. & Weber, C.W. 1988. A review of *Distichlis* spp. for production and nutritional values. In: E.E. Whitehead, Hutchinson, C.F., Timmermann, B.N. &. Varady, R.B. (Eds.), Arid Lands: Today and Tomorrow, Proceedings of an international research and development conference, October 20-25, 1985, Bolder, Colorado: Westview. 809-822 pp.

Zavaleta, A. 1969. Saline and alkaline soils in Peru. In: I. Szabolcs, (Ed.), Symposium on the Reclamation of Sodic and Soda-Saline Soils, Yerevan, Agrochemistry and Soil Science, Budapest 18(Suppl.): 169-180.

CHAPTER 24

HALOPHYTE RESEARCH AND DEVELOPMENT: WHAT NEEDS TO BE DONE NEXT ?

BENNO BÖER

UNESCO Regional Office in the Arab States of the Gulf
Doha, POBox 3945, State of Qatar
b.boer@unesco.org

Abstract. Activities supported by the UNESCO Office in Doha in 2004-2005 UNESCO currently supports activities on water & ecosystems. Since several years, UNESCO supported a number of projects that aimed to enhance the sustainable utilisation of saline water resources. The development of cash crop halophytes, as well as methods for coastal habitat restoration, is believed to play an important role towards sustainable coastal management, and the development of models for sustainable living. This is especially true in these times of fresh water scarcity, population growth, and increased resources consumption.

1. DISCUSSION

In our day, with more than 6 billion people on the planet, and continuous global population pressure, "packing up and moving away is no longer a viable option" (Aronson, 1989), when agricultural land becomes non-productive due to high salt concentrations. Substantial scientific knowledge on halophyte habitat restoration, halophyte functioning, and development has been obtained since Boyko's work into saline irrigation (Boyko, 1968; Boyce, 1964). Aronson's database of salt tolerant plants of the world (Aronson, 1989), and Lieth's publication on "the rational use of high salinity tolerant plants" (Lieth & Al Masoom, 1993) can be regarded as additional important milestones concerning the subject of halophyte research and development. The 1990s and early 2000s brought a number of experiments and new publications into this subject, too many to be listed here. To mention a few, Khan and Ungar's "Biology of salt tolerant plants"(Khan & Ungar, 1995), Choukr-Allah et al.,'s "Halophytes and Biosaline Agriculture (1996), and Hamdy & Lieth's "advanced short course on saline irrigation: halophyte production and utilisation" (Hamdy & Lieth, 1999) where certainly again important scientific documents providing knowledge into halophyte utilisation. More recently, and with the support of UNESCO, Barth & Böer, (2002) compiled information into sabkha ecosystems, and the first volume deals with the sabkha ecosystems of the Arabian Peninsula and adjacent countries. The same authors are

M.A. Khan and D.J. Weber (eds.), Ecophysiology of High Salinity Tolerante Plants, 397-399.

currently developing the second volume on sabkha ecosystems in West & Central Asia, as well as a third volume on sabkha ecosystems in Africa. There are four more volumes planned, on sabkhat in America, East Asia, Australia, and a wrap-up volume on the ecology and utilisation of the sabkhat ecosystems of the world. There are large areas of hyper-saline deserts in the world, and the sabkha ecosystms series aims to provide knowledge on how these systems should be conserved, and how some of the area can potentially be utilised by man.

Salt tolerant plants play increasingly important roles for coastal ecosystem restoration, and for the development of biosaline agriculture. Biosaline agriculture will contribute to redress the pressure on limited freshwater resources, as well as the problem of increasing agricultural areas of salinised soils. Brackish water, saline wastewater, and full strength seawater can be used to irrigate salinity tolerant cash crops on saline soils, provided suitable irrigation and drainage techniques are being applied.

However, it is now important to direct the overall work on halophytes into the future, and to identify the next important milestones towards the sustainable utilisation of saline water resources, as well as plants and soils. A concerted international action is required in order to pursue science based coastal environmental management, and to develop self-financing models and marketable products. It must in the same time not being forgotten, that some marketable products have already been developed, such as *Salicornia bigelovii*, and *Salicornia europaea*, as well as *Aster tripolium*, and these products are being sold as vegetable and salad crops on European markets at comparatively high prices. Of much more value in the Gulf Arab region are *Conocarpus erecta*, and *Conocarpus lanciofolius*, both which are being used as roadside trees. *Sesuvium portulacastrum* was introduced into the Gulf in 1989, and it is now being widely used within almost each major city in the GCC countries, as a salt tolerant replacement for the freshwater dependant bermuda-grass (*Cynodon dactylon*). However, there is currently still a lack of public awareness, and scientific documentation regarding successfully established halophytes.

The establishment of a "World Halophyte Garden" had been suggested as early as 1999 (Böer & Al Ghais, 1999; Böer, 2004). The purpose of a halophytic germplasm collection is to carry out research into the utilisation of saline soil-water conditions. This germplasm collection can assist worldwide when halophytes are required for habitat restoration, as well as for biosaline agricultural research. It is especially important to focus on those halophytes that can tolerate full strength sseawater salinity, such as sea grasses, marine algae, as well as those flowering plants that can produce biomass and reproduce while being continuously irrigated with seawater. It is seawater, that is available in abundance in many dry desert countries, whereas limited resources of freshwater are decreasing rapidly, and countries are being increasingly dependant on seawater desalinisation. Another important milestone that needs to be achieved, and that requires immediate attention, is the development of marketable products based on cash crop halophytes.

In one of the more recent publications, the UNESCO Office in Doha pointed out the importance to focus on marketing, the establishment of an international coordination centre, as well as the establishment of a halophyte garden. During the years 2004/05 the UNESCO Office in Doha is ready to assist activities that aim to achieve one or both of the above mentioned necessary future actions. In particular attention will be given to projects that aim to:

1.1. Establish a World Halophyte Garden "

1.2. Develop marketable cash crop halophytes

2. REFERENCES

Aronson, J. 1989. l. Haloph, - A Data Base of Salt Tolerant Plants of the World. Office of Arid Land Studies. Tucson, Arizona: The University of Arizona. 75 pp.

Barth, H.J. & Böer, B. Eds. 2002. Sabkha Ecosystems Vol. I.: The Arabian Peninsula Region with contributions from Afghanistan, Bahrain, Djibouti, Egypt, Eritrea, Ethiopia, Iran, Jordan, Kuwait, Oman, Pakistan, Qatar, Saudi Arabia, Sudan, United Arab Emirates, and Yemen. In: Lieth, H., Mooney, H.A. & Kratochwil, A. (Eds.), Tasks for Vegetation Sciences, Volume 36: Dordrecht. Netherlands:Kluwer Publisher. 368 pp.

Böer, B. 2004. Halophyte development in the Arabian Peninsula: Recent achievements and prospects. In: Taha, F., Ismail, S. & Jaradat, A. (Eds.), Prospects of saline agriculture in the Arabian Peninsula. Amherst, Massachusetts: Amherst Scientific Publishers. 111-120 pp.

Böer, B. & Al Ghais, S. 1999. A contribution towards the successful propagation of halophytes in harsh environmental conditions.. In: A. Handy & H. Leith (Eds.), Saline irrigation: halophyte production and utilisation. Proceedings. Agadir, Morocco:Symposium Proceedings. 77-95 pp.

Boyce, H. 1964. Principles and experiments regarding irrigation with highly saline and seawater without desalinization. Transactions of the New York Academy of Sciences Series II, 26:1087-1102.

Boyko, H. 1968. Saline irrigation for agriculture and forestry. The Hague, Netherlands: W. Junk. 325 pp.

Choukr-Allah, R., Malcolm, C. & Hamdy, A. Eds 1996. Halophytes and Biosaline Agriculture. New York, New York: Marcel Dekker, Inc. 400 pp.

Hamdy, A. & Lieth, H. 1999. Saline irrigation: halophyte production and utilisation. Proceedings. Agadir, Morocco, April 6-15, 1999. Agadir, Morocco: Symposium Proceedings. 363 pp.

Khan, A. & Ungar, I. 1995. Eds. Biology of salt tolerant plants. Dept. of Botany, Karachi, Pakistan, Pakistan. Karachi, Pakistan: Karachi, Pakistan. 419 pp.

Lieth, H. & Al Masoom, A. 1993. Eds. Towards the rational use of high salinity tolerant plants. Tasks for vegetation science 27/28. Dordrecht, Netherlands:Kluwer Academic Publishers.

Tasks for Vegetation Science

1. E.O. Box: *Macroclimate and plant forms.* An introduction to predictive modelling in phytogeography. 1981 ISBN 90-6193-941-0
2. D.N. Sen and K.S. Rajpurohit (eds.): *Contributions to the ecology of halophytes.* 1982 ISBN 90-6193-942-9
3. J. Ross: *The radiation regime and architecture of plant stands.* 1981 ISBN 90-6193-607-1
4. N.S. Margaris and H.A. Mooney (eds.): *Components of productivity of Mediterranean-climate regions.* Basic and applied aspects. 1981 ISBN 90-6193-944-5
5. M.J. Müller: *Selected climatic data for a global set of standard stations for vegetation science.* 1982 ISBN 90-6193-945-3
6. I. Roth: *Stratification of tropical forests as seen in leaf structure* [Part 1]. 1984 ISBN 90-6193-946-1 *For Part 2, see Volume 21*
7. L. Steubing and H.-J. Jäger (eds.): *Monitoring of air pollutants by plants.* Methods and problems. 1982 ISBN 90-6193-947-X
8. H.J. Teas (ed.): *Biology and ecology of mangroves.* 1983 ISBN 90-6193-948-8
9. H.J. Teas (ed.): *Physiology and management of mangroves.* 1984 ISBN 90-6193-949-6
10. E. Feoli, M. Lagonegro and L. Orlci: *Information analysis of vegetation data.* 1984 ISBN 90-6193-950-X
11. Z. Sv̂esták (ed.): *Photosynthesis during leaf development.* 1985 ISBN 90-6193-951-8
12. E. Medina, H.A. Mooney and C. Vzquez-Ynes (eds.): *Physiological ecology of plants of the wet tropics.* 1984 ISBN 90-6193-952-6
13. N.S. Margaris, M. Arianoustou-Faraggitaki and W.C. Oechel (eds.): *Being alive on land.* 1984 ISBN 90-6193-953-4
14. D.O. Hall, N. Myers and N.S. Margaris (eds.): *Economics of ecosystem management.* 1985 ISBN 90-6193-505-9
15. A. Estrada and Th.H. Fleming (eds.): *Frugivores and seed dispersal.* 1986 ISBN 90-6193-543-1
16. B. Dell, A.J.M. Hopkins and B.B. Lamont (eds.): *Resilience in Mediterranean-type ecosystems.* 1986 ISBN 90-6193-579-2
17. I. Roth: *Stratification of a tropical forest as seen in dispersal types.* 1987 ISBN 90-6193-613-6
18. H.-G. Dässler and S. Börtitz (eds.): *Air pollution and its influence on vegetation.* Causes, Effects, Prophy´laxis and Therapy. 1988 ISBN 90-6193-619-5

Tasks for Vegetation Science

19. R.L. Specht (ed.): *Mediterranean-type ecosystems.* A data source book. 1988 ISBN 90-6193-652-7
20. L.F. Huenneke and H.A. Mooney (eds.): *Grassland structure and function.* California annual grassland. 1989 ISBN 90-6193-659-4
21. B. Rollet, Ch. Hägermann and I. Roth: *Stratification of tropical forests as seen in leaf structure*, Part 2. 1990 ISBN 0-7923-0397-0
22. J. Rozema and J.A.C. Verkleij (eds.): *Ecological responses to environmental stresses.* 1991 ISBN 0-7923-0762-3
23. S.C. Pandeya and H. Lieth: *Ecology of Cenchrus grass complex.* Environmental conditions and population differences in Western India. 1993 ISBN 0-7923-0768-2
24. P.L. Nimis and T.J. Crovello (eds.): *Quantitative approaches to phytogeography.* 1991 ISBN 0-7923-0795-X
25. D.F. Whigham, R.E. Good and K. Kvet (eds.): *Wetland ecology and management.* Case studies. 1990 ISBN 0-7923-0893-X
26. K. Falinska: *Plant demography in vegetation succession.* 1991 ISBN 0-7923-1060-8
27. H. Lieth and A.A. Al Masoom (eds.): *Towards the rational use of high salinity tolerant plants*, Vol. 1: Deliberations about high salinity tolerant plants and ecosystems. 1993 ISBN 0-7923-1865-X
28. H. Lieth and A.A. Al Masoom (eds.): *Towards the rational use of high salinity tolerant plants*, Vol. 2: Agriculture and forestry under marginal soil water conditions. 1993 ISBN 0-7923-1866-8
29. J.G. Boonman: *East Africa's grasses and fodders.* Their ecology and husbandry. 1993 ISBN 0-7923-1867-6
30. H. Lieth and M. Lohmann (eds.): *Restoration of tropical forest ecosystems.* 1993 ISBN 0-7923-1945-1
31. M. Arianoutsou and R.H. Groves (eds.): *Plant-animal interactions in Mediterranean-type ecosystems.* 1994 ISBN 0-7923-2470-6
32. V.R. Squires and A.T. Ayoub (eds.): *Halophytes as a resource for livestock and for rehabilitation of degraded lands.* 1994 ISBN 0-7923-2664-4
33. T. Hirose and B.H. Walker (eds.): *Global change and terrestrial ecosystems in monsoon Asia.* 1995 ISBN 0-7923-0000-0
34. A. Kratochwil (ed.): *Biodiversity in Ecosystems: Principles and case studies of different complexity levels.* 1999 ISBN 0-7923-5717-5
35. C.A. Burga and A. Kratochwil (ed.): *Biomonitoring: General and applied aspects on regional and global scales.* 2001 ISBN 0-7923-6734-0

Tasks for Vegetation Science

36. H.-J. Barth and B. Böer (eds.): *Sabkha Ecosystems*. Volume I: The Arabian Peninsula and Adjacent Countries. 2001 ISBN 1-4020-0504-0
37. R. Ahmad and K.A. Malik (eds.): *Prospects for Saline Agrigulture*. 2002 ISBN 1-4020-0620-9
38. H. Lieth and M. Mochtchenko (eds.): *Cashcrop Halophytes Recent Studies*. 10 Years after the Al Ain Meeting. 2003 ISBN 1-4020-1202-0
39. To be Published.
40. M.A. Khan, and D.J. Weber (eds.): *Ecophysiology of High Salinity Tolerant Plants*. 2006 ISBN 1-4020-4017-2